UPDATED

Keith Johnson

New Physics for You

Updated Edition for All GCSE Examinations

Nelson Thornes

This edition published in 2011 by: Nelson Thornes Ltd, Delta Place, 27 Bath Road, CHELTENHAM, GL53 7TH, United Kingdom

ISBN 978 1 4085 0922 7 11 12 13 14 15 / 10 9 8 7 6 5 4 3 2

A catalogue record for this book is available from the British Library

Page make-up by Tech-Set with additional typesetting by Fakenham Photosetting

Printed in Croatia by Zrinski

Website:

The website at **www.physicsforyou.co.uk** gives you details of exactly which pages in this book you need to study for your particular GCSE examination course.

Make sure you visit this website and print out the correct sections.
They will show you:

- which topics you need to learn for your particular examination, and
- which page numbers to read in this book.

'What is the use of a book,' thought Alice, 'without pictures or conversations?'

Lewis Carroll, *Alice in Wonderland*

Everything should be made as simple as possible, but not simpler.

Albert Einstein

There is no higher or lower knowledge, but one only, flowing out of experimentation.

Leonardo da Vinci

I do not know what I may appear to the world, but to myself I seem to have been only a boy playing on the seashore, and diverting myself in now and then finding a smoother pebble or a prettier shell than ordinary, while the great ocean of truth lay all undiscovered before me.

Sir Isaac Newton

Introduction

Physics For You is designed to introduce you to the basic ideas of Physics, and show you how these ideas can help to explain the world in which we live.

This book is based on successful earlier editions of the same name, but new pages and extra questions have been added to cover the latest requirements of the new GCSE Examinations.

Physics For You has been designed to be interesting and to help you to pass your exams, whether you are using it for a Physics course or as part of a Core Science or Additional Science course.

The book is carefully laid out so that each new idea is introduced and developed on a single page or on two facing pages. Words have been kept to a minimum and as straightforward as possible. Pages with a red band in the top corner are the more difficult pages and may be left out at first.

Throughout the book there are many simple experiments for you to do. A safety sign: ⚠ means your teacher should give you further advice (for example, to wear safety glasses).

Each important fact or new formula is printed in **heavy type** or is in a box. There is a summary of important facts at the end of each chapter.

At the back of the book there is advice for you on practical work, key skills, ideas and evidence in science, careers, revision and examination techniques, as well as help with mathematics.

Questions at the end of a chapter range from simple fill-in-a-missing-word sentences (useful for writing notes in your notebook) to more difficult questions that will need some more thought.
In calculations, simple numbers have been used to keep the arithmetic as straightforward as possible.

At the end of each main topic you will find a section of further questions taken from actual GCSE examination papers.

Throughout the book, cartoons and rhymes are used to explain ideas and ask questions for you to answer. In many of the cartoons, Professor Messer makes a mistake because he does not understand Physics very well. Professor Messer does not think very clearly, but I expect you will be able to see his mistakes and explain where he has gone wrong.

Here I would like to thank my wife, Ann, for her constant encouragement and her help with the many diagrams and cartoons.

I hope you will find Physics interesting as well as useful. Above all, I hope you will enjoy **Physics For You**.

Keith Johnson

Professor Messer gets in messes,
Things go wrong when he makes guesses.
As you will see, he's not too bright,
It's up to you to put him right.

The small . . .

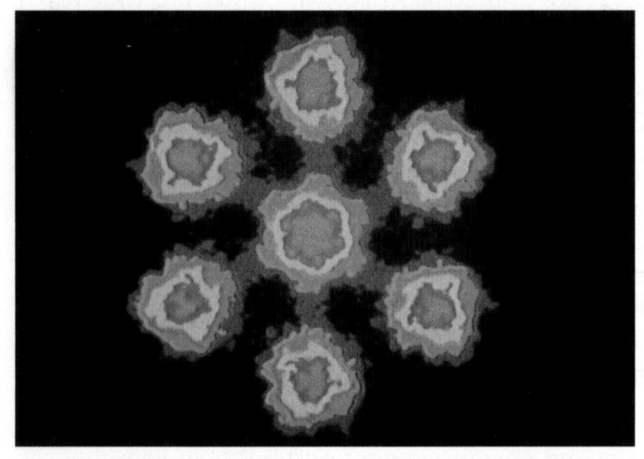

Seven atoms in a uranyl microcrystal, photographed with an electron microscope and false colour added. They are magnified 100 million times.

. . . and the large.

The Andromeda galaxy. It contains about 100 000 million stars, and the distance across it is over 100 000 light-years (10^{21} metres).

Contents

Basic ideas

Heat (thermal energy)

Mechanics

Earth Physics and Astronomy

Waves: Light and Sound

Electricity and Magnetism

Nuclear Physics

Extra sections

Revision details at **www.physicsforyou.co.uk**

chapter 1

How Science works

The main purpose of this book is to help you gain scientific knowledge and understanding.
But you also need to know 'How Science works', and how scientists work.

You will learn a lot of this by planning and then doing your own experiments and investigations, to collect data (see the opposite page).

'Practical skills are important'

How scientists work
The purpose of Science is to find out how the Universe works.
We are trying to explain the world in which we live.
We do experiments and investigations to observe and collect evidence, rather like detectives investigating a crime.

From the evidence scientists try to form a theory or 'thought-model', to explain the evidence.
(For example, the kinetic theory of molecules on page 16.)

A good theory or model is one which can be used to make a prediction, which we can test by experiment.
If the experiment contradicts it, then the theory is modified or changed to a new theory that fits the facts ...until new evidence disproves this theory, and so on.
In this way, human understanding of our world has developed step by step. This is how Science works.

As Albert Einstein said:
"No amount of experimentation can ever prove me right; but a single experiment may prove me wrong."

The history of Science shows how scientific ideas have changed step by step (see pages 371–379).

In the modern world, scientists usually work in teams and share their ideas with other teams, by publishing them in books or on the internet (see page 370 for more details).
It is vital that their evidence is reliable and valid. It is then for society as a whole to make decisions based on that evidence.
For example, what to do about global warming (see page 107).

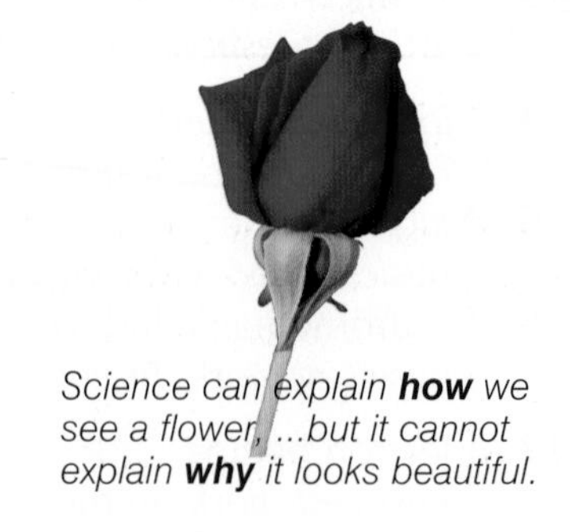
*Science can explain **how** we see a flower, ...but it cannot explain **why** it looks beautiful.*

Of course there are many questions that Science cannot answer yet (eg. how to cure all cancers). And there are questions that Science can never attempt to answer (eg. religious questions).

The opposite page gives a summary of the key words used in Physics investigations.
There are more details about **How Science works** on pages **358–365**.

Read through the box below, and discuss any difficult words in your group or with your teacher. See also the Glossary below. Make sure that you understand all of the words.

Science is a powerful tool for answering certain questions. To do this we need to **plan** an investigation to collect **data**. We can do this by **observing** and **measuring**. Usually we are trying to find a link between 2 **variables**.

When planning an investigation we always try to make it a **fair test**. We need to ensure that the data we collect are **reliable** and **valid**. As well as the primary data that we collect, we can also use **secondary data**.

For some investigations we can use **ICT**, either to take measurements (using a **sensor**) or to **model** the situation.

When we have collected data, we need to present it clearly, in order to see any **patterns** in the data. We can use **tables**, **bar-charts** and **line-graphs** to display the data.

On a line-graph we can draw a **line of best fit**. This lets us see any **anomalous** data that do not fit the pattern. The line of best fit may allow us to **draw a conclusion**.

We should always **evaluate** our work. This means:
– evaluating the investigation to see if it could be improved,
– evaluating our data to see if they are **reliable** and **valid**.

The main steps in an investigation

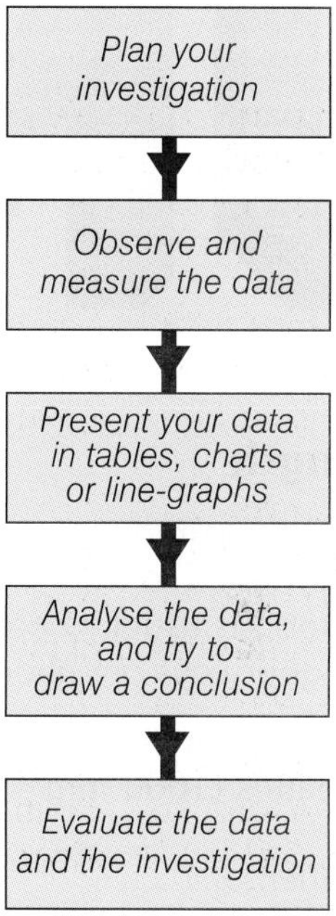

Plan your investigation
↓
Observe and measure the data
↓
Present your data in tables, charts or line-graphs
↓
Analyse the data, and try to draw a conclusion
↓
Evaluate the data and the investigation

Glossary

Data: a series of measurements, used as evidence.

Variables These are things that you can vary or change during your investigation. There are 3 main types:
- **independent** (or input) variable. This is the thing that ***you*** decide to change.
- **dependent** (or outcome) variable. This is the variable that changes as a result. It is the variable that you measure.
- **control** variables. These are all the variables that must not change, so that it is a fair test.

For example: **Variables**
When you stretch an elastic band,
- the independent variable is the force that you apply,
- the dependent variable is the length of the elastic (which you can measure).

Reliable evidence is data that we can trust. If someone else did the same experiment would they find the same evidence? Your evidence will be more reliable if you repeat the readings.

For example: **Reliable**
Measuring the time of a pendulum more than once, and taking an average, gives you more reliable data.

Valid evidence is data that measures what you intended, and is directly relevant to your investigation.

For example: **Valid**
Measuring the volume of food is not valid evidence of the amount of energy in it.

Secondary evidence is data collected by someone else. You may find it in a book or on the internet, but you should always check to see if it is reliable and valid.

For example: **Secondary data**
Data on road safety published by a car manufacturer ...but it may be biassed.

chapter 2

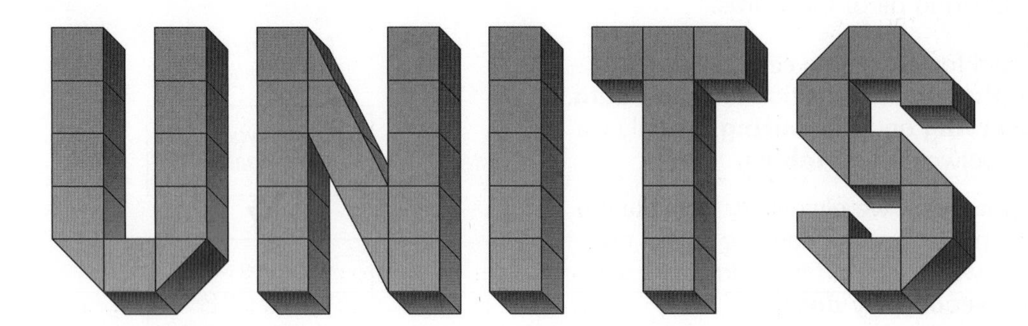

▶ Length

Length is measured in a unit called the ***metre*** (often shortened to m).
A door knob is usually about 1 metre from the ground; doorways are about 2 m high.

We often use centimetres (100 cm = 1 metre) or millimetres (1000 mm = 1 metre).

Experiment 2.1

a) Look at a metre rule. Which marks are centimetres and which are millimetres?
b) It is useful to know the length of your hand-span. Mine is 22 cm (0.22 m); what is yours?
c) Use the metre rule to measure
   – the length of your foot
   – your height.
   Write down your answers in mm and also in m.

To measure short lengths very accurately we can use vernier calipers or a micrometer.

When measuring in Physics we try to do it as accurately as we can.
Professor Messer is trying to measure the length of a block of wood with a metre rule but he has made at least six mistakes.

How many mistakes can you find?

Experiment 2.2
Measure the length of a block of wood taking care not to make any of the Professor's mistakes.

▶ Mass

If you buy a bag of sugar in a shop, you will find the ***mass*** of sugar marked on the bag. It is written in ***grams*** (g) or in ***kilograms*** (kg). 'Kilo' always means a thousand, so 1 kilogram = 1000 grams.

In Maths and in Physics, a 'k'
Means a thousand of whatever you say
For grams and for metres
And even, for teachers,
The size of their annual pay.

The mass of this book is about 1 kilogram.
People often get confused between mass and weight, but they are ***not*** the same (see pages 65 and 68).

Experiment 2.3 ⚠
Lift some masses labelled 1 kg, 2 kg, 5 kg and 1 g.

'kilo' is not the only prefix:

Mega (M)	= 1 million	=	1 000 000
kilo (k)	= 1 thousand	=	1 000
centi (c)	= 1 hundredth	=	$\frac{1}{100}$
milli (m)	= 1 thousandth	=	$\frac{1}{1000}$
micro (μ)	= 1 millionth	=	$\frac{1}{1\,000\,000}$
nano (n)	= 1 thousand-millionth	=	$\frac{1}{1\,000\,000\,000}$

▶ Time

In Physics, time is always measured in ***seconds*** (sometimes shortened to s).
You can count seconds very roughly, without a watch, by saying at a steady rate: ONE (thousand) TWO (thousand) THREE (thousand) FOUR . . .

Experiment 2.4 ⚠
Use a stopclock or stopwatch to measure the time for a complete swing of a **pendulum** (see page 99) or the beating of your heart.
What is the time for 100 of your heartbeats?
What is the time for one heartbeat?
By how much does it change if you run upstairs?

All the other units you will meet in this book are based on the metre, the kilogram and the second. They are called ***SI units***.

Very large and small numbers

For very large or very small numbers, we sometimes use a shorthand way of writing them, by counting the number of zeros (see also page 391).
For example:

a) 1 million = 1 000 000 (6 zeros) = 10^6

b) 2 million = 2 000 000 = 2×10^6

c) 0.000 001 = $\frac{1}{1\,000\,000}$ (1 millionth) = 10^{-6}

In this shorthand way, write down:
one thousand, one thousandth, 10 million, one hundredth, 3 million, 30 thousand.

Approximate length of time in seconds	Events
10^{18}	Expected lifetime of the Sun
10^{17}	Age of the Earth
10^{15}	Time since the dinosaurs lived
10^{13}	Time since the earliest human
10^{10}	Time since Isaac Newton lived
10^{9}	Average human life span
10^{7}	A school term
10^{5}	One day
10^{0}	One second
10^{-2}	Time for sound to cross a room
10^{-7}	Time for an electron to travel down a TV tube
10^{-8}	Time for light to cross a room
10^{-11}	Time for light to pass through spectacles
10^{-22}	Time for some events inside atoms

chapter 3

I wish I could find some use for all that energy!

Energy can exist in different forms, as you can see in the cartoon.

People get their energy from the **chemical energy** in their food.

Cars run on the chemical energy in petrol. A firework in the cartoon has chemical energy which it transforms to **thermal energy (heat)** and **light** and **sound** energy when it explodes.

Some forms of energy are called ***potential*** energy.
One kind of potential energy is the **elastic energy** (also called **strain** energy) stored in the stretched elastic of a catapult.
The bucket over the door also has some stored potential energy, called **gravitational potential energy**. When the bucket falls down, this gravitational energy is transferred to movement energy.

The moving pellets from the catapult and the moving people all have movement energy, called **kinetic energy**.

The television set is taking in **electrical energy** and transferring it to thermal energy and light and sound energy.

Another form of energy is **nuclear energy**, which is used in nuclear power stations.

These forms of energy are shown in the diagram on the opposite page (see also page 98).

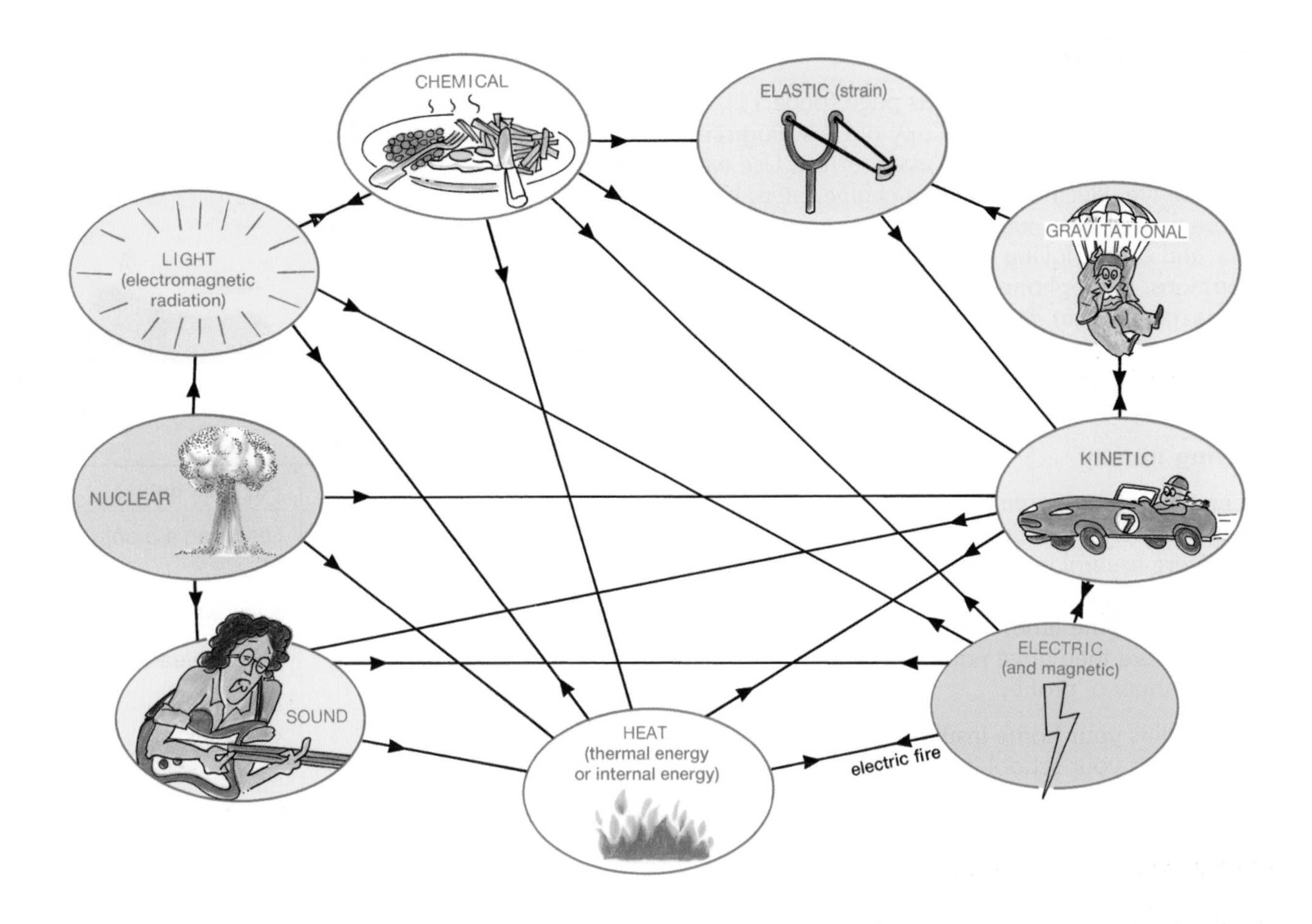

The connecting lines on the diagram show the different ways that energy can be changed from one form to another.

See if you can decide what the energy changes are in the following objects.

For example:
A firework changes ***chemical*** energy to ***thermal*** energy, ***light*** and ***sound*** energy.
Copy and complete these sentences.

1. A TV set changes .... energy to .... energy.
2. A match changes .... energy to .... energy.
3. A light-bulb changes .... energy to .... energy.
4. A catapult changes .... energy to .... energy.
5. A falling bucket changes .... energy to .... energy.
6. An electric fire changes .... energy to .... energy.
7. A human body changes .... energy to .... energy.
8. A microphone changes .... energy to .... energy.
9. An atomic bomb changes .... energy to .... energy.
10. A car engine changes .... energy to .... energy.

You can learn more about energy changes and transfers in chapter 16.

Energy changes

Taking in chemical energy

In the diagram on the previous page (page 11), one energy change has been labelled 'electric fire'. Copy out the diagram into your book and then add the correct label to every arrow. Use words from the following list: coal fire, electric fire, steam engine, atom bomb (on four arrows), car engine, battery, loudspeaker, dynamo, very hot object, friction, bow and arrow, falling parachutist, cricket ball rising in the air, vibrations, microphone, thermocouple, solar cell, solar panel, plants, glow-worm, fluorescent lamp, girl landing on a trampoline, hanging a weight on a spring, an arm muscle tightening, sound-absorbing material, electroplating.

Saving money

Energy costs money and we use large amounts of energy/money each day.
There are several ways of saving energy/money in your home (and making your home more comfortable at the same time). The table shows the time taken before they have paid for themselves and start to show a 'profit'.

Method	Payback times*
Lagging the hot-water cylinder	Less than a month
Draught excluders	A few weeks
Lagging the loft (see page 43).	About 3 years
Wall cavity insulation	4–7 years
Double glazing in windows	About 10 years

*all these times are shorter if you get a government grant

How well is your home insulated?
How well is your school insulated? Write a list of recommendations.

Energy is measured in units called ***joules***.
The joule is a small unit. To lift this book through a height of 10 cm needs about 1 joule.
When you walk upstairs you use over 1000 joules.

The diagram below shows the energy (in joules) involved in different events.
(Remember: 10^5 = number with 5 zeros = 100 000 and $10^{-5} = \frac{1}{100\,000}$)

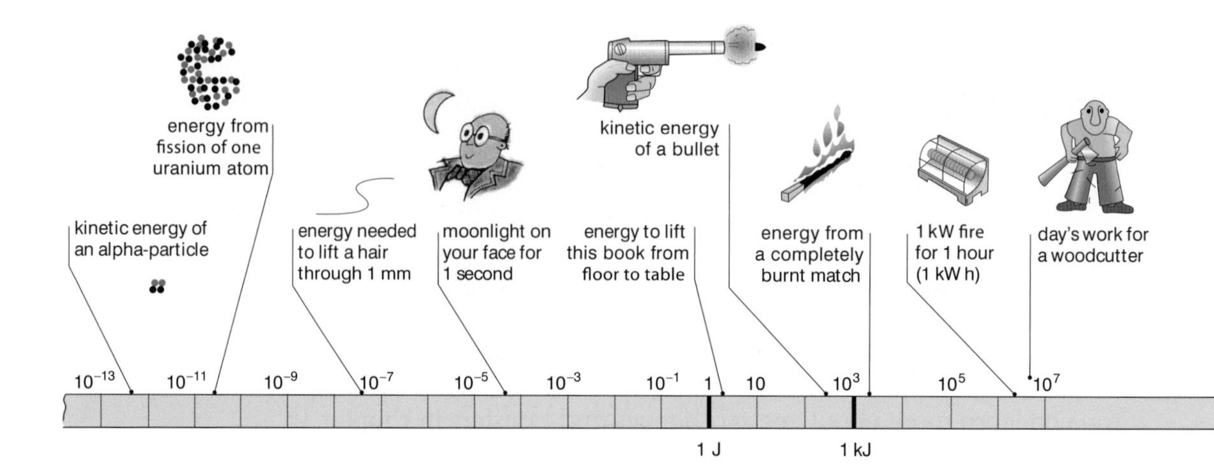

▶ The energy crisis

We are only just beginning to realise fully that our planet Earth is a spaceship with ***limited*** food and fuel (and we are taking on more passengers each year as the population increases).

Our supplies of energy cannot last for ever.

Oil and ***natural gas*** will be the first to disappear. If the whole world used oil at the rate it is used in America and Europe, our oil supplies would end in about 4 years! They are ***non-renewable***.
As it is, the world's oil supplies might last for about 40 years.

How old will you be then? In what ways will your life be different without oil (and therefore without petrol and plastics)?

Natural gas will last a bit longer – perhaps 60 years.
Coal will last longer – perhaps 300 years with careful mining.

Nuclear energy might help for a while – but it causes problems due to the very dangerous radioactive waste that is produced (page 350). Also, each power station lasts only about 30 years and is difficult to dismantle because of the radioactivity.

We ***waste*** huge amounts of energy. It takes over 5 million joules of energy to make one fizzy-drink-can and we throw away 700 million of them each year! Making paper and steel uses particularly large amounts of energy, but very little is recycled.

We ***must*** find new ways of obtaining energy. The Sun's energy is free but it is not easy to capture it. Governments are looking for new sources of energy (see next page). They will probably not give enough energy for the future. Our main hope is that the H-bomb ***fusion*** process (page 156) will eventually be controlled.

Power stations are wasteful (see p. 104). The overall efficiency (from power station to your home) is about 25%. Three-quarters of the energy is entirely wasted! The wasted energy could be used to heat nearby homes.

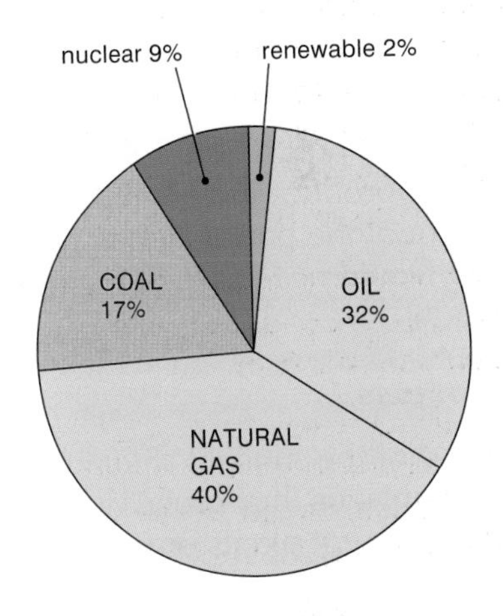

Energy sources of Britain at present. What will happen when the fossil fuels (oil, gas, coal) run out?

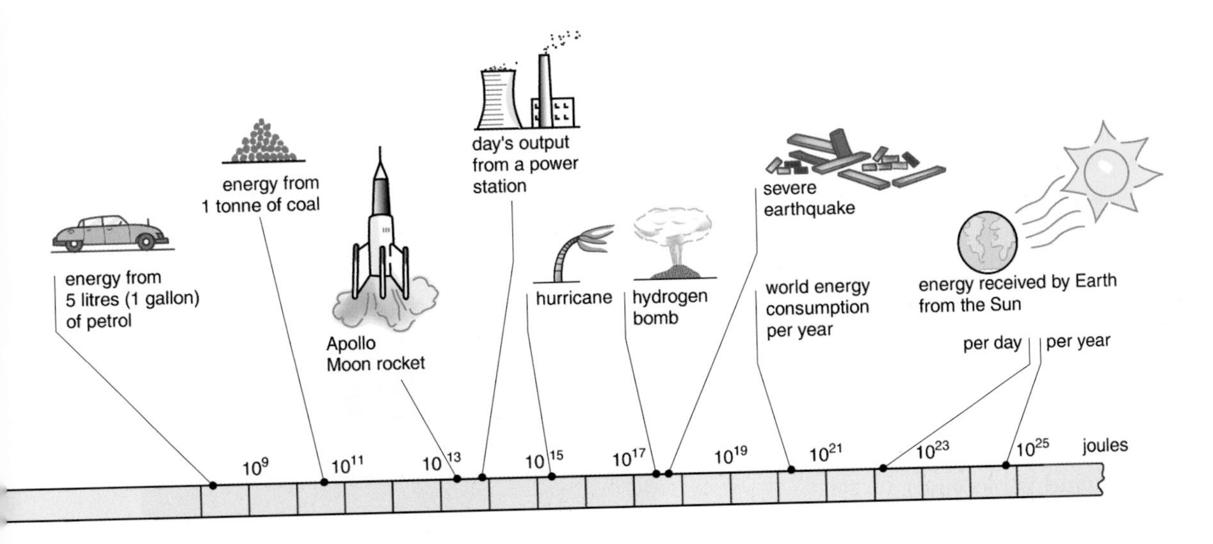

▶ Renewable sources of energy

Some sources of energy are ***renewable***. They are not used up like coal or oil.
Physicists and engineers are working hard to develop better machines to use these sources of energy.

Source of energy	Original source
Solar	Sun
Biomass	Sun
Wind	Sun
Waves	Sun
Hydro-electric	Sun
Tides	Moon
Geothermal	Earth

Solar energy

The Earth receives an enormous amount of energy directly from the Sun each day, but we use very little of it. Some homes have ***solar panels*** on the roof (see page 50). In hot countries solar ovens can be used for cooking (see page 48) and for producing electricity in ***solar thermal towers***.

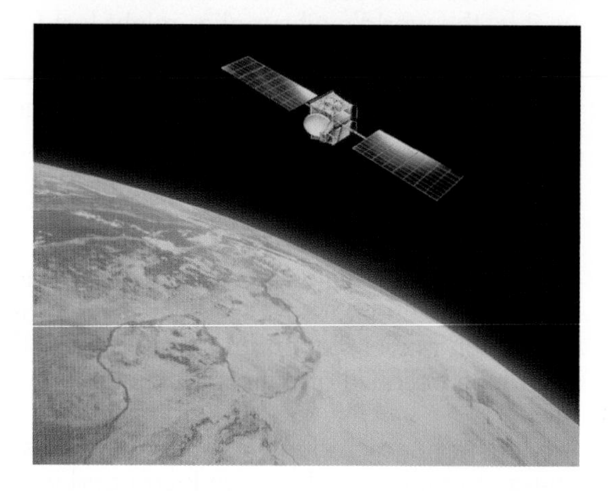

Space ships and satellites use ***solar cells*** to convert sunlight into electricity. You may have seen calculators powered by solar cells.

Covering part of the Sahara Desert with solar cells would produce energy but would be very expensive. To equal the power of one modern power station you would need 40 square kilometres of solar cells.

Biomass

Some of the sunlight shining on the Earth is trapped by plants, as they grow. We use this ***biomass*** when we eat plants or when we burn wood.

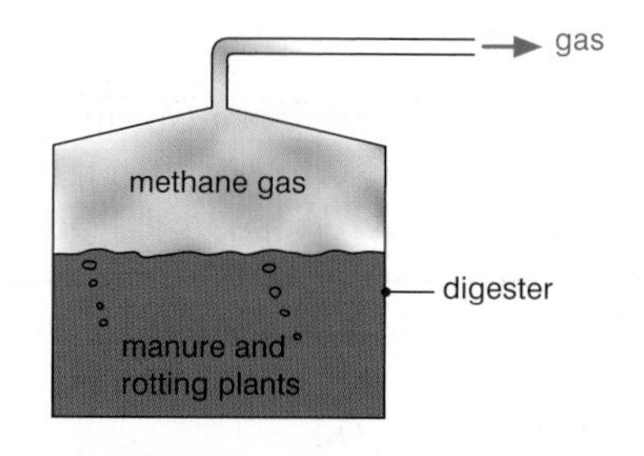

In Brazil they grow sugar cane and then use the sugar to make alcohol ('bio-ethanol').
The alcohol is then used in cars, instead of petrol.

Rotting plants can produce a gas called methane which is the same as the 'natural gas' we use for cooking. If the plants rot in a closed tank, called a ***digester***, the gas can be piped away and used as fuel for cooking. This is often used in India.

Wind energy

This energy also comes from the Sun, because winds are caused by the Sun heating different parts of the Earth unequally.

Modern 'wind-generators' are very efficient but it takes about 2000 very large wind turbines to provide as much electricity as one modern power station (and only if the wind is blowing).

Wave energy

Waves are caused by the winds blowing across the sea. They contain a lot of free energy.

One method of getting this energy is to use large floats which move up and down with the waves.
The movement energy can be converted to electricity. However, we would need about 20 kilometres of floats to produce as much energy as one power station.

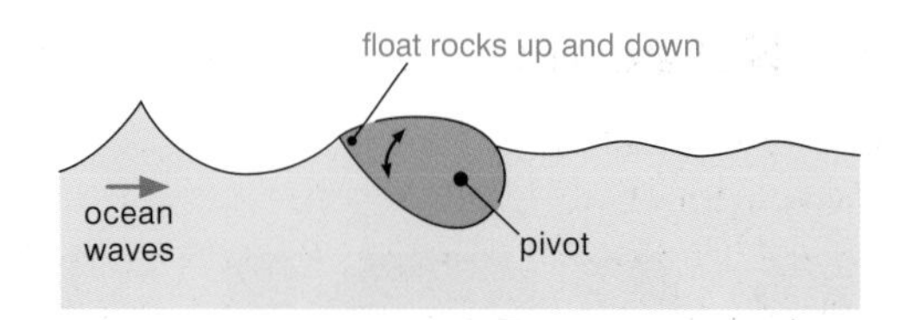

Hydro-electric energy

Dams can be used to store rain-water, and then the falling water can be used to make electricity (see experiment 16.5 on page 101).

This is a very useful and clean source of energy for mountainous countries like Norway and China.

The same idea can be used to store energy from power stations that cannot easily shut down. At night, when demand is low, spare electricity can be used to pump water up to a high lake. During the day, the water can be allowed to fall back down, to produce electricity when it is needed.

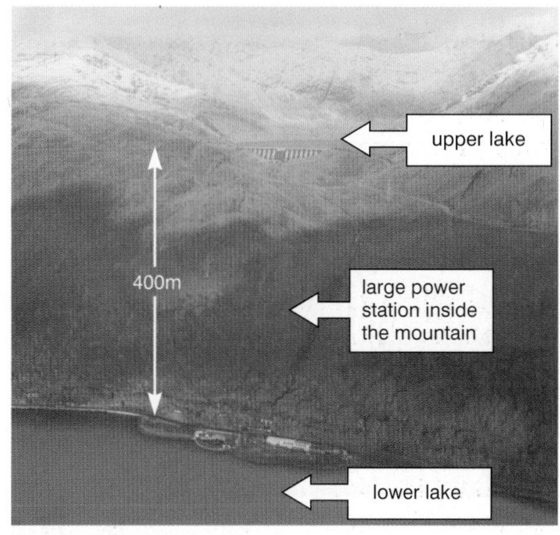

A pumped storage scheme in Scotland

Tidal energy

As the Moon goes round the Earth it pulls on the seas so that the height of the tide varies.

If a dam is built across an estuary, it can have gates which trap the water at high tide. Then at low tide, the water can be allowed to fall back through the dam and make electricity (see experiment 16.5).

A tidal power station in France

Geothermal energy

The inside of the Earth is hot (due to radioactivity, see chapter 39). In some parts of the world (like New Zealand) hot water comes to the surface naturally. In other countries cold water is pumped down very deep holes and steam comes back to the surface.

Summary

Energy exists in several different forms:
chemical, electric, magnetic, kinetic, potential (elastic and gravitational), sound, nuclear, electromagnetic radiation (including light) and thermal energy (internal energy or heat).

Energy can be changed from one form to another.

Some sources of energy are non-renewable and will be used up: coal, oil, gas, nuclear.
Other sources are renewable: solar, biomass, wind, wave, hydro-electric, tidal and geothermal.
See also chapter 16, pages 103–106.

chapter 4

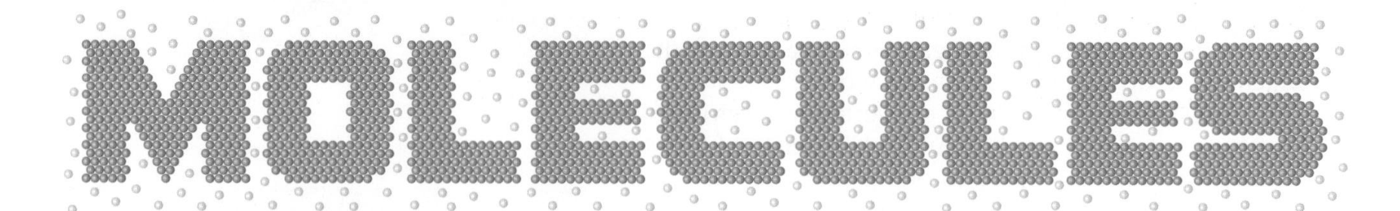

Do you know that as you are reading this sentence you are punched on the nose more than 100 billion billion times!!?

This is because the air is made up of billions and billions of tiny particles called ***molecules*** – rather like the title at the top of this page is made up of particles. But the molecules are so small that we cannot see them even through a powerful microscope. Since each molecule is very small and light, it does not hurt as much as when a big fist hits your nose.

▶ High speed gas

The molecules hitting your face are moving very fast. Their average speed is about 1000 m.p.h. (1600 km/h), but with some moving much faster and some slower.

The billions and billions of air molecules around you are bouncing and colliding with each other at this very high speed. It's just as though the room is full of tiny balls bouncing everywhere and filling the whole room.

If the air is cooled down, the molecules have less energy and move more slowly.
If the air is heated up, the molecules have more energy and so move faster (and hit your nose harder).

This idea about molecules is called the ***kinetic theory.***

▶ Three states of matter

Every substance can exist in three states: ***gas, liquid*** and ***solid.***

The air in this room is a gas. But if it is cooled down it can become liquid or even solid.

In the same way, water can be a solid (ice), a liquid, or a gas (steam):

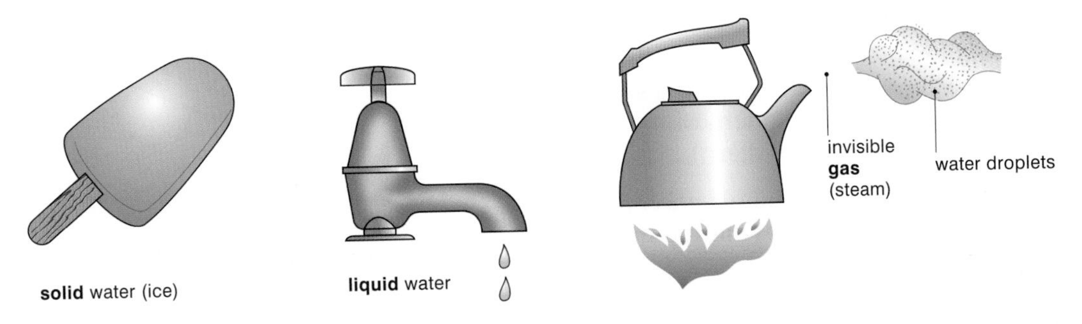

In a **gas**, the particles are moving very quickly.

In a **liquid**, the particles are moving more slowly and are held together more closely by forces between the particles. This is why a large volume of gas ***condenses*** into a smaller volume of liquid.

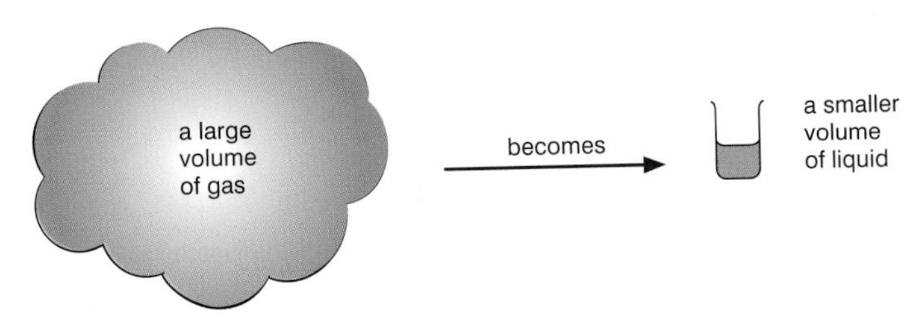

These forces between particles make the liquid have a definite size, although it can still change its shape and flow.

In a **solid**, the forces between particles are so strong that the solid has a definite shape as well as a definite size.

There are two kinds of forces between molecules:
– ***attractive*** forces if molecules try to move apart
– ***repulsive*** forces if molecules try to move closer.
Normally these forces balance.
If you try to stretch a solid (and so pull the particles apart) you can feel the attractive forces.
If you squeeze a solid (and so push its particles closer together) you can feel the repulsive forces.

These forces are ***electric*** forces (see page 241).

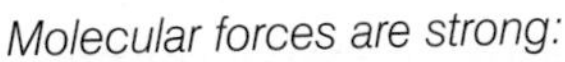
Molecular forces are strong:

▶ Experiments

You will need some marbles (to act as 'molecules') and a flat box or tray with vertical sides.

Experiment 4.1 A solid
Place the marbles in the tray. Tilt the tray slightly so that the 'molecules' collect at one end.

You can see that the group of 'molecules' has a definite size and shape. This is imitating a ***solid*** (a very cold solid). Sketch the pattern of the particles (a regular ***crystalline*** pattern).

The particles in a solid are continually vibrating. If the solid gets hotter, its particles vibrate more.

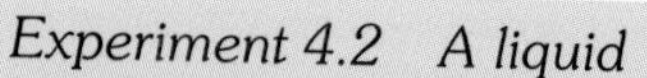

Experiment 4.2 A liquid
Now rock the tray gently from side to side.
You can see that the group of 'molecules' now has more energy and does not have a definite shape. This is imitating a ***liquid***.

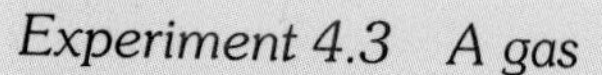

Experiment 4.3 A gas
Next, with the tray flat on the table, shake the tray to and fro and from side to side to give the 'molecules' a lot more energy.
You can see that the 'molecules' fill all the available space in the tray.
This is imitating a ***gas***.

Heat up the gas by shaking the tray faster.

Experiment 4.4 Brownian movement
No-one has ever seen an air molecule, but here is an experiment to help us believe that they exist.

A microscope is used to look into a small glass box which contains some smoke as well as air molecules.

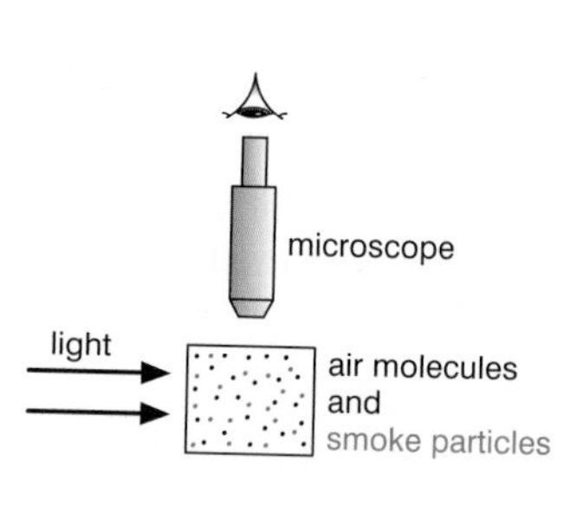

In a bright light, the smoke particles show up as bright specks which move in a continuous and jerky random movement.
This is called ***Brownian movement***.

It happens because the smoke particles (like your nose) are being punched by molecules. Because the smoke particles are so light, they twist and jerk, as you can see through the microscope.

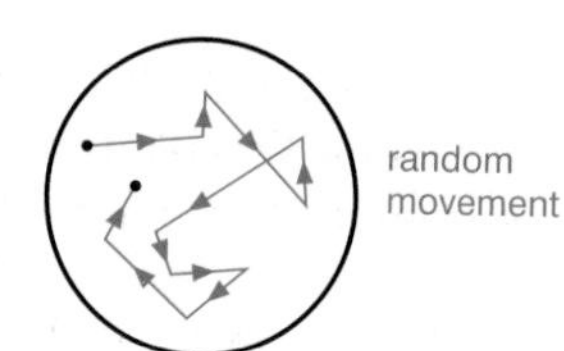

Experiment 4.5
While shaking your marbles tray to imitate a gas, drop into it a ***larger*** marble (to act as a particle of smoke).
Watch the random movement of the 'smoke particle' as it is punched by the 'molecules'. This is Brownian movement.

▶ Diffusion

When food is being cooked in the kitchen, it can be smelt in other rooms in the house, even when there are no draughts to move the air.

This is because molecules from the food are moving around at high speed, and after billions of collisions eventually reach all parts of the house. This is called ***diffusion***.

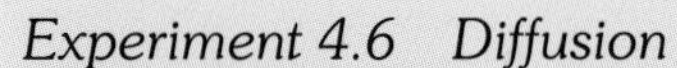

Experiment 4.6 Diffusion
While shaking your marbles tray to imitate a gas, drop into one corner a coloured marble (acting as a 'smell molecule'). Watch it while you keep shaking. You will see that eventually the 'smell molecule' will diffuse to any part of the tray.

Draw a rough sketch of the random movement of the 'molecule'.

Experiment 4.7 Diffusion in a liquid ⚠
Fill a tall beaker with water and leave it for a while so that it becomes still and at room temperature.
Carefully drop in a single crystal of potassium permanganate. When this dissolves it makes the water purple.
Leave the beaker where it will not be touched or knocked for several weeks and observe it.

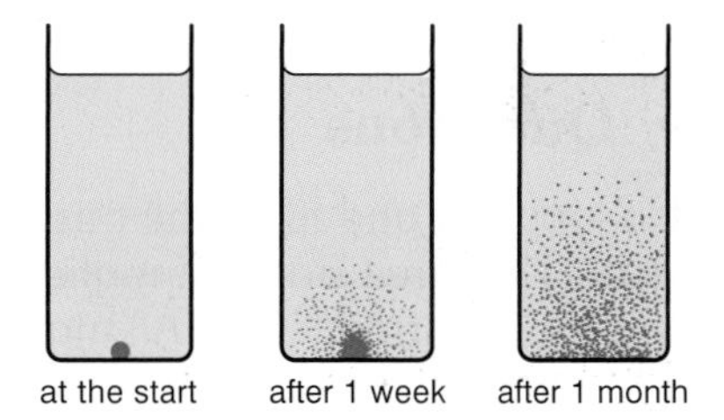

It takes a long time for liquids to diffuse. Why is this?

Experiment 4.8 Diffusion in a gas ⚠
Your teacher may be able to show you the diffusion of some bromine. This is a gas that you can see (it is brown), but it is very poisonous so special apparatus and care must be used. When the tap is opened, the bromine runs down and starts diffusing.

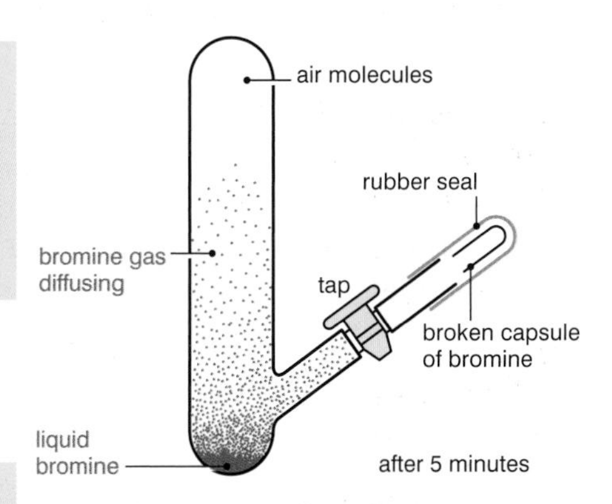

Why is the diffusion in this experiment faster than in the previous experiment?

Experiment 4.9 Diffusion into a vacuum ⚠
Your teacher may be able to repeat the bromine experiment but with the main tube completely empty of air molecules (by using a vacuum pump to take out all the molecules).
What happens this time as the tap is opened to let in the bromine? (Don't blink or you might miss it.)

Because there are no air molecules to get in the way, this shows you the speed of bromine molecules – it is about 500 m.p.h. (800 km/h) at room temperature!

Summary

All substances can exist in 3 states ***solid***, ***liquid*** and ***gas***. All substances are made of particles (***atoms*** and ***molecules***).

Molecules can ***diffuse*** through other molecules.

If the substance is heated, the molecules have more energy and move faster.

Brownian movement of smoke particles helps us to believe in molecules.

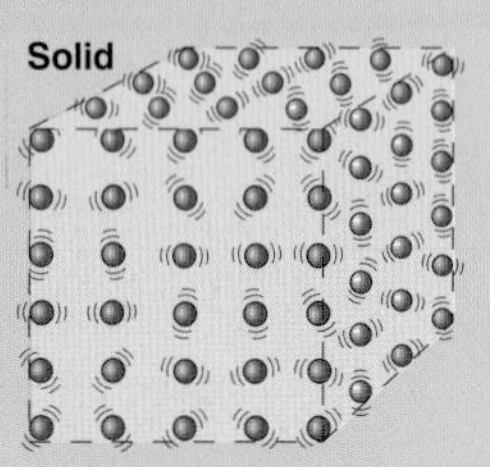

Particles in regular rows, vibrating about fixed positions. Strong forces of attraction keep it a definite shape.

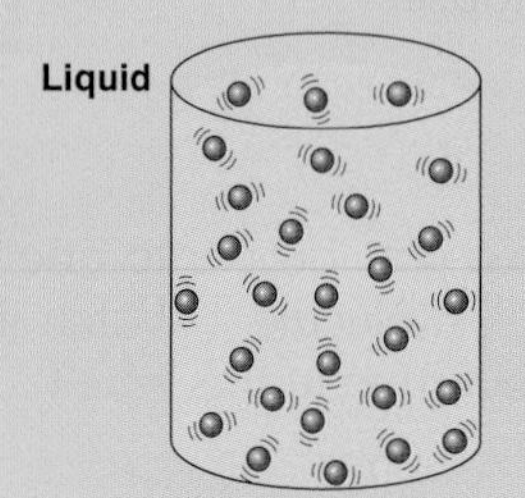

Particles can move around, but still held loosely together by forces of attraction.

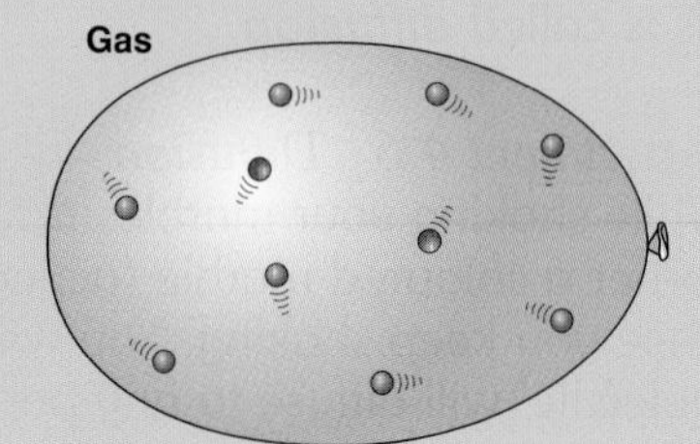

Particles are far apart, moving at high speed, and colliding with the walls of the balloon.

▶ Questions

1. Copy out and fill in the missing words:

a) The three states of matter are . . . . , ... and . . . . . All matter consists of tiny . . . . travelling at . . . . speed.

b) A solid has a definite size and a definite . . . . because the forces between its . . . . are very . . . . . A liquid has weaker forces between its . . . . and so does not have a definite . . . . .

c) If a substance is cooled down, its molecules move at a . . . . speed. When some molecules pass through some other molecules, we call it . . . . . The twisting and jerking of smoke particles when they are hit by air . . . . is called . . . . .

2. State whether each of these is describing a solid, a liquid, or a gas:

a) A fixed size and shape.
b) No definite size or shape.
c) Particles vibrate about a fixed position.
d) A fixed volume but no definite shape.
e) The particles move at high speed.
f) Almost no attraction between particles.

3. Explain why perfume can be smelled some distance away from the person wearing it.

4. Explain why

a) solids have a definite shape but liquids flow
b) solids and liquids have a fixed size but gases fill whatever container they are in.

5. a) Explain what is meant by Brownian movement and how it helps us to believe in molecules.
b) What change would you expect to see in the movement if the air was cooled down?

6. Copy and complete the rhyme and explain why you do not see this happening:

A rather small student called Brown,
Was asked why he danced up and down,
He said "Look, you fools,
It's the air . . . . ,
They constantly knock me around."

7. A tin can containing air is sealed. If it is then heated, what can you say about:

a) the average speed of the molecules,
b) how often the molecules hit the walls of the can,
c) how hard the molecules hit the walls,
d) the air pressure inside the can?

Further questions on page 61.

chapter 5

Expansion

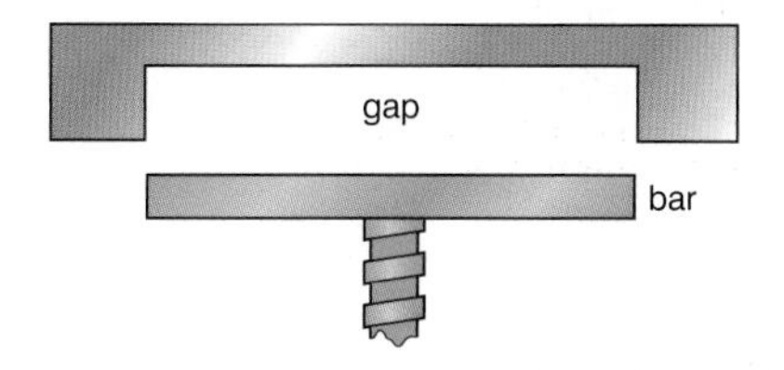

Experiment 5.1
This metal bar will ***just*** fit into the gap when they are both cold. Heat the bar with a Bunsen burner. Does it fit the gap now?

What happens when it cools down?

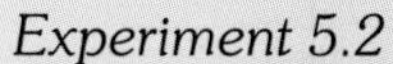

Experiment 5.2
This metal ball will ***just*** slip through the metal ring when they are both cold.

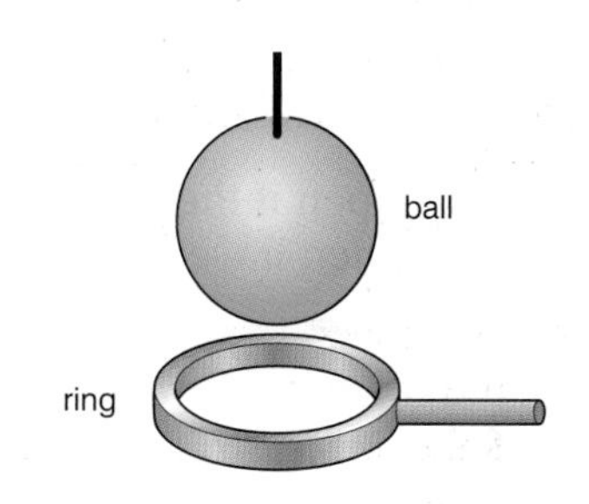

Does it go through when the ball is heated?

When objects get hotter they grow bigger. We say they ***expand***.
When objects cool down they get smaller. We say they ***contract***.
It is difficult to see this change in size because it is so small.

Experiment 5.3 ⚠
This thick bar is heated with a Bunsen burner and then the big nut is tightened.

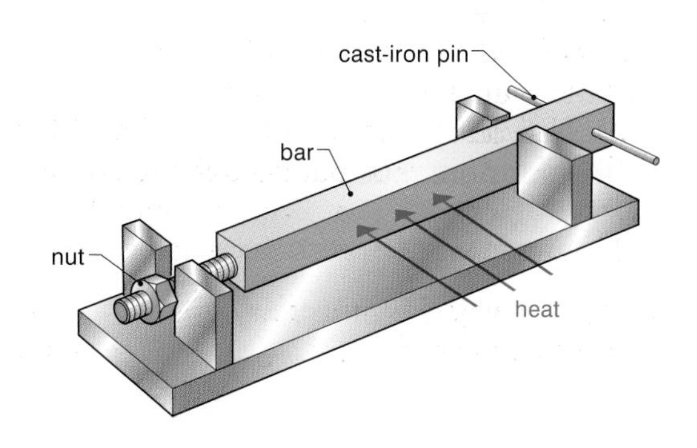

What happens to the cast-iron pin when the bar cools down? (Do not stand too near.)

This shows that:
– when bars ***contract*** you get big ***pulling*** forces.
– when bars ***expand*** you get big ***pushing*** forces.

▶ Expansion can cause trouble

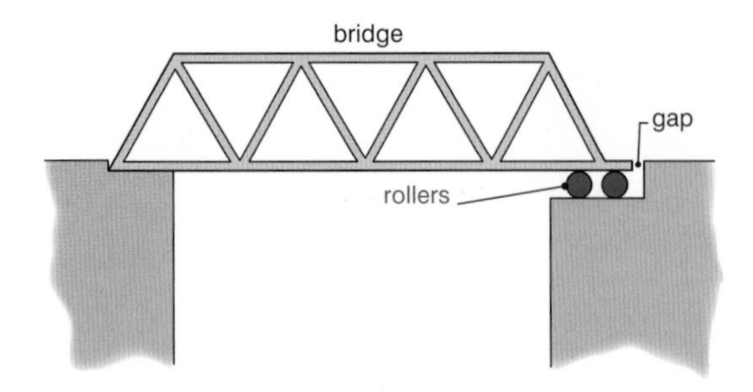

Bridges are often made of big steel bars. When they get hotter, they get longer.
The Forth Railway Bridge is more than 1 metre longer in summer than in winter.

Bridges are usually put on rollers so that they can expand and contract without causing damage.

There must be an ***expansion gap*** in the road at the end of a bridge. Have you seen one?

What would happen if there was no gap?

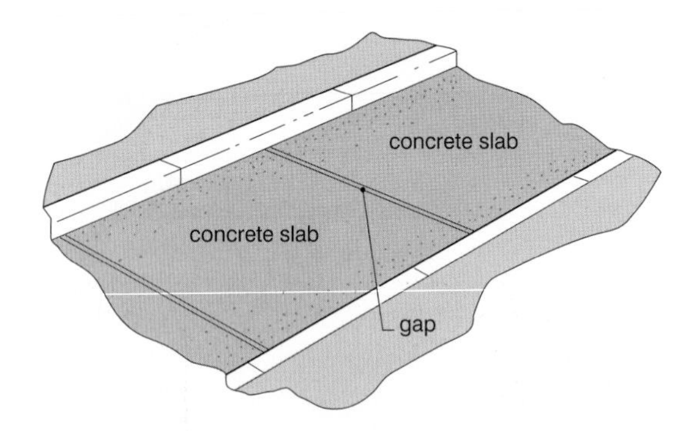

Roads are often made of large concrete slabs. They also expand. There are ***expansion gaps*** between the slabs, filled with a soft substance which can be squeezed easily in hot weather.

If you look carefully, you may find similar expansion gaps in the floor of your school.

Telephone wires look like this in summer:
What will they look like in winter?

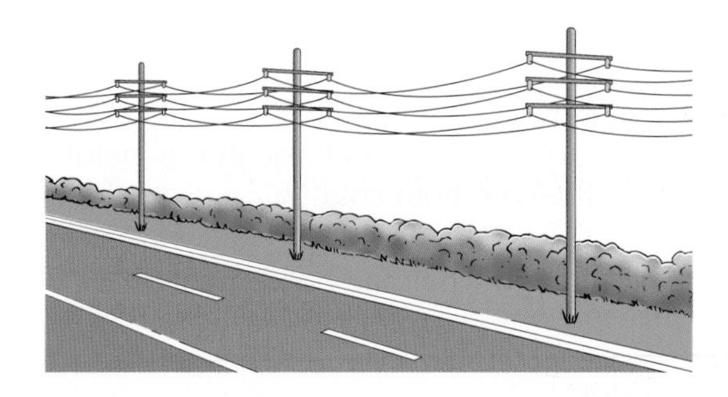

What would happen if the wires were made tight in summer and then it went cold?

Thick glass beakers and milk-bottles sometimes crack if boiling water is put in them. This is because the inside tries to expand while the outside stays the same size.

'Pyrex' glass beakers do not crack because 'pyrex' does not expand as much as ordinary glass.

Steel railway lines need expansion gaps. Sometimes the gaps are at the end of each length of rail (you can hear the clicks as the wheels go over them). Nowadays many rails are welded together and the gaps are much farther apart.

A railway mechanic was sent
To lay down some track across Kent.
He forgot to design
Some gaps in the line,
And when it got hot, it went bent.

▶ Physics at work: Using expansion

Experiment 5.4 ⚠

Look at a ***bi-metallic strip***. It is made of two metal strips (often brass and iron) which are placed side by side and then riveted or welded together.

Heat it gently with a Bunsen burner. What happens?

When it is heated, brass expands more than iron and so the strip bends with brass on the outside because it is longer.

Which way would it bend if you made the bi-metallic strip very cold?

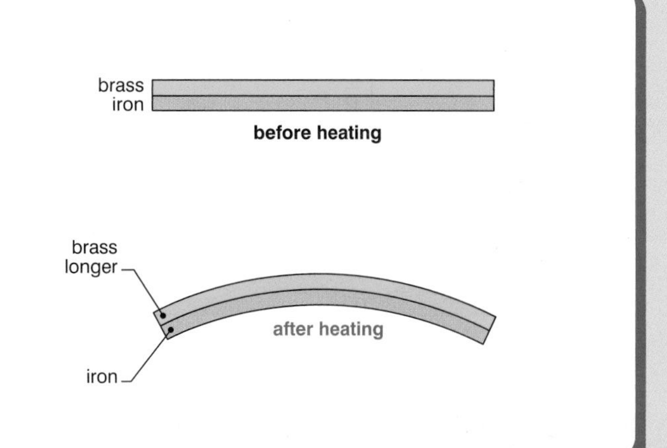

A fire alarm

Experiment 5.5 ⚠

The diagram shows how you can make a fire-alarm.

Hold the bi-metallic strip in a clamp, with the brass on top.

Heat the bi-metallic strip gently. What happens?

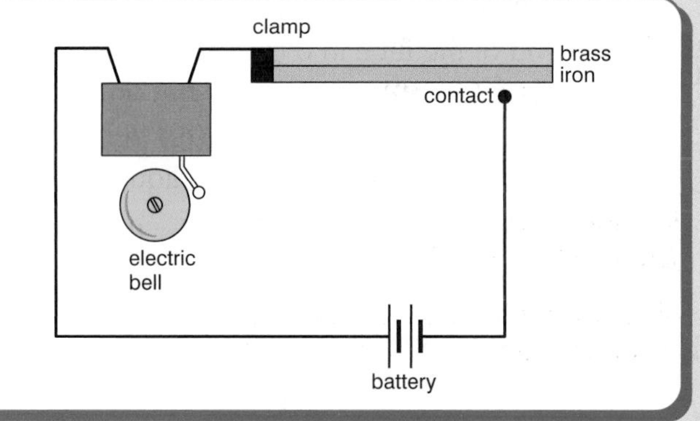

Electric thermostat

A ***thermostat*** is used to keep something at the same temperature, without getting too hot or too cold.
In the diagram, the electric current is flowing through the contact and the bi-metallic strip to the heater.

What happens if it gets too hot?
What happens later when it cools?

This is used in electric irons, to keep the iron at the correct temperature.

Which way would you turn the control knob if you wanted a cooler iron, for ironing nylon?

See also the circuit-breaker on page 272.

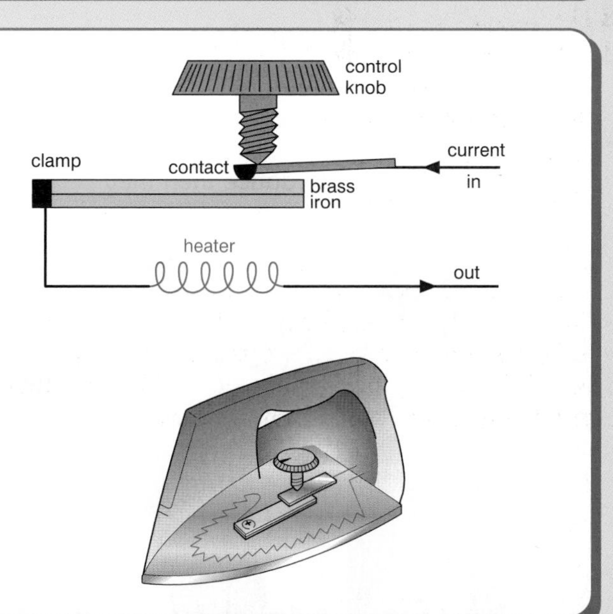

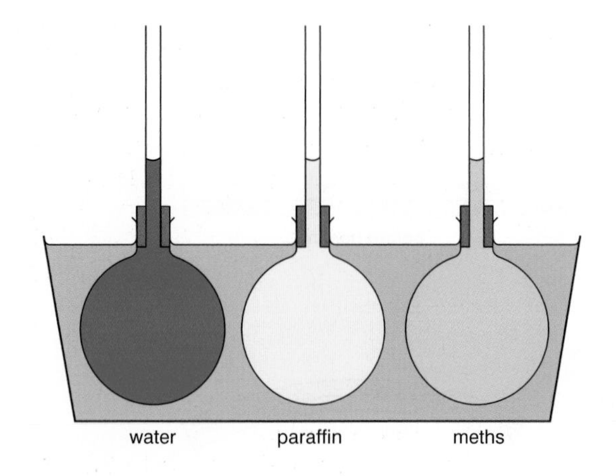

▶ Expansion of liquids

Experiment 5.6 ⚠
We need three glass bottles or flasks which are all the same size and have narrow tubes coming out of the top.

Fill them with three different liquids up to the same level as in the diagram. Put them in a dish and then fill the dish with hot water.

What happens? Which liquid expands the most? Do liquids expand more or less than solids?

Liquids expand more than solids.

▶ Expansion of gases

Experiment 5.7
Get a glass flask with a rubber bung and a tube as in the diagram: Dip the end of the tube in a beaker of water and then warm the flask with your hands.

What do you see? What is expanding?
Do gases expand more or less than liquids?

Now take your hands off and wait for the flask to cool down.
What happens? Why?

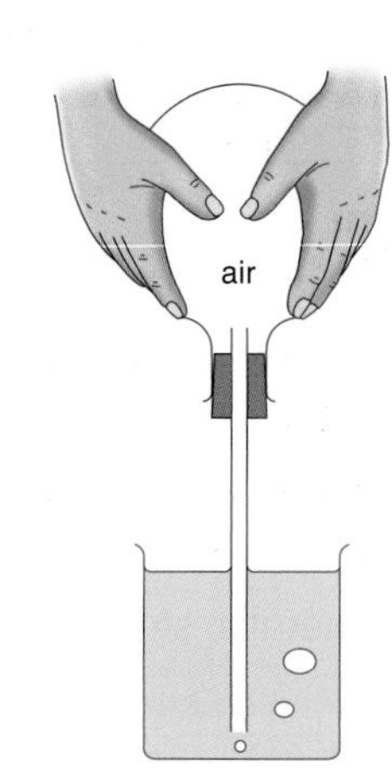

Gases expand more than liquids.
Expanding gases are used in rockets (see page 160).

▶ Explaining expansion

If a group of people are just standing still then they can stay close together and they don't take up much space.
But if the people start to dance then they take up more space – the group expands.

In the same way, if a substance is heated, the molecules start moving more and so they take up more space – the substance expands.

Notice that the molecules themselves do not get any bigger – but as a group they take up more space.

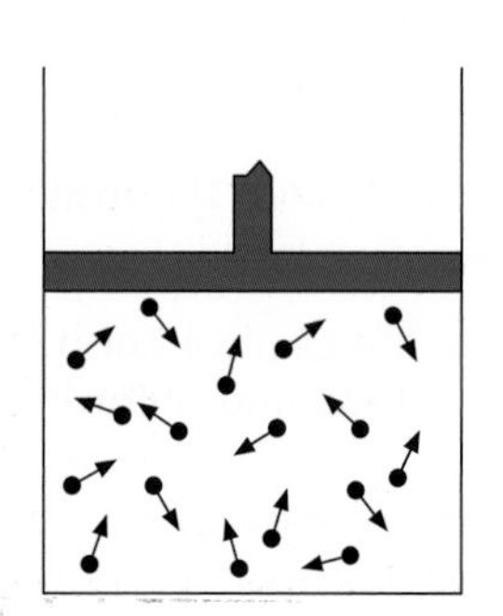
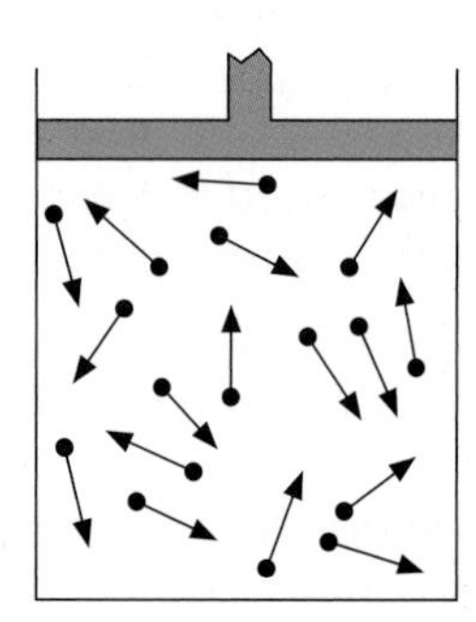

Summary

When objects are heated, they expand and push with large forces. Expansion gaps are needed in bridges, roads, etc.

Liquids expand more than solids.
Gases expand more than liquids.

A bi-metallic strip is made of two different metals fixed together. When heated, it bends (with the metal with the bigger expansion on the outside).

Substances expand because the molecules are moving more and take up more space.

▶ Questions

1. Copy out and fill in the missing words:
  a) When an object is heated it . . . . and when it cools it . . . . . If we try to prevent it from contracting, it exerts a very large . . . . . If a bi-metallic strip made of brass and iron is heated, it . . . . because . . . . expands more than . . . . and the . . . . goes on the outside of the bend.
  b) Liquids expand more than . . . . but less than . . . . .
  c) When things are heated, they get bigger because the . . . . are moving more.

2. a) If a bottle was filled with a liquid, tightly sealed and then heated, what might happen?
  b) Why does a bottle of lemonade always have space between the top of the liquid and the cap?
  c) Why might a beach-ball burst if left in the Sun?
  d) Why can a dented table-tennis ball often be mended by warming it in hot water?
  e) Why must you never put sealed bottles or 'aerosol' spray cans on a fire?

3. Can you explain this cartoon?

4. a) Draw a labelled diagram of a fire alarm and explain how it works.
  b) How could you change it to a frost alarm?

5. Here is a circuit often used in cars:

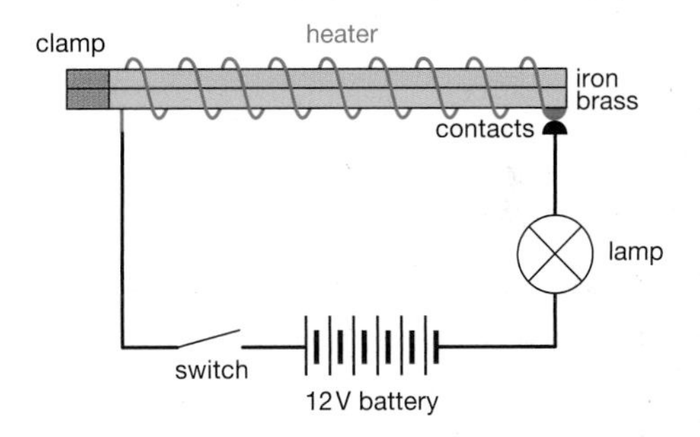

  a) When the switch is connected, what happens to the light and the heating coil?
  b) What happens to the bi-metallic strip?
  c) What happens then to the light and the heater?
  d) What happens then to the bi-metal strip?
  e) What happens next?
  f) What would you see?
  g) Where is this used on a car?

Further questions on page 61.

chapter 6

Thermometers

Thermometers are used to measure ***temperature***.
Temperature is not the same thing as the total energy in an object.

To understand this, let's compare a white-hot 'sparkler' firework with a bath-full of warm water:

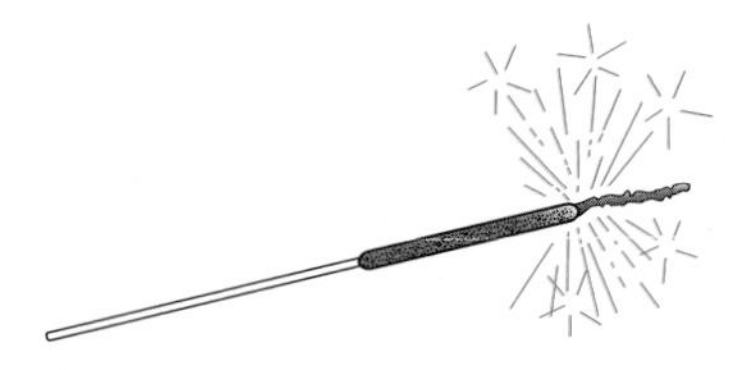

The tiny 'sparkles' are at a very high temperature but contain little energy because they are very small

The water is at a lower temperature but it contains more internal energy because it has many more molecules

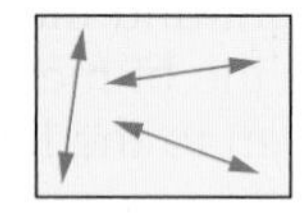

Each sparkle contains a few molecules vibrating at very high temperature but with little total energy

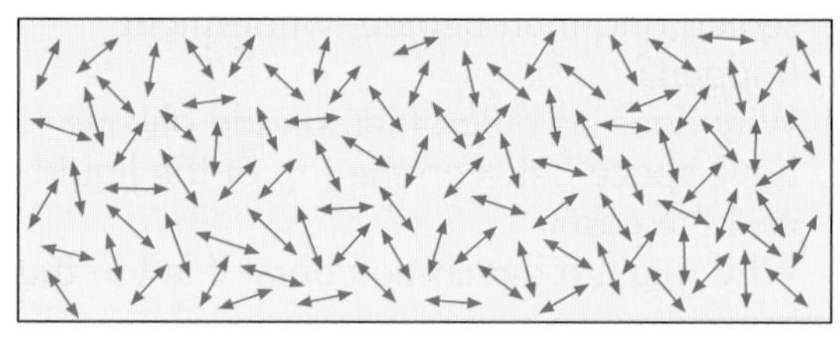

The bath has many molecules, each vibrating at low temperature but with more total energy

You can see that the bath has more energy.
Its molecules have more ***internal energy***.
To raise the temperature of an object you must give more energy to its molecules.
Heat energy travels from a hot object to a cold object.

A laboratory thermometer

This uses the expansion of a liquid to measure temperatures accurately. Look at the diagram opposite. If the liquid gets warmer, it expands along the tube, where there is a scale of numbers marked °C or ***degrees Celsius*** (or degrees centigrade). There is a vacuum (no air) above the liquid so it can move easily along the tube.

To make the thermometer ***sensitive,*** with a long scale, it should have a narrow-bore tube and a large bulb.

To make the thermometer ***quick-acting,*** it should have a bulb made of thin glass so that the heat can get through easily.

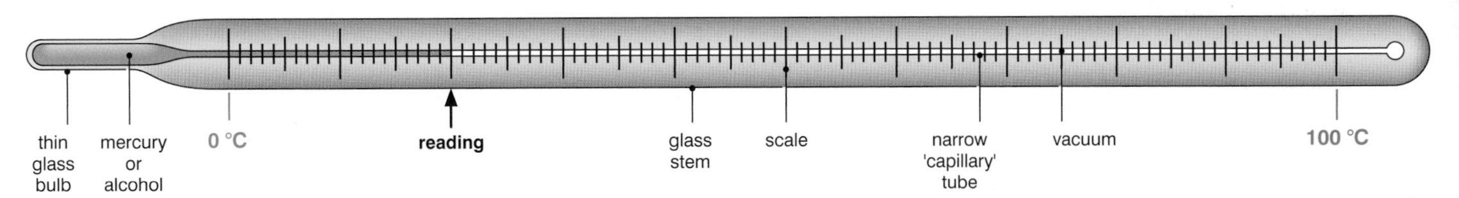

What is the reading on the thermometer in the diagram? (This is the temperature of a warm room.)

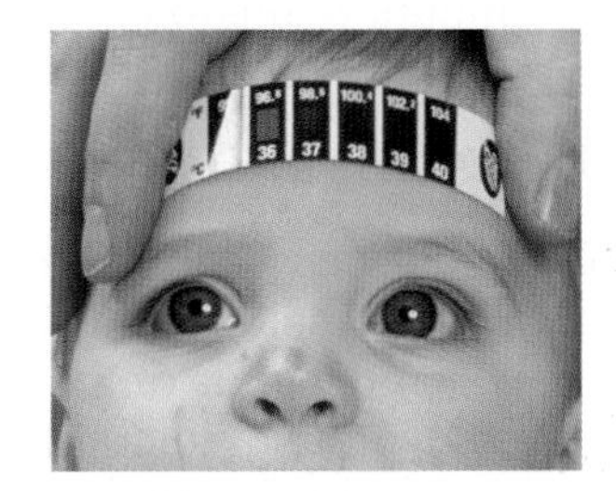

A liquid-crystal thermometer

Choice of liquids

Mercury is used because a) it expands evenly as the temperature rises and b) it is a good conductor of heat. Unfortunately it freezes solid if it is used in very cold places.
Mercury is poisonous and care must be taken if the glass breaks.

Alcohol is often used instead, because a) it can be used at very low temperatures and b) its expansion is six times greater than mercury. Unfortunately it cannot be used in very hot places because it boils at a lower temperature than mercury.

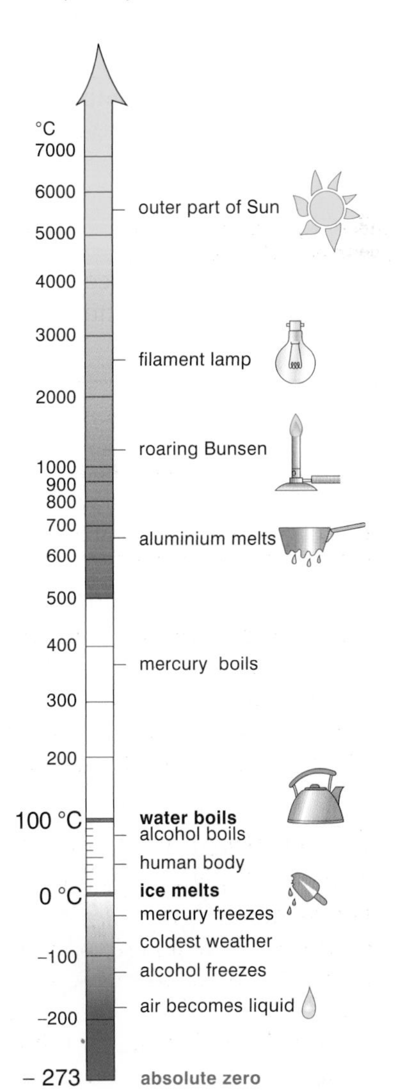

Temperature sensors (electronic)

In your experiments you may use a temperature sensor.
This uses a ***thermistor*** (see page 319). An interface circuit is used to connect it to a computer (see page 314). The computer stores the data (as a ***data-logger***) and displays it for you.

A clinical thermometer is used by a doctor or nurse to measure the temperature of your body. A modern clinical thermometer has a thermistor in the probe and a digital readout to display the result:

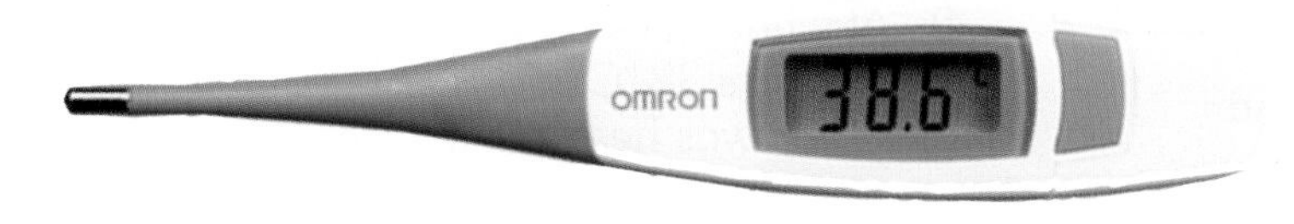

Absolute zero

As substances get colder, gases condense into liquids and liquids freeze into solids. If the temperature continues to get colder and colder, the molecules vibrate less and less until eventually they have their lowest possible energy. This happens at the coldest possible temperature, at −273 °C, called ***absolute zero***. (See also page 31.)

Scientists often measure temperatures on the ***Kelvin scale***, which begins at absolute zero but increases just like the Celsius scale. This means 0 °C becomes 273 kelvin (written 273 K) and 100 °C = 373 K.

On a hot day, the temperature is 27 °C – what is this in kelvin?

Summary

The temperature of an object depends upon how much its particles are vibrating.

Internal energy (thermal energy) is the total amount of energy in an object.

Heat energy travels from a hot object to a cold object.

Absolute zero is the coldest possible temperature at −273 °C. Temperature in kelvin = °C + 273.

▶ Questions

1. Copy out and fill in the missing words:
a) Thermometers are used to measure . . . . which depends on how much the . . . . are vibrating. Internal energy is the amount of . . . . in an object.
b) The coldest possible temperature is . . °C, called . . . .

2. Copy out and complete the rhyme:
Thermometers were often made
With marks described as c . . . .
But modern times have seen a fuss
To change the name to C . . . .

3. Copy out and place ticks in the table to show which liquid is better in each case:

	Mercury	Alcohol
Expands more evenly		
Expands more		
A better conductor of heat		
Useful at higher temperatures		
Useful at lower temperatures		

4. Explain the difference between heat and temperature.

6. Describe and explain the features of a thermometer which will make it
a) sensitive, b) quick-acting.

7. a) Convert 10 °C to kelvin.
b) Convert 300 K to °C.
c) What is your body temperature in kelvin?

8. Professor Messer was taken ill. The table shows how his temperature varied:

Time (hours)	0	1	2	3	4	5	6	7	8	9	10
Temperature (°C)	37	39	40	40	40	39.5	38.3	37.7	37.3	37	37

a) Plot a graph of his temperature (for the range 36–40 °C) against time (in hours).
b) What was his highest temperature?
c) When did this happen?
d) When do you think an ice-bath was used to cool him?
e) For how long was his temperature above normal?
f) At what times do you think his temperature was 38 °C?
g) When was he cooling fastest?

5. *Professor Messer blows his top. Tell him why he's such a flop:*

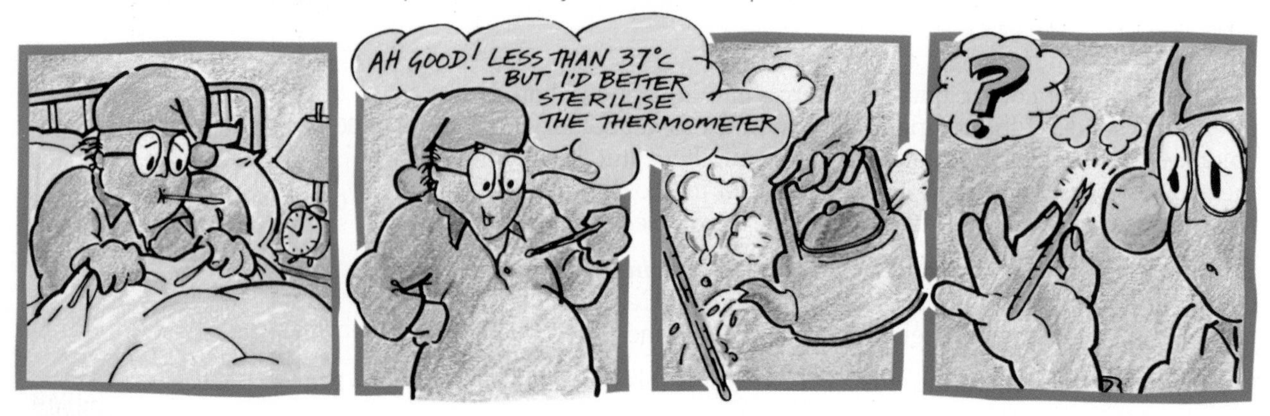

AH GOOD! LESS THAN 37°C - BUT I'D BETTER STERILISE THE THERMOMETER
?

chapter 7

The GAS Laws

When investigating the behaviour of gases, we must consider ***three*** varying quantities: **pressure**
volume
temperature.

In the following experiments we always keep one of these variables constant and investigate the other two variables.

Experiment 7.1
Keeping the temperature constant – **Boyle's Law**
Use the apparatus in the diagram to investigate how the **volume** of air depends on the **pressure** applied (with the temperature kept constant).

Use the vertical scale to find the **volume** of the trapped air.
What is the volume of air in the diagram?
What is the volume of air in your apparatus?

The pressure on the trapped air is measured using a Bourdon pressure gauge (as shown here).
What is the pressure on the gauge in the diagram?
What is the pressure in your apparatus?
Why is the pressure not zero? (See page 80.)

Now use the pump to exert more pressure and squeeze the air. This will make the trapped air slightly warmer, so wait a minute to let it cool back to room temperature.

Now measure the new pressure and volume and put your results in a table.
Use the pump again, to get more results.

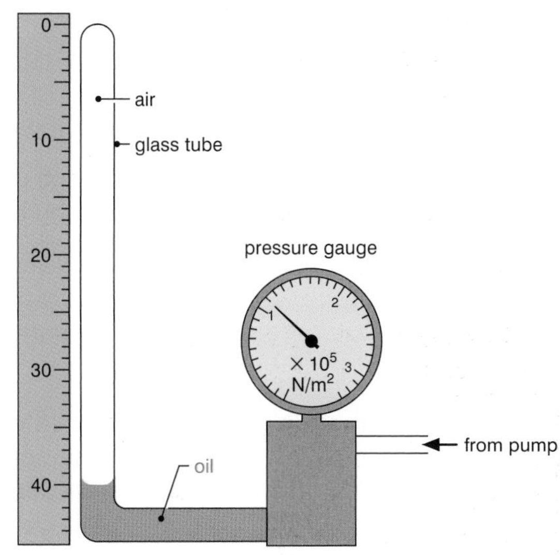

Some possible results:

Pressure p	Volume V	$p \times V$
1.1	40	44
1.7	26	44
2.2	20	
2.6	17	

Then calculate $\boldsymbol{pV}$ (pressure × volume) for each set of results. What do you notice?

This is called **Boyle's Law** after its discoverer, Robert Boyle:

For a fixed mass of gas, at constant temperature, pV = constant or $p_1 \times V_1 = p_2 \times V_2$

Did you notice that if p is doubled, V is halved?
If p increases to 3 times as much, V decreases to $\frac{1}{3}$rd. This means:

volume is *inversely proportional* to pressure, or $V \propto \frac{1}{p}$

See Example 1 on page 33.

▶ Charles' Law

Experiment 7.2 ⚠
Keeping the pressure constant – **Charles' Law**
In this experiment you will investigate how the **volume** of air changes as the **temperature** changes (with the pressure kept constant).

In the usual apparatus, the air is held inside a glass capillary tube by a short length of concentrated sulphuric acid (mercury can be used but the acid dries the air to give better results).

The length **y** of the trapped air is a measure of the volume of the air.

Hold the glass tube on to a ruler with rubber bands. Move the tube until the bottom of the trapped air is opposite the zero mark of the ruler.

An alternative is to use a plastic syringe which is lubricated with silicone oil and sealed by a rubber tube and a clip so that it is half-full of air.

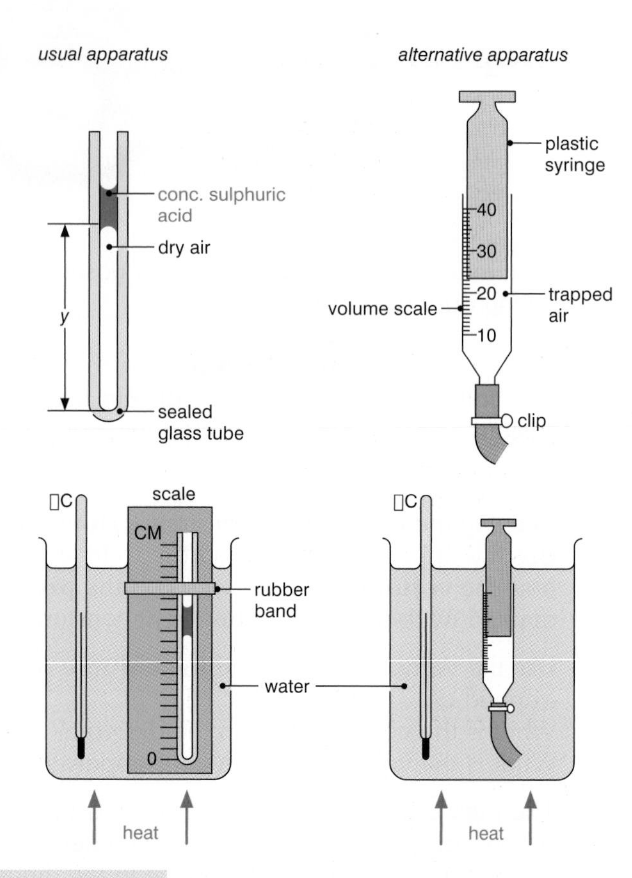

Put your apparatus in a tall beaker of cold water with a thermometer. Wait until the trapped air is the same temperature as the water and then measure the **volume** and the **temperature**.

Then heat the water until it is about 20 °C hotter. Again, wait before taking the readings of volume and temperature.
Repeat this at other temperatures until the water boils.
Put your results in a table.

At constant pressure:

Volume	Temperature (°C)

Then plot a graph to show how the **volume** varies with **temperature** (at constant pressure).

You used air in your experiment but other gases give the same result.

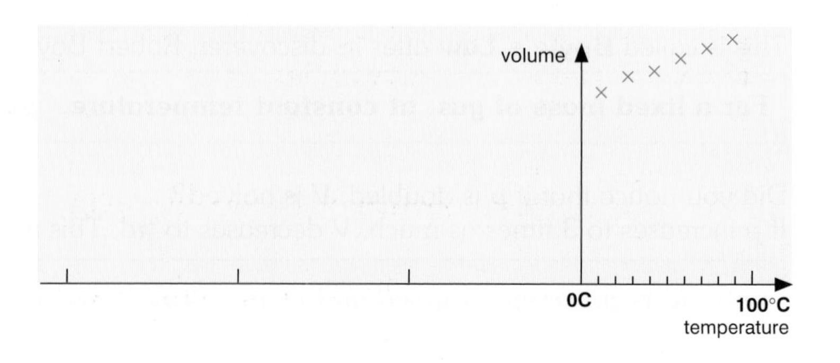

Through the points on your graph, draw the ***best*** straight line (it helps to hold the paper near your eye and look along the crosses to see the best line):

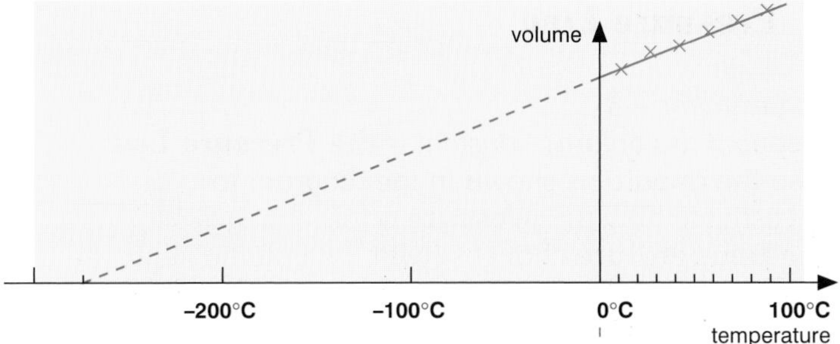

Because the graph is a ***straight*** line it means that the gas expands ***uniformly*** with temperature (when the pressure is constant).

What happens to the volume when the gas is cooled? From your graph find the temperature at which it seems the volume would become zero. Mark it on your graph.

Although the gas would liquefy before cooling to this temperature, this experiment suggests that there is a limit to how cold objects can be. This is called the **absolute zero** of temperature (see also page 27). Careful experiments show that absolute zero = −273 °C.

In science, we often measure temperature on the **Kelvin scale**. This begins at absolute zero and increases just like the Celsius scale.
This means that 0 °C becomes 273 kelvin (written 273 K) and 100 °C becomes 373 K.
Your body temperature is 37 °C – what is this in kelvin?

The kelvin temperature of a gas is directly proportional to the average kinetic energy of its molecules.

Redraw your graph (or re-label the temperature axis) so that it shows the absolute temperature, in kelvin:

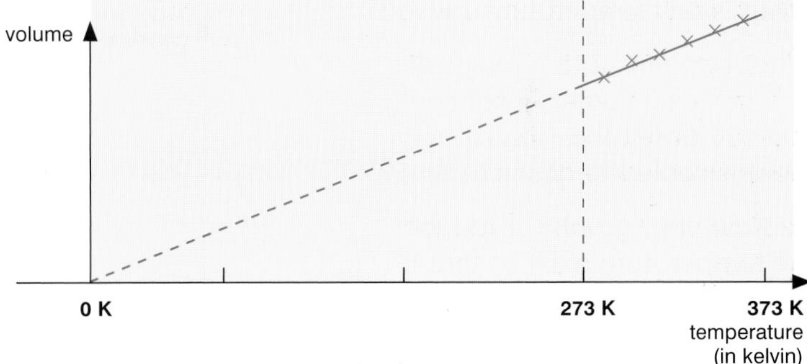

Since the graph is a ***straight line*** now passing ***through the origin*** it means that:

> **For a fixed mass of gas, at constant pressure, volume *(V)* is directly proportional to the absolute temperature *(T)***

or, in symbols: $\boldsymbol{V \propto T}$

$\therefore\ \boldsymbol{V = \text{constant} \times T}$ (see page 390)

$$\therefore\ \boxed{\frac{V}{T} = \textbf{constant}}$$ (if the pressure is constant)

This is called **Charles' Law.**
T must be the absolute temperature, measured in kelvin.

▶ Pressure Law

Experiment 7.3 ⚠

Keeping the volume constant – **the Pressure Law**

Use the apparatus shown in the diagram to investigate how the **pressure** of air changes as the **temperature** changes (with the volume kept constant).

The **pressure** is measured using a Bourdon pressure gauge as shown. What is the pressure on the gauge in the diagram?

What is the **pressure** in your apparatus?
What is the **temperature** of your apparatus?

Now heat the water by about 20 °C. Then wait for several minutes to allow the air in the flask to reach the temperature of the water.
Take the readings of pressure and temperature.
Repeat this at several temperatures until the water boils. Put your results in a table.

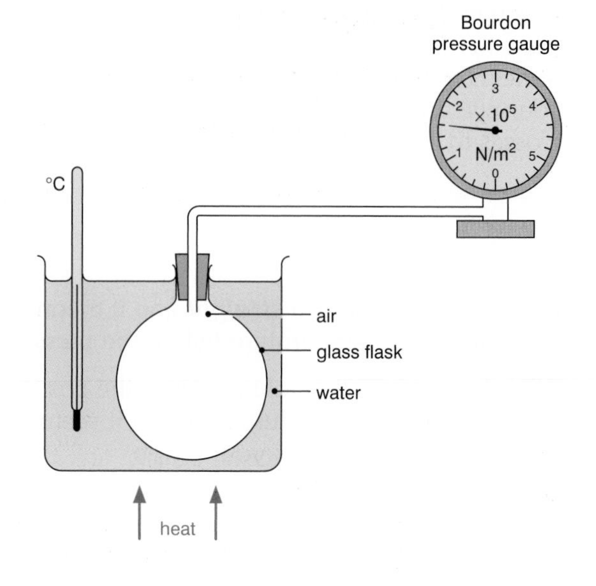

At constant volume:

Pressure $\times 10^5$ (N/m²)	Temperature (°C)	Absolute temperature (K)

Plot a graph of pressure : temperature in °C.
Compare it with the graph on the previous page.
According to your graph, at what temperature would the pressure become zero? How does this compare with the accepted value?
What is this temperature called?

What happens to the molecules of a gas as it cools? What could you say about the molecules if the gas cooled to absolute zero?

Redraw your graph (or re-label the temperature axis) so that it shows
pressure : ***absolute*** temperature:

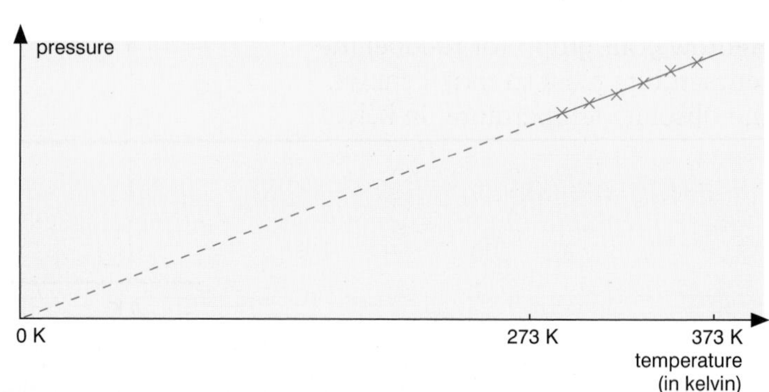

Since the graph is a ***straight line*** passing through the ***origin***, it means:

> **For a fixed mass of gas, at constant volume,**
> **pressure (p) is directly proportional to the absolute temperature (T)**

or, in symbols: $\boldsymbol{p \propto T}$

$\therefore \quad \boldsymbol{p = \text{constant} \times T}$

$$\therefore \quad \boxed{\frac{p}{T} = \textbf{constant}} \quad \text{(if the volume is constant)}$$

As before, ***T*** must be in kelvin.

This apparatus can be used as a thermometer: ***a constant volume gas thermometer.***
The dial of the pressure gauge can be marked out in °C.

▶ The gas equation

From the last three experiments, we have 3 equations:

pV = **constant (Boyle's Law)** $\frac{V}{T}$ = **constant (Charles' Law)** $\frac{p}{T}$ = **constant (Pressure Law)**

These 3 equations can be combined into one, called the **ideal gas equation:**

$$\frac{pV}{T} = \text{constant}$$

If a fixed mass of gas has values p_1, V_1 and T_1, and then some time later has values p_2, V_2 and T_2, then the equation becomes:

$$\frac{p_1V_1}{T_1} = \frac{p_2V_2}{T_2}$$

Example 1

A deep-sea diver is working at a depth where the pressure is 3.0 atmospheres. He is breathing out air bubbles. The volume of each bubble is 2 cm^3. At the surface the pressure is 1.0 atmosphere. What is the volume of each bubble when it reaches the surface?

Assuming the temperature is constant, so Boyle's Law applies:

Formula first: $p_1 \times V_1 = p_2 \times V_2$

Then numbers: $3.0 \times 2 = 1.0 \times V_2$

$\therefore$ Volume of bubble $= \underline{6\ \text{cm}^3}$

Note that p_1 and p_2 must have the same unit. Similarly with V_1 and V_2.

Example 2

A bicycle pump contains 70 cm^3 of air at a pressure of 1.0 atmosphere and a temperature of 7 °C.
When the air is compressed to 30 cm^3 at a temperature of 27 °C, what is the pressure?

What do we know? $p_1 = 1.0$ atmosphere | $p_2 = ?$
$V_1 = 70\ \text{cm}^3$ | $V_2 = 30\ \text{cm}^3$
$T_1 = 273 + 7 = 280$ K | $T_2 = 273 + 27 = 300$ K

Then the formula: $$\frac{p_1V_1}{T_1} = \frac{p_2V_2}{T_2}$$

Put in the numbers: $$\therefore \quad \frac{1 \times 70}{280} = \frac{p_2 \times 30}{300}$$

$$\therefore \quad p_2 = \frac{70 \times 300}{30 \times 280} = \underline{2.5 \text{ atmospheres}}$$

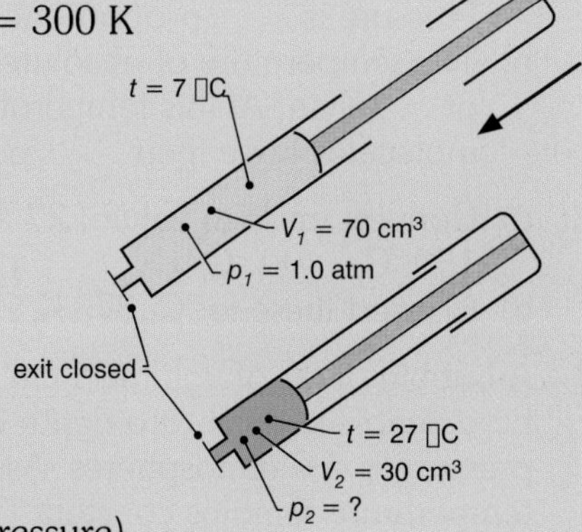

Note 1. p_2 has the same unit as p_1. Similarly with V_2 and V_1. T_1 and T_2 must be in kelvin.

Note 2. If questions refer to **s.t.p.** (**s**tandard **t**emperature and **p**ressure) this means T = 273 K (0 °C) and p = 760 mm Hg (see page 80).

The gas laws and the kinetic theory

We saw on page 16 that air pressure is caused by the bombardment of billions of molecules.

Boyle's Law
If the volume of a fixed mass of gas is made smaller then the molecules will be closer together. If the volume is halved, then the number of molecules per cm^3 will be doubled, and so the pressure will be doubled (Boyle's Law).

Absolute zero
As the temperature falls, the molecules have less energy. Absolute zero is the temperature at which molecules have their lowest possible energy.

The Pressure Law
If the temperature rises, then the molecules have more energy and move faster.
If the volume of the container is kept constant then the molecules hit the walls harder and more often and so the pressure rises (the Pressure Law).

Charles' Law
If the pressure is to be kept constant even though the molecules begin to move faster, then the volume must increase. This lets the molecules move farther apart so that the pressure can stay the same (Charles' Law).

Summary

For a fixed mass of gas in each case:

Boyle's Law. At constant temperature: $p \propto \frac{1}{V}$ or $pV = \textbf{constant}$ or $p_1 V_1 = p_2 V_2$

Charles' Law. At constant pressure: $V \propto T$ or $\frac{V}{T} = \textbf{constant}$ or $\frac{V_1}{T_1} = \frac{V_2}{T_2}$

Pressure Law. At constant volume: $p \propto T$ or $\frac{p}{T} = \textbf{constant}$ or $\frac{p_1}{T_1} = \frac{p_2}{T_2}$

T must be in kelvin (= 273 + temperature in °C)

Gas equation: $\frac{pV}{T} = \textbf{constant}$ or $\frac{p_1V_1}{T_1} = \frac{p_2V_2}{T_2}$

▶ Questions

1. Copy and complete these sentences:
a) Boyle's Law: for a . . . . mass of gas, at constant . . . . , ... × . . . . is constant. Pressure is . . . . proportional to . . . . .
b) The temperature of absolute zero is . . °C or . . kelvin. At this temperature the molecules have their . . . . possible . . . . .

2. a) Convert these to kelvin: 27 °C, −3 °C, 150 °C, −90 °C.
b) Convert these to °C: 373 K, 200 K, 1000 K.

3. A bubble of air of volume 1 cm^3 is released by a deep-sea diver at a depth where the pressure is 4.0 atmospheres. Assuming its temperature remains constant ($T_1 = T_2$), what is its volume just before it reaches the surface where the pressure is 1.0 atmosphere?

4. A sealed syringe contains 60 cm^3 of air at a pressure of 1.0 atmosphere. The piston is pushed in slowly until the volume is 40 cm^3. What is the pressure now?

5. The air in a bicycle tyre has a volume of 1000 cm^3 and a pressure of 2.5 atmospheres. If the air is released (at the same temperature) so its pressure is 1.0 atmosphere, what is the volume of the air?
A bicycle pump has a volume of 100 cm^3 and is now used to pump up the tyre. After 1 stroke of the pump, what is the pressure in the tyre?

6. A mass of gas has a volume of 380 cm^3 at a pressure of 560 mm Hg and a temperature of 7 °C. What is its volume at s.t.p.?

Further questions on page 61.

chapter 8

Measuring HEAT

We have seen that the moving molecules of an object have ***internal energy***. When an object is heated, its internal energy increases. Like all other forms of energy, internal energy is measured in a unit called the ***joule*** (usually written J).

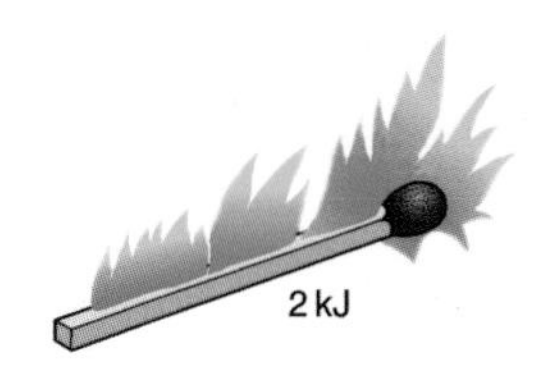

A joule is a small unit of energy – if a match burns up completely, it produces about 2000 J. This can also be written as 2 kJ where
1 kJ = 1 kilojoule = 1000 joules.
For larger amounts of energy we use MJ where
1 MJ = 1 megajoule = 1 million joules (1 000 000 J).

Your body needs energy – energy to keep warm (at 37 °C) and energy to work and play. You get this energy from the fuel that you eat.

Can you work out the rhyme?

Little Jack Horner,
Sat in a corner,
Feeling so chilly and cool.
He said, "I should eat,
And so produce . . . . ,
The unit of which is a . . . . ."

Too many joules, too few joules

Your body needs a total of about 10 MJ per day.
Lying asleep, your body uses only about 4 kJ every minute, but for making your bed or dancing you use about 20 kJ every minute.

The table below shows some energy values (called ***calorific values***), measured in megajoules per kilogram of substance.

Butter	32 MJ/kg	Eggs	7 MJ/kg	Natural gas	55 MJ/kg
Sugar	16	Potatoes	4	Petrol	47
Meat	12	Fish	3	Coal	30
Bread	10	Fruit	2	Wood	15
Ice cream	9	Carrots	2	Dynamite	6

If you want to lose weight, which foods should you avoid?
Which food would you take on an expedition to the Arctic?

Did you hear about the cat burglar who stole the family joules?
He used them for (h)eating!

▶ Heating water

Experiment 8.1
Measure out 1 kg of cold water into a large beaker.
Take the initial temperature of the water accurately.
Put an electric heater in the water and connect it to a 12-volt supply through a ***joulemeter***, which will measure how much energy we give the water.

Note the joulemeter reading and switch on.
When you have given the water 10 000 J, switch off, stir gently and note the highest temperature that is reached.
What is the rise in temperature of the water?

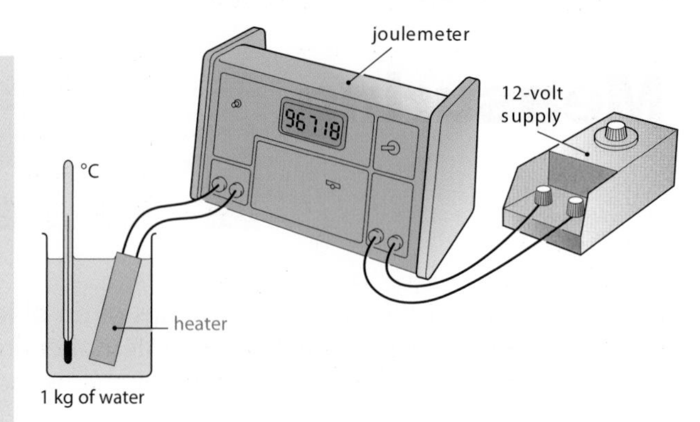

(If you have not got a joulemeter, you can use a 50-watt heater switched on for 200 seconds.)

Experiment 8.2
If you gave twice as much energy (20 000 J) to the same 1 kg of water, what would be the temperature rise?
Predict and then try it.

Is it exactly what you expected? Does it double?
Would you get a better result if you wrapped cotton wool round the beaker?

Do other substances need as much energy as water?

Experiment 8.3
Put your themometer and heater into a 1 kg block of aluminium. Give it the same amount of energy (10 000 J) as in experiment 8.1, and find the temperature rise.
Is it the same as when you had 1 kg of water?

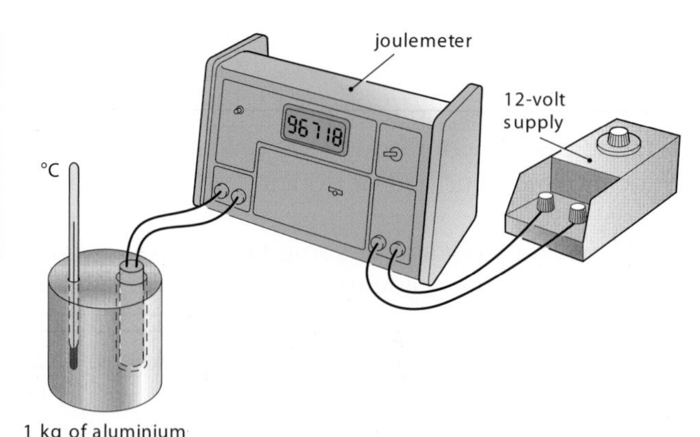

If the temperature rise is 5 times as much, it means that aluminium needs only $\frac{1}{5}$th as much energy as water (to give the same temperature rise to the same mass).

Experiment 8.4
If you have time, repeat this experiment with other solids and liquids and compare the results.

We find that different substances have different appetites or capacities for the internal energy of the molecules (see p. 26). Water is very greedy. It needs 5 times as much energy as the same mass of aluminium to produce the same temperature rise.

Different objects have different ***thermal capacities***.

▶ Specific heat capacity

If you calculate how much energy 1 kg of water needs to become 1 °C hotter, you will find it needs about 4200 J. This number is called the ***specific heat capacity*** of water.

The specific heat capacity of a substance is the amount of energy (in joules) that is needed to raise the temperature of 1 kg of the substance by 1 °C.
Its unit is written J/kg °C or J/kg K.

Here are some values of this number – notice that the number for aluminium is about $\frac{1}{5}$th of the number for water.

Water	4200	J/kg °C	Ice	2100	J/kg °C
Meths	2500		Aluminium	880	
Paraffin	2200		Sand	800	
Mercury	140		Copper	380	

All these numbers are for 1 kg and 1 °C.
The energy needed to raise the temperature of 2 kg through 1 °C would be twice as much, and the energy needed to raise 2 kg through 3 °C would be three times as much again – that is, six times as much.
So to calculate the energy needed, we multiply like this:

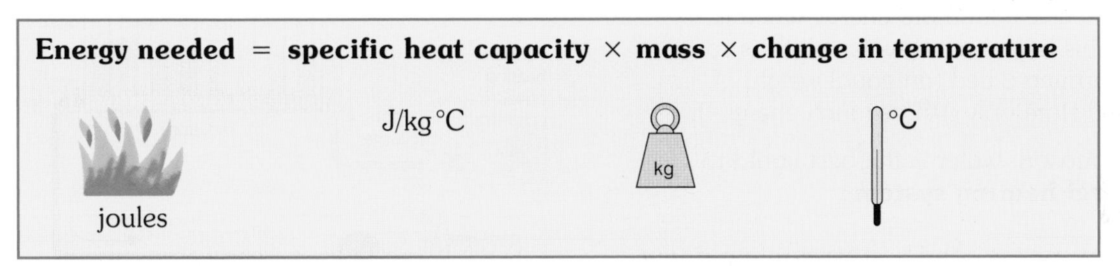
Energy needed = specific heat capacity × mass × change in temperature

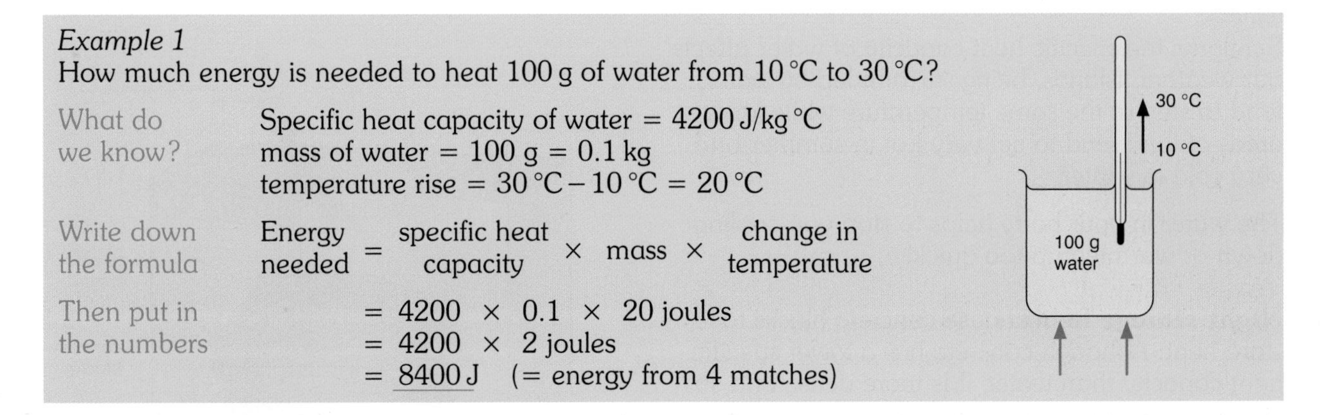
Example 1
How much energy is needed to heat 100 g of water from 10 °C to 30 °C?

What do we know?
Specific heat capacity of water = 4200 J/kg °C
mass of water = 100 g = 0.1 kg
temperature rise = 30 °C – 10 °C = 20 °C

Write down the formula
Energy needed = specific heat capacity × mass × change in temperature

Then put in the numbers
= 4200 × 0.1 × 20 joules
= 4200 × 2 joules
= 8400 J (= energy from 4 matches)

In a similar way, using the same formula, you can work out ***your*** values for specific heat capacity from the results of your experiments.

If you met a girl called Julie, would you expect her to be full of energy?

Example 2
A 2 kW (2000 W) electric heater supplies energy to a 0.5 kg copper kettle containing 1 kg of water. Calculate the time taken to raise the temperature by 10 °C. (Use values from the table on page 37.)

2 kW = 2000 W = 2000 J/s = 2000 joules in each second (see also page 111).
∴ Energy supplied in t seconds = $(2000 \times t)$ joules

Assuming no energy is lost to the surroundings:

Energy lost by hot object (heater) = **Energy gained by cold objects** (water and kettle)

$$2000 \times t = \left(\text{mass} \times \text{specific heat capacity} \times \text{rise in temp.}\right)_{\text{of water}} + \left(\text{mass} \times \text{specific heat capacity} \times \text{rise in temp.}\right)_{\text{of copper}}$$

$$2000 \times t = (1 \times 4200 \times 10)_{\text{water}} + (0.5 \times 380 \times 10)_{\text{copper}}$$

$$2000 \times t = (42\,000) + (1900) = 43\,900$$

$$\therefore t = \underline{22 \text{ seconds}}$$

▶ Storing heat energy

You could see from the table that water has easily the highest specific heat capacity. It needs more energy to heat it up. It stores more energy when it is hot, and so it gives out more energy when it cools down. This is why a hot-water bottle is so effective in warming a bed (an equal weight of mercury would store only $\frac{1}{30}$th as much energy!).

For the same reason, water is the best liquid to use in a **central heating system**:

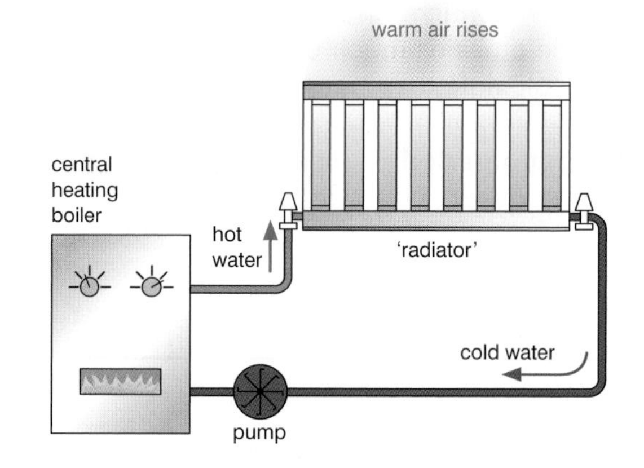

In a similar way, water is the best for ***cooling down*** the engines in cars and factories (see page 44).

Similarly, the specific heat capacity of water affects our weather. Islands, being surrounded by water, tend to stay at the same temperature while large areas of land tend to get very hot in summer and very cold in winter.

The water in your body helps to stop you cooling down or warming up too quickly.

Night storage heaters use concrete blocks to store heat. Although concrete has a lower specific heat capacity than water, it is more dense and so the same mass takes up less space in your house. The concrete blocks are heated by electric coils using cheaper night-time electricity. This stored energy is then released slowly during the day.

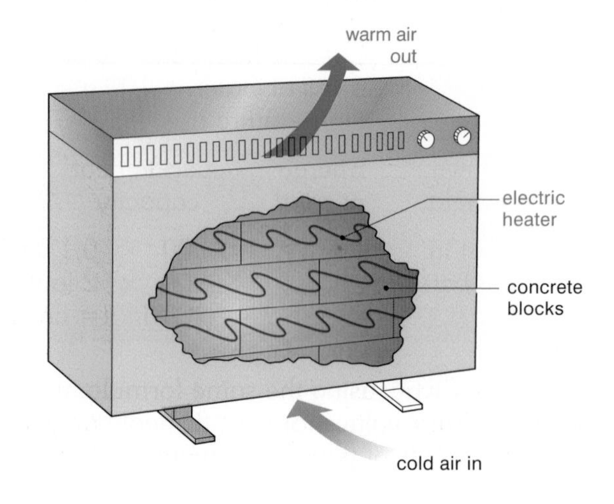

Summary

Thermal energy, like all other forms of energy, is measured in joules.
1 kJ = 1000 joules.
1 MJ = 1 000 000 joules.

Energy needed to change temperature = specific heat capacity × mass × change in temperature

▶ Questions (Use the tables on pages 35 and 37 if necessary.)

1. Copy out and complete :
a) The unit of energy is the . . . . .
1 kJ stands for one . . . . .
1 MJ stands for one . . . . .
b) The specific heat capacity of a substance is the amount of . . . . needed to raise one . . . . of the substance through . . °C.
The unit of specific heat capacity is . . . .
The specific heat capacity of water in these units is . . .
c) The formula for the energy needed to change the temperature of a substance is:
d) Energy lost by . . . . object = energy . . . . by . . . . object.

2. *Jack Spratt could not eat fat,*
His wife could not eat lean.
While Jack ate fruit or carrots,
His wife would lick ice-cream.
Two kilograms they each put in
And one grew fat and one grew thin.
I ask you now to calculate
How many joules of heat each ate.

3. Professor Messer fills his car. How many faults can you find with his idea (there are at least 4)?

4. Calculate the amounts of energy needed to change the temperature of
a) 2 kg of water by 5 °C
b) 500 g of water by 4 °C
c) 100 g of aluminium from 20 °C to 30 °C
d) 200 g of copper from 60 °C to 10 °C

5. A 2 kg block of iron is given 10 kJ of energy and its temperature rises by 10 °C.
What is the specific heat capacity of iron?

6. At the sea-side, the Sun shines down equally on the sand and the sea. Assuming they both absorb the same amount of energy, why is the sand hotter than the sea?

7. Professor Messer has an outdoor swimming pool, which contains 100 000 kg of water.
a) He needs to heat the water from 15 °C to 20 °C. How much energy does this need?
b) How much would this cost if the electricity board charge 3p for 1 MJ of energy?
c) On a hot summer day the Sun shines for 10 hours and the water absorbs solar energy at a rate of 20 kW. How much energy is absorbed that day?
d) By how much does the Sun heat up the pool?

chapter 9

CONDUCTION, CONVECTION and RADIATION

There are ***three*** ways in which heat can be transferred.

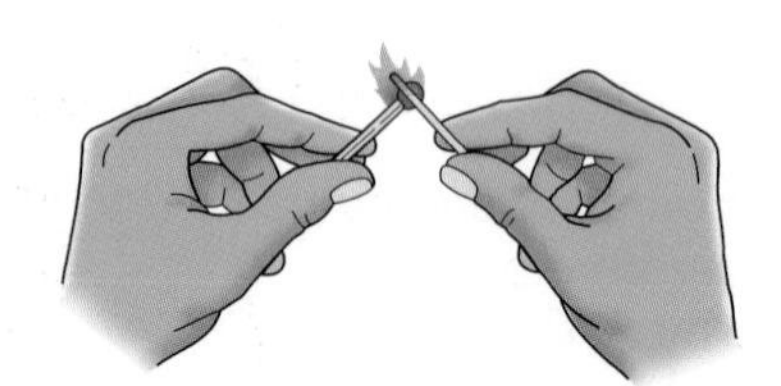

Experiment 9.1 **Conduction** ⚠
Get a piece of stiff copper wire about the same length as a match. Strike the match and hold the copper wire in the flame.

What happens?
Does the energy get to your hand quicker through wood or through copper?

We say that copper is a better ***conductor*** than wood. The energy has travelled from atom to atom through the copper.

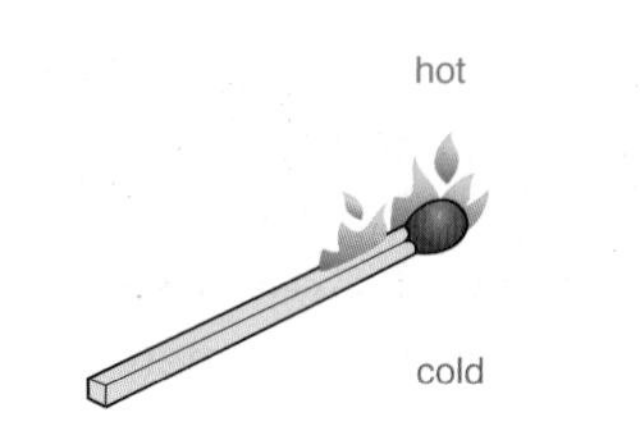

Experiment 9.2 **Convection** ⚠
Hold your hand over and then under the flame of a match.

What do you notice?

The hot air expands and then rises.
We say the heat is ***convected*** upwards.

Radiation

The thermal energy of the Sun is ***radiated*** to us in the same way that the light reaches us from the Sun.

Here is a 'model' to help us see the difference between the three ways. Three ways of getting a book to the back of the class:

1. ***Conduction***: a book can be passed from person to person – just as heat is transferred from atom to atom.
2. ***Convection***: a person can walk to the back of the class carrying the book. This is the way hot air moves in convection, taking the energy with it.
3. ***Radiation***: a book can be thrown to the back of the class rather like the way energy is radiated from a hot object.

CONDUCTION

> *Experiment 9.3*
> Get some rods (all the same size) of different substances – for example, copper, iron and glass. Rest them on a tripod and fix a small nail near one end of each rod, using vaseline as 'glue'.
>
> Heat the other ends of the rods equally with a Bunsen burner:
>
> What happens? How many minutes does it take for the first and second nail to drop off?

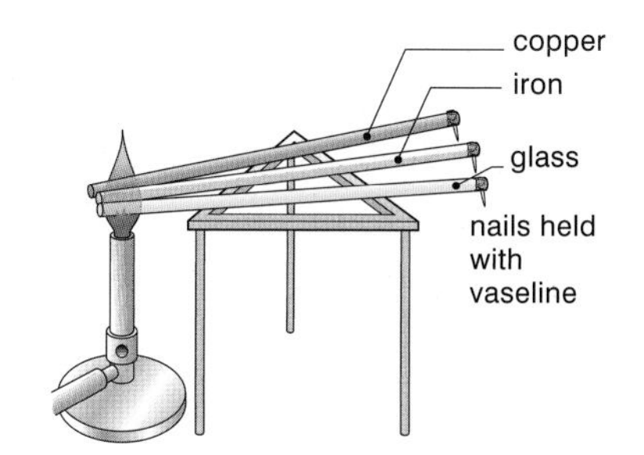

All metals are good conductors of heat.
Copper is a very good conductor.
Glass is not a good conductor of heat – it is an ***insulator***.

Pans for cooking are usually made with a copper or aluminium bottom and a plastic handle. What would happen if the handle was made of copper?

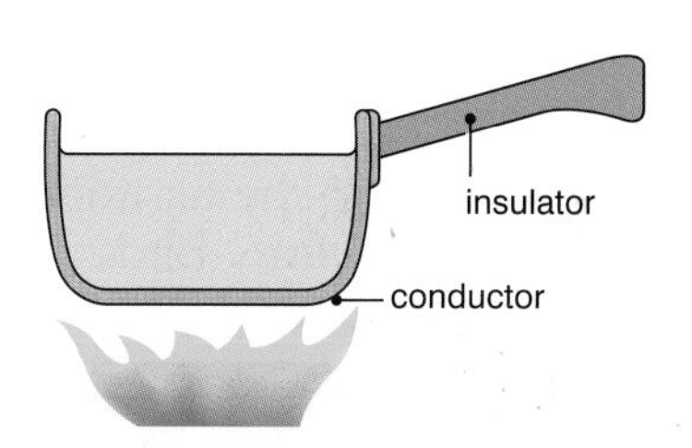

The heat energy is conducted from the hot end to the cold end.
It is transferred from atom to atom.
At the hot end the atoms are vibrating a lot:
This vibration is gradually passed along to the other atoms as they bump into each other.

This happens in all substances, but in a metal something extra happens.
A metal has many free-moving electrons.
These free electrons can travel through the metal, transferring the energy more quickly.

Insulators do not have these free electrons.
This explains why all metals are good conductors of heat (and electricity).

> *Experiment 9.4* ⚠
> Hold a large test-tube of water at the bottom and aim a gentle Bunsen flame just below the water surface until the water boils at the top, without hurting your hand.
>
> Is water a good conductor or a poor conductor?

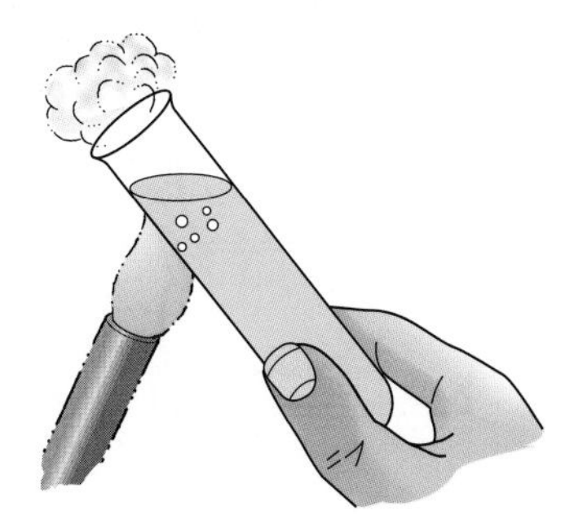

Most liquids are poor conductors of heat.

▶ Insulators

Is air a conductor or an insulator?

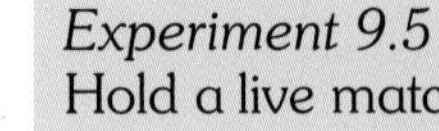

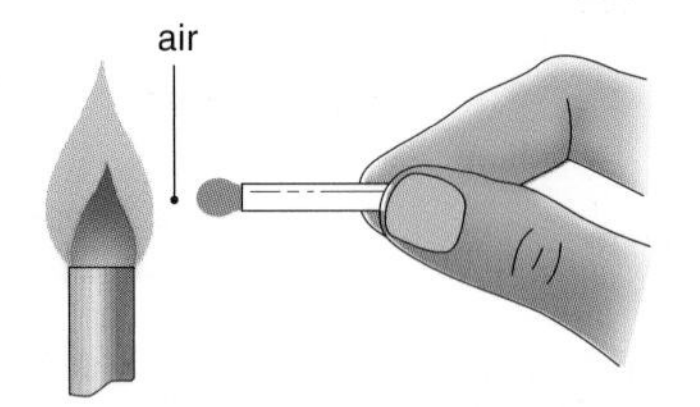

Experiment 9.5 ⚠

Hold a live match about 1 cm away from a very hot Bunsen flame.

Does the match get hot enough to burst into flame?

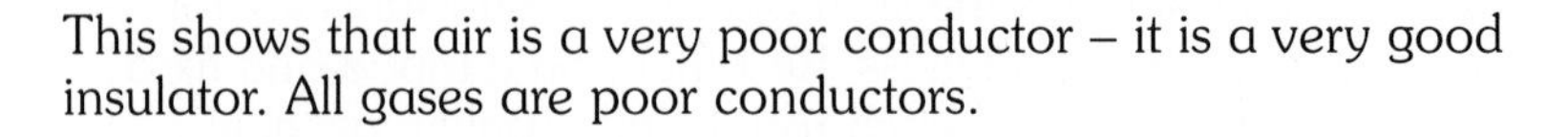

This shows that air is a very poor conductor – it is a very good insulator. All gases are poor conductors.

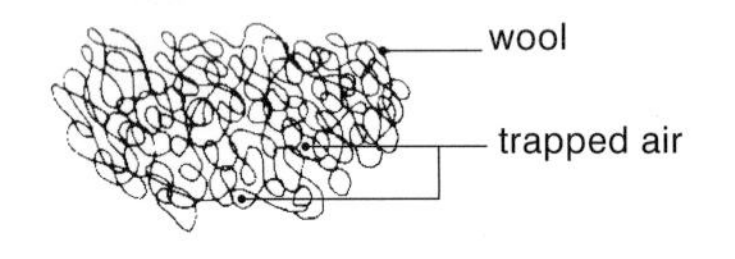

Many insulators contain tiny pockets of trapped air to stop heat being conducted away.

For example, wool feels warm because it traps a lot of air.

The air trapped in and between our clothes and blankets keeps us warm.

In the same way, the air trapped in fur and feather keeps animals warm. Birds fluff up their feathers in winter to trap more air.

A refrigerator has insulation material round it to keep it ***cold***. The insulation reduces the amount of heat conducted to the inside from the warmer room.

Pipes and hot-water tanks should be lagged with insulation material to reduce the loss of energy.

▶ Physics at work: Keeping houses warm

Heating your house is expensive. The owners of this house pay £1000 a year for energy – look where the money goes!

There are many ways in which you can save money by insulating your home:

Which of these ways applies to your home?

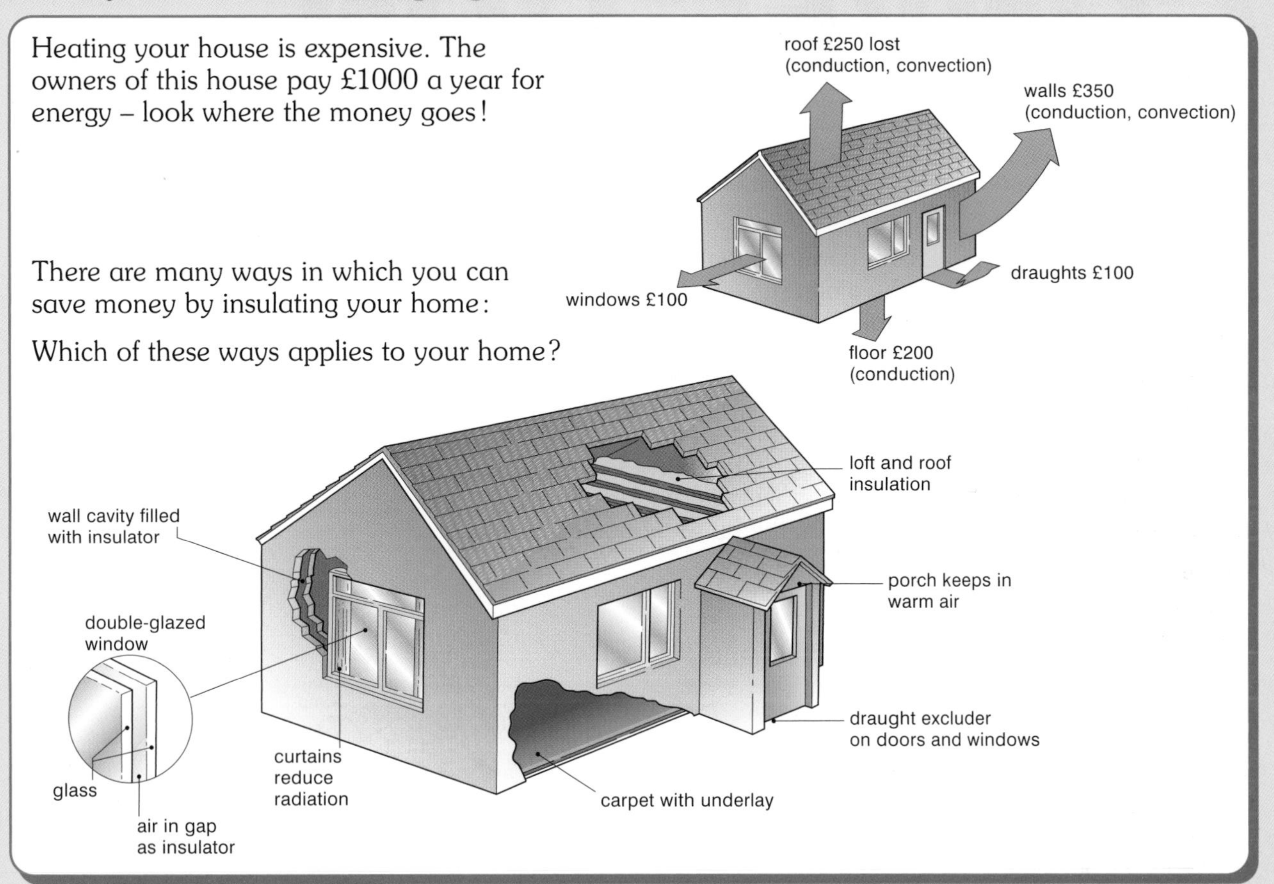

U-values

Architects calculate the heat loss from a house by using ***U-values***:

Example

An uninsulated roof measures 10 m by 10 m. How much heat is lost through the roof if the temperature inside the house is 20 °C and outside it is 5 °C?

What do we know? U-value of roof (*see* table) = 2.0
area of roof = 10 m × 10 m = 100 m^2
temperature difference = 20 − 5 = 15 °C

Then formula: Heat lost per second (W) = U-value × area × temperature difference

Then numbers: = 2.0 × 100 × 15

= <u>3000 joules per second</u>

U-values	(in W/m^2 °C)
Roof, tiled, no insulation	2.0
Roof, tiled, insulated	0.4
Wall with air cavity	1.5
Wall with insulation in cavity	0.5
Window, single glazed	5.6
Window, double glazed	3.0

Now calculate how much heat is lost if the roof is insulated.

CONVECTION

Experiment 9.6 ⚠
Put a single crystal of potassium permanganate into a beaker of water near to the side, as in the diagram.
Warm the water near the crystal with a small Bunsen flame.

The crystal colours the water so that you can see the movement of the ***convection currents*** in the water. The warm water rises in the same way that warm air rises in a flame.
This is because the warm water has expanded, and so it is ***less dense*** than the cold water. The molecules are farther apart.
The cooler water is denser and falls to the bottom until it is heated.

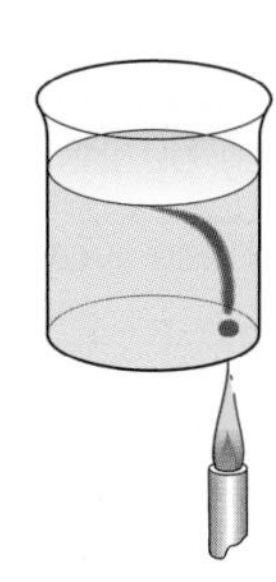

Car engines are cooled by convection currents in the water pipes:
A pump is often used to help the water to circulate. This is 'forced convection'.

Water is a very good substance to carry the unwanted heat away from the engine to the 'radiator'.
The 'radiator' is a ***heat exchanger*** where the hot water gives up its energy to the air.

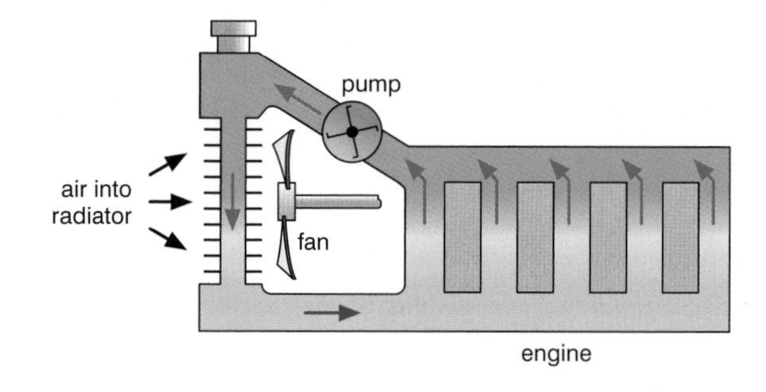

The diagram shows a simple domestic hot-water system as it might be in your home:

Convection currents take energy from the boiler up to the storage tank which is connected to the hot taps.

Where would you add lagging to insulate this system?
Why is this important for the future of our planet? (See page 107.)

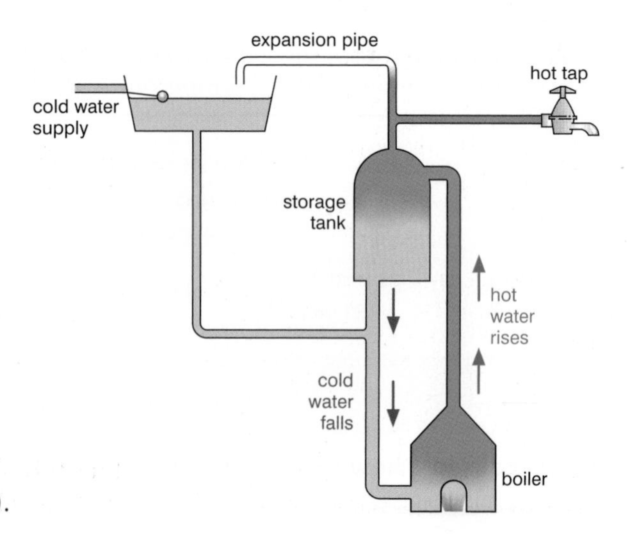

Convection currents are also used in central-heating systems and by night-storage heaters (see page 38).
There are convection currents inside Earth (page 146).

Experiment 9.7 ⚠
Investigate convection currents in a model room like the one in the diagram, using smoke so that you can see how the air moves.

Hold the smouldering paper near the window, before and then after lighting the fire.

What do you notice?

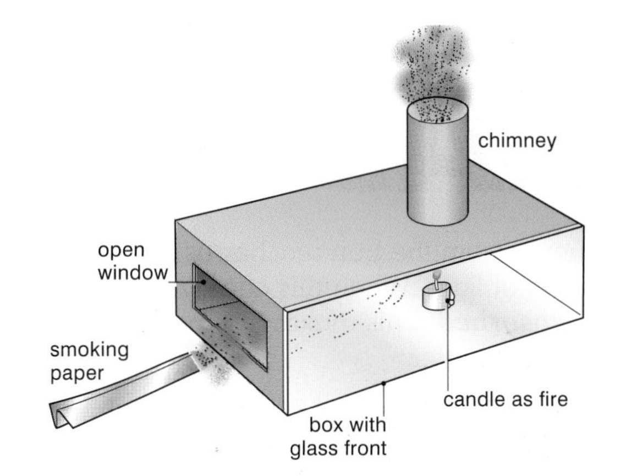

This shows that fires and chimneys help to ventilate rooms.

Experiment 9.8
Draw a spiral on paper as shown. Cut it out and hang it from cotton. Use it to investigate convection currents near fires and windows.
It will rotate as convection currents move past it.

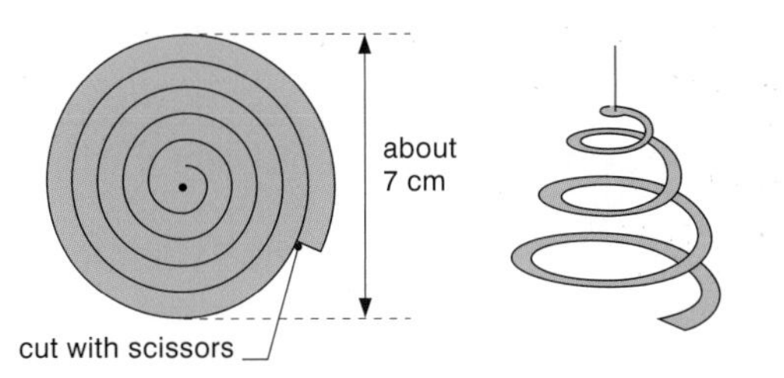

The Sun can cause very large convection currents which we call ***winds***.

In the day-time the land warms up more than the sea. The warm air rises over the land and cool air falls over the sea. So we feel a sea breeze:

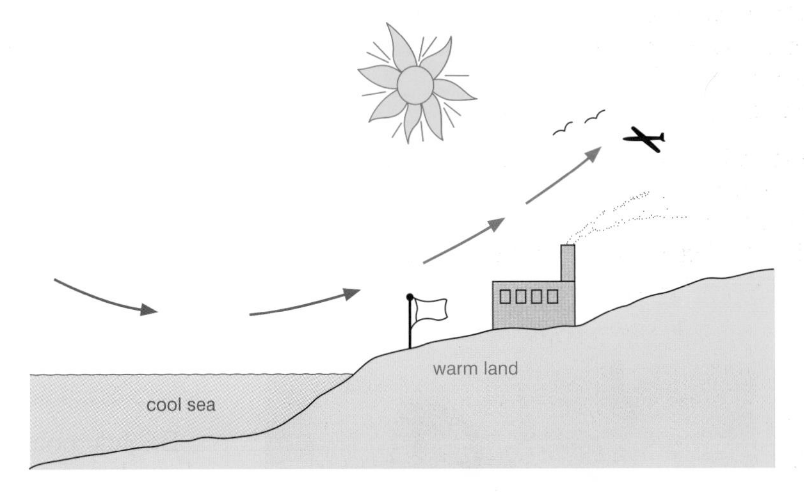

Rising convection currents over the land are used by glider pilots to keep their planes in the air.

Can you draw a diagram of what happens at night when the land cools quicker than the sea?

A rather small boy called Maguire,
Sat next to a roaring hot fire.
Life is full of surprises:
He found hot air rises,
And he floated up higher and higher!

Hot-air balloons rise in the air in the same way as convection currents. The hot air inside is less dense than the air outside.

RADIATION

Energy from the Sun reaches us after travelling through space at the speed of light. When this energy hits an object, some of it is taken in or ***absorbed***. This makes the molecules vibrate more – and so the object is hotter.

All objects take in and give out energy (as radiation) all the time. Different objects **emit** (give out) different amounts of radiation, depending on their ***temperature*** and their ***surface***.

Experiment 9.9
Get two cans of equal size, one that is polished and shiny on the outside, and one that is black and dull. Each one should have a lid.
Put a thermometer in each, and pour in ***equal*** amounts of hot water. Let them cool, side by side. Stir the water and take their temperatures every minute.

Which can cools down more quickly?
Which can is losing energy more quickly?

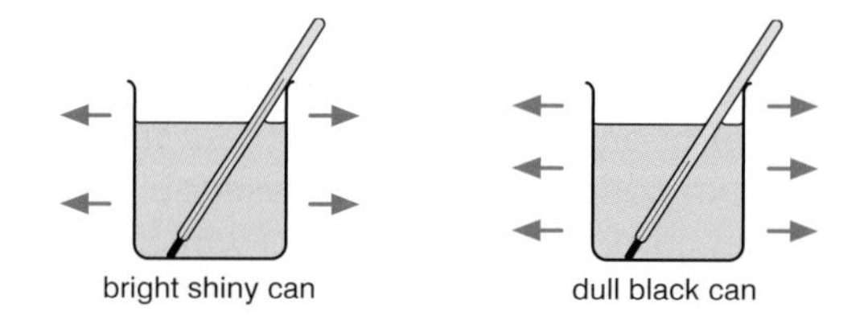

A dull black surface loses energy more quickly – it is a good radiator.
A bright shiny surface is a poor radiator.

The hotter an object, the more energy it radiates.

Brightly polished kettles and teapots do not lose much energy by radiation – they keep warm.

Marathon runners need to keep warm at the end of the race. The shiny blanket reduces radiation (and convection and evaporation).

The cooling fins on the back of a refrigerator, in a car radiator and on a motor bike engine should be dull black so that they will radiate away more energy.

Central heating 'radiators' are not usually painted dull black because they lose most of their energy by convection and should be called 'convectors'.

If different surfaces give ***out*** different amounts of energy, do different surfaces take ***in*** different amounts of energy?

Experiment 9.10
Use the same two cans as in experiment 9.9, but this time pour in equal amounts of cold water. Place the two cans in full sunlight or place them equal distances from an electric fire.
Stir and take the temperatures every minute.

Which can heats up more quickly?
Which can is absorbing energy more quickly?

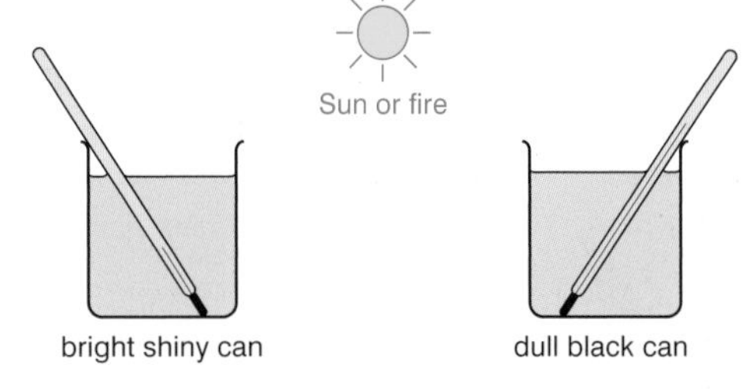

A dull black surface is a good absorber of radiation (as well as a good radiator). It takes in and gives out a lot of radiation.
A bright shiny surface is a poor absorber of radiation – it reflects the radiation away.

The bigger the temperature difference between an object and its surroundings, the faster the heat is transferred.

In hot countries, people wear bright white clothes and paint their houses white to reduce absorption of energy from the Sun.

In the same way, petrol storage tanks (and sometimes factory roofs) are sprayed with silver paint to reflect the Sun's rays:

A fire-fighting suit is bright and shiny so that it does not take in a lot of energy and burn the fire-fighter:

▶ Infra-red rays

Energy to heat us up travels from the Sun at the speed of light, just like the light rays. The rays which cause the most heating are called ***infra-red rays***. These are like light rays but have a ***longer wavelength*** than the light we use to see with (see page 209).

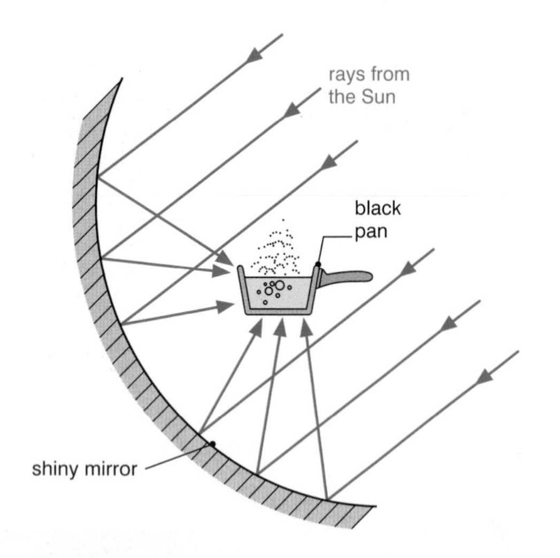

Solar ovens (solar furnaces) are used in hot countries. Like visible light rays, infra-red rays are reflected by mirrors. A large curved (concave) mirror is used to collect the rays from the Sun, to focus them on to a kettle or a pan for cooking, or on to a solar thermal tower to make electricity.

(A TV 'dish' works in the same way – it collects radio waves from a satellite and passes the signals to your TV set; see page 182.)

In an electric fire, a mirror is used in the opposite way, to reflect rays out to the room.

Experiment 9.11
On a sunny day, use a concave mirror to heat a spoonful of water or to light a match.
Why does it help if you first blacken the head of the match with a pencil?

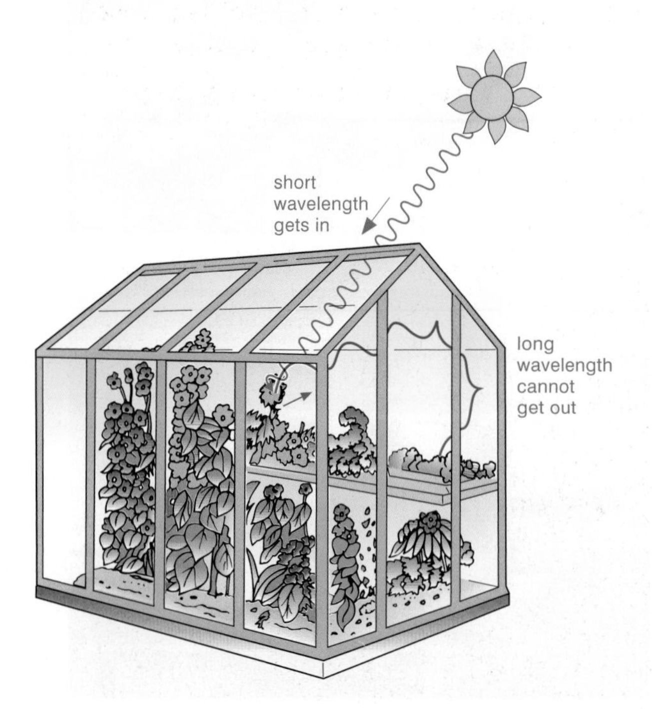

Greenhouses

The inside of a greenhouse is warmer than the outside because the rays from the very hot Sun have a short wavelength which can get through the glass. These rays are absorbed by the plants which get warmer and also radiate infra-red rays. However, because the plant is not very hot, the rays have a longer wavelength and cannot get through the glass. So energy is radiated in, but cannot easily radiate out again.

A '***greenhouse effect***' applies to planet Earth. Carbon dioxide and other gases emitted by factories and power stations act like the glass in the greenhouse. This may be making the Earth hotter and affecting our weather (see page 107).

The bowl of my goldfish, (called Egbert by me),
Was left in the Sun for an hour or three.
It absorbed infra-red, Till Egbert was dead,
So we all ate boiled Egbert for tea.

▶ The vacuum flask

A vacuum or 'Thermos' flask will keep tea hot (or keep ice-cream cold).
It does this by reducing or stopping conduction, convection and radiation.

It is a double-walled glass bottle. In the space between the two walls, both pieces of glass are coated with shiny bright 'silvering' and the air is pumped out to form a vacuum.

A ***vacuum*** is used because it stops energy transfer, by stopping conduction and convection.
The ***silvering*** on one glass wall reduces radiation of energy and the silvering on the other glass wall reflects back any infra-red rays that may have been radiated.

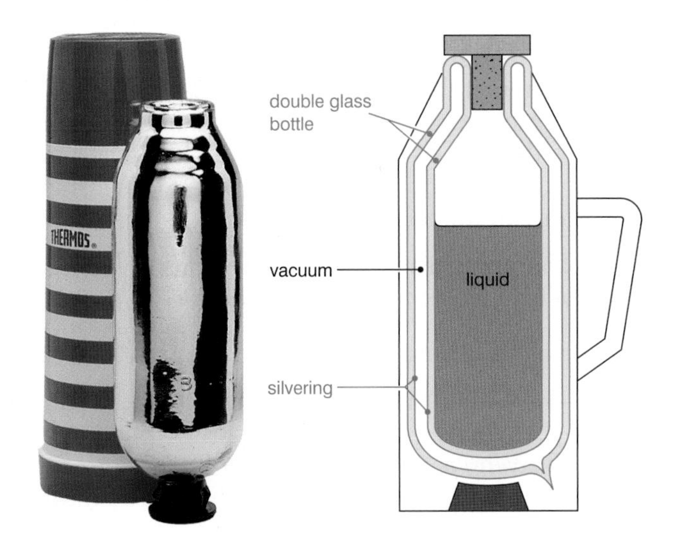

Summary

Conduction

The energy is passed from one vibrating atom to the next.
All metals are good conductors.
Water is a poor conductor.
Air is a very poor conductor.

Convection

Hot liquids and gases expand and rise while the cooler liquid or gas falls (a convection current).

Radiation

Energy is transferred by infra-red rays which can travel through a vacuum at the speed of light.

Dull black surfaces are good radiators and good absorbers.
Shiny bright surfaces are poor radiators and poor absorbers – they reflect the infra-red rays.

The hotter the object, the more energy it radiates.

Professor Messer "RAYS-ES" the temperature !

Physics at work: Infra-red radiation

Solar heating

Although the Sun's energy is free, it is not easy to make use of it. One way is to fit **solar panels** to the roof of a house, in order to provide hot water. The hot water can be used for washing or for central heating radiators.

The photograph shows some houses with solar panels: Which way do you think is North?

Here is a simple design for a solar panel:

1. Why is the main surface dull black?
2. Why is it covered with glass? (See page 48.)
3. Why is the back insulated?
4. Should the pipes inside the box be made of copper or plastic?
5. Why is the pipe inside the box made as long as possible?
6. Why should the pipe CD be kept as short as possible?
7. Where would you add insulation?
8. Why does the water move through the pipes?
9. Why is the storage tank placed at a higher level than the solar panel?
10. Why would an electric immersion heater usually be added? (See page 264.)

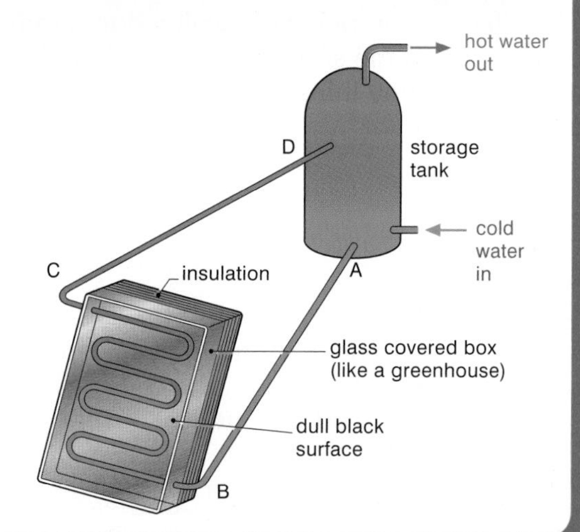

Thermography

Special photographs which are taken using infra-red rays are called ***thermographs***. Thermographs can be used by doctors because diseased parts of the skin are often hotter (and show up white or red on the thermograph).

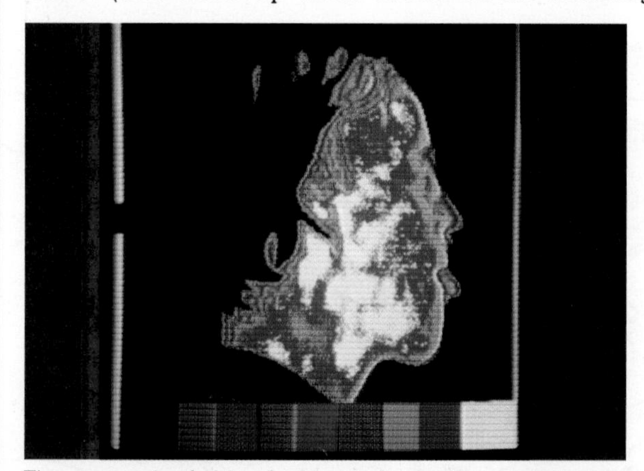

Thermograph of side of a human head, tongue out

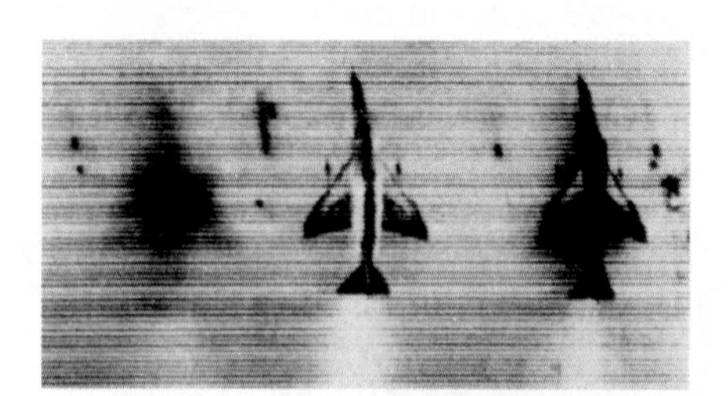

An infra-red photograph of an airfield.
The plane at the right has warm fuel in its wings.
The plane in the centre has its engines running.
At the left, a 'heat shadow' shows where a plane has left the airfield.

▶ Physics at work: Keeping warm

When string vests are worn under clothes, they trap pockets of air and keep you warm. Why are string vests no good by themselves?

Why are the eskimos wearing their coats with the fur side inwards?

Why does it help to have fur at the neck, wrists and bottom?

New-born babies are sometimes wrapped in a blanket coated with shiny aluminium. This reflects the infra-red rays and keeps baby warm.

Why do mountain-rescue teams also carry some of these shiny blankets?

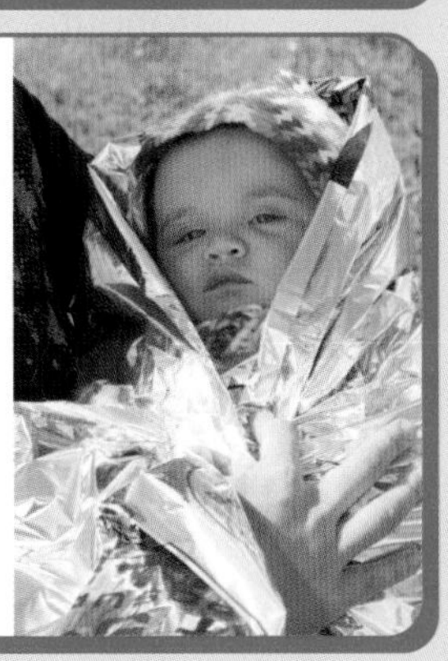

The latest padding for anoraks uses Physics to reduce conduction, convection *and* radiation:

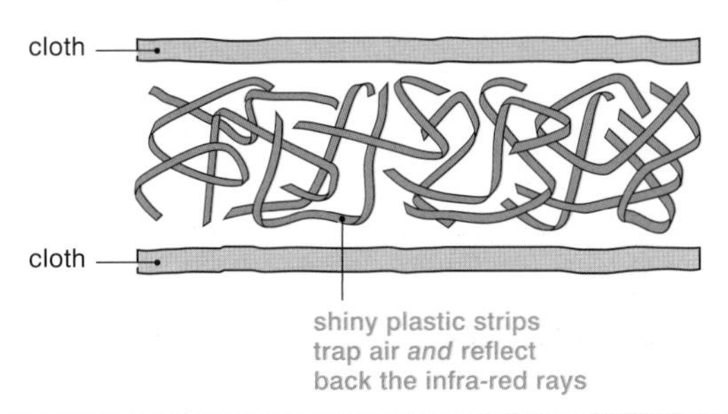

cloth
cloth
shiny plastic strips trap air *and* reflect back the infra-red rays

Physicists have found ways to show the energy we lose from our bodies.

Here the ***Schlieren*** method shows us the convection currents round a person's body:

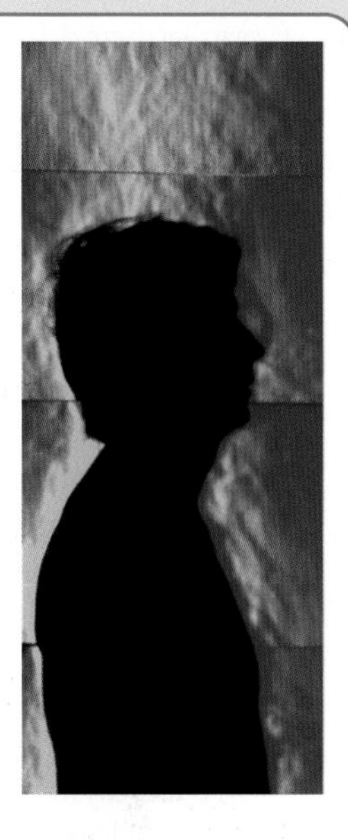

Here is a photograph taken using infra-red rays:
This thief was walking away from the camera in total darkness!
You can see that more energy was lost from his warm bare hands and where his clothes were tight (at the shoulders and trousers).

Why does it help to have loose clothes?

Things to do:

1. Devise an investigation to compare the insulation of two anoraks. You can have a beaker of hot water placed in the open sleeve of an anorak. What else would you need? List all the things you would do to ensure this was a ***fair*** test. ⚠
2. A competition for the class: using only a yoghurt pot and anything that will fit ***inside*** it, who can keep an ice-cube longest? Can you beat 3 hours?

▶ Questions

1. Copy out and fill in the missing words:
a) Thermal energy travels through the bottom of a pan by . . . . . The energy is passed from one vibrating atom to the next. All metals are good . . . . . Plastics, water and air are poor . . . . (good . . . .).
b) . . . . currents can form when liquids and gases are heated. The cold fluid . . . . and the hot fluid . . . . .
c) Energy can travel through empty space by . . . . rays, which can be . . . . by mirrors like light rays. Dull black surfaces are . . . . radiators and . . . . absorbers. Shiny, bright surfaces are . . . . radiators and . . . . absorbers.
d) A vacuum flask uses silvering to cut down heat transfer by . . . . and uses a vacuum to cut down heat transfer by . . . . and . . . . .

2. Explain the following:
a) Copper or aluminium pans are better than iron ones.
b) A fur coat is warmer if worn inside out.
c) Sheets of newspaper can be used to keep ice-cream cold, and keep chips hot.
d) In winter fat people keep warmer than thin people.
e) A carpet feels warmer to bare feet than lino or concrete.
f) Two thin blankets are usually warmer than one thick one.
g) Eskimos build their igloos from snow not ice.

3. Explain seven ways by which energy loss from a house can be reduced.

The Professor's ***roof*** loses £250 of energy per year. It can be insulated for £400 and would then lose £50 per year.
The ***walls*** lose £350 per year, would cost £750 to insulate and would then lose £100 per year. Discuss what he should do.

4. a) Why do flames go upwards?
b) Why do a chimney and a fire help to ventilate a room?
c) Why would you crawl close to the floor in a smoke-filled room?
d) Why is the element at the bottom in an electric kettle?
e) Why is the freezer compartment at the top of a refrigerator?
f) Why is there often a sea breeze during the day and a land breeze at night?
g) How can a 500 kg glider stay up in the air?
h) How have convection currents in the Earth moved the continents? (See p. 146)

5. Explain, with a diagram, how a domestic hot-water system works.

6. a) Why does a shiny teapot stay hotter than a dull brown teapot?
b) Why are fire-fighting suits made of shiny material?
c) What would probably happen if an astronaut's spacesuit was dull black?
d) Why is it more likely to be frosty on a clear night than on a cloudy night?
e) Why is the back of a refrigerator painted black?
f) If pieces of black paper and white paper are laid on snow in sunshine, what is likely to happen?
g) What might be the effect of sprinkling black soot over the ice in the arctic or the antarctic?

7. Explain why plants are warmer in a greenhouse than outside. How does the 'greenhouse effect' affect the Earth?

8. Explain, with a diagram, how a vacuum flask can a) keep tea hot and b) keep ice cold.

9. Transistors or 'microchips' that may get hot inside a computer often have a metal '***heat sink***' clipped to them.
a) Why is it made of aluminium?
b) Why has it got large 'fins'? c) Why is it black?

Further questions on page 62.

chapter 10

Changing State

SOLID to LIQUID

Before a solid can change state into a liquid, it must be given some energy.

In fact to melt 1 kg of ice at 0 °C into 1 kg of water at 0 °C, we must give it 340 000 joules of energy.

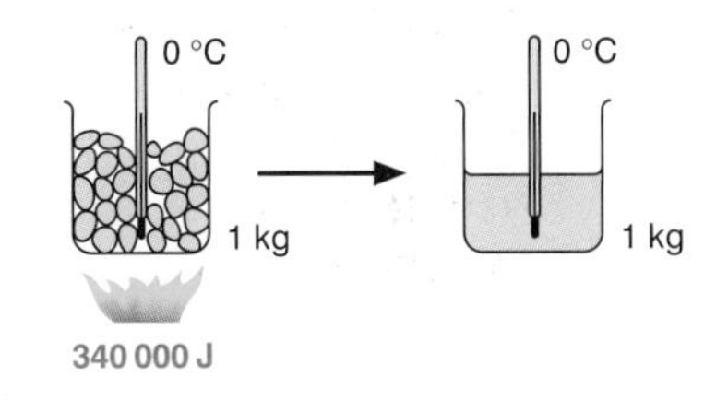

This energy is called 'hidden heat' or ***latent heat*** because it does not make it any warmer – it is still at 0 °C.

In an opposite way: to make 1 kg of ice at 0 °C, a refrigerator would have to take 340 000 J ***away from*** 1 kg of water at 0 °C.

This number – 340 000 J/kg of water – is called the ***specific latent heat of fusion of water***. (Fusion means melting.) Other substances have different values.

The specific latent heat of fusion of a substance is the amount of energy (in joules) needed to melt 1 kg of the solid to liquid without changing the temperature. Its unit is J/kg.

Ice cubes cool down your drink because the ice needs a lot of latent heat in order to melt

Molecules

The latent heat energy is given to the molecules in a solid, so that they can move more freely as in a liquid.
In ice there are strong forces between the molecules, so it needs a lot of energy to melt it.

▶ Calculating how much energy is needed

If 1 kg of ice at 0 °C needs 340 000 J to melt, 2 kg would need twice as much and 3 kg would need three times as much. We multiply like this:

> **Energy needed to melt ice at 0 °C = 340 000 × mass**
> joules (J/kg) kg

> *Example*
> How much energy is needed just to melt 100 g of ice at 0 °C?
>
> What do we know? Mass of ice = 100 g = 0.1 kg
> specific latent heat = 340 000 J/kg
>
> Formula: **Energy needed = 340 000 × mass**
>
> Put in the numbers: = 340 000 × 0.1 joules
> = 34 000 J (= energy from about 20 matches)

> *Experiment 10.1*
> Put 100 g of ice in a beaker with a heater connected to a joulemeter.
>
> Find how many joules are needed to just melt the ice without making it hotter.
>
> When you have found the number of joules for 0.1 kg of ice, calculate the number for 1 kg of ice and see if it agrees with the accepted value of 340 000 J/kg. If it is different, suggest reasons.

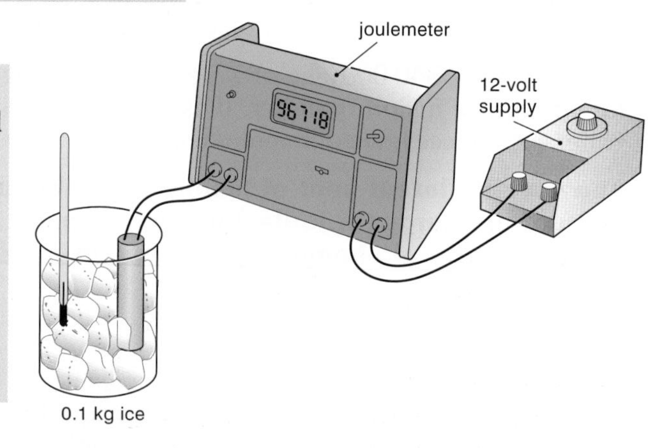

> *Experiment 10.2 To find the melting point of a substance* ⚠
> Put some of the substance (eg. hexadecanol) in a test tube and melt it by putting the test tube in a beaker of boiling water.
>
> Then take out the test tube and let it cool down, using a sensor or thermometer to take the temperature every $\frac{1}{2}$ minute.
>
> Plot a graph of your results (use a computer if possible).
>
> From your graph, work out the melting point of the substance.
>
> Using the idea of latent heat, explain what is happening during the level part of the graph.

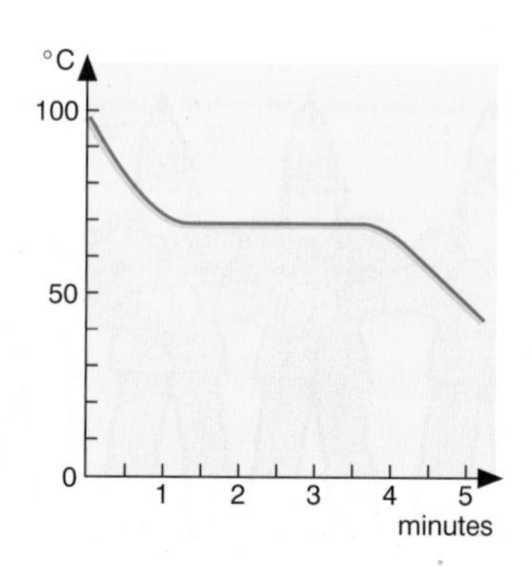

Changing State

LIQUID to GAS

When water is heated by a Bunsen burner, it reaches 100 °C and then stays at that temperature as the water changes into steam.

The energy from the Bunsen is needed as latent heat – the energy is needed to overcome the forces between the molecules (and also to push back the surrounding air molecules).

In fact, to boil 1 kg of water at 100 °C into steam at the same temperature, we must give it 2 300 000 joules of energy.

In an opposite way, if 1 kg of steam at 100 °C condenses to hot water at 100 °C, it will give up 2 300 000 J of energy.

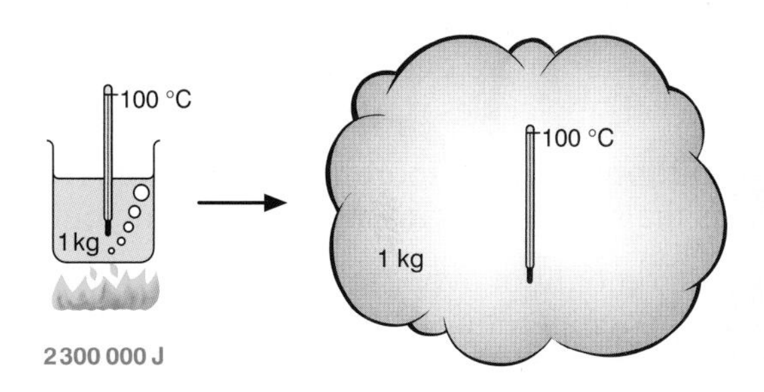

This number – 2 300 000 J/kg of water – is called the ***specific latent heat of vaporisation of water***. Other substances have different values.

The specific latent heat of vaporisation of a substance is the amount of energy (in joules) needed to boil 1 kg of the liquid to gas without changing the temperature. Its unit is J/kg.

Experiment 10.3 ⚠
You can do an experiment to find the number for water after reading question 8 at the end of this chapter.

To calculate the energy needed to boil a different mass of water, we use a formula like the last one:

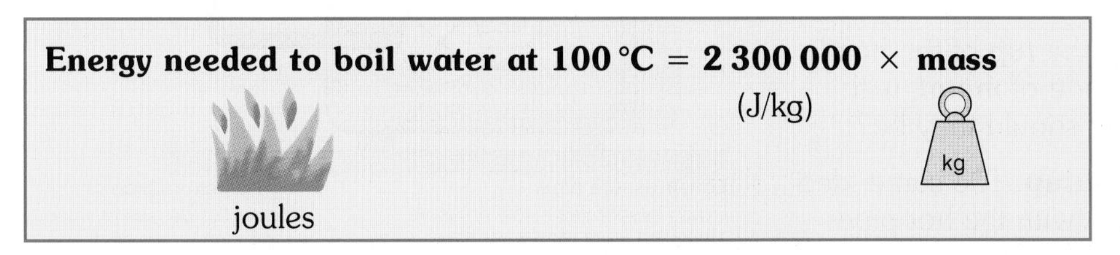

Energy needed to boil water at 100 °C = 2 300 000 × mass
joules
(J/kg)
kg

How much energy would be needed to boil 2 kg of water at 100 °C?

▶ Evaporation

A liquid can change state without boiling. Even on a cool day, a rain puddle can dry up. We say the water is ***evaporating***.

> *Experiment 10.4*
> Put some methylated spirit on the back of your hand.
> What do you feel as it evaporates away?

How can you tell the wind direction by holding up a wet finger?
Why do you often feel cold after getting out of the swimming baths?
Why is it dangerous for you to be out on the hills in damp clothes?

Evaporation causes cooling

In each case the liquid takes energy from your body, so that the liquid can change state by evaporating.
The faster (more energetic, hotter) molecules leave, and the slower (cooler) molecules stay behind until your body warms them.

Your body sweats more in hot weather. This is so that you keep cool as the sweat takes in latent heat from your body and evaporates.

The amount of evaporation depends on:

- the temperature of the liquid,
- the surface area of the liquid,
- whether there is a wind or draught of air,
- how dry the air is.

When the water evaporates, it becomes a gas called ***water vapour***.
Water vapour in the air can ***condense*** back to liquid water, to form clouds or fog.

The refrigerator

An electric motor pumps a liquid (a 'refrigerant') from a low-pressure pipe to a high-pressure pipe. From there it squirts through a narrow hole, back into the low-pressure pipe, and round again:

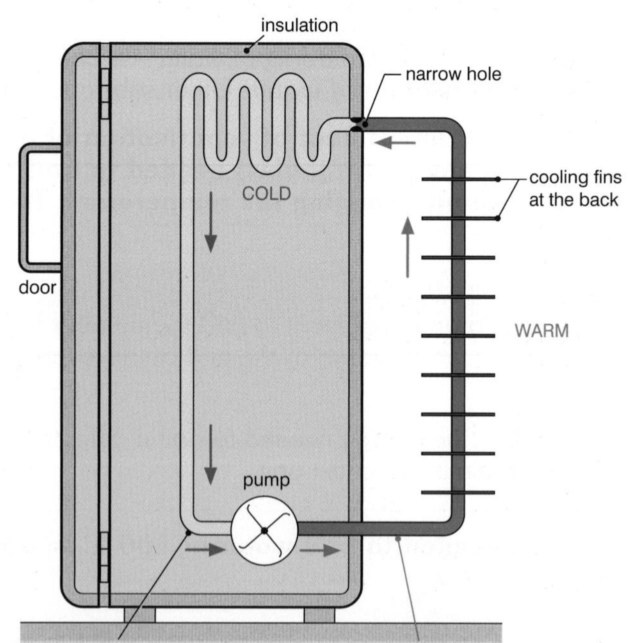
insulation
narrow hole
cooling fins
at the back
COLD
WARM
door
pump
Low-pressure pipe
Liquid evaporates in low pressure.
Takes in Latent Heat from pipe,
so pipe is cold.
High-pressure pipe
Vapour condenses to liquid
under high pressure.
Gives out Latent Heat to pipe,
so pipe is warm.

Why is the cold pipe put at the ***top*** of the box?
Why are the cooling fins made of ***metal***, with a ***large*** area? What ***colour*** should they be?

The refrigerator is a **heat pump**. The same idea can be used to heat houses (with the hot pipes inside the house and the cold pipes outside).

▶ Changing the melting point in 2 ways

Experiment 10.5   Adding impurities
Put some ice in a beaker and sprinkle salt on it.
Stir until the ice is melting, and then take its temperature.
What do you find?

When water is not pure, it melts at a temperature below 0 °C.

Impurities lower the melting point of water.

Why is salt put on icy roads by the Council?
Why do motorists spray 'de-icer' liquid on their windscreens?

Mrs Messer: How do you make anti-freeze?
Professor: Hide her coat!

Experiment 10.6   Increasing the pressure
Press two ice cubes together as hard as you can, and then release the pressure. What has happened?

Squeezing causes the ice to melt, but when you release the pressure it re-freezes and 'glues' the ice cubes together.
We make snowballs the same way. This shows that when the pressure is increased, ice melts at a temperature below 0 °C.

Increased pressure lowers the melting point of water.

▶ Changing the boiling point in 2 ways

Experiment 10.7   Adding impurities
Put some salt or other impurity into water and heat it until it boils. Measure the boiling point. What do you find?

Impurities raise the boiling point of water.

Why do potatoes cook faster if you put salt in the water?

Experiment 10.8   Pressure (Teacher demonstration) ⚠
Put some hot (but not boiling) water in a strong flask and use a pump to reduce the air pressure. What happens?

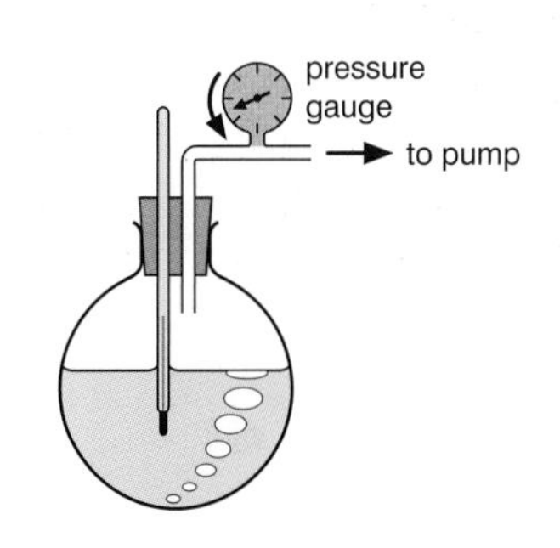

Reducing the pressure lowers the boiling point of water.
This is because the molecules find it easier to escape from the liquid when the pressure is less.

As you go higher in altitude, atmospheric pressure decreases.
At the top of Mount Everest water boils at only 70 °C.

In an opposite way, increasing the pressure raises the boiling point of water.
This is used in a ***pressure cooker***, where the water boils at about 120 °C, so food cooks quicker.

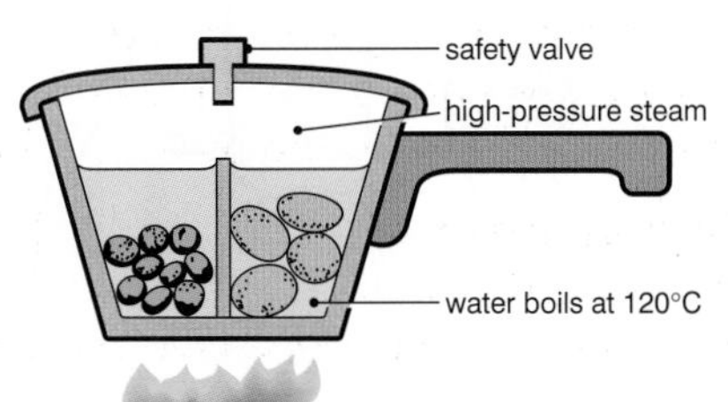

How to get water from the ground – even in the desert

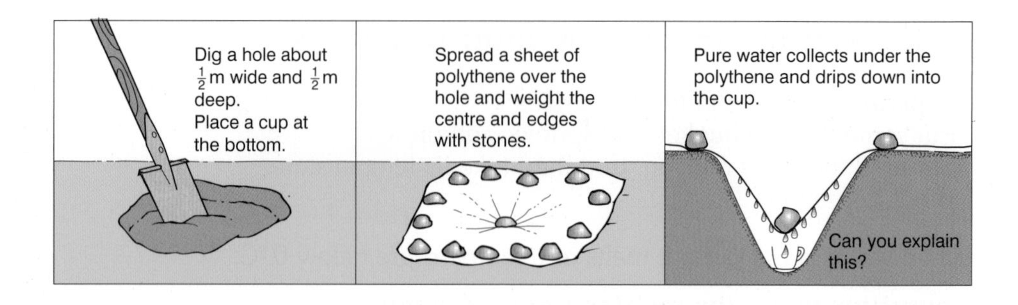

Dig a hole about $\frac{1}{2}$ m wide and $\frac{1}{2}$ m deep.
Place a cup at the bottom.
Spread a sheet of polythene over the hole and weight the centre and edges with stones.
Pure water collects under the polythene and drips down into the cup.
Can you explain this?

Summary

Substances must be given latent heat before they can melt or boil. This energy is needed to overcome forces between the molecules.

The specific latent heat of fusion (melting) of a substance is the amount of energy (in joules) needed to melt 1 kg of the solid to the liquid, without changing the temperature.
Its unit is J/kg.

The specific latent heat of vaporisation (boiling) of a substance is the amount of energy (in joules) needed to boil 1 kg of the liquid to the gas, without changing the temperature.
Its unit is J/kg.

Evaporation depends on the temperature, the dryness of the air, the wind and the surface area.

Impurities or a change of pressure affect the melting point and the boiling point of water as the diagram shows:

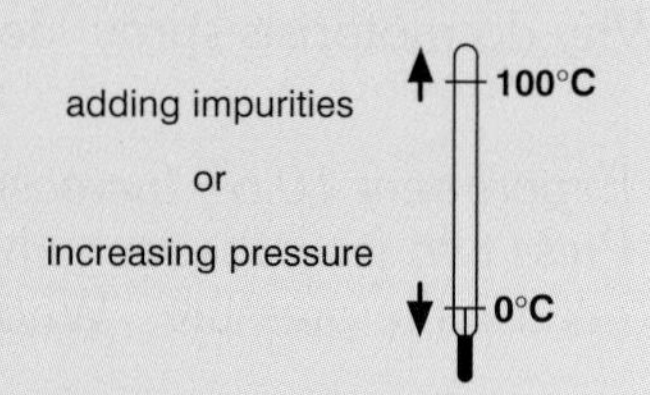

Energy need to change state = specific latent heat × mass

joules J/kg

When liquids evaporate, they cool down. This is used in the refrigerator.

Professor Messer's boiling mad!
He feels quite sure that he's been had.
The mercury just won't go high,
Please work it out and tell me why:

I'LL JUST CHECK THIS THERMOMETER IN BOILING WATER

DRAT! IT READS ONLY 90°C – IT MUST BE A FAULTY ONE....
....I'LL HAVE TO THROW IT AWAY AND JUST GO AND BUY A NEW ONE

▶ Questions

1. Copy out and complete:

a) A solid must be given . . . . heat before it can melt into a liquid, and a liquid must be given . . . . heat before it can boil into a gas. The energy is used to overcome the forces between the . . . . so that they can move more freely.

b) The specific latent heat of fusion of a substance is the amount of . . . . (measured in . . . .) needed to melt . . kg of a solid into a . . . . without changing the . . . . . Its unit is . . . . / . . . . .

c) The formula is:

Energy needed to change state = specific . . . . heat × . . . .

d) Impurities . . . . the melting point of water. Increased pressure . . . . the melting point of water.

e) Impurities . . . . the boiling point of water. Increased pressure . . . . the boiling point of water.

f) Evaporation causes cooling because only the . . . . (hotter) molecules escape, leaving the slower (. . . .) molecules in the liquid.

2. Explain the following, using the idea of molecules where possible.

a) Lakes are more likely to freeze than the sea.

b) Salt and sand are sprinkled on roads in winter.

c) Motorists put 'anti-freeze' in the cooling system in winter.

d) Snowballs cannot be made in very cold weather.

e) Icebergs take a long time to melt.

3. When water freezes, it ***expands*** (unlike most other substances). How does this explain:

a) water pipes often bursting in cold weather,

b) not finding out about them immediately?

4. What factors affect the rate of evaporation of a liquid? How would you investigate this?

5. Explain the following, using the idea of molecules where possible:

a) A steam burn is worse than a hot-water burn.

b) A pressure cooker cooks food faster.

c) A car cooling system is usually pressurised.

d) An astronaut needs a pressurised space-suit all round his body.

e) It is difficult to make tea on Mount Everest.

f) Wet clothes feel cold.

g) Tea is cooled more rapidly by blowing on it.

h) Athletes should put on a tracksuit soon after finishing a race.

i) A swimming pool is colder on a windy day.

6. Explain, with a labelled diagram, how a refrigerator works.

7. If the specific latent heat of ice is 340 000 J/kg and the specific latent heat of steam is 2 300 000 J/kg, calculate

a) the energy needed to melt 2 kg of ice at 0 °C

b) the energy needed to melt 500 g of ice at 0 °C

c) the energy needed to boil 3 kg of water at 100 °C

d) the energy needed to boil 100 g of water at 100 °C.

8. An electric kettle (marked 3 kW) produces 3000 joules of heat every second. It is filled with water, weighed and switched on.
After coming to the boil, it is left on for a further 80 seconds and then switched off.
It is found to be 100 g lighter.
Calculate the specific latent heat of steam.
Suggest reasons why your answer does not agree with the accepted value.

9. If the specific latent heat of ice is 340 000 J/kg, and the specific latent heat of steam is 2 300 000 J/kg and the specific heat capacity of water is 4200 J/kg K, calculate the heat needed to change 2 kg of ***ice*** at 0 °C to ***steam*** at 100 °C.

Further questions on page 63.

▶ Energy resources (see also page 142)

1. The list below gives some sources of energy.

Natural Gas *Oil* *Coal* *Nuclear*
Hydro (Water) *Wind* *Wave* *Solar*
Geothermal *Tidal*

a) i) Name **two** of these sources which are used to generate electricity on a large scale in the United Kingdom. [2 marks]

ii) When electricity is generated in power stations much of the input energy is 'wasted'. State **one** reason why electrical energy is so useful that we accept this energy 'loss'. [1]

b) Coal is a fossil fuel.

i) Give **one** disadvantage of coal. [1]

ii) Name **one other** fossil fuel in the list above. [1]

c) Solar energy is a renewable source of energy.

i) Explain what is meant by the term renewable energy. [1]

ii) Name **one other** renewable energy source in the list above. [1]

iii) Explain why solar energy is not likely to be used to generate electricity on a large scale in the United Kingdom. [1]

d) i) What is geothermal energy? [1]

ii) Explain briefly how geothermal energy can be extracted. [2] (AQA)

2. Fossil fuels, geothermal energy and nuclear fuels are examples of non-renewable energy resources.

Construct a table of **renewable** energy resources.

Your table should:

- refer to **four** different renewable energy resources;
- for **each**, explain briefly how it is used as a source of useful energy;
- for **each**, explain briefly **one** advantage and **one** disadvantage;

Your table should **not**:

- give any advantage or disadvantage more than once;
- give a disadvantage which is the opposite of the advantage you have given for that resource. [17] (AQA)

3. a) Explain clearly the difference between renewable and non-renewable sources of energy. [1]

b) Give **two** examples of

i) renewable energy sources, [2]

ii) non-renewable energy sources [2]

c) Give **two** reasons why it is necessary to cut down the use of non-renewable energy sources. [2] (WJEC)

4. a) Explain how a named renewable energy source is used to produce electricity. [2]

b) Briefly discuss **one** environmental problem associated with the products of large scale combustion of coal to generate electricity. You should identify the problem and describe the implications for the environment. [3] (Edex)

5. The map shows an industrial region (shaded).

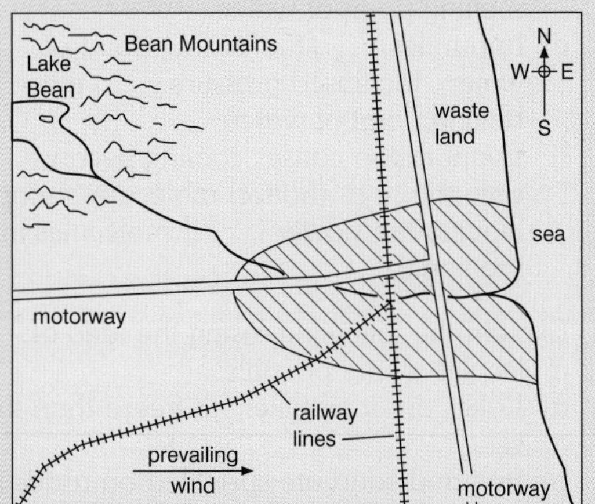

The prevailing wind is from the west. There is a nearby mountainous area, from which a river flows through the region. The major road and rail links are shown.
A power station is to be built to supply electrical energy to the region. The energy will be for a range of domestic and industrial uses. The choice is between a coal fired power station, wind turbines and a hydroelectric scheme. Three local groups each support a different opinion. Choose which option you would support and justify your choice by making reference to the financial, social and environmental implications of your choice compared with those of the alternative systems. [8] (AQA)

 Note: although a question may be marked (AQA) etc, it will often be relevant to other examinations as well.

▶ Molecules, Expansion

6. a) The properties of materials can be explained by using a particle model of matter.
   i) What are the 3 states (phases) of matter?
   ii) Describe, briefly, how the particles (molecules) are arranged in the 3 states (phases) of matter. [6]

b) Brownian motion is one piece of evidence in support of this model of matter.
   i) Draw a ***simple*** diagram of apparatus used to observe Brownian motion.
   ii) Describe and explain what is observed with the apparatus. [6] (NI)

7. Use the idea of particles to explain why
a) the air in a car tyre exerts pressure on the walls of the tyre. [1]
b) the pressure inside the tyre increases when the tyre is pumped up. [1]
c) the pressure inside the tyre is higher at the end of a journey. [4]

8. In terms of the forces of attraction between the particles, the particle spacing and their motion,
a) describe and explain the difference in density between liquids and gases, [4]
b) explain why gases diffuse more quickly than liquids, [4]
c) describe and explain the difference in compressibility between solids and gases, [6]
d) describe and explain the change in volume that occurs on boiling. [5]

9. The diagram below shows a section through a gas oven thermostat.

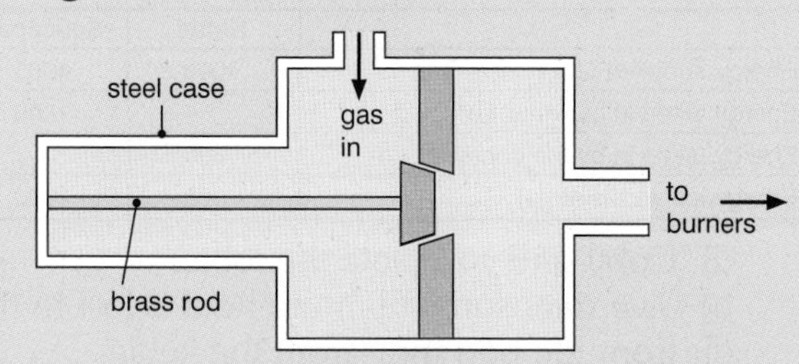

Explain how the thermostat controls the flow of gas to the burners. [3]

▶ Gas Laws

10. The diagram shows apparatus in which a fixed mass of air was compressed in a calibrated syringe, which was approximately half full of air at atmospheric pressure and 17 °C.

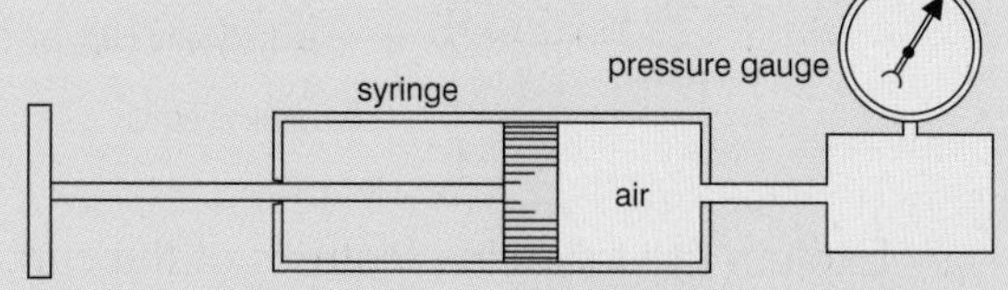

Corresponding values of volume and pressure of the trapped air are shown in the table.

Pressure (kPa)	50	60	75	90	105	120
Volume (m^3)	0.000 48	0.000 40	0.000 32	0.000 27	0.000 23	0.000 20
$\frac{1}{\text{volume}}$ (m^{-3})		2500		3704		5000

a) Copy and complete the table by calculating values for $\frac{1}{\text{volume}}$. [3]
b) On graph paper, plot a graph of pressure on the y-axis against $\frac{1}{\text{volume}}$ on the x-axis. [3]
c) What relationship between pressure and volume of the trapped air can be deduced from your graph? Explain your answer. [3]

11. A tank of helium gas is brought from an outside storeroom to a warm laboratory.
a) Explain, in terms of its molecules,
   i) how the gas exerts a pressure. [3]
   ii) why the pressure of the gas increases with temperature. [2]

b) The temperature and pressure of the helium are recorded as the gas warms up to room temperature:

Temperature (°C)	Pressure (kPa)
8.0	107.5
10.5	108.5
12.0	109.0
14.0	110.0
18.0	111.5
23.5	113.5

   i) Which is the dependent variable? [1]
   ii) Draw a graph of pressure against temperature. [3]
   iii) Use your graph to find the pressure of the helium at 0 °C. [2]
   iv) Explain why the pressure of the helium at 0 °C is not 0 kPa. [1]
   v) At what temperature should the pressure theoretically be 0 kPa? [1]

(Edex)

▶ Conduction, convection, radiation

12. The diagram shows a cup used to keep drinks hot.

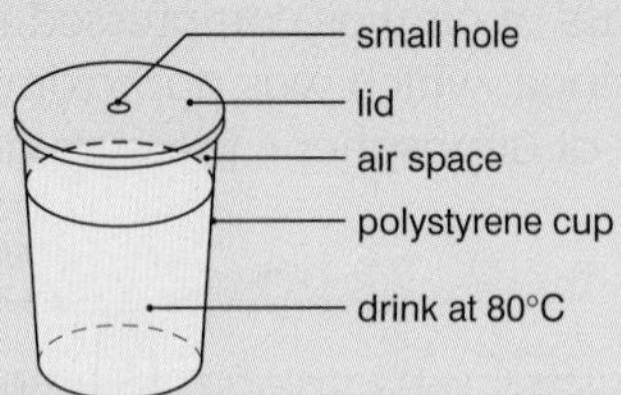

Explain the following features of the cup.
Use **some** of these words in your answers:

Conductor *Conduction* *Convection*
Evaporation *Expansion* *Insulator*
Pressure *Radiation*

a) Polystyrene is used as material for the cup. [2]
b) The lid helps to keep the drink hot for a longer time. [2]
c) There is a small hole in the lid. [2]

13. An isolated community uses coal and oil as their main sources of energy. They decide to put solar panels on their roofs and use small windmills on a hill to generate electricity.

a) They now have four sources of energy:
coal oil solar wind
i) Name **one** of these which is renewable.
ii) Explain clearly what a **renewable source of energy** means. [2]
b) Give **two** reasons why the community have decided to use solar panels and windmills. [2]
c) i) Why are the windmills put on a hill-top?
ii) Why are the solar panels put on a south-facing roof? [2]
d) The houses of the community were not insulated and lost heat in several ways. The chart below shows the percentage of heat lost from different areas.
i) Which part of the house is in most need of insulation? [1]
ii) What is the percentage of heat lost through the walls? [1]

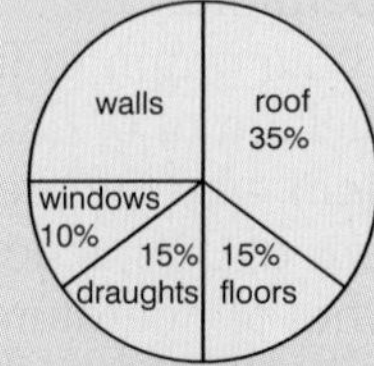

iii) Suggest what could be done to reduce heat loss through the floors. [1] (OCR)

14. Jack hopes to improve the insulation in his house by adding an extra layer of loft insulation. Suggest ways in which he could test, by experiment,

a) if the extra layer really does improve the insulation of the house, [3]
b) if it is cost effective. [2]

15. The table gives information about ways of reducing energy loss from a house:

Method of reducing energy loss	Cost of fitting	Annual saving
Draught-proofing	£50	£50
Hot-water-tank jacket	£20	£15
Loft insulation	£200	£50
Temperature controls on radiators	£100	£20

a) Which method saves money by preventing the house becoming too warm? [1]
b) Which method reduces energy loss by the smallest amount? [1]
c) Which method pays for itself soonest? [1]
d) What is the payback time on loft insulation? [1] (AQA)

16. Some water is being heated in a metal pan on the hotplate of an oven.

a) Explain, in terms of the particles in the metal, how heat energy is transferred through the base of the pan. [2]
b) Energy is transferred through the water by convection currents. Explain what happens to cause a convection current in the water. [3]
c) Some energy is transferred from the hotplate to the air by *thermal radiation*. What is meant by *thermal radiation*? [1]
(AQA)

17. A kettle and a saucepan contain the same volume of water. 4000 J of heat energy was given to the water in each container.

	Kettle	Saucepan
Energy supplied (J)	4000	4000
Energy taken in by water (J)		1800
Energy taken in by the container (J)	600	
Energy lost to the air (J)	800	1000

a) Copy and complete the table. [2]
b) Give a reason why more heat is lost to the air from the pan than from the kettle. [1]
c) Explain how the table shows that the kettle is more efficient for heating water. [1]
(WJEC)

18. *Describe and explain* **two** ways by which heat loss from a building may be reduced. [4] (WJEC)

19. An uninsulated house at 20 °C loses energy in each minute as shown:

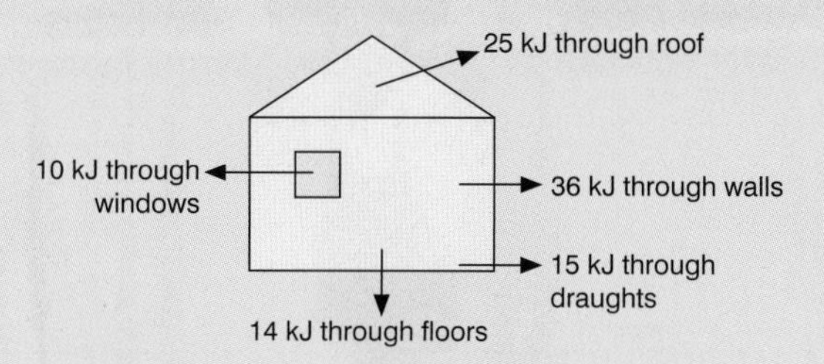

a) What is the total energy lost per minute?
b) 80% of roof losses can be saved by insulating the loft,
50% of window losses can be saved by double glazing,
$66\frac{2}{3}$% of wall losses can be saved by cavity wall insulation,
60% of draught losses can be saved by draught excluders.
Calculate the savings achieved by
i) insulating the loft and fitting draught excluders,
ii) double glazing and cavity insulation, and
iii) comment on your **two** answers. [4] (WJEC)

20. This question is about solar panels (page 50).
a) What is the purpose of the following:
i) the insulation behind the absorber panel? [1]
ii) having the absorber panel painted black? [1]
iii) having a glass cover on the top of the panel? [1]
b) i) Name suitable materials for making the absorber panel and water-ways. [2]
ii) Give your reasons for the choice of such materials. [2] (Edex)

21. This question is about a vacuum flask.
a) Give **two** features of the flask which reduce heat loss by conduction. [2]
b) Give **one** feature of the flask which reduces heat loss by radiation. [1] (AQA)

22. A saucepan and water are heated on a hotplate from 20 °C to 100 °C.
The hotplate provides 250 000 J of energy.
168 000 J of energy is used to heat the water, and 28 800 J is used to heat the pan.
a) How much energy is lost to the surroundings? [1]
b) Explain why the pan and hotplate method is not a very efficient way of boiling water. [2] (WJEC)

23. Hannah makes a cup of coffee. She notices that the kettle has a power rating of 2000 W. It takes 225 seconds to heat 1 kg of water from 20 °C to 100 °C.
a) Calculate the energy transferred by the kettle, using the equation:
energy = power × time [2]
b) Calculate the energy absorbed by the water. [The specific heat capacity of water = 4200 J/kg/°C] [3]
c) The energy absorbed by the water is ***less*** than the energy transferred by the kettle. Explain why. [1] (OCR)

▶ Change of state

24. During rain, pools of water form on a road. After the rain has stopped, the water slowly evaporates.
a) Explain the process of evaporation, by writing about molecules. [3]
b) State **two** weather conditions which make this process happen more quickly. [2] (OCR)

25. a) Use the molecular theory to explain why evaporation causes cooling.
b) State **two** ways by which the rate of evaporation of a liquid may be increased.
c) State **two** ways in which boiling is different from evaporation. [2,2,2]

26. Jack boils some water in a kettle. The lid of the kettle rattles up and down. Explain the following in terms of the movement of the water molecules.
a) The liquid water changes into steam. [2]
b) There is an upward force on the lid of the kettle. [2]
c) The lid of the kettle falls again, after it has risen. [3]

Parachutists have two forces acting on them.
One is the force upwards due to the air resistance.
What is the other force?

The metal weights have two forces acting on them.
One is their weight (the pull of gravity downwards).
What is the other force?

The rope has two forces acting on it.
One is the rock pulling upwards.
What is the other force?

The ice-skater has two forces on her.
One is the man pushing upwards.
What is the other force?

chapter 11

Pushes and pulls are ***forces***. Whenever we are pushing or pulling, lifting or bending, twisting or tearing, stretching or squeezing, we are exerting a force.

Experiment 11.1
Exert some push and pull forces on different objects including a ball, a spring, a piece of plasticine and an elastic band. What do you notice? Make a list of the things that forces can do to the objects.

Forces can
a) change the speed of an object
b) change the direction of movement of an object
c) change the size or shape of an object.

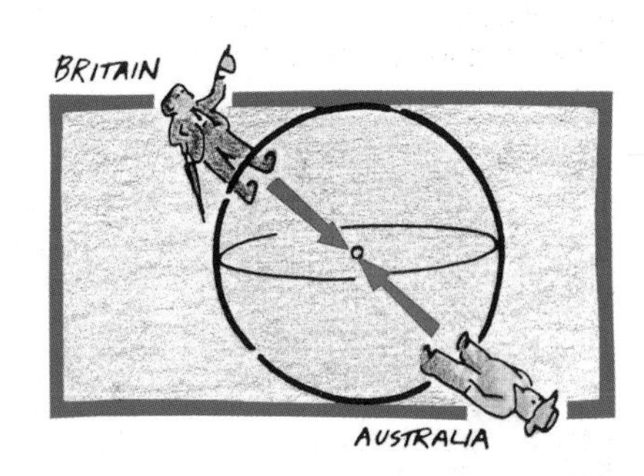

Weight is a very common force. Weight is the force of gravity due to the pull of the Earth. This is the force that is pulling you downwards now. It acts towards the centre of the Earth.

An apple fell on Newton's head,
Moved downwards from the tree,
And that was when Sir Isaac said:
"The force is gravity."

The weight of an object varies slightly at different places on the Earth.

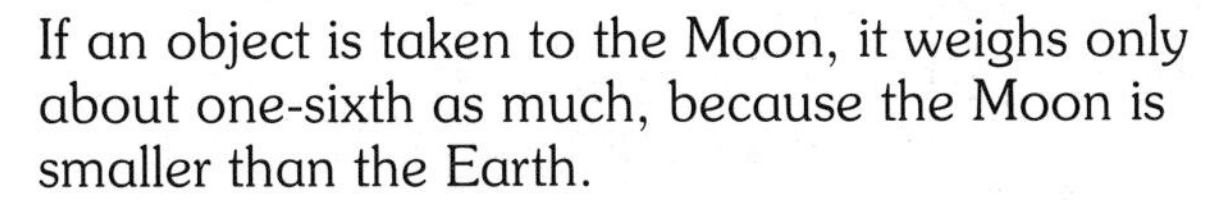

If an object is taken to the Moon, it weighs only about one-sixth as much, because the Moon is smaller than the Earth.

In outer space, well away from any planets or stars, objects become weightless.

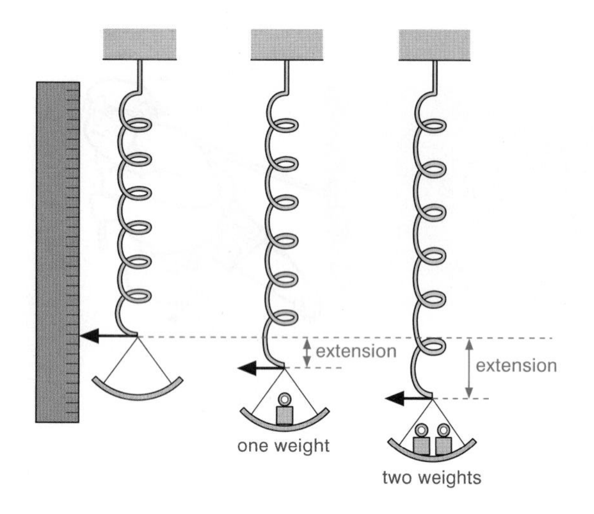

Stretching a spring

Experiment 11.2
Find out how forces change the length of a spring. Hang a spring from a retort stand and fix a pointer at the bottom of the spring so that it points to the scale of a ruler which is clamped vertically.

You will need several ***equal*** weights. Hang one weight on the spring and measure the ***extension*** of the spring.

Then add another weight so that there are two pulling down on the spring. Again measure the extension (from the original position).
Increase the load by adding more weights and measure the extension each time.

What do you notice?
If you double the load, does it double the extension?
If you put on three times the load, do you get three times the extension?

This means that:

The extension is directly proportional to the stretching force.

This is called **Hooke's Law**. This law also applies to the stretching of metal wires, and girders in bridges. It is important to engineers.

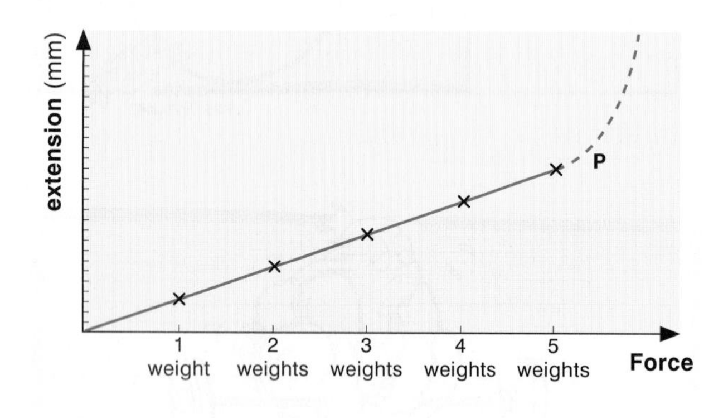

From your results, plot a graph of extension against the stretching force (the load).

What shape is the graph?

A straight line through the origin of the graph confirms that the extension is ***directly proportional*** to the stretching force (see page 390).

What happens with very heavy loads?
Hooke's Law only applies to the straight part of the graph (up to the ***limit of proportionality***).

The point P is called the ***elastic limit***. If a spring is taken beyond this limit, it will not return to its old shape. It is permanently ***deformed***.

Force = **spring constant** × **extension**
(N) (N/m) (m)

Experiment 11.3 ⚠
Use the method of experiment 11.2 to investigate the stretching of:
a) A thin copper wire. (Wear safety glasses for these experiments.)
b) A rubber band. What do you find?

▶ Measuring forces

A ***spring balance*** uses a spring to measure the weight of an object or to measure the strength of any pulling force. The marks on the scale are equally spaced because of Hooke's Law.

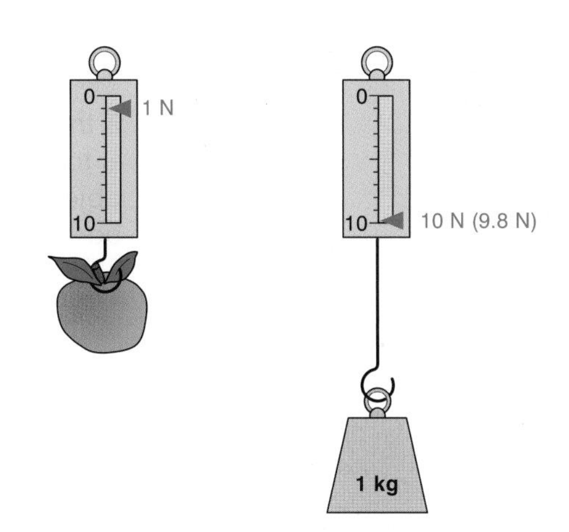

Forces are measured in units called ***newtons*** (often written N), named after Sir Isaac Newton.

An apple weighs about 1 newton (1 N).
The weight of this book is about 10 newtons (10 N).

> *Experiment 11.4*
> Use a spring balance to find the weight of a 1 kg mass.

A mass of 1 kg (here on Earth) weighs almost 10 newtons. (To be more exact, 1 kg on Earth weighs 9.8 newtons.)
A mass of 2 kg weighs 20 N, and so on:

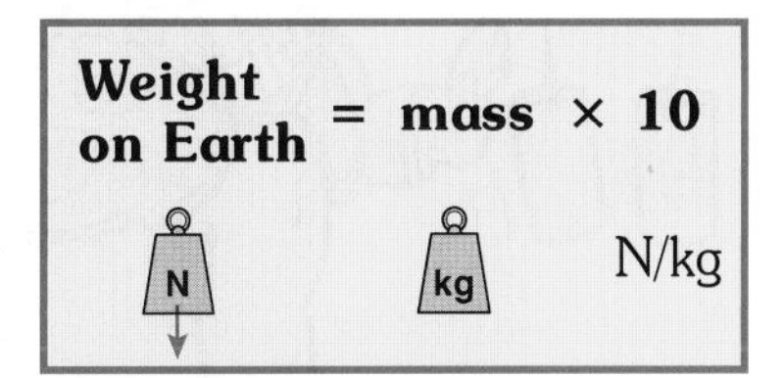

How much would a 1 kg mass weigh on the Moon, if the pull of gravity there is about one-sixth of the pull of gravity here?

> *Experiment 11.5*
> Use a weighing machine to find the weight of your body in newtons.

Your weight changes slightly from place to place on the Earth – at the North Pole you would weigh about 3 newtons heavier (like having 3 extra apples in your pocket).

How much would you weigh on the Moon?

How many press-ups could you do on the Moon?

> *Experiment 11.6*
> Use a force-meter (a very strong spring balance) to measure the strength of your arm muscles when you are pushing and when you are pulling.

Here is a weighty question from Professor Messer: does 1 newton of lead weigh more than 1 newton of feathers?

▶ Mass and inertia

If this book (which has a weight of about 10 newtons here on Earth) is taken into outer space, well away from the Earth or any other object, it will become weightless. But it would still be the same book with the same ***mass***.

The mass of an object is the amount of matter in it. It is measured in ***kilograms*** (kg). The mass of this book is about 1 kg.

When an object is stationary, it needs a force to make it move. The bigger the mass, the bigger the force needed to start it moving. We say that masses have ***inertia***, a reluctance to start moving.

Experiment 11.7
Hang two tins from long pieces of string. Fill one with wet sand and leave the other empty. Try pushing the cans.

Which is the harder to push?
Which has the larger inertia?
Which has the larger mass?

Since this experiment depends on the ***mass*** of the object (not the weight), you would feel just the same effect on the Moon or out in space – if you pushed a large mass you would feel a large amount of inertia.

Experiment 11.8
Place a piece of smooth card on top of a milk bottle and place a coin on top. Flick the card away with your finger.

What happens?
Why does the coin stay when the card moves?

Explain the Physics of this cartoon:

In a similar way, moving objects need a force to ***stop*** them moving. Their inertia tends to keep them moving.

> *Experiment 11.9*
> Pull aside the cans used in experiment 11.7 and when they swing back, stop them with your hand.

Which is the harder to stop?

Passengers in a car have a lot of inertia and so they need seat-belts. If the car stops suddenly, the people will tend to keep on moving (through the windscreen), unless the seat-belts exert large forces to stop them. See also page 138.

Your inertia can kill you!

What happens when the car turns a corner? Your body, because of its inertia, will tend to travel straight on. You can feel your body sway as the car turns the corner, but fortunately your seat exerts a force on you and this pulls you round the corner with the car.

Cars are designed to crumple safely in crashes

If you are standing on a bus, what happens to you when the bus:
a) starts moving?
b) stops moving?
c) turns a corner to the left?

Sir Isaac Newton stated all this in **Newton's First Law of Motion:**

> **If the forces on a mass are *balanced* (no resultant force), then**
> - **if it is at rest, it stays at rest**
> - **if it is moving, it keeps on moving at a constant speed in a straight line.**

There is more about forces in chapter 14.

Professor Messer puts his skates on. Explain the Physics of this cartoon:

▶ Circular motion

A snooker ball travels in a ***straight line***. This is because of Newton's 1st Law (page 69).

On an icy road, with no friction, a car 'skids' ***in a straight line:***

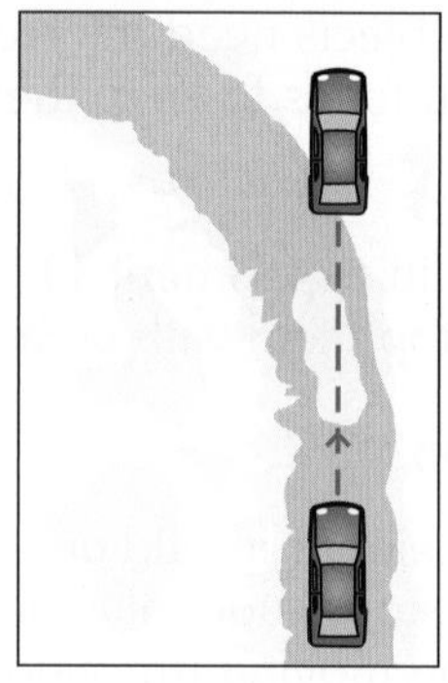

On an icy road

However on a dry road, the car can move in a curve. This is because there is a friction force between the tyres and the road.
To turn a corner or move in a curve, a force has to be applied.

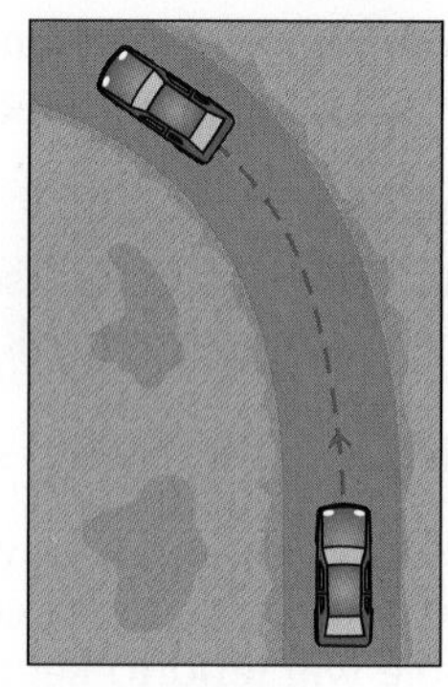

On a dry road

Experiment 11.10
Fasten a 'conker' or a cork to a piece of string, and whirl it round your head in a circle:
Can you feel the force in the string?

Newton's First Law says that the conker will continue to move in a straight line, unless a force is applied. The force in this case is due to your hand pulling in the string and the conker.

This force is acting ***towards the centre*** of the circle and is called **the centripetal force**:

You can also feel this force when you are on a roundabout.

If there is no centripetal force (if the string breaks) then the object moves in a straight line:

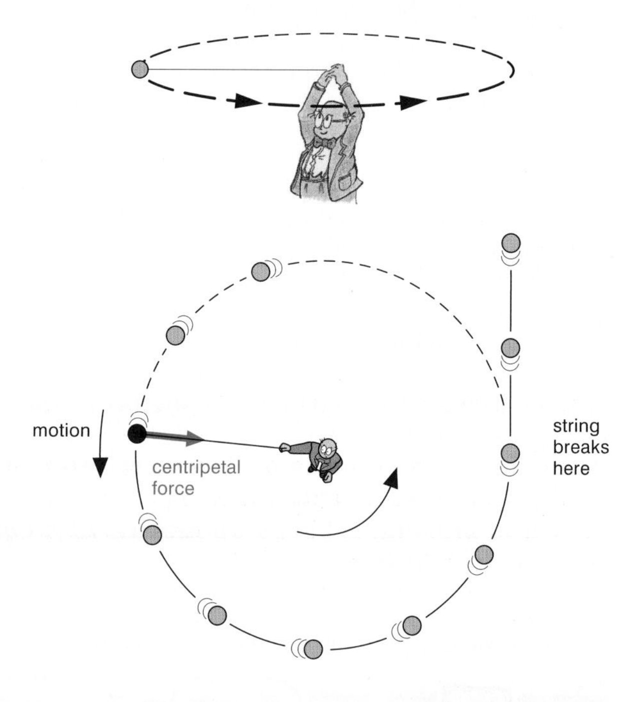

Strange as it may seem, although the conker is moving at a steady speed in the circle, it is ***accelerating***.
The conker is turning at a steady ***speed***, but its direction is changing all the time. This means that its ***velocity*** is changing (see page 122).
And this means that the conker is accelerating all the time, towards the centre of the circle!
The centripetal force is the resultant force needed to cause this acceleration (see page 130).

An object moving in a circle needs a ***bigger*** centripetal force if:

- the **mass** of the body is ***bigger***,
  (A lorry needs a bigger force than a car.)
- the **speed** of the object ***increases***,
  (A fast car needs a bigger force than a slow one.)
- the **radius** of the circle is ***smaller***.
  (A tight corner needs more force than a gentle curve.)

He is pulling on the string to provide the centripetal force

A spin-drier

In a spin-drier, the clothes spin very fast in a metal drum. They are held 'in orbit' by the force from the wall of the drum.
However there are holes in the wall. A drop of water which is next to a hole has nothing to keep it in orbit, and so it travels ***in a straight line*** and leaves the drum.

What do you think happens to a piece of dust on a CD when you start to play it?

The Earth in orbit

The Earth orbits the Sun in an almost circular path: What provides the centripetal force here?

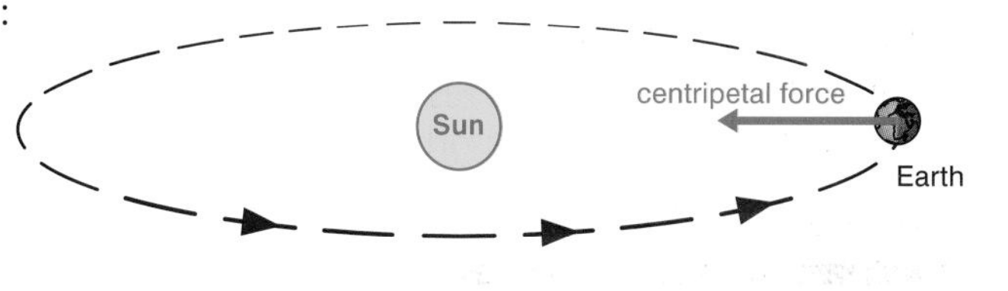

The gravitational pull of the Sun on the Earth provides the centripetal force which keeps us in orbit.
In the same way, the Moon is kept in orbit round the Earth by the gravitational pull of the Earth.
Artificial satellites are kept in orbit round the Earth by gravity (see pages 152, 154).
What keeps a comet in orbit (page 153)?

In a similar way, in an atom, the electric force of attraction between a proton (+) and an electron (−) keeps the electron in orbit (see page 242).

Look at the people on the fairground ride:

What is providing the centripetal force?
What happens if the ride spins faster?
What happens if one of the chains snaps?

Summary

Forces can
a) change the speed of an object
b) change the direction of movement of an object
c) change the size or shape of an object.

Hooke's Law: The extension of a spring is proportional to the force pulling it (up to the limit of proportionality).

Weight is the force due to the pull of gravity. It can be measured in newtons (N) by a spring balance and varies from place to place.

Mass is the amount of matter in an object. It can be measured in kilograms (kg) and does not change from one place to another.
A mass of 1 kg here on Earth weighs about 10 N (9.8 N).

Newton's First Law:
Every mass stays at rest or moves at constant speed in a straight line unless a resultant force acts on it.

To move in a circular orbit, a centripetal force is needed (acting towards the centre).

▶ Questions

1. Copy out and fill in the missing words:
a) When a spring is pulled, the . . . . is proportional to the . . . . (called . . . . Law).
b) Weight is the force of . . . . on an object due to the pull of the . . . . .
c) Weights and other forces are measured in . . . . .
d) Here on Earth, the pull of gravity on a mass of 1 kg is . . newtons.
e) Weight is measured by a . . . . balance and . . . . from place to place.
f) Newton's First Law says: every mass stays at . . . . or moves at constant . . . . in a . . . . line unless a resultant . . . . acts on it.
g) To move in a . . . . orbit, an object must have a . . . . force on it.

2. An object has a mass of 4 kg. What is its weight (in newtons) here on Earth?

3. An astronaut has a mass of 60 kg.
a) What is her weight here on Earth?
b) What is her weight on the Moon?
c) What is her mass on the Moon?

4. A spring is 20 cm long when a load of 10 N is hanging from it, and 30 cm long when a load of 20 N is hanging from it. Draw diagrams and work out the length of the spring when
a) there is no load on it
b) there is a load of 5 N on it.

5. In a spring experiment, the results were:

Load (N)	0	1	2	3	4	5	6	7
Length (mm)	50	58	70	74	82	90	102	125
Extension (mm)								

a) What is the length of the spring when unstretched?
b) Copy and complete the table.
c) Plot a graph of extension : load.
d) One of the results is wrong. Which is it? What do you think it should be?
e) Mark the elastic limit on your graph.
f) What load gives an extension of 30 mm?
g) What would be the spring length for a load of 4.5 N?

6.

Should Professor Messer pull the tablecloth quickly or slowly?

Explain the Physics of this trick, using the correct scientific words.

7. An engineer needs to know how far a long steel beam will sag under a load. The table shows some results:

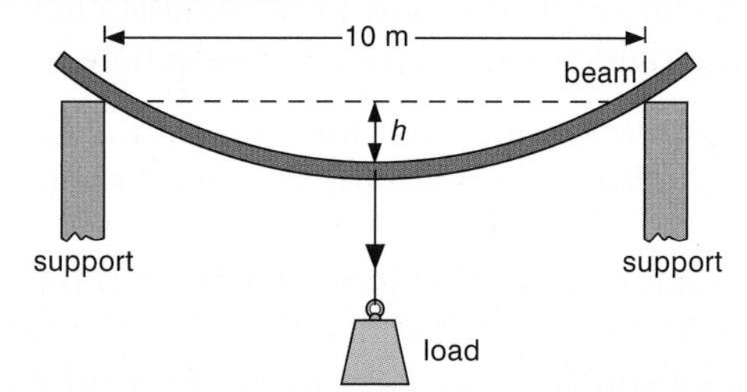

Load (N)	1000	2000	3200	4400	5200	6500
Sag, *h*(cm)	2.0	4.0	6.6	8.8	10.4	13.4

a) Plot a graph of the sag, ***h***, against load.
b) One of the measurements of ***h*** is wrong. Which is it? What do you think it should be?
c) What is the sag for a load of 4500 N?
d) What load would give a sag of 52 mm?
e) Would a longer beam sag more or less? Sketch its graph on the same axes.

8.

Could the convict escape this way?

Could he lift the ball more easily in a prison on the Moon? Could he jump higher in a prison on the Moon?

9. What provides the centripetal force for:
a) a car turning a corner?
b) the icy rocks in Saturn's rings?
c) the rim of a bicycle wheel?

Further questions on page 140.

▶ Physics at work: Building Bridges

Experiment 1 Bending a beam
Put a ruler as a bridge between two books. Then press down gently on the middle so that the 'beam' bends.
What must be happening to the molecules of the ruler?

In the lower part of the ruler the molecules are pulled farther apart (as the ruler bends). This part of the ruler is in ***tension***.

In the upper part of the ruler the molecules are pushed closer together. This part of the ruler is in ***compression***.

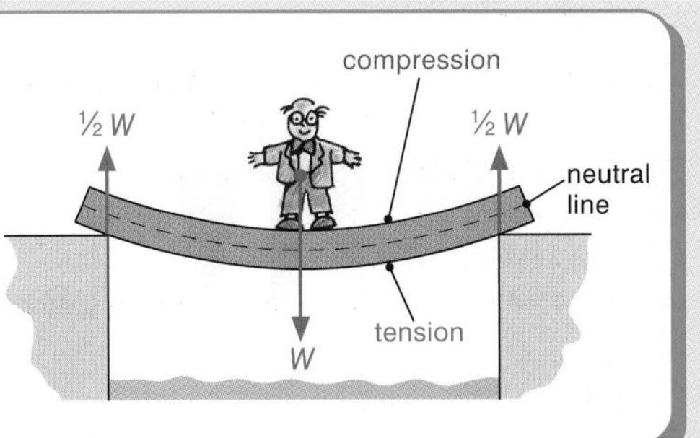

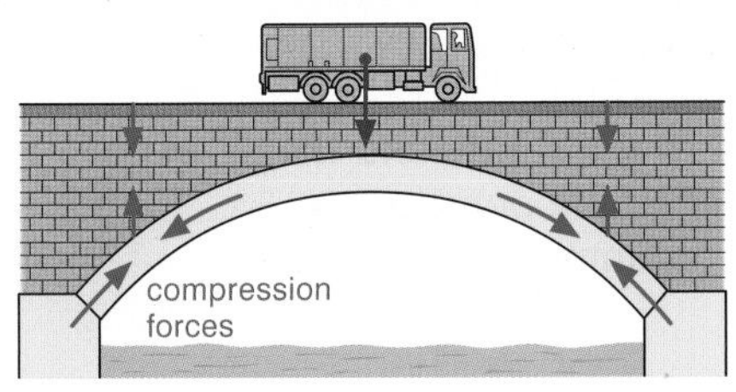

*An **arch** bridge has only compression forces. It can be safely built from concrete (or stone or brick) because they are strong in compression.*

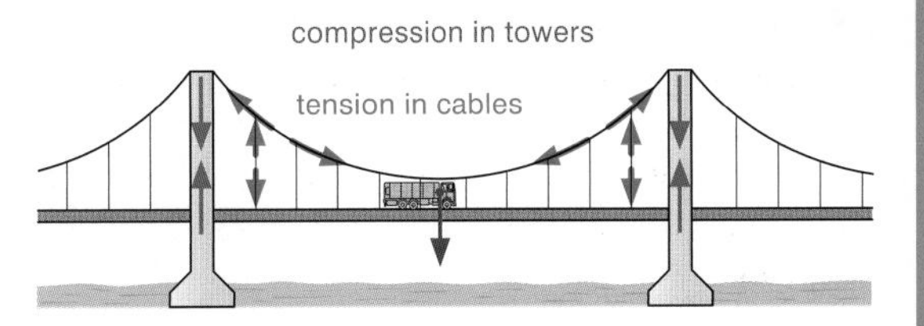

*A **suspension** bridge. Here the towers (in compression) can be made from concrete, but the cables (in tension) are made from steel.*

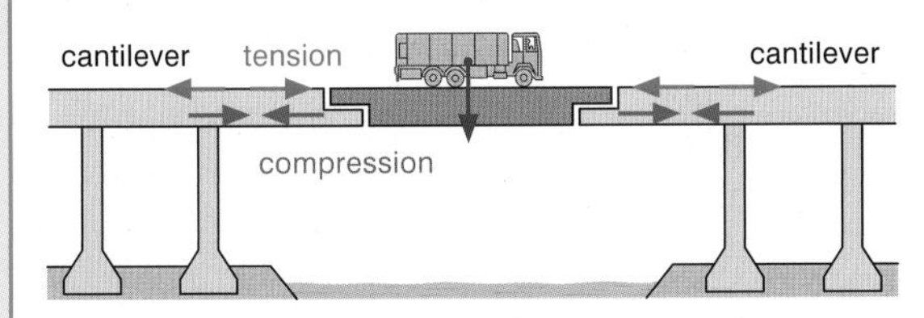

*A **cantilever** bridge is an extension of a simple beam bridge (see above). The cantilevers allow a wider gap to be bridged.*

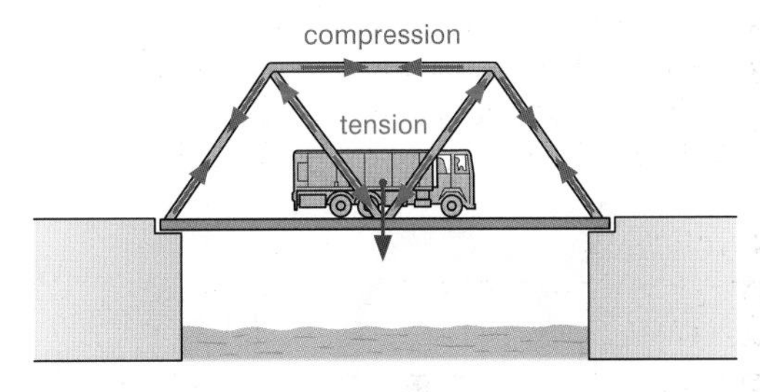

*A **girder** bridge is a strengthened beam bridge. It uses steel triangles to make it rigid. Longer bridges need more triangles (see also page 22).*

Survey your local bridges: what are they made of; which parts are in tension, or in compression?

Experiment 2 Testing girders
For girders you can use drinking straws, with pins to fasten them (or use balsa wood, or wooden 'spills', Lego, or Meccano).
Use girders to build these shapes:

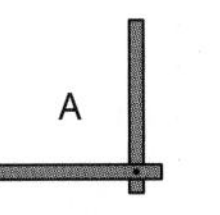

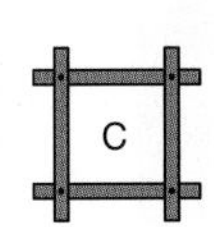

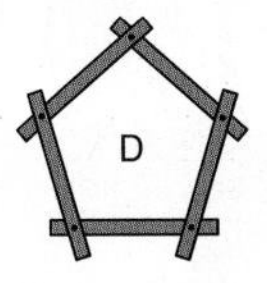

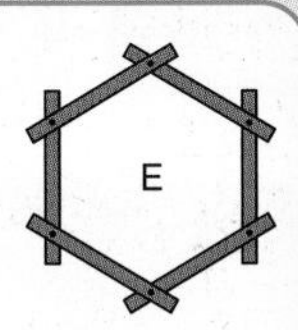

B is much more rigid than the others. Why? How can you add extra girders to make them more rigid?

Experiment 3 Building a girder bridge: a competition
Task: use your girders to build a bridge to cross a 10 cm gap.
You can use up to 10 straws but imagine each one costs £1000. You can test each bridge (until it collapses) by hanging a paper cup from it and gently adding weights or sand to it.
How can you ensure your tests are ***fair***? Who can build the cheapest bridge to support a 500 g mass?
Who can build the strongest bridge for £10 000?

chapter 12

Which is heavier – iron or wood? Many people say iron – and yet an iron nail is lighter than a wooden tree!
What people mean is that iron and wood have different ***densities***.
To measure density, we need to measure the mass of a definite volume of the substance. In fact:

$$\textbf{Density} = \frac{\textbf{mass}}{\textbf{volume}}$$

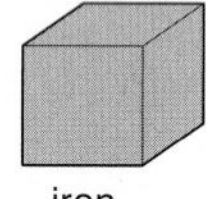

1 m^3 of iron
mass = 8000 kg
$\therefore$ density = 8000 kg/m^3

iron

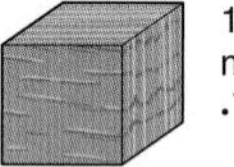

1 m^3 of oak wood
mass = 700 kg
$\therefore$ density = 700 kg/m^3

wood

If the mass is measured in **kg** (kilograms) and the volume in $\mathbf{m^3}$ (cubic metres), the density is measured in $\mathbf{kg/m^3}$ (kilograms per metre cubed).

Sometimes the mass is measured in **g** (grams) and the volume in $\mathbf{cm^3}$ (cubic centimetres) so the density is measured in $\mathbf{g/cm^3}$ (grams per centimetre cubed).

Here are the densities of several substances:

Substance			**Density**	
Solid	**Liquid**	**Gas**	**kg/m^3**	**g/cm^3**
Gold			19 000	19
	Mercury		14 000	14
Lead			11 000	11
Iron			8 000	8
	Water		1 000	1
Ice			920	0.92
	Petrol		800	0.80
		Air	1.3	0.0013

What do you notice about the numbers in the table?

Polystyrene has a very low density

Example
An engineer needs to know the mass of a steel girder which is 20 m long, 0.1 m wide and 0.1 m high. (Density of steel = 8000 kg/m^3)

Calculate the volume first:

$$\text{Volume of girder} = \text{length} \times \text{width} \times \text{height} = 20\,\text{m} \times 0.1\,\text{m} \times 0.1\,\text{m} = 0.2\,\text{m}^3$$

Formula:

$$\textbf{Density} = \frac{\textbf{mass}}{\textbf{volume}}$$

Then put in the numbers:

$$8000 = \frac{\text{mass}}{0.2\,\text{m}^3}$$

$$\therefore \text{ mass} = 8000 \times 0.2 = \underline{1600\,\text{kg}}$$

▶ Measuring density

Experiment 12.1
Find the density of a **solid** in three stages:

1. Find the mass of the solid (for example, a stone) by using a top-pan balance.
2. Find the volume of the solid by using a measuring cylinder. Remember to put your eye at the right level, and to read the ***bottom*** of the meniscus.
3. Use the formula: $\text{density} = \frac{\text{mass}}{\text{volume}}$

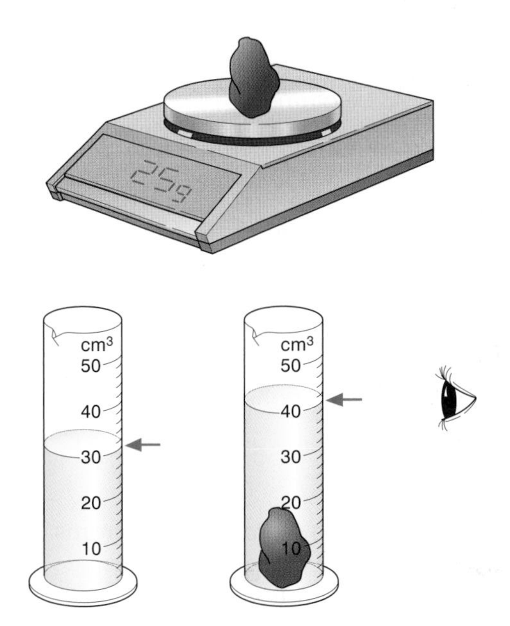

What is the density of the stone shown here?

Geologists need to measure the density of rocks. It helps them to identify the kind of rock, and where it comes from.

Milk inspectors and beer inspectors measure the density of the liquid to see if it has been watered down.

Experiment 12.2
Find the density of **water** in five stages:

1. Find the mass of a dry empty beaker using a beam balance or a top-pan balance.
2. Use a measuring cylinder to pour exactly 100 cm^3 of water into the beaker.
3. Now find the mass of the beaker with the water in it.
4. Work out the mass of the water from the results of steps 1 and 3.
5. Use the formula:

   $\text{density} = \frac{\text{mass}}{\text{volume}}$ to calculate the density of water.

What unit is your result measured in?
How does your result compare with the value shown opposite?
What would your result be in units of kg/m^3?

Experiment 12.3
Repeat experiment 12.2 to find the density of another liquid.

Can you find Professor Messer's mistake? What is the correct answer?

Summary

1. Density $= \frac{\text{mass}}{\text{volume}}$

Changing round the formula (see page 390):

2. Volume $= \frac{\text{mass}}{\text{density}}$

3. Mass = volume × density

Density of water = 1000 kg/m^3 = 1 g/cm^3.

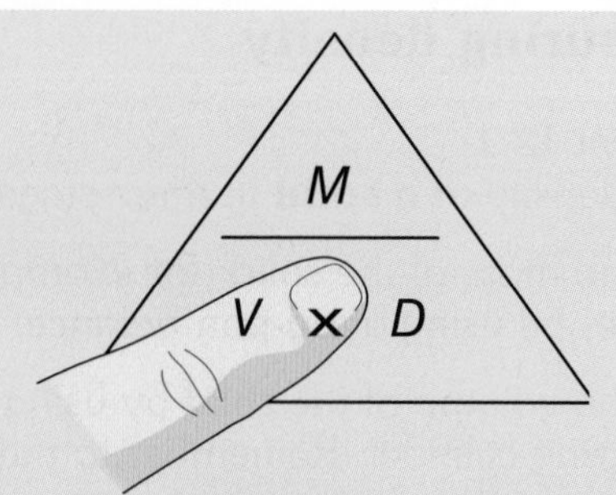

To find the one you want, cover up that letter in the triangle and the remaining letters show you the formula.

▶ Questions

1. An object has a mass of 100 grams and a volume of 20 cubic centimetres.
What is its density?

2. An object has a mass of 40 000 kilograms and a volume of 5 cubic metres.
What is its density?

3. An object has a volume of 3 cubic metres and a density of 6000 kilograms per cubic metre. What is its mass?

4. Copy and complete this table.

Object	Density (kg/m^3)	Mass (kg)	Volume (m^3)
A		4000	2
B	8000		4
C	2000	1000	
D		2000	4

a) Which object has the greatest mass?
b) Which has the smallest volume?
c) Which objects could be made of the same substance?
d) Which object would float on water?

5. A water tank measures 2 m × 4 m × 5 m.
What mass of water will it contain?

6. An object has a mass of 20 000 kilograms and a density of 4000 kilograms per cubic metre. What is its volume?

7. A stone of mass 30 grams is placed in a measuring cylinder containing some water. The reading of the water level increases from 50 cm^3 to 60 cm^3. What is the density of the stone?

8. Professor Messer wants to load some bricks into his van. There are 1000 bricks, and when stacked neatly they measure 2 m by 1 m by 1 m.
a) What is the volume of the stack?
b) What is the volume of one brick?
c) If the density of brick is 2500 kg/m^3, what is the mass of the stack?
d) If his van's maximum load is 1000 kg, how many bricks can he load?

9. The density of air is 1.3 kg/m^3. What mass of air is contained in a room measuring 2.5 m × 4 m × 10 m? How does this compare with your mass?

10. Describe, in detail, with diagrams, how you would find the density of a rock.

11. Professor Messer has been offered a 'gold medallion' by a woman in the market.
a) How (in detail) could he test it?
b) He found its mass was 550 g and its volume was 50 cm^3. What do you think? (See the table on page 74.)
c) What is the connection with Archimedes in ancient Greece?

Have you started to revise? See page 382.

chapter 13

You can push a drawing pin into a piece of wood – but you cannot push your finger into the wood even if you exert a larger force. Why? What is the difference between a sharp knife and a blunt knife?

The difference in each case is a difference of ***area*** – the point of the drawing pin and the edge of the sharp knife have a small area.

A force acting over a small area gives a larger **pressure**. Pressure is ***force per unit area***, or

$$\textbf{pressure} = \frac{\textbf{force} \text{ (in newtons)}}{\textbf{area} \text{ (in square metres)}}$$

Its unit is newtons per square metre (N/m^2). This unit is also called the pascal (Pa), named after Blaise Pascal, who investigated air pressure.

Why do eskimos wear snow-shoes?

Example 1
An elephant weighing 40 000 N stands on one foot of area 1000 cm^2 ($= \frac{1}{10} \text{ m}^2$). What pressure is exerted on the ground?

Formula first: $\text{Pressure} = \dfrac{\text{force}}{\text{area}}$

Then put in the numbers: $= \dfrac{40\,000 \text{ N}}{\frac{1}{10} \text{ m}^2}$

$= 40\,000 \times 10 \quad \text{N/m}^2$

$= \underline{400\,000 \text{ N/m}^2}$

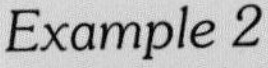

Example 2
What is the pressure exerted by a girl weighing 400 N standing on one 'stiletto' heel of area 1 cm^2 ($= \frac{1}{10\,000} \text{ m}^2$)?

Formula first: $\text{Pressure} = \dfrac{\text{force}}{\text{area}}$

Then put in the numbers: $= \dfrac{400 \text{ N}}{\frac{1}{10\,000} \text{ m}^2}$

$= 400 \times 10\,000 \quad \text{N/m}^2$

$= \underline{4\,000\,000 \text{ N/m}^2}$ (ten times bigger!)

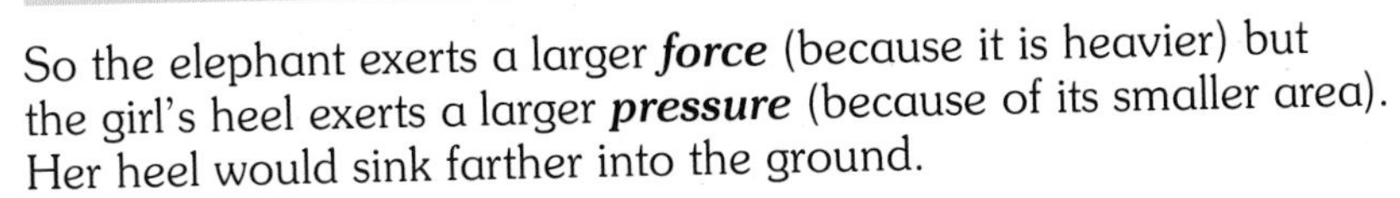

So the elephant exerts a larger ***force*** (because it is heavier) but the girl's heel exerts a larger ***pressure*** (because of its smaller area). Her heel would sink farther into the ground.

Why do camels have large flat feet?

Sir jumps quickly to his feet
He's got the point (– he's got a scar!)
The pressure, acting on his seat,
Is force per unit areaaaaaagh!

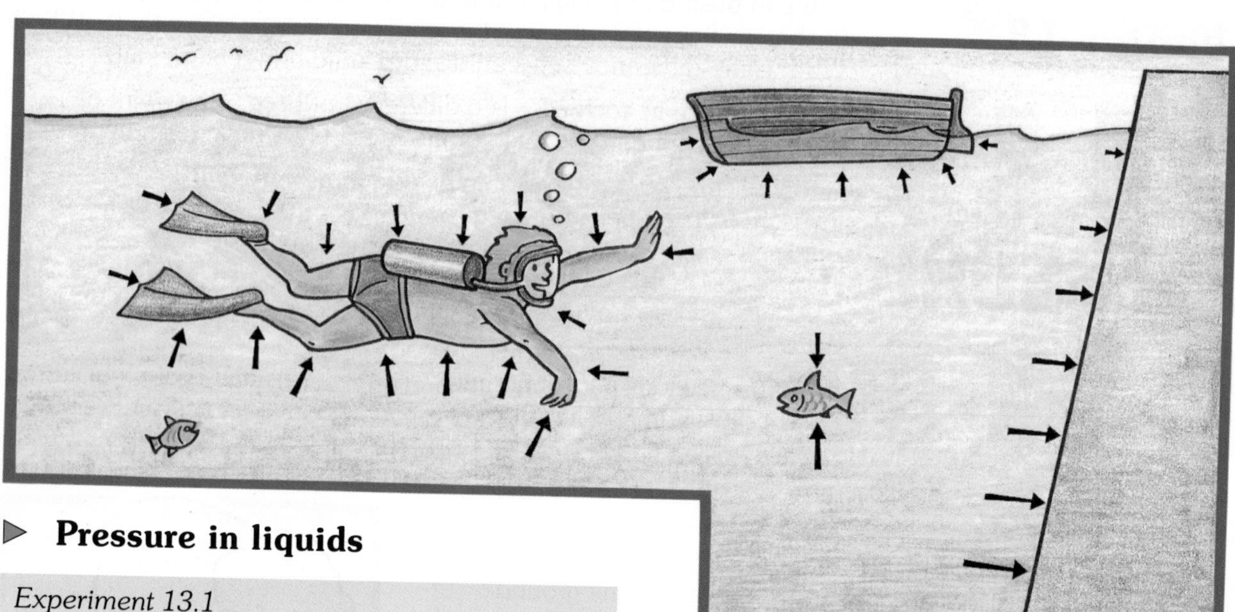

Pressure in liquids

Experiment 13.1
Fill a polythene bag with water.
Use a pin to poke several holes in the plastic while you squeeze the bag gently. What do you see?

Squeezing the top of the bag causes the water to squirt out. This means that:
Pressure is transmitted throughout the liquid.

Also, the water squirts out in ***all*** directions. This means that:
Pressure acts in all directions.
In the picture above you can see that the red arrows are acting in ***all*** directions on the diver – up and down, and left and right.

Experiment 13.2
Use a tall can with three equal sized holes drilled in the side. Fill it with water. What happens?

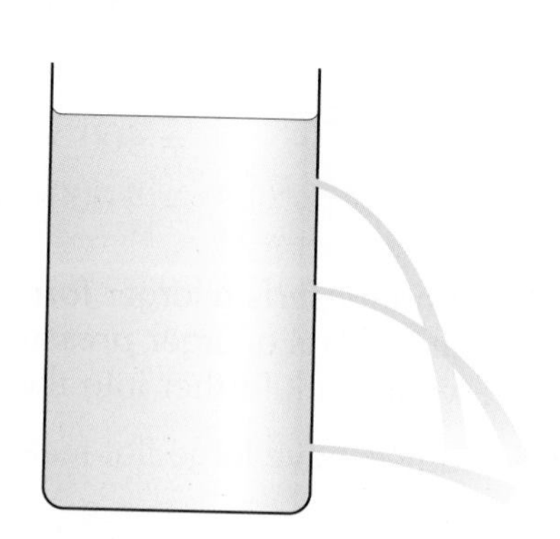

Where does the water squirt out fastest?
Where is the pressure greater – at the top hole or at the bottom hole?
Pressure increases with depth.

As well as depth, the pressure also depends on:
– the density of the liquid,
– the pull of gravity, g = 10 N/kg (see page 67).

In fact you can calculate the change in pressure by:

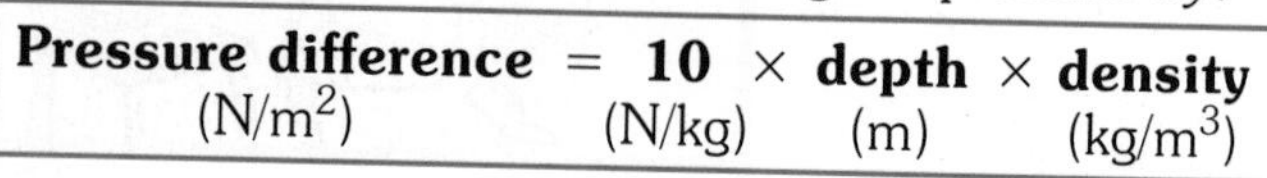

$$\underset{(\text{N/m}^2)}{\textbf{Pressure difference}} = \underset{(\text{N/kg})}{\mathbf{10}} \times \underset{(\text{m})}{\textbf{depth}} \times \underset{(\text{kg/m}^3)}{\textbf{density}}$$

▶ Hydraulic machines

The pressure in a liquid can be used to work machinery.

The diagram shows a narrow syringe A, connected by a tube to a wider syringe B. They are filled with a liquid.

What would happen if you pushed on piston A?

The pressure is transmitted from one piston to the other, so piston B moves out a little bit. The pressure is the ***same*** at both ends: at piston A it is caused by a ***small*** force acting over a small area. At piston B it exerts a **larger** force acting over the larger area. The force has been magnified.

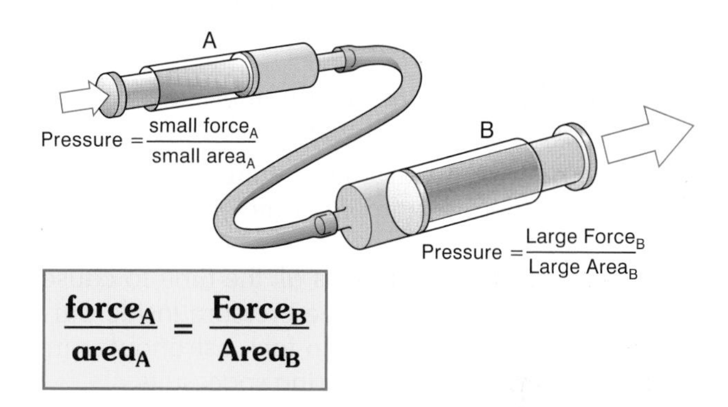

A
B
Pressure = small forceA / small areaA
Pressure = Large ForceB / Large AreaB

$$\frac{\text{force}_A}{\text{area}_A} = \frac{\text{Force}_B}{\text{Area}_B}$$

Hydraulic disc brakes

Experiment 13.1 showed that pressure is transmitted throughout a liquid. This idea is used in a car, to apply equal forces to all four wheels, and to magnify the force on the brake.

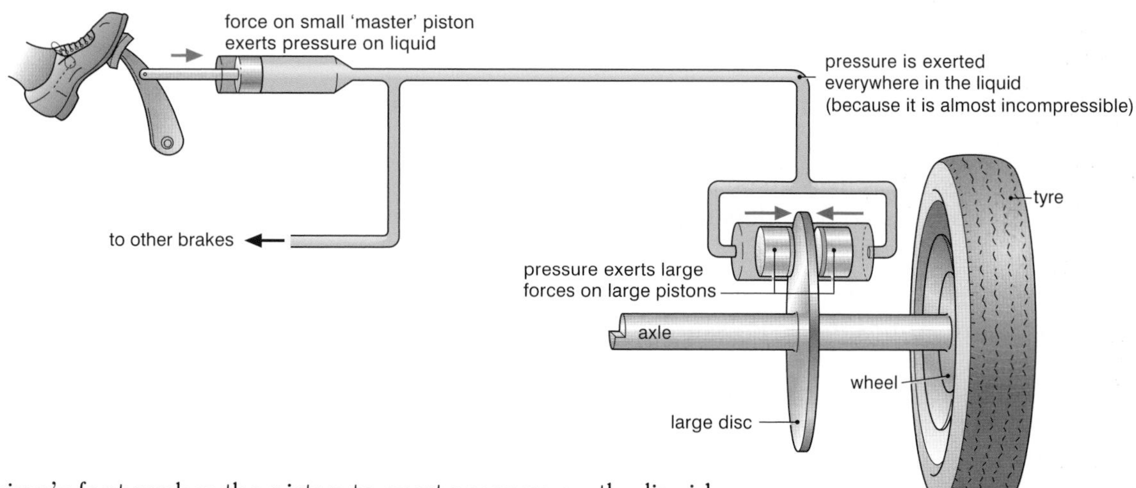

force on small 'master' piston exerts pressure on liquid
pressure is exerted everywhere in the liquid (because it is almost incompressible)
to other brakes
tyre
pressure exerts large forces on large pistons
axle
wheel
large disc

The driver's foot pushes the piston to exert pressure on the liquid. This pressure is transmitted to pistons on each side of a large disc on the axle. The pressure makes the pistons squeeze the disc (like the brakes on a bicycle) to slow down the car. Exactly the same pressure is applied to the other brakes on the car.

If the pistons at the disc have ***twice*** the area of the master piston, they will each exert ***twice*** the force that the driver applies with her foot. The force is magnified by the increased area of the pistons.

Other hydraulic machines use the same principle – for example, a hydraulic car jack, and the moving arms on this mechanical digger:

▶ Atmospheric pressure

We are living at the bottom of a 'sea' of air called the atmosphere, which exerts a pressure on us (just as the sea squeezes a diver).

> *Experiment 13.3 (Teacher demonstration)* ⚠
> Remove the air from the inside of a can by connecting it to a vacuum pump. (Alternatively, the air can be removed by boiling some water in the can so that the steam drives out the air before a bung is used to seal the can.) What happens?

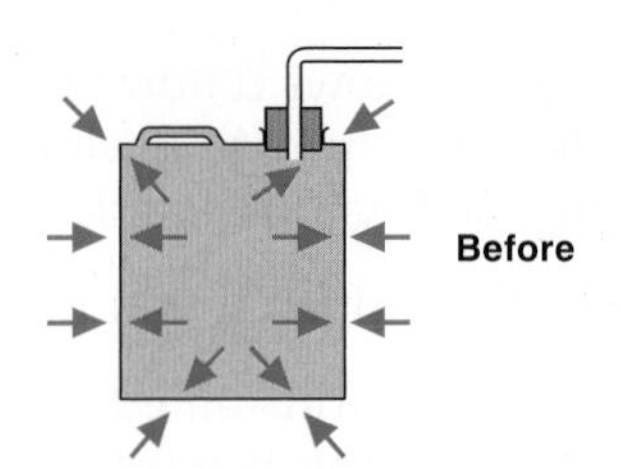

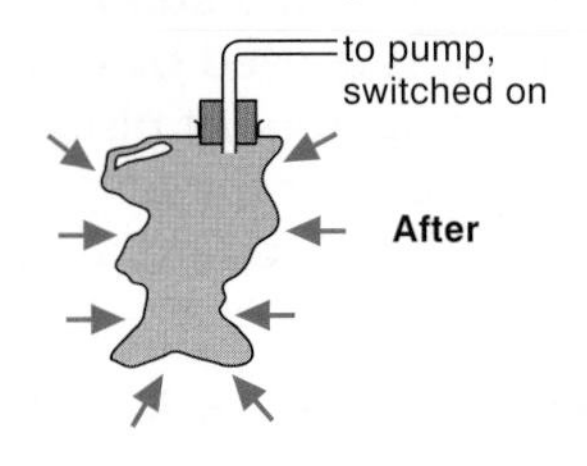

Before the pump is switched on, molecules are hitting the outside and the inside of the can with equal pressure.

After the pump is switched on, there are almost no molecules inside the can and the pressure of the molecules outside the can crushes it!

Molecules of air are hitting us all the time to cause this pressure (see page 16). Fortunately the pressure inside our bodies is equal to this, so we are not crushed. In fact, astronauts must keep this air pressure round them by wearing spacesuits.

The pressure exerted on us by the atmosphere is about 100 000 newtons per square metre (100 kN/m^2)!

The mercury barometer

A mercury barometer can be made by filling a long glass tube with mercury, and then turning it upside down in a bowl of mercury.

The column of mercury is held up by the air pressure. As the air pressure varies from day to day (depending on the weather), the height of the mercury varies.

The distance to measure is shown in the diagram. A height of 760 mm is called ***Standard Atmospheric Pressure*** (written 760 mm Hg).

vacuum (no air molecules)

mercury

h

air pressure

If you suck on a straw, the drink moves up for the same reason – atmospheric pressure pushes the liquid up into your mouth.

A manometer pressure gauge

A manometer is a U-tube containing a liquid, usually water.

> *Experiment 13.4*
> Use a manometer to measure the pressure of the gas supply.

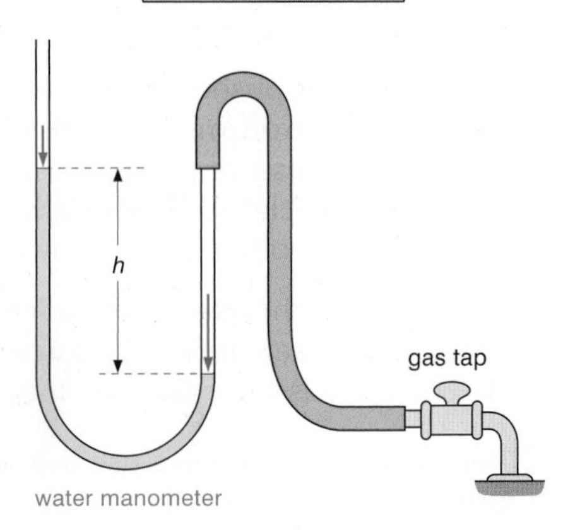

The pressure of the gas changes the levels of the water in the manometer. The pressure can be found (in 'centimetres of water') by measuring the height marked ***h***.

Summary

Pressure (in N/m^2) = $\frac{\text{force (in N)}}{\text{area (in m}^2\text{)}}$

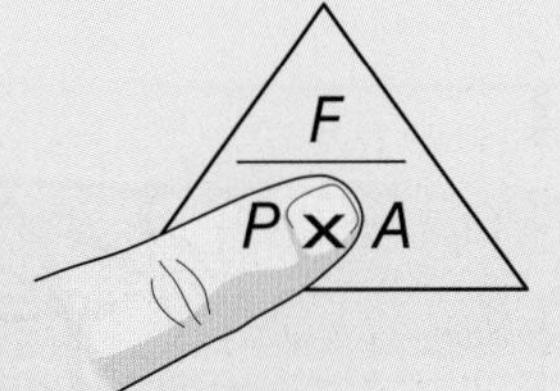

Pressure is transmitted throughout a fluid and acts in all directions.

Pressure in a fluid increases with depth.

Pressure is used in hydraulic machines.

Atmospheric pressure is about 100 000 N/m^2 (or about 760 mm Hg on a mercury barometer) but depends on the weather and the altitude.

▶ Questions

1. a) Pressure = ÷÷÷. Its units are . . . . per square . . . . (or . . . .).
b) In fluids (liquids and gases), pressure acts in . . . . directions, and pressure . . . . as the depth increases.
c) In a hydraulic machine, the . . . . is the same throughout. The larger piston exerts the . . . . force.
d) Air pressure can be measured by a mercury . . . . . The . . . . height of the mercury is usually about . . mm.

2. A box weighs 100 N and its base has an area of 2 m^2. What pressure does it exert on the ground?

3. If atmospheric pressure is 100 000 N/m^2, what force is exerted on a wall of area 10 m^2?

4. Use your Physics to explain the following:
a) Stiletto heels are more likely to mark floors.
b) Eskimos wear snowshoes.
c) It is useful for camels to have large flat feet.
d) Tractors have large tyres, bulldozers have caterpillar tracks and heavy lorries may need eight rear wheels.
e) It hurts to hold a heavy parcel by the string.
f) It is more comfortable to sit on a bed than on a fence.
g) A ladder would be useful if you had to rescue someone from an icy pond.
h) An Indian fakir can lie on a bed of nails if there is a large number of nails.

5. Use your Physics to explain the following:
a) You can fill a bucket from a downstairs tap quicker than from an upstairs tap.
b) Deep sea divers have to wear very strong diving suits.
c) A dam is thicker at its base than at the top.
d) A hole in a ship near the bottom is more dangerous than one nearer the surface.
e) A giraffe must have a stronger heart than a human.
f) A barometer will show a greater reading when taken down a coal mine.
g) Aeroplanes are often 'pressurised'.
h) Astronauts wear spacesuits.

6. Referring to the table of densities on page 74, which of those substances will a) sink in water b) float on water c) sink in mercury d) float on mercury?
A solid has a mass of 2000 kg and a volume of 4 m^3. Will it sink or float in water?

7. In a hydraulic brake, a force of 500 N is applied to a piston of area 5 cm^2.
a) What is the pressure transmitted throughout the liquid?
b) If the other piston has an area of 20 cm^2, what is the force exerted on it?

8. Explain, with diagrams, how each of the following uses atmospheric pressure:
a) a drinking straw b) a syringe
c) a rubber sucker d) a vacuum cleaner
e) an altimeter f) a rubber plunger used for clearing blocked sinks.

Further questions on page 140.

chapter 14

▶ Friction

Friction is a very common force. Whenever one surface slides over another, friction always tries to oppose the movement. Friction is often a nuisance, because it wastes energy.

Experiment 14.1
Put your hands together and rub hard. What do you notice? This waste of heat energy reduces the efficiency of machines.

Reducing friction

1. The slide in the park is polished smooth so that you can slide down easily.
   Through a microscope, even polished surfaces look rough, so there is always some friction (as the lumps on one object catch and stick to the lumps on the other object):

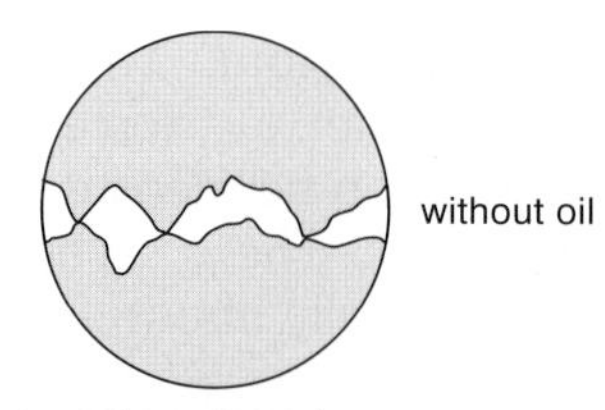

views through a microscope

2. Friction can be reduced by ***lubricating*** with oil:

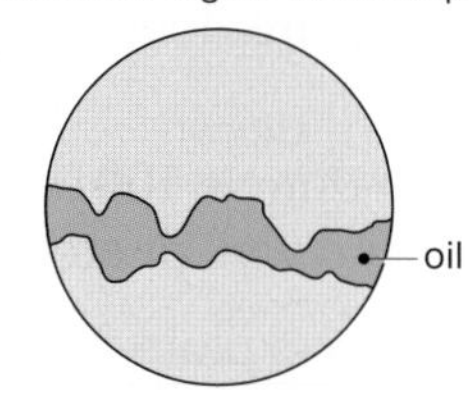

If workers puff that rough is tough Reduce the toil by adding oil

3. Another way of reducing friction is to separate the surfaces by air. This is how a hovercraft works.
4. A fourth way of reducing friction is to have the object ***rolling*** instead of sliding.
   This is what happens with ball bearings.
5. A fifth way: boats, cars, planes and rockets are ***streamlined*** to reduce friction with water or air. Even so, rockets get very hot when they enter the Earth's atmosphere.
   Dolphins are streamlined to reduce friction:

Advantages of friction

Although it is often a nuisance, friction is also very useful. Friction allows you to pick up this book with your hands.
Our lives depend on the friction at the brakes and tyres of cars and bicycles.
Air friction (drag) slows down the parachute of a falling man so that he can land safely.

You are able to walk only because of friction with the floor (try walking on wet ice!).
Knots in string and the threads in your clothes are held together by friction.
Nails and screws are held in wood by friction.
What things do you think would happen if there was suddenly no friction at all in this room?

▶ Physics at work: Friction and Road safety

Car-drivers and cyclists rely on friction to stop.

When a driver has to brake, it takes time for him to react. In that fraction of a second, the car can travel many metres.
This is called the **thinking distance**.

A driver's reaction time is usually about 0.7 s.
It is slower if he has taken alcohol or other drugs, or if he is tired. See the experiment on page 132.

The **braking distance** is the distance the car will travel ***after*** the brake is pressed.

The braking distance ***in***creases,

- If the car is travelling faster. The brakes have to transfer the car's kinetic energy (to heat). The formula for KE has the speed ***squared*** (p. 109). So, if the car goes at three times the speed, it has ***nine*** times the energy and so the braking distance is ***nine*** times as long!
- If the car is heavier (because it has more KE).
- If the road surface is wet or smooth. On a wet road the braking distance is about twice as long! Also the car may skid. Special high-friction surfaces are often used before traffic-lights.
- If the car is poorly maintained – with worn brakes or worn tyres. The grooves in tyres must be at least 1.6 mm deep. The grooves are carefully designed to clear away water on a wet road.

Example
A car is travelling at 20 m/s (45 mph).
The driver has a reaction time of 0.7 seconds.
How far does he travel ***before*** he starts to brake?

$$\textbf{speed} = \frac{\textbf{distance travelled}}{\textbf{time taken}} \quad \text{(see page 122)}$$

$\therefore$ distance travelled = speed × reaction time
= 20 m/s × 0.7 s

$\therefore$ Thinking distance = 14 m (see chart below)

Too much kinetic energy

Old and new

The total stopping distance = thinking distance + braking distance

What patterns can you find in this chart:

Shortest stopping distances *on a dry road, with good brakes, and a good reaction time.*

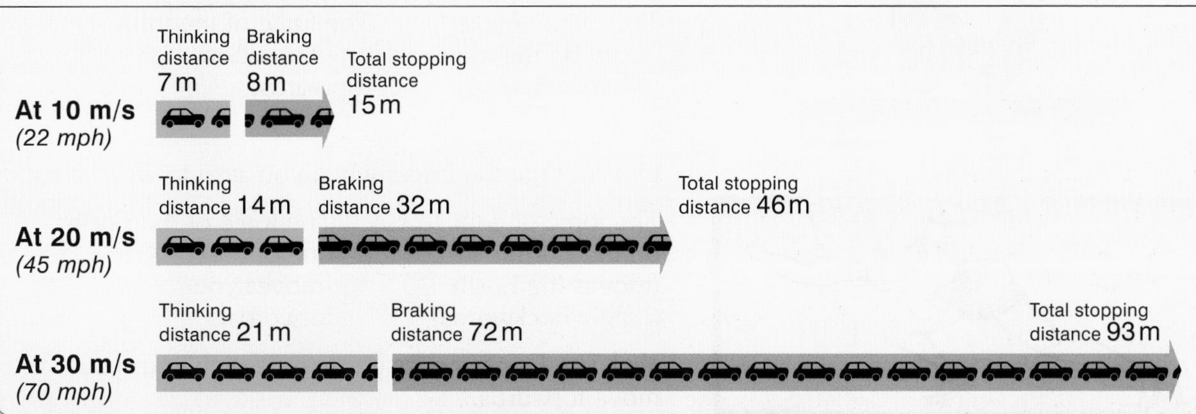

▶ Newton's Third Law

The two teams are having a tug-of-war contest. If the rope is not moving, what can you say about the forces exerted by the two teams?

Experiment 14.2
Use two spring balances to have a tug of war:

What do you notice about the readings on the balances?

Newton noticed that forces were ***always*** in pairs and that the two forces were ***always*** equal in size but opposite in direction. He called the two forces ***action*** and ***reaction***. **Newton's Third Law of Motion** is:

The action force and the reaction force are equal and opposite.

In the tug-of-war picture, if the rope is not moving, then:

the force of team A on team B	=	the force of team B on team A

You can see that the way we get the sentence describing the second force is by ***changing round the words from the first sentence***. We use the same method in the following examples.

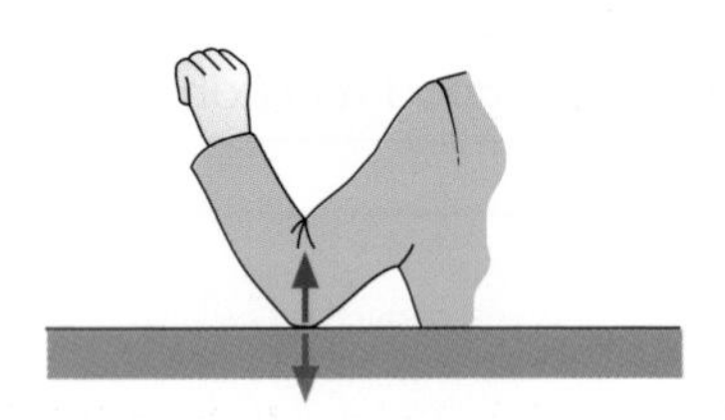

a) Rest your elbow on the table. What is the pair of forces here?

The force of your elbow on the table (downwards)	=	The force of the table on your elbow (upwards)

b) What are the forces when you start to walk or run?

The force of your feet on the Earth (moves the Earth slightly backwards)	=	The force of the Earth on your feet (moves you forwards).

Yes! The Earth moves backwards slightly as you move forwards.

c) Now consider a block of wood resting on a table. There are four forces here, in ***two*** pairs:

First pair (red)

The force of the block on the table (downwards) = The force of the table on the block (upwards)

These forces act where the block touches the table.

Second pair (blue)

The force of gravity of the Earth pulling down on the block (this weight in effect acts at the centre of the block – see p. 92) = The force of gravity of the block pulling up on the Earth (this force in effect acts at the centre of the Earth)

When the block is on the table, these four forces are equal and ***balanced***. But if the block is allowed to fall to the floor, only the second pair of forces exists and as the block moves downwards, the Earth moves slightly upwards!

Experiment 14.3 ⚠

Get a plastic water rocket and follow the instructions to prepare it. What happens when you release it and let the compressed air force out the water?

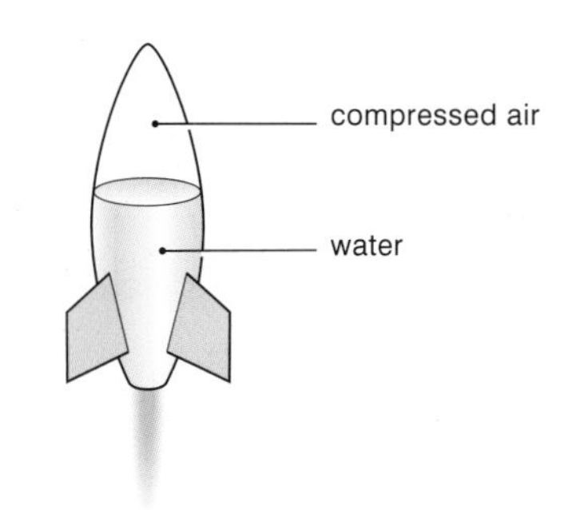

Copy and complete this equation:

The force of the rocket on the water (downwards)	=	The force of the . . . . on the . . . . (. . . .wards)

This is why rockets and jet engines move forward (see page 138 and page 114).

Experiment 14.4

Blow up a balloon and release it. What happens?

Why? (See page 160.) Write down an equation for the pair of forces.

When a gun is fired, why does it recoil?
Write down an equation for the pair of forces (see the cartoon on page 160).
Why does the bullet move farther than the gun?

▶ Adding forces

If you want to move this book to another place you will have to apply a force. ***Two*** things about the force are important: the **size** of the force and the ***direction*** of the force.

We show the direction of a force by an arrow and show the size of a force by the length of the arrow drawn to a chosen scale – for example, 1 cm for 1 newton.

5 N

5 cm

Scale: 1 cm for 1 N (1 newton)

Because both ***size*** and ***direction*** are important, force is called a **vector** quantity. Some other vector quantities are: displacement, velocity, and acceleration (see page 122).

Other quantities are called **scalars**. Scalars have only a size, no direction. For example: temperature, money, mass, and volume.

Money is a scalar
Direction is not important

Adding scalars

For scalar quantities, 2 + 2 = 4, always.

Adding vectors

For vector quantities, like forces, 2 and 2 ***sometimes*** equals 4!
For example:

a) Forces of 2 N and 2 N acting in the same direction, add up to give a **resultant force** of 4 N:

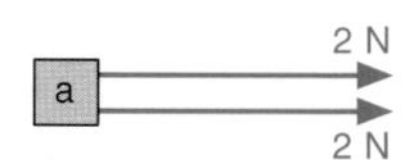

b) Forces of 2 N and 2 N acting in opposite directions cancel out to give ***no*** resultant force at all: These forces are ***balanced***.
Newton's First Law applies (see page 69).

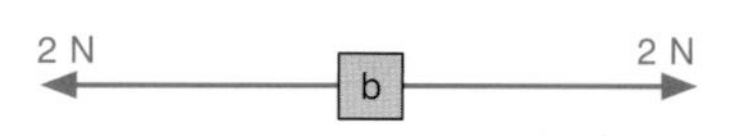

c) What would be the resultant force of 5 N opposed by 2 N?
The 2 N force cancels out part of the 5 N force. How much is left?

d) (More difficult) When two forces are not in the same straight line, the resultant force can be found by drawing a ***parallelogram of forces***:

The diagonal line shows the direction and size of the resultant force.

How big is the resultant force in the diagram? How can you describe its direction accurately?

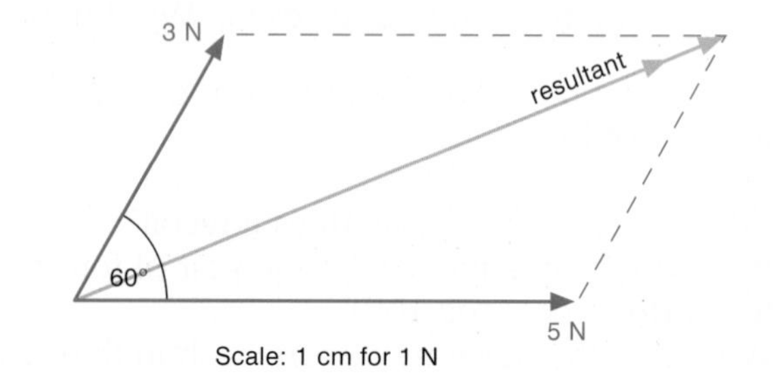

Scale: 1 cm for 1 N

▶ Free-body force diagrams

To see the effect of forces on an object it is best to draw ***only*** the forces on that **one** object.

For example, the diagram shows all the horizontal forces on the moving car:

It shows ***only*** the forces on the car, not the forces on the road. It shows:

- a ***driving force***, where the tyre is pushed forward by the road,
- a ***friction force***, due to air resistance (drag) and perhaps the car's brakes. It is usually shown by a single arrow.

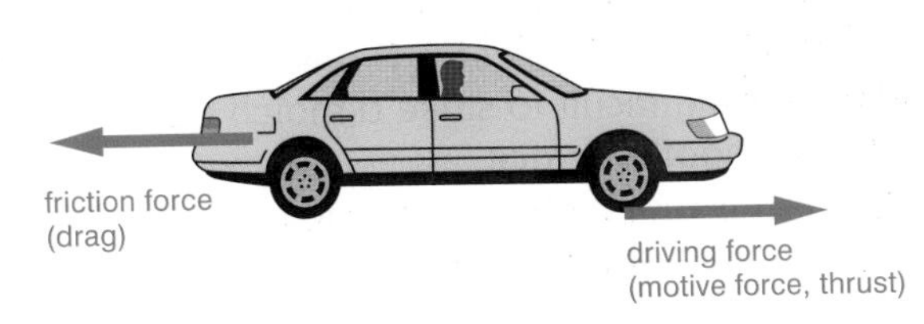

A car travelling on a horizontal road

This is called a free-body force diagram

Look how these forces affect the moving car:

The driving force is bigger than the friction force, so the car accelerates

The forces are equal, so the car travels at constant speed (Newton's First Law, page 69)

The driver brakes, so the car slows down (decelerates)

More examples:

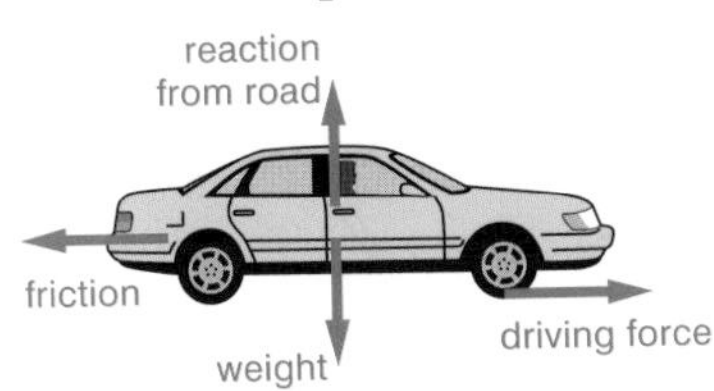

This shows the vertical forces on the car as well

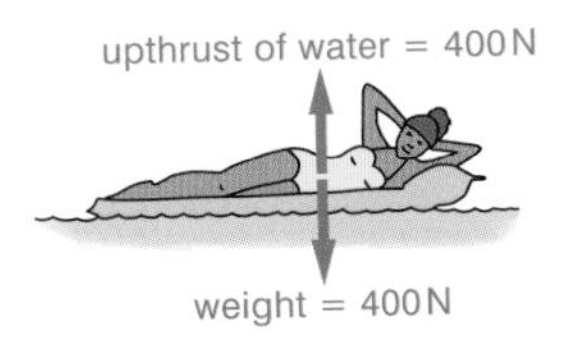

She floats because the forces are balanced

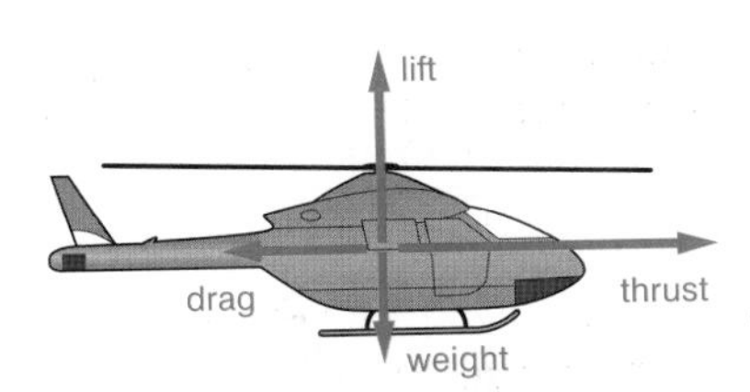

Can you see why this helicopter is moving up and forward?

The diagram shows a model boat, with 4 forces on it:

The Weight W is 5 newtons. This is balanced by the Upthrust U of the water. How big is force U?

The Thrust T is 10 N, while the Friction force F is 6 N. What is the resultant force on the boat? Which way does it move? What happens to its speed?

What happens to F as the boat moves faster? What happens when F and T become equal?

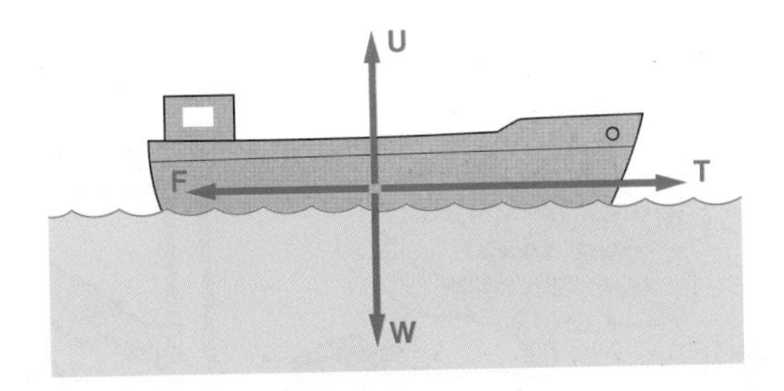

Summary

Friction always opposes movement.
Vectors have size and direction.
Scalars have size only.

Forces are always in pairs.
Newton's Third Law: the action and reaction forces are equal and opposite.

▶ Questions

1. Use your Physics to explain the following:
- a) Walking on wet ice is difficult.
- b) It is difficult to strike a match on a smooth surface.
- c) It is more difficult to pull a boat on the beach than in the sea.
- d) Wet floors and wet roads are dangerous.
- e) Cars are likely to skid on loose gravel.
- f) Aborigines light fires by rubbing pieces of wood together.
- g) Spaceships get hot when they return to Earth.
- h) Sliding down a rope can burn you.
- i) Speedboats have sharper bows than barges.
- j) Racing cyclists wear smooth tight clothes.

2. One night Professor Messer dreamed that all friction suddenly disappeared. Write a story about what might happen to the Professor. Why did his dream turn into a nightmare?

3. The table shows the air resistance on a bicycle at different speeds:

Speed (m/s)	0	1	2	3	4	5
Friction (N)	0	10	30	70	130	200

- a) Draw a smooth graph of this data.
- b) How fast can the cyclist travel if she can exert a forward push of 180 N?
- c) When the cyclist crouches lower, she travels faster. Explain why.

5. For each of the following forces, describe the reaction, giving its direction and stating what it acts on.
- a) the push (north) of a boot on a football
- b) the push (west) of a crashing car on a wall
- c) the push (backwards) of a swimmer on the water
- d) the pull of gravity on a falling apple.

6. Explain the following:
- a) A gun recoils when it is fired.
- b) Fire-fighters have to brace themselves when aiming a fire hose.
- c) You, by yourself, can move our planet.
- d) An astronaut is drifting away from his spaceship. How can he return, using only an aerosol spray?

7. The diagram shows a firework rocket:
As it flies through the air, there are 3 forces on it.
- a) Which 3 arrows show the 3 forces?
- b) Copy the diagram with these 3 forces.
- c) Label the 3 arrows, using these words:
  weight ***thrust***
  air resistance (drag)
- d) What can you say about these forces when the rocket is just taking off?
- e) Why does the rocket come back down?

4. Explain this cartoon – how many pairs of forces can you find?

More questions on page 140.

▷ Physics at work: Friction with the air

Car bodies are designed so that the air-flow is as smooth as possible.
They are ***streamlined*** to reduce the **drag** or **air resistance** caused by friction.

A car in a wind-tunnel. Smoke-streams show it is well-designed to keep the air-flow smooth.

The size of the drag force depends on:

- The shape of the object. Racing-cars and boats are streamlined to reduce drag.
- The area of the object. A larger parachute is slower.
- The speed of the object. There is more friction at high speed. This limits a car or a parachute to its **terminal velocity** (see the diagram on the right).
- The fluid. There is more friction in water than in air. Water is more ***viscous***.

A **parachute** is designed to have a lot of drag.

As a parachutist falls down, he transfers his gravitational potential energy to kinetic energy and to heat (which warms up the air).

Here is a **speed–time graph** for the sky-diver shown opposite (see also page 124).

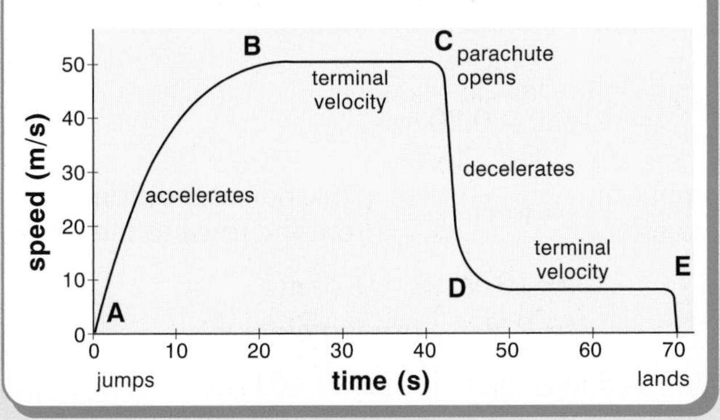

A sky-diver jumps out of a helicopter.
He uses air resistance to land safely:

At the start, there is only one force on the sky-diver. This is his **weight**. The weight of this sky-diver is 700 N.

This ***unbalanced*** or ***resultant*** force makes him accelerate.

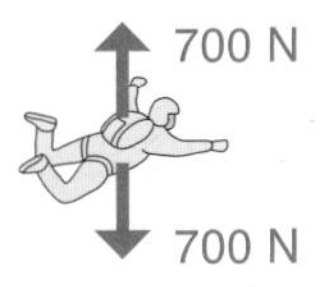

As he travels faster, the friction force (air resistance or drag) Increases.

Eventually, the 2 forces are ***equal*** (***balanced***, so no resultant force). He stops accelerating and travels at a constant speed. (See Newton's First Law, page 69.)
This speed is called his **terminal velocity.**

When his parachute opens, the air resistance increases.
The resultant force is now **up**wards.
This makes him slow down (decelerate), until . . . .

. . . . the 2 forces are ***equal*** again.
Because the 2 forces are balanced, (so no resultant force) he now travels at a constant speed.
This is his new terminal velocity.

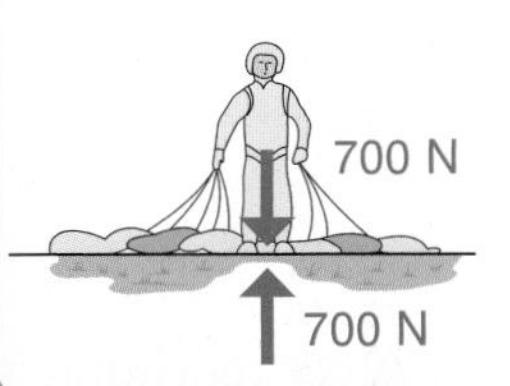

When he hits the ground, it pushes up to make him decelerate quickly.

When he stands on the ground, the ground pushes up on his feet.
The upward force is ***equal*** to his weight (so no resultant force).

chapter 15

Turning Forces

If you want to undo a very tight nut on your bicycle, would you choose a short spanner or a long spanner?

Why is the handle on a door placed a long way away from the pivot? Can you close a door easily by pushing near the pivot? Who would win if a strong man pushed ***at*** the pivot and a small boy pushed the opposite way at the handle? Try it.

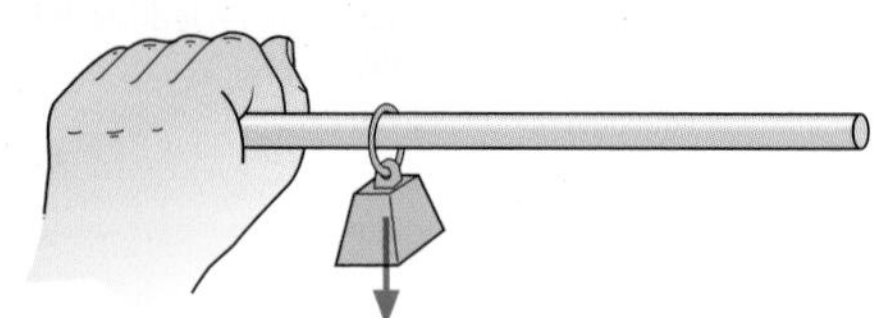

> *Experiment 15.1* ⚠
> Hold a rod horizontally and hang a weight on the rod near your hand.
> Move the weight to different distances from your hand and try to keep the rod horizontal.
>
> What happens?
> Now try it with a heavier weight.

You can feel that the turning effect on your hand depends on the ***size*** of the force (the weight) and the ***distance*** from your hand.
The turning effect of a force is called a **moment** or a **torque**.
It is calculated by:

Moment of a force	=	**force** (newtons)	×	**perpendicular distance (from the force to the pivot)** (metres)

The distance used is always the ***shortest*** (perpendicular) distance.
Moments are measured in newton-metres (often written N m).

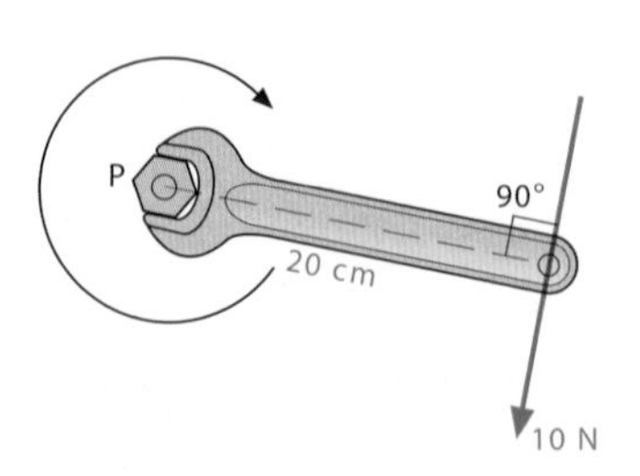

Example
What is the turning effect of the force in the diagram, about the nut at point P?

Perpendicular distance from the force to P = 20 cm = 0.20 m

Formula first: Moment of the force = force × perpendicular distance from the force to the pivot

Then numbers: = 10 N × 0.20 m

= <u>2 N m</u> turning ***clockwise***.

A car mechanic might need to apply a torque of 40 N m to a nut.

You know that a see-saw can be balanced even when the two people have different weights, by sitting them at different distances from the pivot.

When this see-saw is balanced and is not moving (***in equilibrium***), the moment of the girl (a clockwise moment) must ***equal*** the moment of the man (an anti-clockwise moment).
This is called the **principle of moments**:

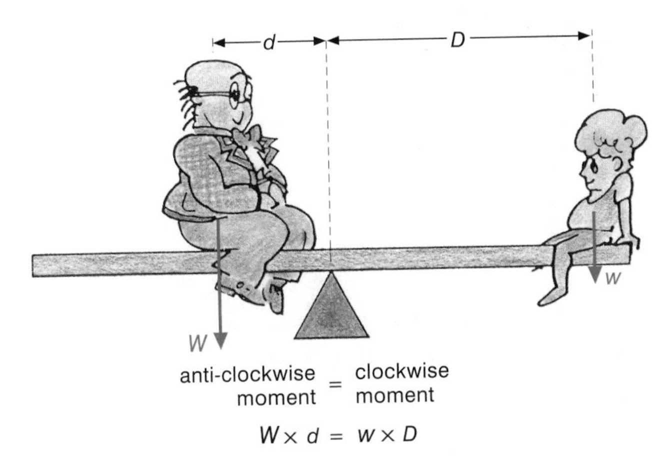

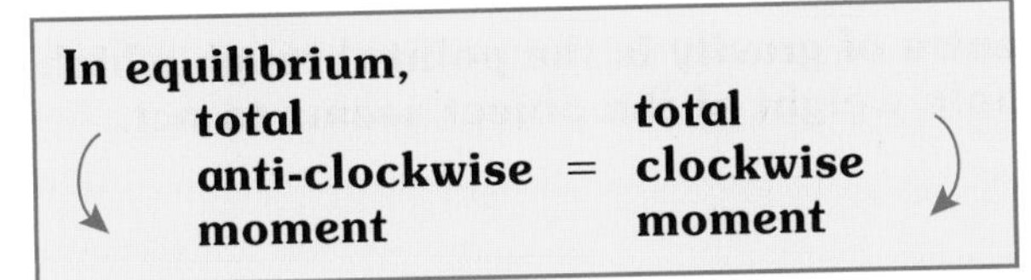

In equilibrium,
total anti-clockwise moment = total clockwise moment

You can check this by an experiment:

Experiment 15.2
Get a metre rule with a hole drilled at its centre; push a large pin through and clamp it to a retort stand. If the rule is not balanced, weight the higher end with plasticine until it balances.
You need weights of 1 newton (a 100 g mass will do) and 2 newtons (a 200 g mass will do) to hang from loops of string.

Adjust the weights until the rule balances. Measure the distances ***d*** and ***D*** and note them in the table. Repeat with these and other weights at different distances. Repeat with two weights on one side (see the example below).

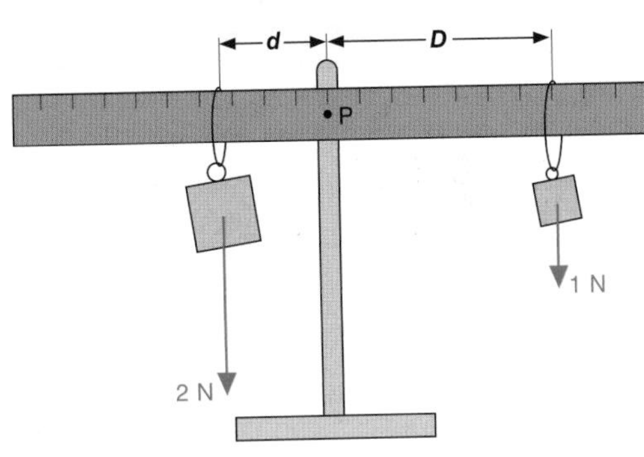

Anti-clockwise

Weight W (N)	Distance d (cm)	Moment $W \times d$
2	20	40
2	15	

Clockwise

Weight w (N)	Distance D (cm)	Moment $w \times D$
1	40	40
1		

What do you find about the anti-clockwise moments and the clockwise moments?

Example
If the ruler in the diagram is balanced, what is the weight ***W***?

Equation first: In equilibrium, total anti-clockwise moments = total clockwise moments

Then put in numbers (see diagram):

$$W \times 25 = (4 \times 15) + (1 \times 40)$$
$$W \times 25 = 60 + 40$$
$$W \times 25 = 100$$
$$W = \underline{4 \text{ newtons}}$$

▶ Centre of mass (centre of gravity)

What part of a ladder would you rest on your shoulder in order to carry it most comfortably?

Although the force of gravity pulls down on every little molecule, the weight of the ladder seems to act at its centre. This point on the ladder, where it balances, is called the **centre of gravity**, or **centre of mass**.

The centre of gravity is the point through which the whole weight of the object seems to act.

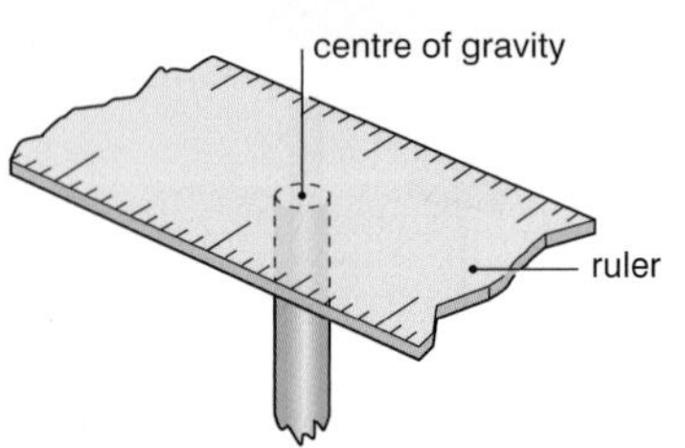

Experiment 15.3
Find the centre of gravity of a ruler by moving it until it balances on the end of a pencil.

Where is the centre of gravity of a ruler?
Where is the centre of gravity of a plank?

Use a similar experiment to find the centre of gravity of a book, a retort stand or a hammer.

Experiment 15.4
Tie a small weight to a piece of string to form a ***plumbline***. Let it hang down from your hand.

Why does it always hang vertically?
What can a plumbline be used for?

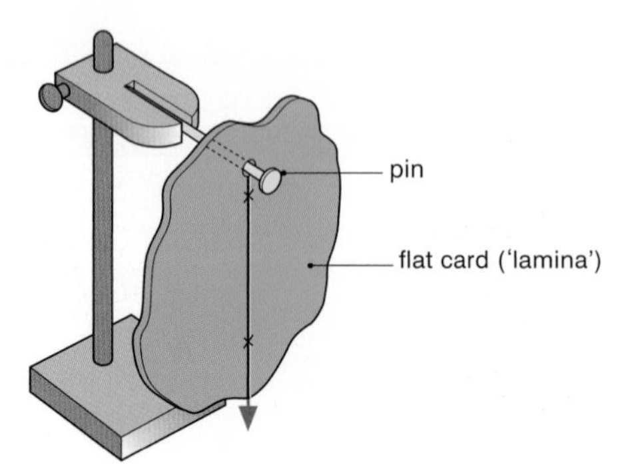

Experiment 15.5
Find the centre of gravity of a piece of flat card (a '***lamina***') using a plumbline.
Let the card hang ***freely*** from a large pin held in a retort stand. Hang a plumbline from the same pin.

Mark the position of the plumbline by two crosses on the card. Join the crosses with a ruler.

Just as the plumbline hangs with its centre of gravity vertically below the pivot, so also will the card. This means that the centre of gravity of the card is ***somewhere*** on the line you have marked. How can you find where it is on this line?

To find just where the centre of gravity is on this line, re-hang the card with the pin through ***another*** hole and again mark the vertical line. The only point that is on ***both*** lines is where they cross, so this point must be the centre of gravity.

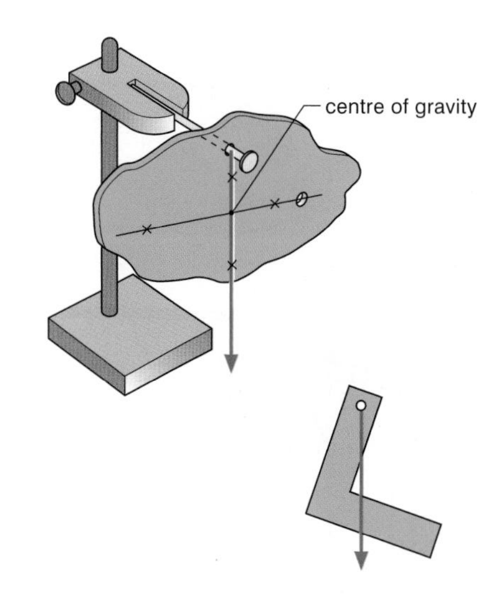

How can you check that this point really is the centre of gravity?

Experiment 15.6
Repeat experiment 15.5 with an L-shaped piece of card. Where does the centre of gravity seem to be?

▶ Stability

This plumbline is said to be in ***stable equilibrium*** because if you push it to one side, it returns to its original position. It does this because when you push it to one side its centre of gravity rises and gravity tries to pull it back to its lowest position.

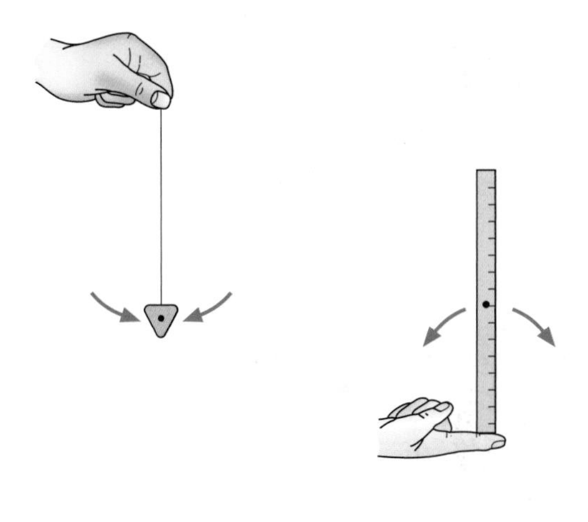

If you carefully balance a ruler vertically on your finger it is in ***unstable equilibrium***, because if it moves slightly, its centre of gravity falls and keeps on falling down.

A billiard ball on a perfectly level table is in ***neutral equilibrium***, because if it is moved, its centre of gravity does not rise or fall.

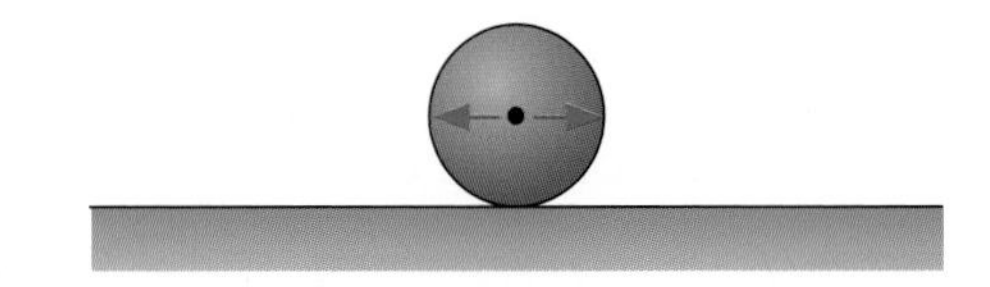

Is a tightrope walker in stable or unstable equilibrium?

If the line of action of the weight (***W***) lies outside the base of the object, there is a turning moment, and the object tends to fall over:

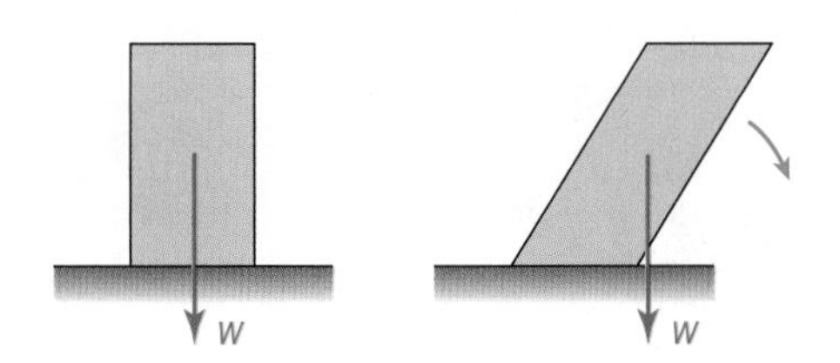

▶ Stable and unstable objects

Most of the objects we use each day are in stable equilibrium.
How can objects be made more stable?

> *Experiment 15.7*
> Place a box or matchbox upright on a rough piece of wood (A). Slowly tilt the wood until the box topples over. Note the angle of the wood.
>
> Raise the centre of gravity by weighting the top of the box with plasticine or sliding up the matches (B). Tilt the wood again. Is the box more stable or less stable?
>
> Place the box in other positions and see when it topples.

To make the box more stable, should it have
a) a high or a low centre of gravity?
b) a narrow or a wide base?

Why is a rowing-boat less stable if you stand up?

Which is more stable, a racing car or a double-decker bus? How could the bus be made more stable? Why are passengers not allowed to stand upstairs?

How are these objects made stable?

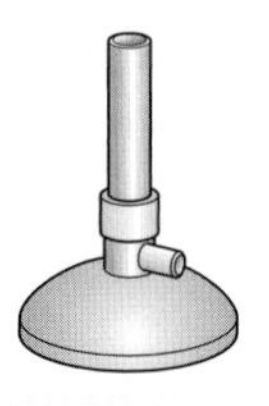

Bunsen burner

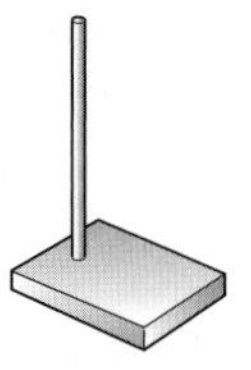

retort stand

wine glass

Discuss the Physics of these cartoons:

Where do you think her centre of gravity is?

A dangerous stunt! What kind of stability is this?

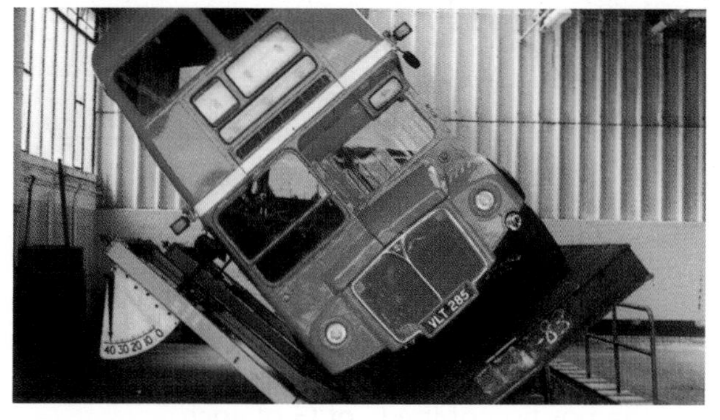

A bus being tested for stability.
Where do you think its centre of mass is?

Doing the 'Fosbury Flop'. She is keeping her centre of gravity as low as possible – why?

Summary

Moment of a force (newton-metre) = force (newtons) × perpendicular distance from the force to the pivot (metres)

Principle of moments:

In equilibrium,

total anti-clockwise moments = total clockwise moments

The ***centre of gravity*** is the point through which the whole weight of the object seems to act. An object can be in stable, unstable or neutral equilibrium depending on the position of its centre of gravity. To be more stable an object needs a low centre of mass and a wide base.

Discuss the Physics of these cartoons:

I'LL PUT ALL THESE HEAVY THINGS IN THE TOP DRAWER

HOW CAN SHE DO THAT UP THERE BLIND-FOLDED ?

I THINK THE DOOR LOOKS NICER WITH THE HANDLE AT THIS SIDE !

▶ Questions

1. Copy and complete these sentences.
a) The turning effect (or . . . .) of a force is equal to the force multiplied by the . . . . distance from the . . . . to the . . . . . Its unit is . . . . .
b) The principle (or law) of moments states that, in . . . . , the total . . . . are equal to the total . . . .
c) The centre of gravity (centre of . . . .) is the point through which the whole . . . . of the object seems to act.
d) A stable object should have a . . . . centre of . . . . and a . . . . base.

2. Use your Physics to explain the following:
a) A mechanic would choose a long spanner to undo a tight nut.
b) A door handle is placed well away from the hinge.
c) It is difficult to steer a bicycle by gripping the centre of the handlebars.

3. A mechanic applies a force of 200 N at the end of a spanner of length 20 cm. What moment is applied to the nut?

4. The first two diagrams show rulers balanced at their centres of gravity.
What is the weight of: a) X? b) Y?

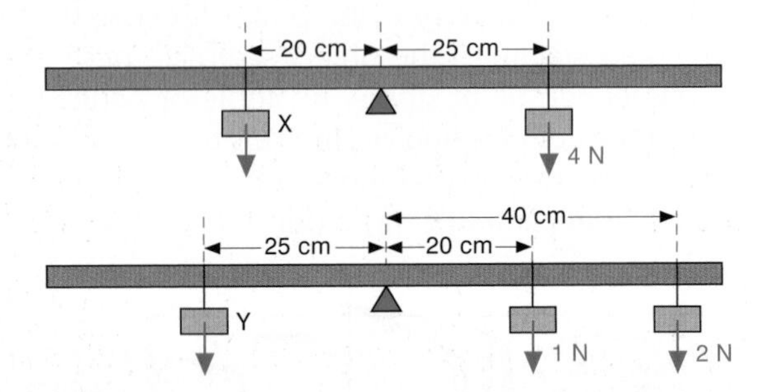

c) What is the weight of the ruler, Z?

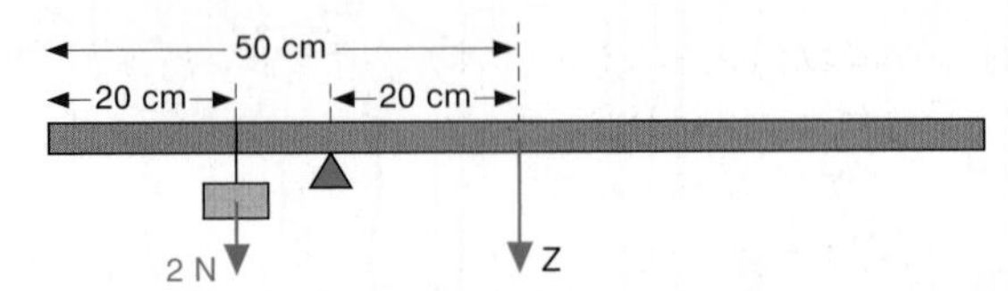

5. A crane driver has a chart in his cab which tells him how far he can safely extend the jib of his crane:

Size of load (tonnes)	Maximum length of jib (m)	Moment (load × length)
10	12	
20	6	
30	4	
40	3	

a) Copy and complete the table.
b) What do you notice? Why is this?
c) What length of jib could he have with a load of 24 tonnes?

6. A, B, C and D are pieces of hardboard. Copy the diagrams and mark with a cross the position of the C of G of each.

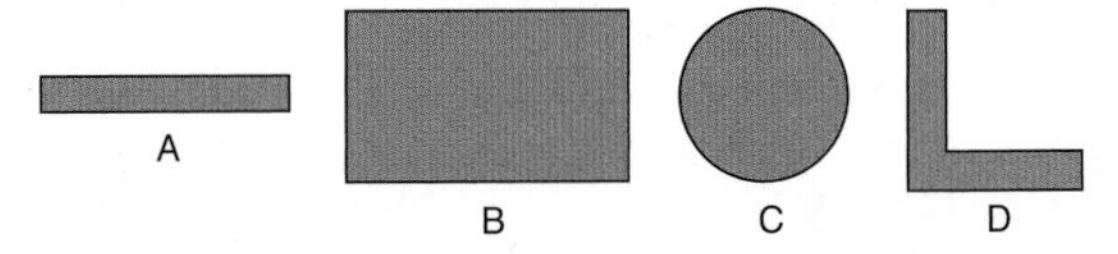

7. Sketch diagrams and mark the approximate positions of the centres of mass of
a) a hammer b) a boomerang
c) a tea cup d) a human being.

8. Explain the following:
a) A Bunsen burner has a wide heavy base.
b) Racing cars are low, with wheels wide apart.
c) A boxer stands with his legs well apart.
d) A wine glass containing wine is less stable.
e) Lorries are less stable if carrying a load.

There was an old teacher called Grace,
Who often fell flat on her face.
The reason you see,
Was her C of G,
Its place was too high for her base.

Further questions on page 140.

chapter 16

Work, Energy and Power

If the man in the picture is moving the van, we say he is doing ***work***.
He is doing work only if there is ***movement*** against an ***opposing*** force.
The opposing force is often friction (as in the picture), or gravity.
The man can only do this work if he has some ***energy***.

If the man pushes with a stronger force or moves it a longer distance he will do more work. In fact,

Work done	=	**force** (newtons)	×	**distance moved** (metres)

The distance used in this formula must be the distance moved ***in the direction of the force***.

If the force is in newtons and the distance is in metres then the work is measured in ***joules***, often written J (see page 35).

1 joule is the work done when a force of 1 newton moves through 1 metre (in the direction of the force).

Example
A man lifts a brick of mass 5 kg from the floor to a shelf 2 metres high. How much work is done?

The opposing force in this case is the ***weight*** of the brick.
Here on Earth a mass of 1 kg weighs 10 newtons (see page 67).
∴ a mass of 5 kg weighs 50 newtons.

Formula first: **Work done** (J) = **force** (N) × **distance moved** (m)

Then put in the numbers: = 50 N × 2 m
= <u>100 joules</u>

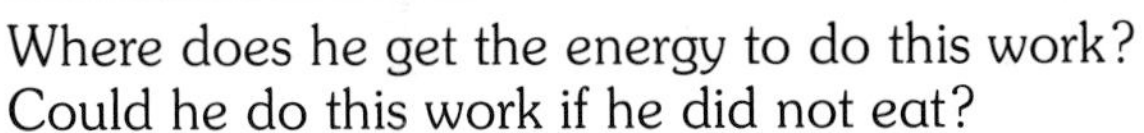

Where does he get the energy to do this work?
Could he do this work if he did not eat?

If a man pushes a van against a friction force of 300 newtons for a distance of 10 metres how much work does he do?

Large amounts of work may be measured in kilojoules or even megajoules (see page 35).

▶ Forms of energy

We have seen already (on page 10) that there are several forms of energy.
We have studied **thermal** energy and found that it is really the movement energy of the molecules.

All moving objects have movement energy, called **kinetic** energy. A heavy lorry travelling fast has a lot of kinetic energy (and if you get in the way it will use some of this energy to damage you).

Some forms of energy are called **potential** energy.

When an object is lifted to a higher place, it is given more **gravitational** ***potential energy***. This is stored energy which it can give out if it falls down.

A second kind of potential energy is **elastic** ***potential energy*** (or strain energy) which is stored in the elastic of a catapult or in a stretched bow.

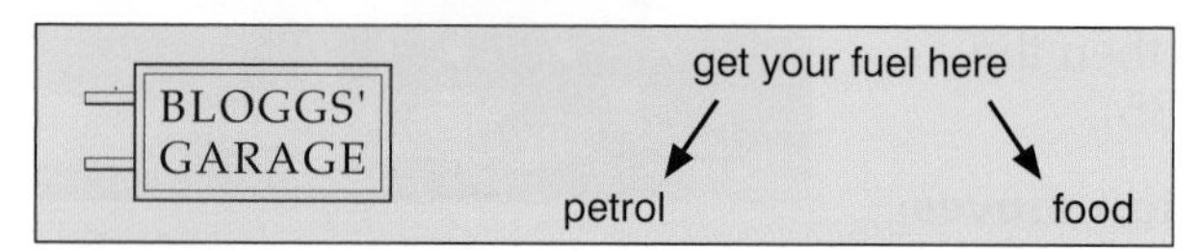

Chemical *energy is energy stored in food and other fuels. Your body gets its energy from food you eat.*

Sound *energy and* ***light*** *energy are common types of energy.*

Electrical *energy is common and useful. Like other forms of energy, it can be very dangerous.* ***Magnetic*** *energy is always connected with electrical energy.*

Nuclear *energy is stored in the centre (or nucleus) of an atom. The Sun, like a hydrogen bomb, runs on nuclear energy.*

Changing energy from one form to another

We have seen already (page 11) that energy can be changed from one form to another. When this happens, the ***amount*** of energy stays the ***same.*** We say it is '***conserved***'. This is because energy cannot be made or destroyed. See also page 102.

This fact is called:

> **the Principle of Conservation of Energy:**
> energy can be changed from one form to another, ***but it cannot be created or destroyed.***

▶ Energy and work

Energy is the ability to do work.
The amount of work that is done tells us how much energy has been transferred from one form to another.

Work done = Energy transferred

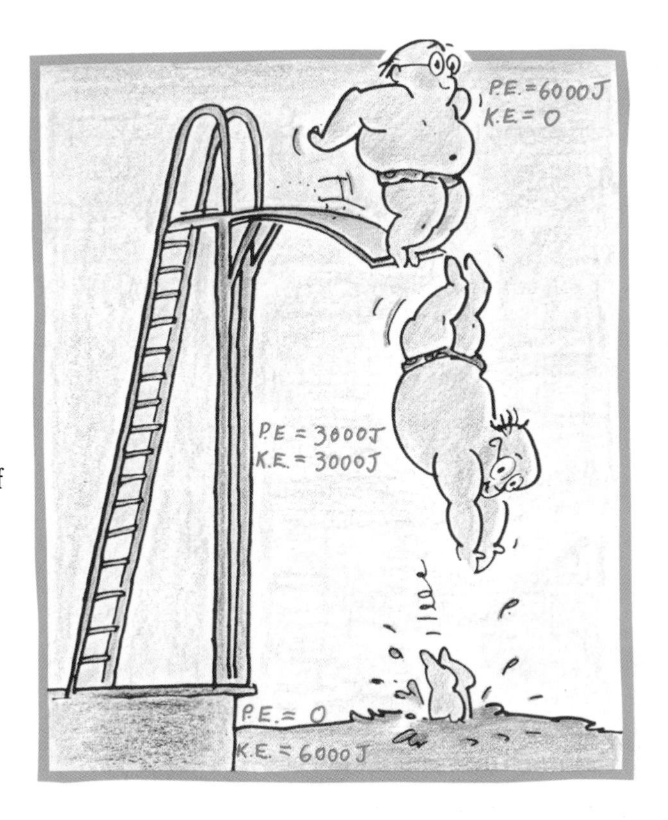

Working against gravity

In the example on page 97 we calculated that the work done by the man in lifting the brick to the shelf was 100 joules. This means that 100 joules of his chemical (food) energy was transferred to 100 joules of gravitational potential energy.

Falling under gravity

The diver in the diagram uses 6000 joules of his chemical energy to climb to the top where he has 6000 J of gravitational potential energy (PE) but no kinetic energy (KE).
As he falls down, his potential energy is transferred into an ***equal*** amount of kinetic energy.
The total amount of PE + KE is always 6000 J.

The pendulum

When a pendulum swings to and fro, its energy is constantly changing from potential energy to kinetic energy and back again.

This energy is gradually transferred to heat by friction with the air. This heat is 'low-grade' energy and we cannot make use of it.
All energy eventually becomes low grade.

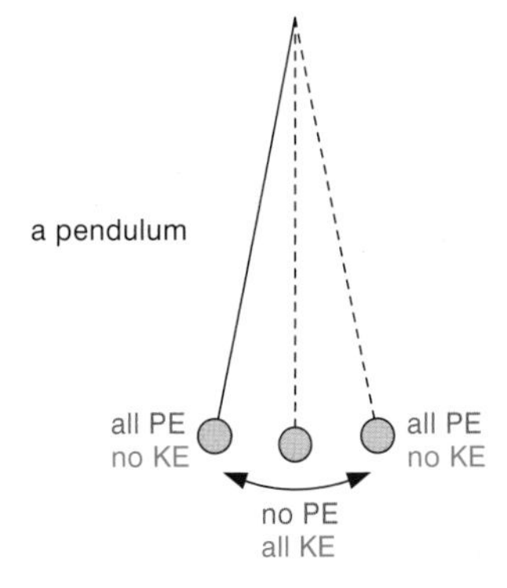

What are the energy transfers in these examples?

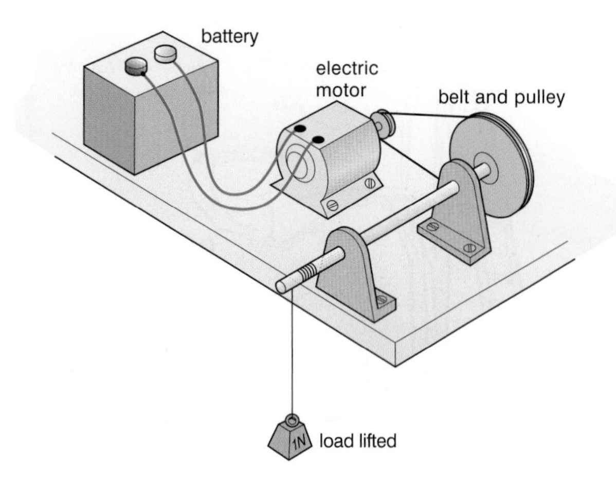

▶ Energy experiments

Experiment 16.1
Arrange an electric motor, as in the diagram, so that it can turn an axle from which a load is hanging.

What happens when you connect the motor to a battery?

What are the energy transfers here?

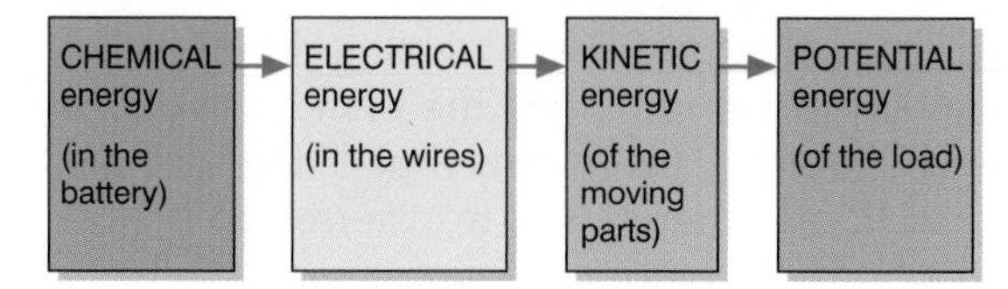

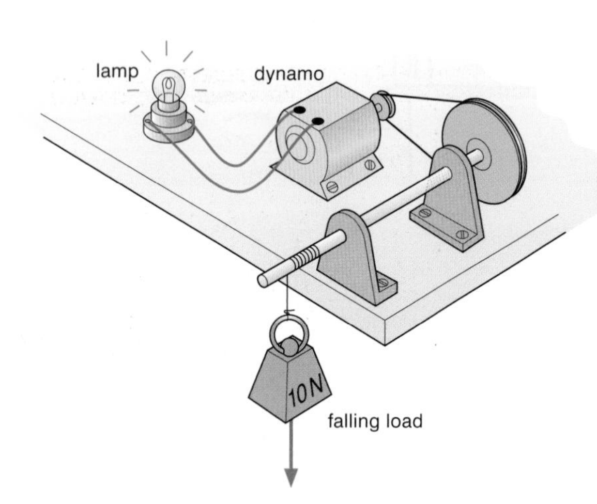

Experiment 16.2
Disconnect the battery and connect a lamp in its place. Start with a heavy load raised high and let it fall to drive the 'motor' which will now act as a ***dynamo*** and produce electricity to light the lamp.

What are the energy transfers here?

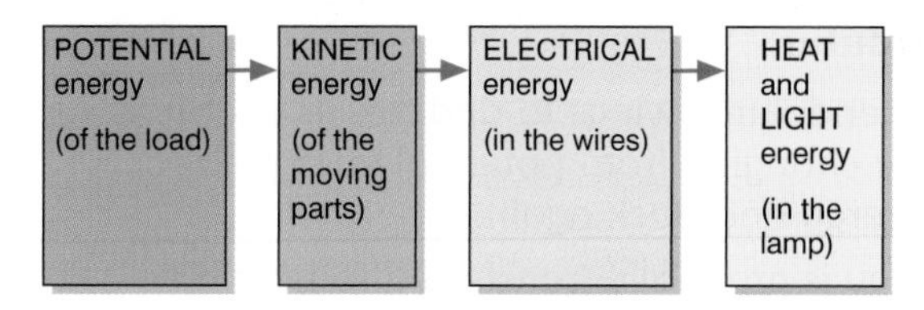

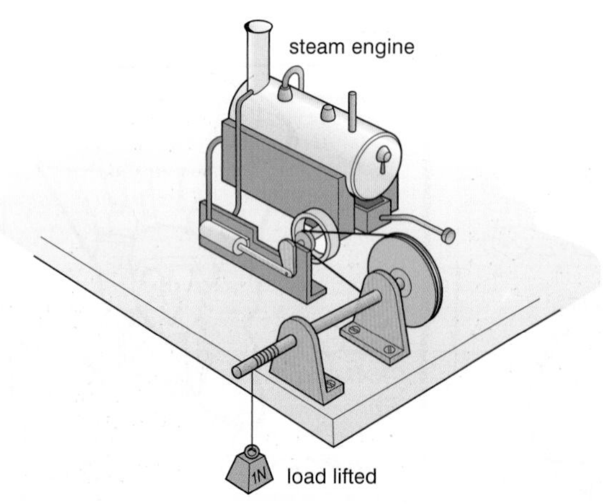

Experiment 16.3 ⚠
Use a model steam engine to lift a load.

What are the energy transfers here?

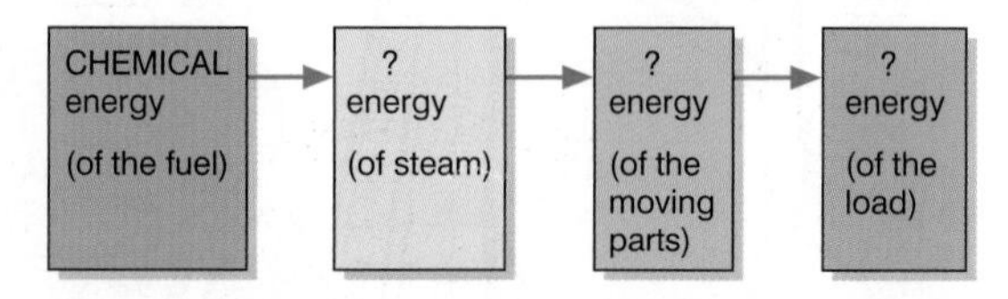

Measure the weight of the load and the height it is lifted.
Then calculate the work done on the load.

Experiment 16.4 ⚠
Build a model power station using a steam engine and a dynamo. Connect them to a lamp to represent all the lights in your home.

What are the energy transfers here? (See also p. 104.)

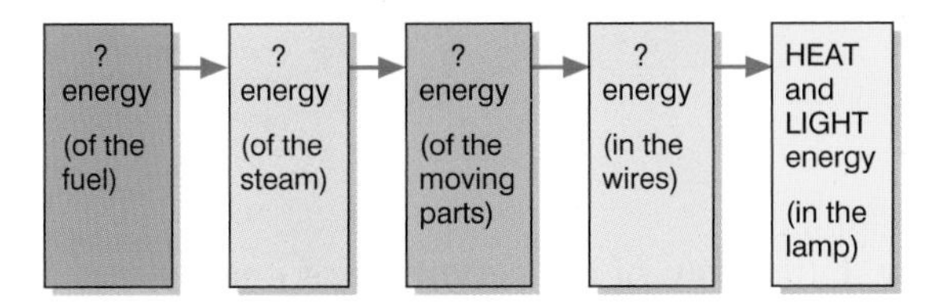

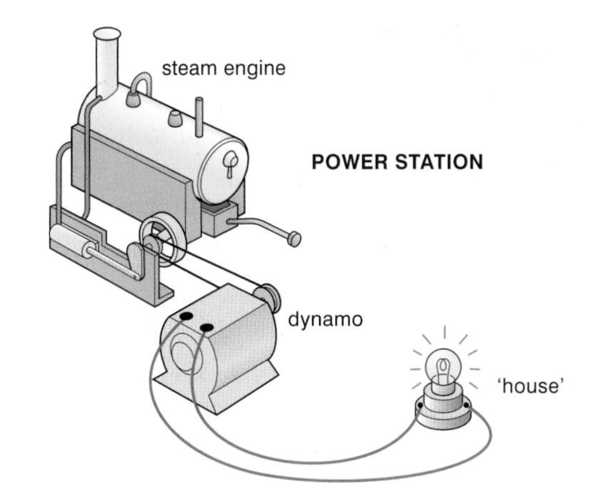

Experiment 16.5
Build a model hydroelectric power station using the energy of falling water to turn a turbine and drive a dynamo.

What are the energy transfers here?

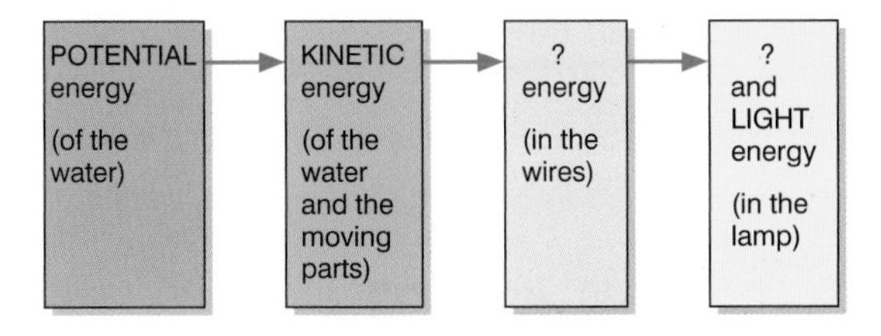

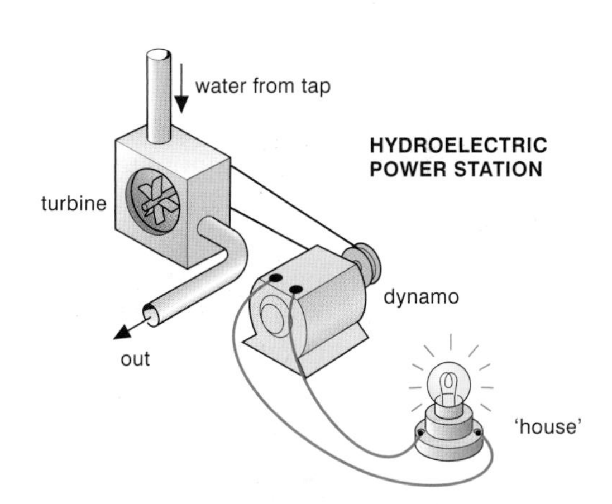

▶ Sources of energy

Where does our energy come from? The answer is that most of it comes from the ***Sun!***
Study the different parts of this diagram and you will see how.

heat and light energy
evaporation and convection currents raise potential energy of water (for hydroelectric power)
plants grow in sunlight (chemical energy)
animals eat plants
unequal heating causes winds for sailboat and wind-mills
plants decay and eventually form coal and oil (chemical energy)
tides are caused by the Moon
oil and natural gas
geothermal energy comes from radioactivity in the rocks
coal

SAVE IT

▶ Energy Transfer Diagrams

When a torch is switched on, it transfers chemical energy (in the battery) to heat and light energy (in the bulb).
We can show this on an **Energy Transfer Diagram** (also known as a **Sankey** diagram):

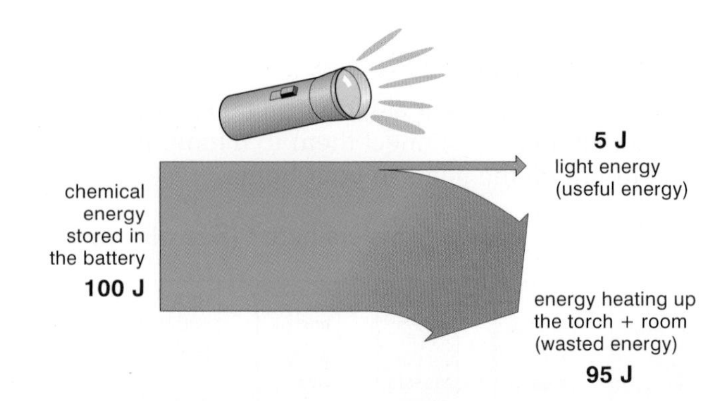

The ***thickness*** of each arrow is drawn to scale to show the amount of energy.

Notice that the total amount of energy after the transfer is the ***same*** as the amount before.
We say the energy is 'conserved'.
This is the first law of energy:

Law 1 (the law of conservation of energy)
The total amount of energy is constant.
Energy cannot be created or destroyed.

Although there is the same amount of energy afterwards, ***not all of it is useful***.
Most of the energy just heats up the bulb and then spreads out, heating up the room (by conduction, convection and radiation).
This energy is wasted. We cannot use it.
This is the second law of energy:

Law 2 (the law of spreading of energy)
In energy transfers, the energy spreads out, to more and more places.
As it spreads, it becomes less useful to us.

In the torch above, for every 100 joules of energy input to the bulb, only 5 joules are output as useful light energy. The rest is wasted.
We say the **efficiency** is $\frac{5}{100}$ or 5%.
The definition of efficiency is:

$$\textbf{Efficiency} = \frac{\textbf{useful energy output}}{\textbf{total energy input}} \times \textbf{100\%}$$

The efficiency of a car is less than 25% (see also page 116).

Three-quarters of the chemical energy in the fuel is wasted!

Passenger transport

The Earth's oil is running out (see page 13).
Yet one of the ways we waste a lot of energy (and cause pollution) is in our transport.
Analyse this chart:

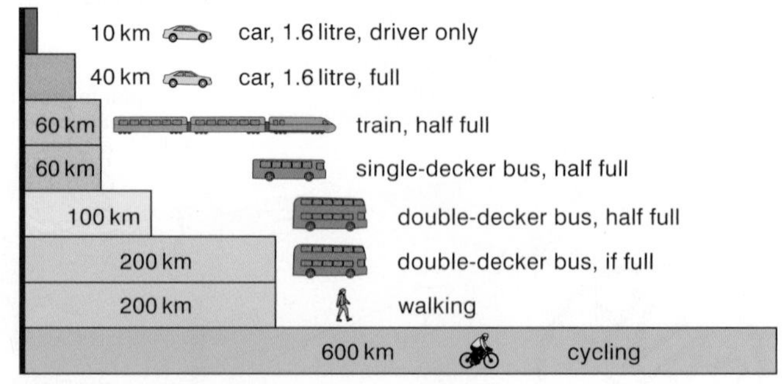

Energy consumption by passenger transport
The distance one passenger can travel using the energy equivalent of one litre of petrol
Source: European Commission

Which is the most efficient form of transport?
Do you use it?

- Discuss the implications of the data in the graph. Try to consider energy resources, people's habits, and their health.
- Write a plan of action to put to the government or your local council.

▶ Calculating efficiency

A modern 'energy-saver' light bulb is a small fluorescent tube (see page 212):
It is 5 times more efficient than an ordinary filament bulb.
Look at its Energy Transfer Diagram:

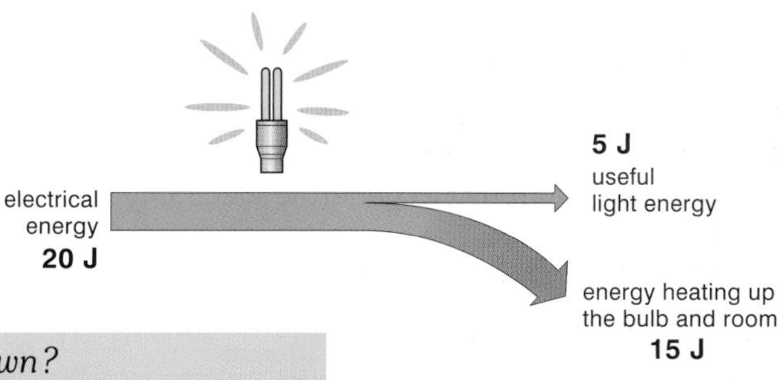

Example 1 What is the efficiency of the bulb shown?

Formula first: $$\textbf{Efficiency} = \frac{\textbf{useful energy output}}{\textbf{total energy input}} \times \textbf{100\%}$$

Then numbers: $$= \frac{5 \text{ joules}}{20 \text{ joules}} \times 100\%$$

$$= \underline{25\%}$$

Experiment 16.6 Efficiency of a motor

You can use the apparatus of experiment 16.1 (page 100) to investigate the efficiency of an electric motor.

To find the useful output, use the example on page 97.

To find the total energy input, use a joulemeter (page 36). (Alternatively, use an ammeter, voltmeter and stopclock and calculate it as in examples (a), (b) on page 266.)

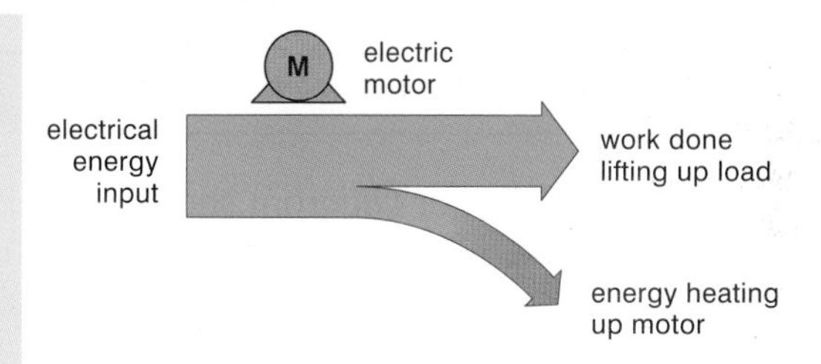

Example 2 A wind-generator (wind turbine)

The kinetic energy of the air hitting the blades of a wind-generator is 2 000 000 joules in each second.
The energy transferred electrically to the power lines is 800 000 joules per second (800 kW).
What is its efficiency?

$$\textbf{Efficiency} = \frac{\textbf{useful energy output}}{\textbf{total energy input}} \times \textbf{100\%}$$

$$= \frac{800\,000 \text{ J/s}}{2\,000\,000 \text{ J/s}} \times 100\%$$

$$= \underline{40\%}$$

If a 800 kW generator can supply electricity for 1000 people, when the wind is blowing, how many would be needed for a city the size of Manchester (1 200 000 people)?

Example 3 A photo-voltaic (PV) solar cell array (page 14)

A solar cell of 4 m^2 has 2000 joules per second (2000 W) shining on it. Its efficiency is just 15%. What is its output?

Useful energy output = 15% of the energy input

$$= \tfrac{15}{100} \times 2000 \text{ joules per second}$$

$$= \underline{300 \text{ joules per second (300 W)}}$$

How many 60 W light bulbs would this light?

Do not confuse these PV solar cells with a solar panel (solar collector) as on page 50

▶ Producing electrical energy

In a power station, energy from a fuel is used to boil water. Then the high-pressure steam is used to turn a turbine, which turns a dynamo to generate electricity:

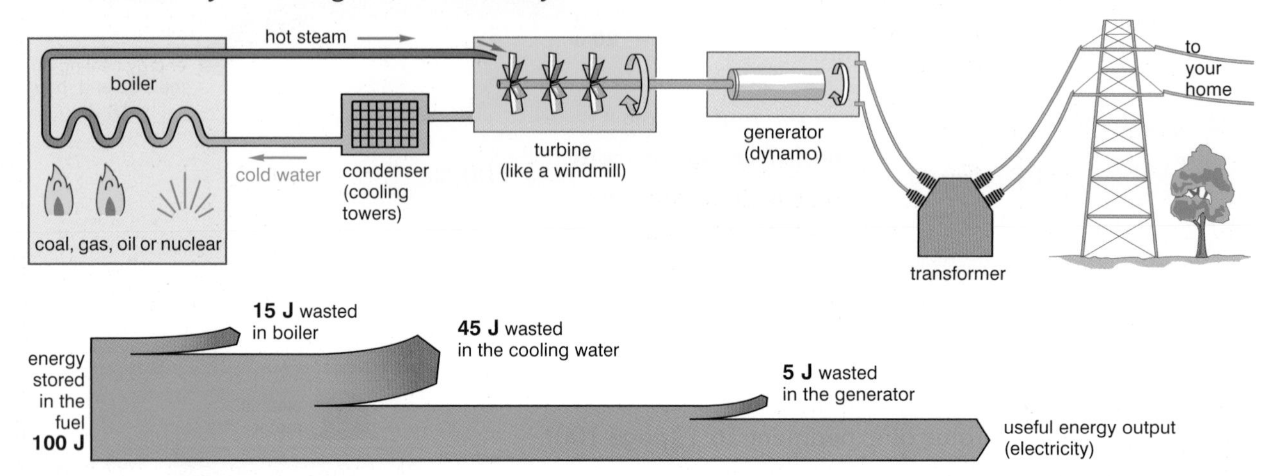

The Energy Transfer Diagram shows what happens to the energy. How much ***useful*** energy is output for each 100 J of fuel energy? What is the efficiency of this power station?

45% of the energy is wasted in the cooling water! Some power stations use this energy to heat local houses. These are called 'combined heat and power' (CHP) schemes. The efficiency of a CHP scheme can be up to 80%.

New gas-powered stations have been made more efficient by using 'combined cycle gas turbines' (CCGT). These use the natural gas to power a jet engine (see page 114) which turns a generator, and then they use the exhaust gases to make steam to drive another generator. This method is more efficient (up to 55%), and they can start up quickly when demand for electricity rises.

Solar cells are expensive but useful in remote sunny areas

In Britain, most power stations burn fossil fuels (coal, gas, oil). Some use nuclear energy (see page 348). They all use the national grid (page 303) for distribution.

The energy to turn the generator can also be got from ***renewable*** resources : the wind, or waves, or from burning biomass, or from falling water (see pages 14–15 and 101). In some countries geothermal energy provides the steam.

However only 3% of the electricity in the UK is made from renewable energy resources! And much of this is made by burning waste materials, or burning the methane gas produced by rubbish dumps.

Waste can be burnt to generate electricity

▷ Physics at work: Supplying electricity

Advantages and Disadvantages of using electricity

Electricity has many ***advantages***: it is very convenient to use in your home, because you can use it with a lot of different appliances, and you don't have to find space to store the electricity. Also it seems clean; it doesn't cause any pollution ***in your home***.

However it has ***disadvantages***: it is dangerous; it cannot be stored to use later (in case there are power cuts); and it needs a complicated network of cables and pylons to connect to your home. And there are other problems that are explained in the boxes below and on the next page:

Reliability of the supply

Fossil fuel and nuclear power stations are ***reliable*** energy sources : they produce electricity whenever it is needed, although they have different start-up times:

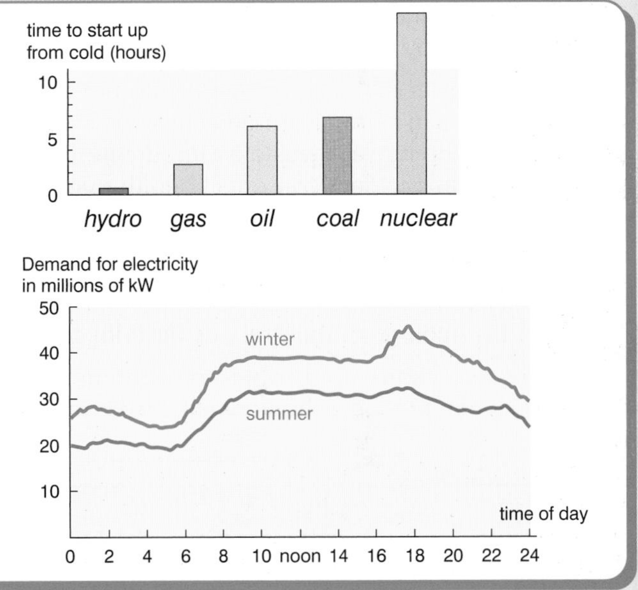

In practice, nuclear power stations normally run all the time, and the others are switched on and off as needed.

Hydro-electric schemes are also very reliable (unless there is a drought) and can start up quickly when there is a demand.

Tidal schemes are reliable, but as the tides vary, so the output may not be at the right time of day for the demand.

Wind power and solar power are unreliable, because they depend on the weather.

Financial costs

The cost per unit of electricity varies but from most of the sources it is about the same:

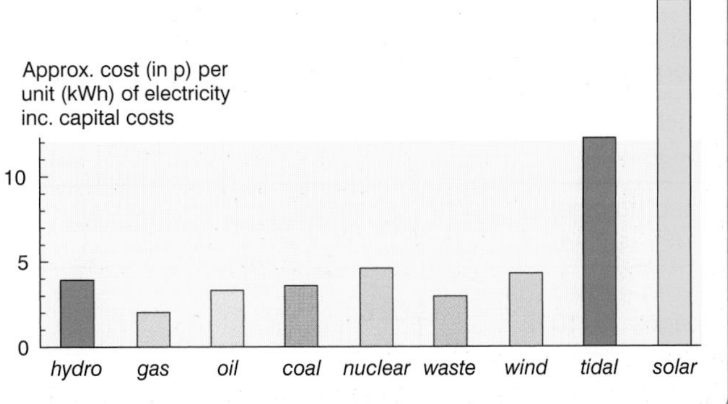

Nuclear fuel is relatively cheap, but the capital costs of building the station are high (p. 349). So are the costs of handling the nuclear waste, and of decommissioning the power station when its useful life has ended.

Solar cells (p. 14) are still the dearest way of producing electricity. Expensive to set up, only 17% efficient, and need the sun to be shining.

▶ Physics at work: Supplying electricity

Effects on the environment

- Burning a fossil fuel (coal, oil, natural gas) always produces carbon dioxide gas (CO_2).

  fuel + oxygen ➡ CO_2 + H_2O + energy

  CO_2 increases the ***greenhouse effect*** (page 48) causing global warming of the Earth. There is no easy way to stop this CO_2 emission, except by burning less fuel.
- Sulphur dioxide (SO_2) is produced by coal-fired and oil-fired power stations. It causes acid rain which harms forests and buildings. The SO_2 gas can be removed, but this makes the electricity 10–20% more expensive.
- Coal-fired stations cause dust and smoke, but these can be removed by electrostatic precipitators (see page 246).
- Nuclear power stations (page 349) do not emit any CO_2 or SO_2, and in normal running cause very little pollution providing the nuclear waste is stored carefully (page 350). However an accident can affect health over a wide area.
- Wind turbines need to be placed on hills or the coast. They only work when it is windy.
- Hydroelectric schemes (page 15) flood large areas of land and affect the ecology of the area, as do tidal schemes (page 15).
- All power stations need cables to distribute the energy, either by unsightly pylons or by very expensive underground cables.

CO_2 (greenhouse effect)
1000
500
0
coal oil gas

SO_2 (acid rain)
10
5
0
coal oil gas

tonnes of gas produced for every million units (kWh) of electricity

The ice caps are melting because of global warming

The table shows the 6 most likely sources for the UK. (Key : the more symbols, the worse it is)

Power station	Cost to build it	Cost of fuel	Air pollution	Cost of electricity	Disadvantages:
Coal-fired, Oil-fired (page 104)	££	££££	!!!!!	££	• emits CO_2, so increases greenhouse effect • emits SO_2 and so causes acid rain • limited fuel available
Gas-fired (page 104)	£	£££	!!	£	• emits CO_2 (but less than coal) • limited fuel available
Nuclear (page 349)	££££	££		£££	• risk of big accident, like Chernobyl • limited fuel; waste needs careful disposal
Wind (turbine) (see page 14)	£££	—	—	£££	• needs many large turbines, noisy, unsightly • unreliable: wind does not blow every day
Hydro-electric (dam) (pages 15, 101)	££££	—	—	£££	• impossible in flat regions • floods a large area, affects ecology
Tidal (barrage) (see page 15)	£££££	—	—	£££££	• needs a place with high tides • affects ecology of the area

▶ Global warming

The Earth's atmosphere acts like a big blanket to keep us warm (at about 15 °C). Without it, it would be like living on the Moon (about −18 °C).

The energy is trapped by the greenhouse gases, including carbon dioxide (CO_2), methane, CFC, nitrogen oxides and water vapour.
Since the industrial revolution, the CO_2 in the atmosphere has increased by 30%.
Over the last 100 years the Earth's temperature has risen by about 0.6 °C and the sea level has risen by about 15 cm. Why is this?

The greenhouse effect (see also page 48).
The Earth absorbs solar radiation and warms up.
It emits infra-red radiation. These longer waves are absorbed by greenhouse gases in the atmosphere.

Look at the 2 graphs and discuss them:

a What do they show?

b What is the effect of the CO_2 axis not starting at zero?

c Do they prove global warming is happening?

d Do they prove that global warming is caused by CO_2 emissions?

e Research to find out what else could be causing global warming. Discuss why the evidence needs to be both ***reliable*** and ***valid***.

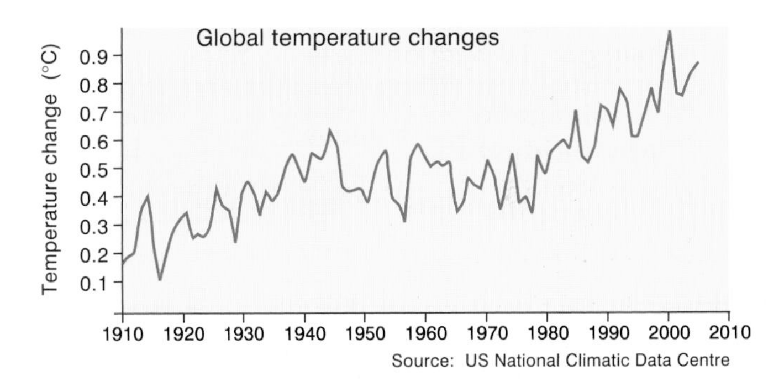

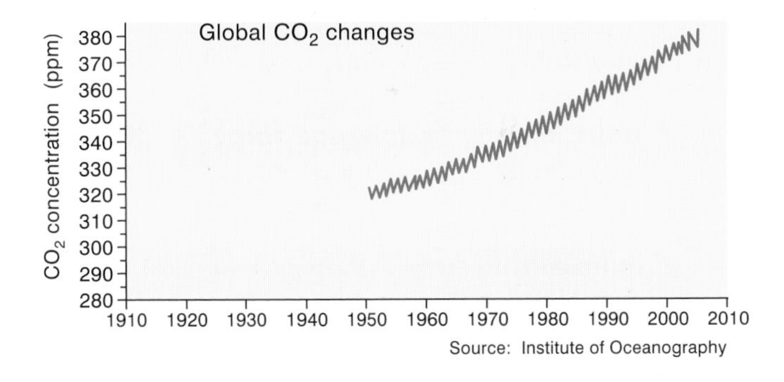

Reducing emissions

f What is the 'Kyoto protocol'? Are you in favour? Why?

g Imagine you live in southern Bangladesh, where land is less than 1 m above sea level. America emits much more CO_2 than any other country. Use your scientific knowledge to write a letter to the US President.

The UK government has a target to reduce CO_2 emissions by 60% by 2050, but progress is slow. Most people are against nuclear power stations (even though they do not emit CO_2 or SO_2). They don't want wind-generators nearby or in the countryside. They prefer cars to buses or trains.

h How do you think the government can change public opinion?

The diagram shows 2 possible futures for the UK:

i Use the information on these 2 pages to discuss the choices. Give reasons for choices you support.

j What is carbon capture technology?

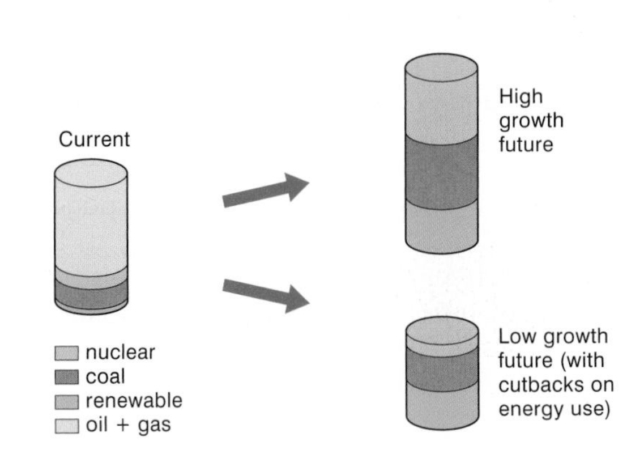

▶ Calculating Potential Energy (PE)

1. Gravitational Potential Energy

This weight-lifter is lifting a mass of 200 kg, up to a height of 2 metres.
We have already seen (on page 97) how to calculate the potential energy of his weights:

gravitational PE = work done so:

change in PE (joules)	=	**weight** (N)	×	**change in height** (m)

But from pages 67 and 131:

weight (in N) = **mass** (in kg) × $\boldsymbol{g}$

where $\boldsymbol{g}$ = **10** here on Earth. So:

change in gravitational PE (joules)	=	**mass** (kg)	×	$\boldsymbol{g}$ (N/kg)	×	**change in height** (m)

$\boldsymbol{g}$ has different values on other planets (see p. 151).

2. Elastic Potential Energy (strain energy)

This is the kind of energy stored in a bow, or in a catapult, or in a spring in a clock.

Energy stored = work done to stretch the bow, so:

Elastic energy (joules)	=	**average force** (newtons)	×	**distance** (metres)

Example

Robin Hood exerts an average force of 100 N in pulling back his bow by 0.5 m. He fires the arrow (mass = 0.2 kg) vertically upwards. How much energy is stored in his bow, and how high does the arrow go?

Elastic energy = average force × distance
= 100 N × 0.5 m = 50 joules

Since this energy is conserved (see page 98), then this must be the kinetic energy as the arrow leaves the bow, and it must also equal the gravitational PE at its highest point:

mass × 10 × height = 50 joules
0.2 × 10 × height = 50
∴ height = 25 metres

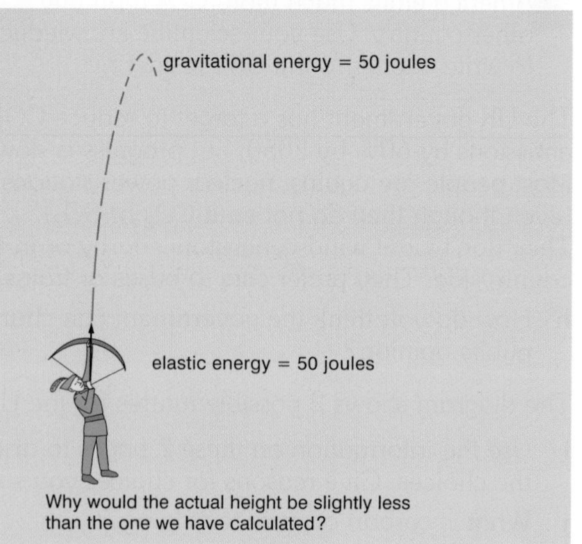

Why would the actual height be slightly less than the one we have calculated?

▶ Calculating Kinetic Energy (KE)

25 000 joules of kinetic energy

A running elephant has more kinetic energy than a running man, because it has more mass.

A racing car has more kinetic energy than a family car because it has a higher speed.

In fact, the formula for the kinetic energy is:

$$\text{Kinetic Energy (joules)} = \frac{1}{2} \times \text{mass (kg)} \times \text{speed squared } (\text{m/s})^2$$

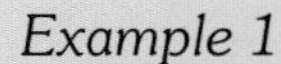

Example 1
An elephant of mass 2000 kg travelling at 5 m/s has
KE = $\frac{1}{2} \times 2000 \times 5 \times 5$ = 25 000 joules.

Example 2
Galileo drops a stone from the leaning tower of Pisa, which is 45 metres high.
At what speed does the stone hit the ground?

The energy is ***conserved*** (see page 102) so, assuming no air resistance:

gravitational energy at the top = **kinetic energy at the bottom**

$$\text{mass} \times 10 \times \text{height} = \tfrac{1}{2} \times \text{mass} \times \text{speed}^2$$
$$10 \times 45 = \tfrac{1}{2} \times \text{speed}^2$$
$$\text{speed}^2 = 10 \times 45 \times 2 = 900$$
$$\therefore \text{speed on impact} = \underline{30 \text{ m/s}}$$

When the stone hits the ground, the kinetic energy is transferred to heat and sound energy.

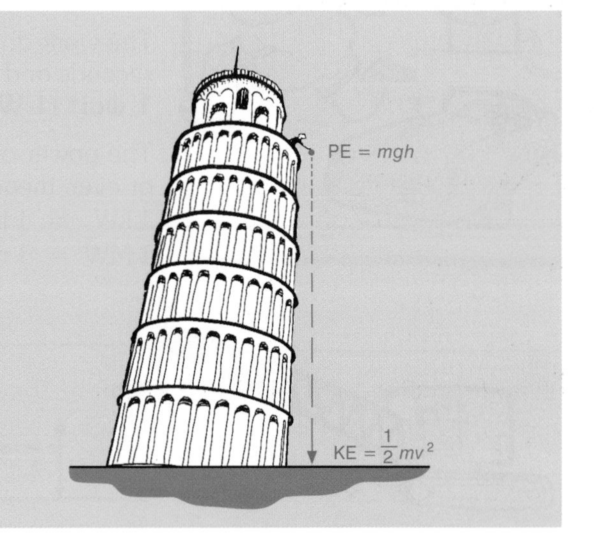

Example 3
A car of mass 800 kg is travelling at 10 m/s.
When the brakes are applied, it comes to rest in 8 m.
What is the average force exerted by the brakes?

The car's KE is transferred to heat the brakes.

From page 99:

Work done = Energy transferred

From the formula on page 97,
and the formula for KE,

$$\text{Force} \times \text{distance moved} = \tfrac{1}{2} \times \text{mass} \times \text{speed}^2$$
$$\text{Force} \times 8 = \tfrac{1}{2} \times 800 \times 10^2$$
$$\therefore \text{Average braking force} = \underline{5000 \text{ newtons}}$$

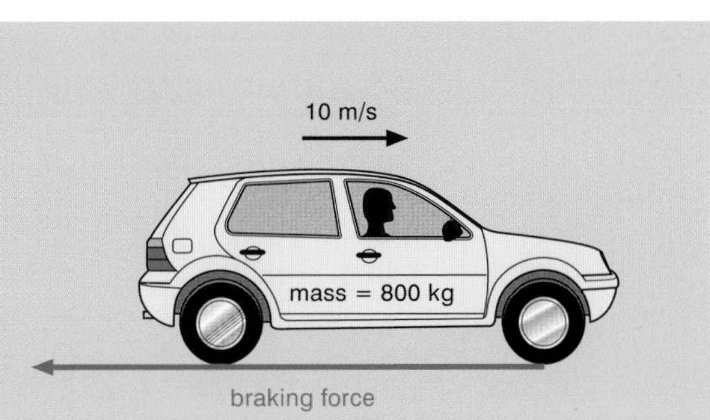

The same method can be used if there is a force *accelerating* an object for a certain distance.

▶ Power

If two cars of the ***same*** weight climb up the ***same*** hill, then they do the ***same*** amount of work. (See p. 97.) But if car A climbs the hill in a shorter time than the other car, we say it has a greater **power**. Power is the ***rate of working*** (the ***rate*** of transferring energy).

$$\textbf{Power} = \frac{\textbf{work done (in joules)}}{\textbf{time taken (in seconds)}} = \frac{\textbf{energy transferred}}{\textbf{time taken}}$$

The work done is measured in joules, the time taken is measured in seconds and so power is measured in ***joules per second*** or ***watts*** (W).
1 watt (1 W) = 1 joule per second

The power of very powerful engines may be measured in kilowatts (kW) or even megawatts (MW).
1 kW = 1 kilowatt = 1000 watts
1 MW = 1 megawatt = 1 000 000 watts.

Example
A crane lifts a load weighing 3000 N through a height of 5 m in 10 seconds
What is the power of the crane?

Calculate it in two parts:

Formula first: **Work done = force × distance moved** (see page 97)

Then put in the numbers: = 3000 N × 5 m
= 15 000 joules

Formula first: $\textbf{Power} = \dfrac{\textbf{work done}}{\textbf{time taken}}$

Then put in the numbers: $= \dfrac{15\,000 \text{ joules}}{10 \text{ seconds}}$
= 1500 W = 1.5 kW

A family car might develop a power of 50 kW, a racing car perhaps 500 kW and a Moon rocket 100 000 MW.

Personal power

How much power can you develop with your legs?

Experiment 16.7 ⚠
You need a flight of stairs to run up, so that you are doing work against the force of gravity (your weight).
First measure your weight in newtons (see page 67).
Then, since you are going to lift your body against the ***vertical*** force of gravity, you must measure the ***vertical*** height of the stairs (in metres).
Then ask someone to use a stop-clock to time how long it takes you to run from the bottom to the top of the stairs.

Now use the method shown on the opposite page to calculate
1. the amount of work you did
2. the power developed by your legs.

Who develops the most power in your class?
What are the energy transfers in this experiment?

VERTICAL DISTANCE
WEIGHT

Wattsisname?
Q. What's Watt?
A. What watt? 'Watt' or 'watt', or what?
Q. Yes, what Watt is the watt named after?
A. James Watt. He knew what was watt!

▶ Electric power

Electric power is also measured in watts. Electric lamps are marked with the electrical power that they transfer.
Look at the writing on a lamp. '60 W' means that the lamp transfers 60 joules per second (from electrical energy to heat and light energy).

60 W

Example
An electric kettle is rated at 2 kW.
How many joules of energy are transferred in 10 seconds?

Power = 2 kW = 2000 W = 2000 joules per second
∴ In 10 seconds, energy transferred to heat = 2000 × 10 joules
= 20 000 joules (= 20 kJ)

2 kW

_ _ _ _ ENERGY CHANGED TO _ _ _ _ ENERGY AND _ _ _ _ ENERGY

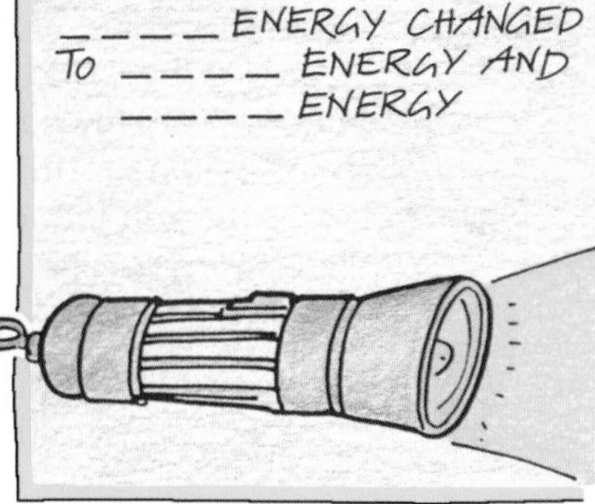
_ _ _ _ ENERGY CHANGED To _ _ _ _ ENERGY AND _ _ _ _ ENERGY

_ _ _ _ ENERGY CHANGED TO _ _ _ _ AND _ _ _ _ AND _ _ _ _ ENERGY

Summary

Work done = force × distance moved = energy changed
(joules) (newtons) (metres)

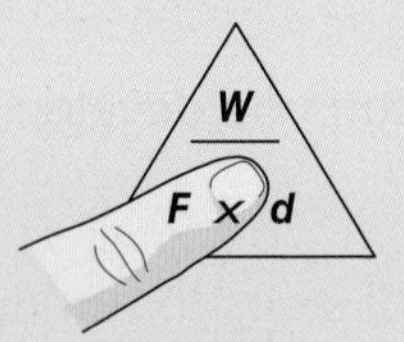

Principle of conservation of energy:
energy can be transformed from one form to another, but it cannot be created or destroyed.

$$\text{Power} = \frac{\text{work done (joules)}}{\text{time taken (seconds)}} = \frac{\text{energy transferred}}{\text{time taken}}$$

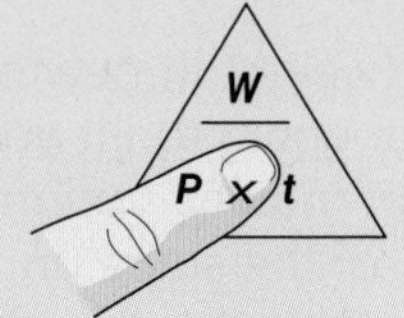

1 watt is a rate of working of 1 joule per second.

Gravitational Potential Energy = $\boldsymbol{mgh}$ joules. Kinetic Energy = $\frac{1}{2}\boldsymbol{mv}^2$ joules.

▶ Questions

1. Copy and complete these sentences.
 a) The work done (measured in . . . .) is equal to the . . . . (in newtons) multiplied by the . . . . moved (in metres).
 b) 1 joule is the work done when a force of one . . . . moves through a distance of one . . . . (in the direction of the . . . .).
 c) The principle of . . . . of energy says that energy can be . . . . from one form to another, but it cannot be . . . . or . . . . .
 d) Work done = ... transferred.
 e) The nine forms of energy are:
 f) The formula for efficiency is:
 g) Power is the rate of doing . . . . .

$$\text{Power (in . . . .)} = \frac{\text{work done (in . . . .)}}{\text{. . . . (in seconds)}}$$

 h) 1 watt equals one . . . . per . . . .

2. A man lifts a parcel weighing 5 newtons from the ground on to a shelf 2 metres high. How much work does he do on the parcel?

3. Draw a labelled Energy Transfer (Sankey) Diagram for each of the following:
 a) a torch
 b) a clockwork toy
 c) a firework rocket
 d) a Bunsen burner
 e) a girl lifting a box on to a shelf
 f) a boy kicking a football
 g) a motor-bike
 h) a hair-dryer
 i) a battery-powered TV set

4. A girl weighing 500 N climbs 40 m vertically when walking up the stairs in an office block. How much work does she do against gravity? What are the energy transfers here?

5. Name the forms of energy stored in:
 a) a slice of bread b) stretched elastic
 c) a torch battery d) a turning flywheel
 e) petrol f) water in a dam.

6. Name the forms of energy at each stage of:
 a) a coal-fired power station
 b) a nuclear power station (see page 349)
 c) a hydroelectric power station.

7. Write an essay about what you think are the most important conclusions that you can draw from the data on pages 102–107.

8. Draw a labelled 'energy chain diagram' to show how the energy in a cheese sandwich comes from the Sun to:
 a) the bread b) the cheese.

9. Professor Messer dreamed one night that the world had suddenly changed so that energy could not be transferred from one form to another. Why did his dream turn into a nightmare?

10. In 1 second, a light bulb transfers 3 joules to light energy and 57 J to heat.
 a) What is the energy input in 1 second?
 b) What is the efficiency?
 c) Draw a Sankey Diagram of it, to scale.

11. What are the energy transfers in these examples?

_____ ENERGY CHANGED TO _____ AND _____ AND _____ ENERGY
_____ ENERGY CHANGED TO _____ AND _____ AND _____ ENERGY
_____ ENERGY CHANGED TO _____ AND _____ AND _____ ENERGY

12. A 100 W light bulb is 5% efficient. Explain carefully what this means.

13. Use the chart on page 83.
a) Plot a graph of the thinking distance data against speed. What do you notice?
b) Plot the braking distance data on the same graph. What do you notice? Can you explain this?
c) Plot the total stopping distance also.
d) From your graph, what is the stopping distance at 15 m/s?

14. A man lifts a weight of 300 N through a vertical height of 2 m in 6 seconds. What power does he develop?

15. A man weighing 1000 N runs up some stairs, rising a vertical height of 5 m in 10 seconds.
a) What power does he develop?
b) What are the energy transfers here?

16. A crane lifts a load of 3000 N through a vertical height of 10 m in 4 seconds. What is its rate of working in a) watts b) kilowatts?

17. What are the energy transfers in these pictures?

18. Use your scientific knowledge to write what you think should be the future energy policy of this country. Discuss the pros and cons of fossil, nuclear, and renewable resources.

19. Write an essay about the evidence for, and the consequences of, global warming.

20. An electric lamp is marked 100 W. How many joules of electrical energy are transformed into heat and light
a) during each second
b) during a period of 100 seconds?

21. A car of mass 1000 kg is travelling at 30 m/s.
a) What is its kinetic energy?
b) It slows to 10 m/s. What is the KE now?
c) What is the change in kinetic energy?
d) If it takes 80 metres to slow down by this amount, what is the average braking force?

22. A girl throws a ball upwards at a velocity of 10 m/s. How high does it go? ($\boldsymbol{g}$ = 10 N/kg)

23. Find out all you can about the work of
a) James Joule b) James Watt.

_____ ENERGY CHANGED TO _____ ENERGY

_____ ENERGY CHANGED TO _____ ENERGY

_____ ENERGY CHANGED TO _____ AND _____ AND _____ AND _____ ENERGY

Further questions on page 142 and page 60.

Physics at work: Heat engines

4-stroke petrol engine

A car or motor-bike uses an ***internal combustion engine***.
In a petrol engine, the petrol vapour is squeezed and then exploded. The chemical energy of the fuel and air is transformed to kinetic energy (and heat, sound). However it is only about 25% efficient.
The 4 steps are:

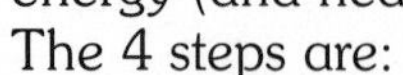

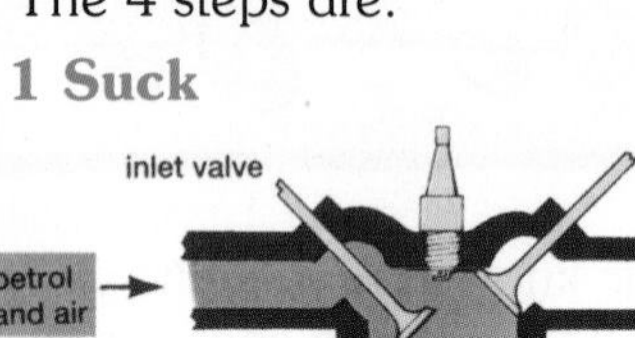

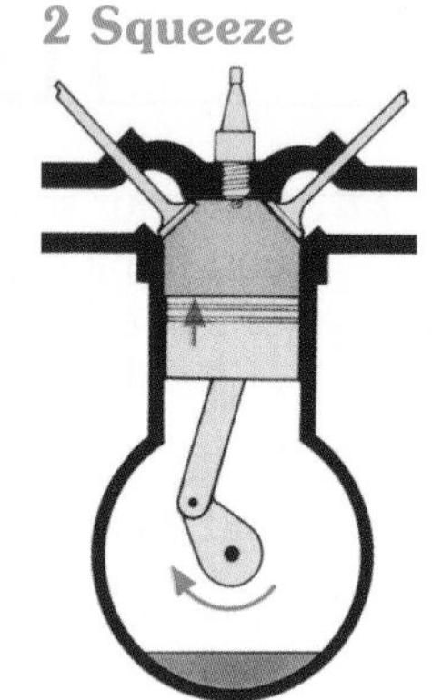

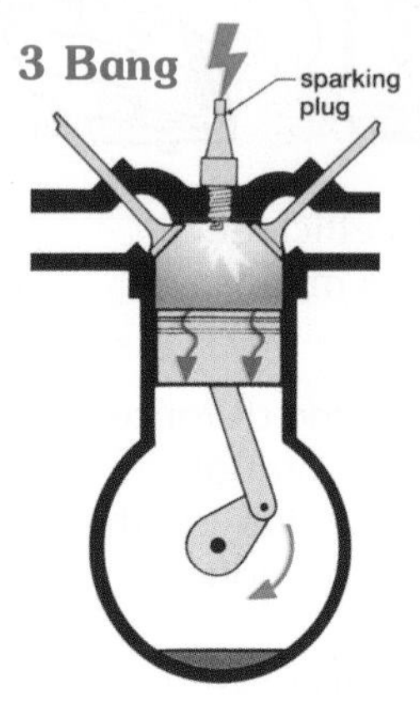

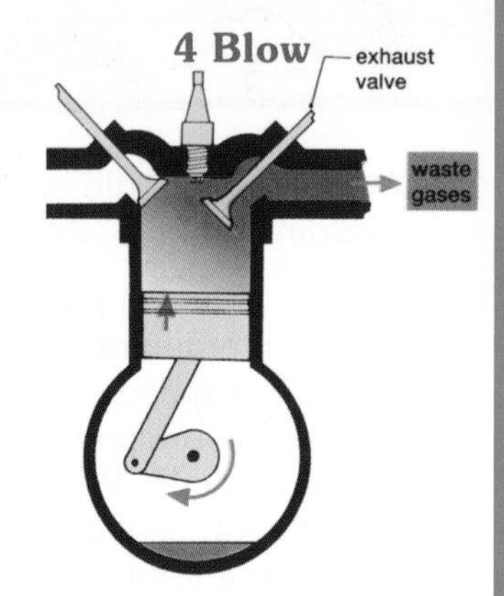

***Induction** stroke*
The inlet valve is open and the piston is moving down. A mixture of petrol vapour and air is sucked in (or injected under pressure).

***Compression** stroke*
The valves close and the piston moves up to squeeze the mixture of petrol and air to about 1/8th of its original volume. So it gets hotter.

***Power** stroke*
An electric spark from the sparking plug ignites the mixture which burns rapidly and expands, forcing the piston down.

***Exhaust** stroke*
The exhaust valve is open and the piston is moving up, to push out the waste gases. The cycle then begins again.

Jet engine

A jet engine also has 4 stages to convert chemical energy to kinetic energy.

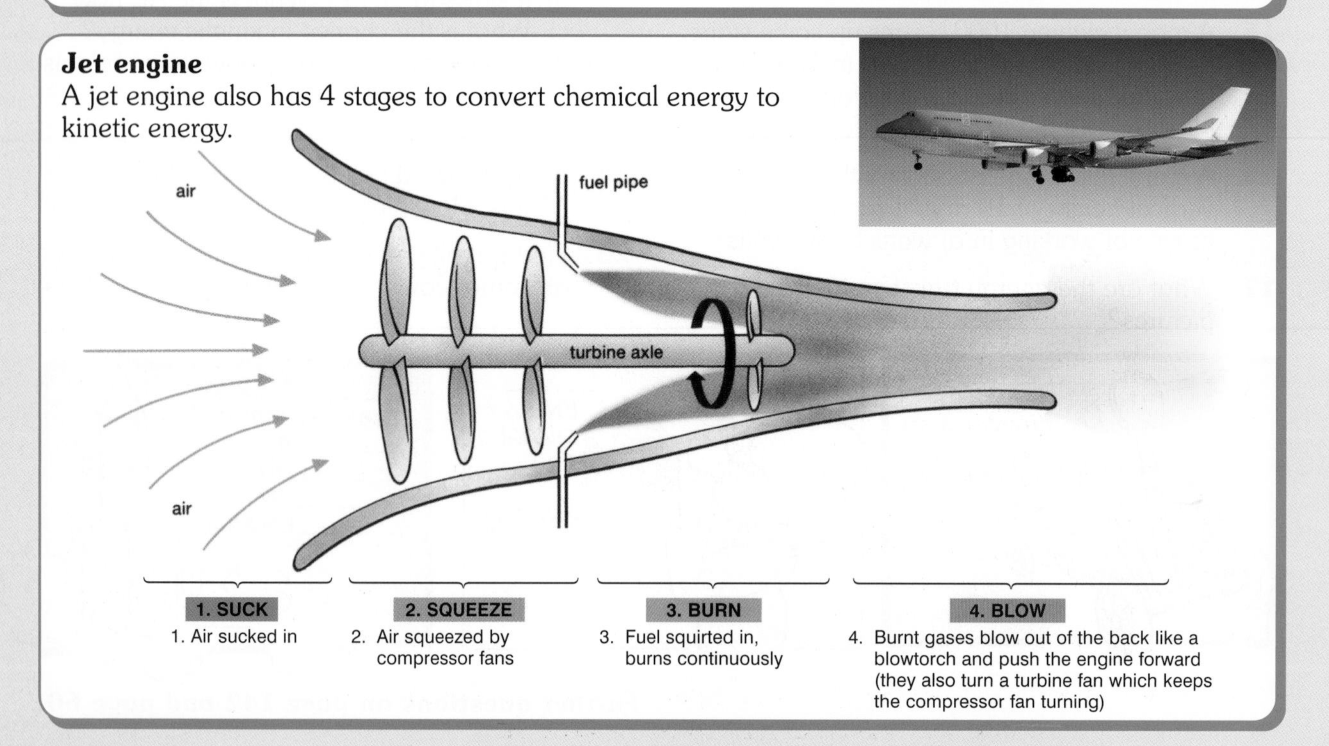

1. Air sucked in
2. Air squeezed by compressor fans
3. Fuel squirted in, burns continuously
4. Burnt gases blow out of the back like a blowtorch and push the engine forward (they also turn a turbine fan which keeps the compressor fan turning)

▷ Physics at work: Cars in the future

Electric cars

Petrol-powered cars cause pollution, so the idea of an electric car seems attractive. The problem is that even large batteries cannot store much energy (compared to a tank of petrol). So an electric car cannot travel far before it needs an electric socket to re-charge. And producing this electricity may cause pollution anyway.

The capacity of a battery is measured in ***ampere-hour*** (A h). 1 ampere-hour means that the battery can deliver a current of 1 amp for 1 hour, or 2 amps for $\frac{1}{2}$ hour, etc.

An electric car for town journeys

Solar-powered cars

It would be nice if we could power cars using sunlight. Unfortunately it is not realistic at present, even in sunny countries, even if the car is covered in solar cells. This is because of the low efficiency of solar cells (see page 103).

Some experimental cars have been built:

Hydrogen fuel-cell cars

A fuel-cell, like a battery, converts chemical energy to electricity. A common fuel-cell used in some new cars uses oxygen (from the air) and hydrogen (stored in tanks under very high pressure).
They can travel up to 300 miles on one tankful of H_2.

The only waste produced by this car is water, so it appears to be pollution-free. However, to produce the hydrogen (from water, see page 273) needs electricity from a power station, which causes pollution unless renewable sources can be used (page 14).

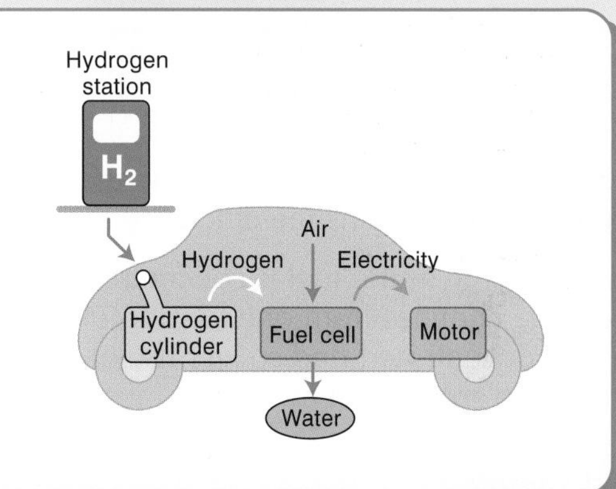

Hybrid cars

A hybrid car uses an electric motor for slow speeds and a 4-stroke petrol engine at higher speeds.
The battery for the electric motor is charged up by a generator (page 298). This is done in 2 ways:

- The generator can be turned by the petrol engine.
- When the car slows down, its kinetic energy is used to turn the generator (and make electricity) instead of being wasted in heating the brakes. This is called '**regenerative braking**'.

The result is an efficient car with less pollution.

A hybrid car : petrol engine + electric motor

Discuss which of the car technologies on this page shows the best way forward.

chapter 17

Machines transfer energy from one form to another.
We know that the total amount of energy put into a machine must equal the total amount of energy output. This is the principle of conservation of energy (see page 98).
However, only ***some*** of the output energy is useful to us. The rest is wasted energy. This affects the ***efficiency*** of the machine.

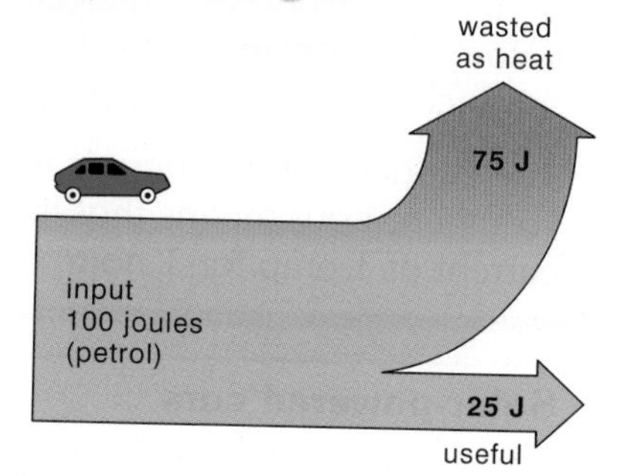

A car is not very efficient. For every 100 joules of chemical energy (petrol) that is put into the car, only 25 joules appear as useful movement energy. The other 75 J are wasted as heat. It is low-grade energy and we cannot use it. The efficiency is calculated by:

$$\textbf{Efficiency} = \frac{\textbf{useful energy output}}{\textbf{total energy input}} \quad \textbf{or} \quad \textbf{Efficiency} = \frac{\textbf{power output}}{\textbf{power input}}$$

Example 1
For this car, the efficiency $= \frac{25}{100} = \underline{0.25}$ or $\underline{25\%}$

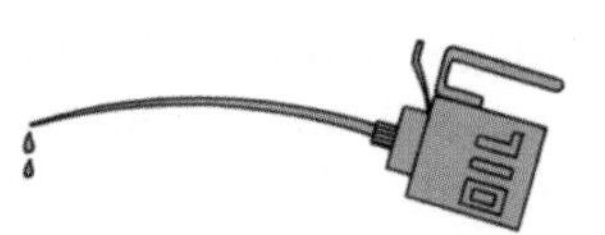

Efficiency will soon decrease
If you forget the oil and grease

Because of friction in a machine there is always some wasted energy. This means the efficiency is **always less than 100%**.

The energy in the fuel is concentrated and useful. Whenever energy changes from one form to another, some of it becomes less concentrated and so less useful. In any machine, energy tends to spread out more and more, into less useful forms.

Example 2
An athlete in a race exerts a force of 80 newtons for a distance of 100 metres, while she uses up 40 000 joules of food energy.
What is her efficiency?

Formula first: Useful energy = work done
= force × distance moved (see p. 97)

Then numbers: = 80 N × 100 m = 8000 joules

Formula: $\text{Efficiency} = \frac{\text{useful energy output}}{\text{total energy output}}$

Then numbers: $= \frac{8000}{40\,000} = \underline{0.20}$ or $\underline{20\%}$

The rest of the energy (32 000 J) appears as useless heat – this is why athletes sweat a lot!

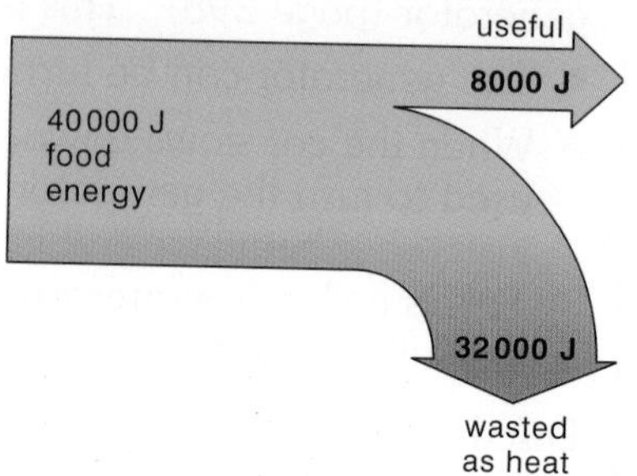

▶ Levers

A lever is a very common machine. It helps us to do work more easily.
Here is a crowbar being used to lever up a large stone. It is like the see-saw on page 91.
The man is 3 m from the pivot, but the stone is only 1 m from the pivot.

He is exerting an ***effort*** force of only 100 N, against a larger ***load*** force of 300 N (the weight of the stone).
The lever is acting as a ***force-magnifier***.

The principle of moments (page 91) shows us why:

$$300\,\text{N} \times 1\,\text{m} = 100\,\text{N} \times 3\,\text{m}$$

To lift the stone by 0.1 m, the man must move $3 \times 0.1 = 0.3$ m
The work done on the stone $= 300\,\text{N} \times 0.1\,\text{m} = 30$ joules
The work done by the man $= 100\,\text{N} \times 0.3\,\text{m} = 30$ joules

This assumes that there is no friction, so the efficiency is 100%.
In practice, if 10% energy is lost by friction, the efficiency = 90%.

Here are some examples of levers:

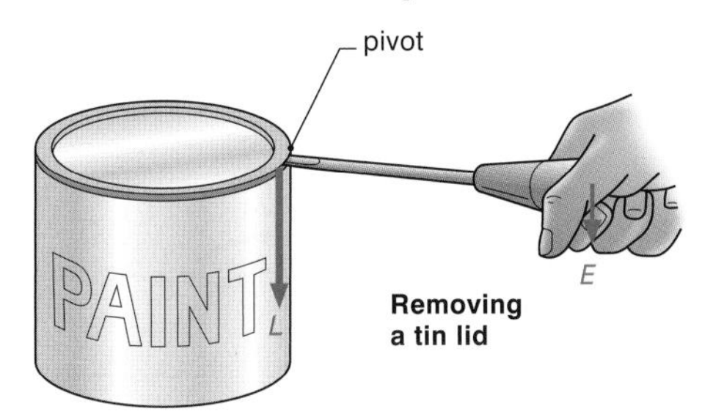

This is a force-magnifier. The effort is farther away from the pivot, just like in the crowbar at the top of the page.

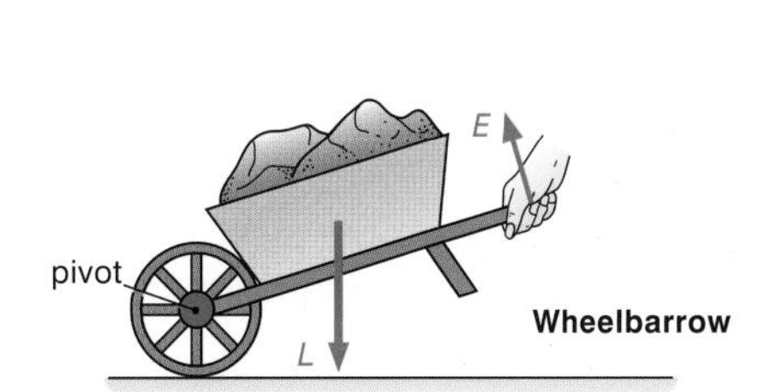

This is a force-magnifier. The effort force needed is about half the load because it is twice as far from the pivot.

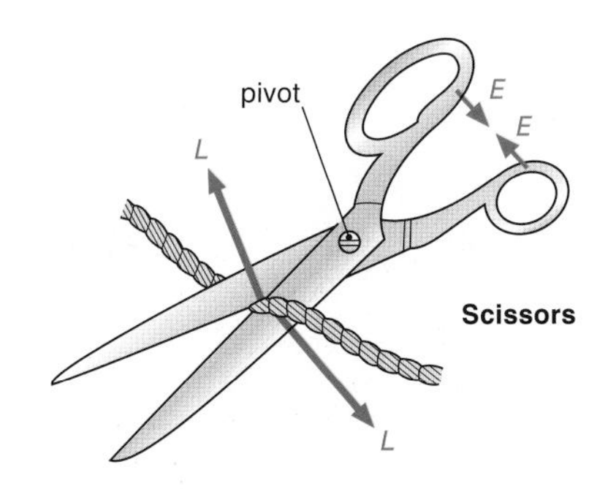

This is a force-magnifier, because the effort is farther from the pivot. What if you try to cut string with the very tip of the scissors?

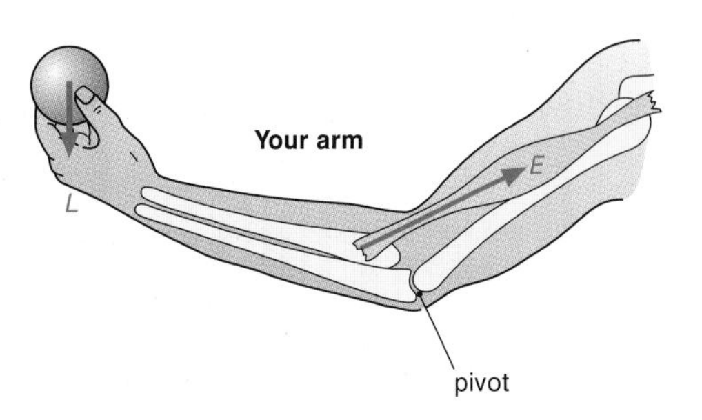

This is not a force-magnifier – it is a distance magnifier. A small contraction of your arm muscle gives a big movement of your hand.

▶ The inclined plane (or ramp)

A ramp, a slope and a hill are examples of inclined planes. They are force-magnifiers.

In the diagram the man is moving a heavy load into the van using a small effort force.
As in all these machines, the smaller force has to go a longer distance – in this case the effort has to move the full length of the slope while the load moves a shorter distance vertically.

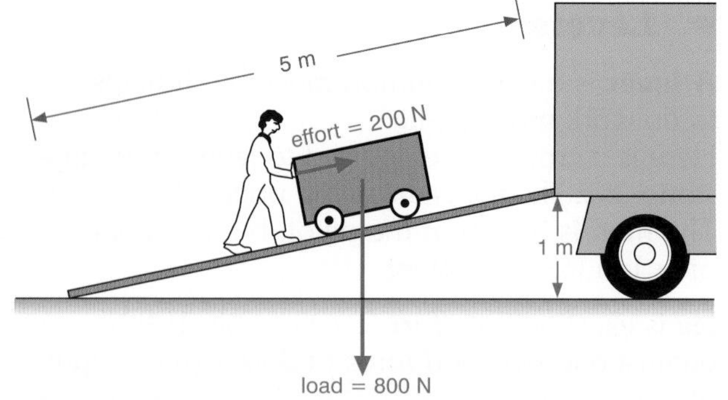

Example
Find the efficiency of the ramp in the diagram.

Useful energy out = work done against gravity (see page 97)
= weight × height lifted = 800 N × 1 m = 800 joules

Total energy in = work done by man
= force × distance moved = 200 N × 5 m = 1000 joules

$$\text{Efficiency} = \frac{\text{useful energy output}}{\text{total energy input}} = \frac{800 \text{ joules}}{1000 \text{ joules}} = \underline{0.80} \text{ or } \underline{80\%}$$

▶ Gears

Gears can be used as force-magnifiers or as distance-magnifiers. It depends on the size of the wheels.

The larger wheel always moves more slowly, but with a larger force. This is a force-magnifier:

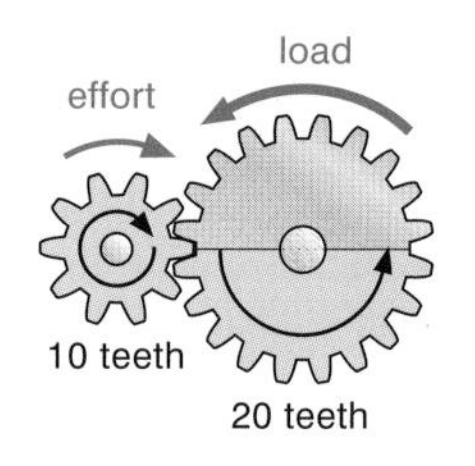

If the wheels are the other way round, the smaller (load) wheel moves faster (but with a smaller force). This is a distance-magnifier and a speed-magnifier:

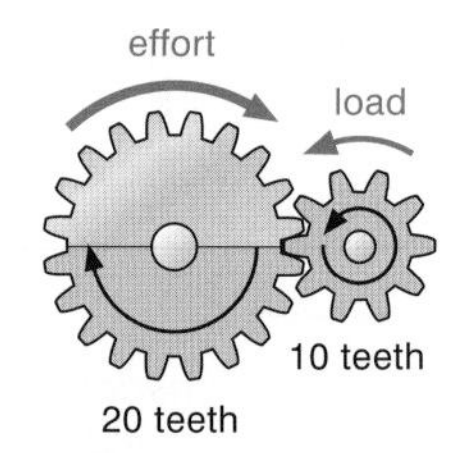

▶ Wheel and axle

This force-magnifier is really a kind of continuous lever.
A small effort on the wheel gives a large turning force on the axle.
The diagrams show some examples:

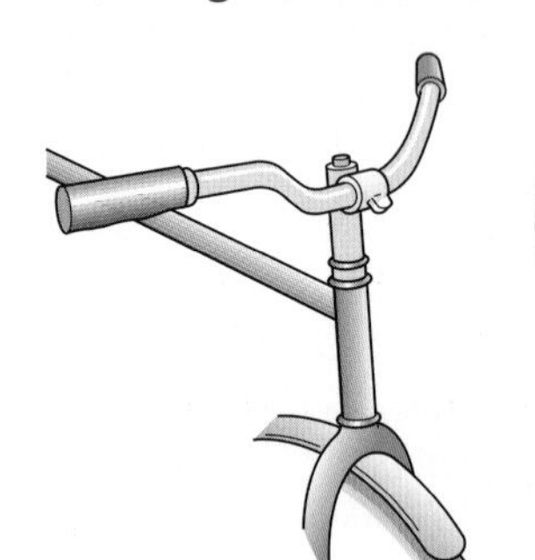

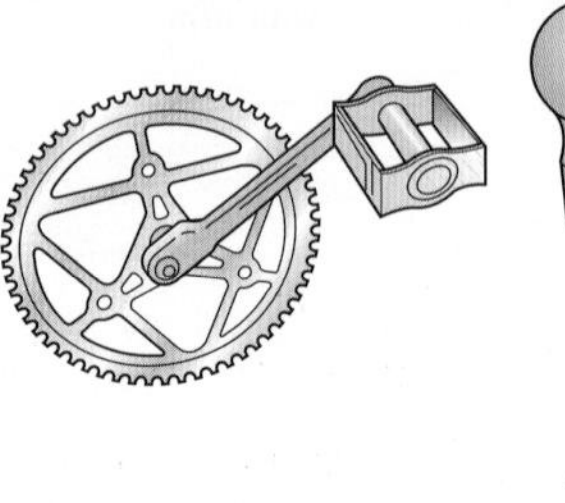

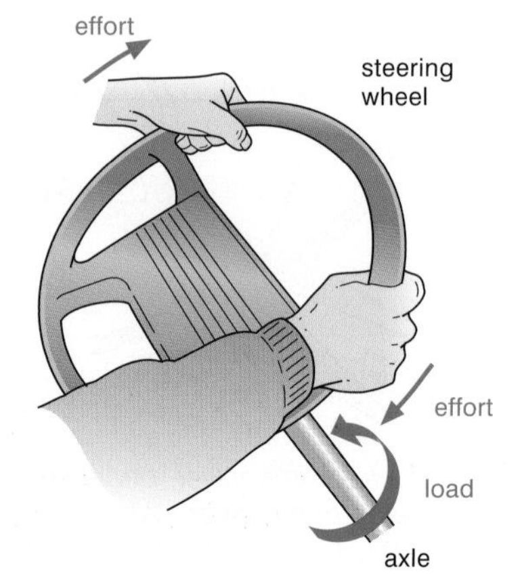

▶ Pulleys

Pulleys are very useful for lifting loads vertically.

Single pulley
Even a single pulley is useful, as it changes the direction of the force. It is easier to pull downwards (with your weight helping you) than it is to pull upwards.

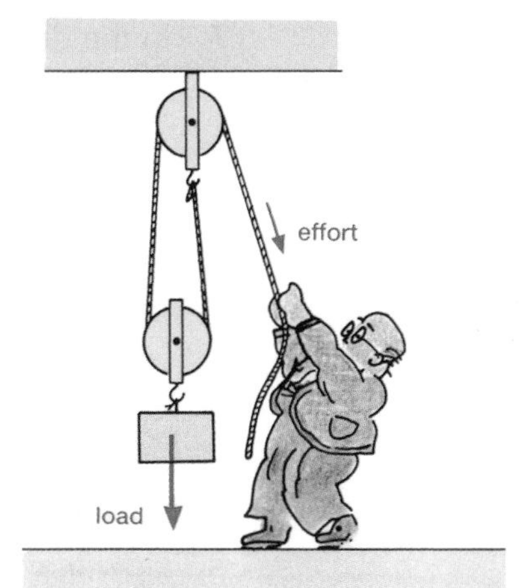

Two pulleys
*Because there are **two** ropes pulling up the load, the effort needed is only about **half** as much.*
As in all these machines, the smaller force has to move a longer distance.

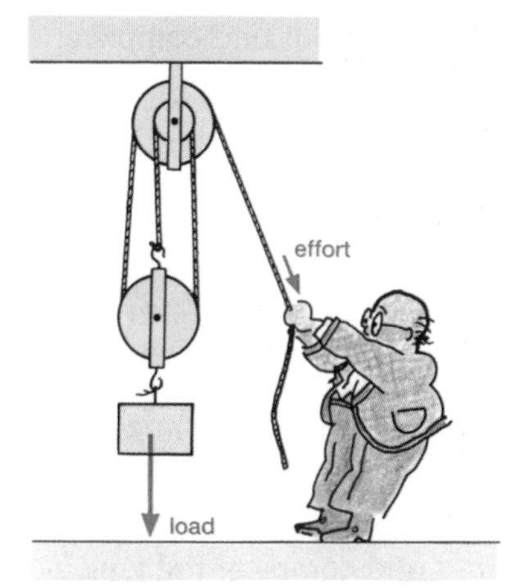

Three pulleys
Three pulleys make it even easier. The efficiency of a pulley system is always less than 100%, because friction transfers some of the energy to heat.

Example
Calculate the efficiency of this pulley system:

Suppose the load is lifted by 1 metre. Each of the four sections of rope has to be shortened by 1 m, so the effort force has to move 4 m.

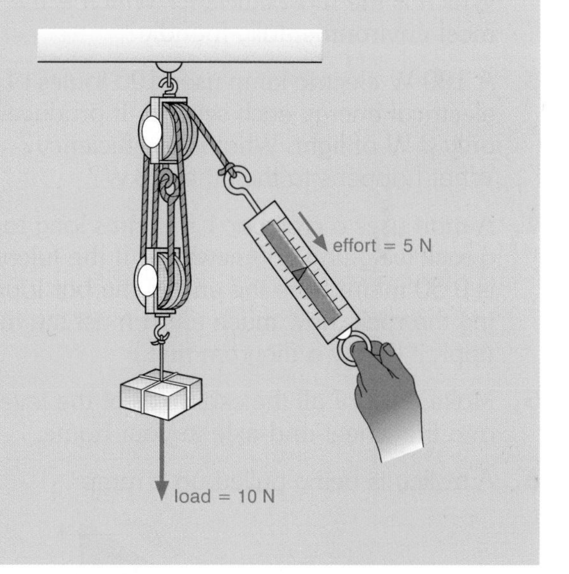

$$\text{Useful energy output} = \text{work done on the load} = \text{weight} \times \text{height lifted} = 10\,\text{N} \times 1\,\text{m} = 10 \text{ joules}$$

$$\text{Total energy input} = \text{work done by the effort} = \text{force} \times \text{distance moved} = 5\,\text{N} \times 4\,\text{m} = 20 \text{ joules}$$

$$\text{Efficiency} = \frac{\text{useful energy output}}{\text{total energy input}} = \frac{10 \text{ joules}}{20 \text{ joules}} = \underline{0.50} \text{ or } \underline{50\%}$$

Summary

$$\text{Efficiency} = \frac{\text{useful energy output}}{\text{total energy input}} = \frac{\text{power output}}{\text{power input}}$$

Machines can be force-magnifiers or distance-magnifiers. Hydraulic (liquid) machines are described on page 79.

Because of friction, the efficiency is always less than 100%.

The smaller force always moves the longer distance.

▶ Questions

1. Copy out and complete:

a) Efficiency $= \dfrac{\text{. . . . energy . . . .}}{\text{. . . . energy . . . .}}$

$= \dfrac{\text{power . . . .}}{\text{. . . . input}}$

b) The energy transferred is calculated by: work done = ... × . . . . moved

c) Because of friction, some . . . . is always wasted. This means that the . . . . is always . . . . than 100%.

2. Here are data for four heat engines. Copy and complete the table:

Engine	Fuel	Input	Useful output	% efficiency
Petrol engine		80 kJ	20 kJ	
Diesel engine		20 kJ	7 kJ	
Steam engine		100 MJ	10 MJ	
Cyclist		500 W	100 W	

Which is the most efficient? Which is the most environmentally-friendly?

3. A 100 W electric lamp uses 100 joules of electrical energy each second. It produces only 2 W of light. What is its efficiency? What happens to the other 98 W?

4. A man uses a crowbar 1.5 metres long to lift a rock weighing 600 newtons. If the fulcrum is 0.50 metre from the end of the bar touching the rock, how much effort must the man apply? (Draw a diagram first.)

5. Make a list of all the examples of the lever and the wheel-and-axle in your home.

6. A trolley is being pulled up a ramp:

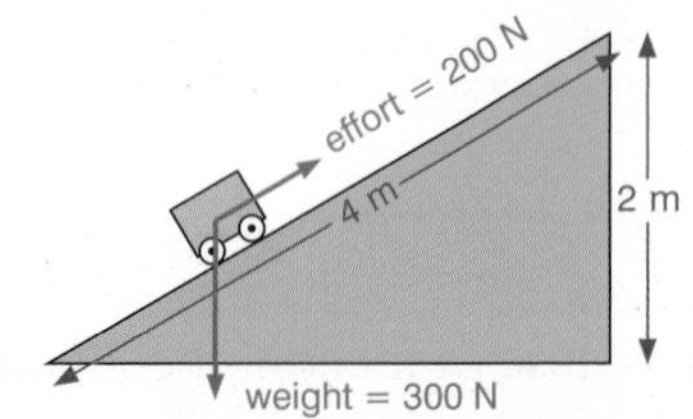

Calculate:
a) the work done on the load (see page 97)
b) the work done by the effort
c) the efficiency of the machine.

7. In the diagram, gear A is rotating at 30 revolutions per second.

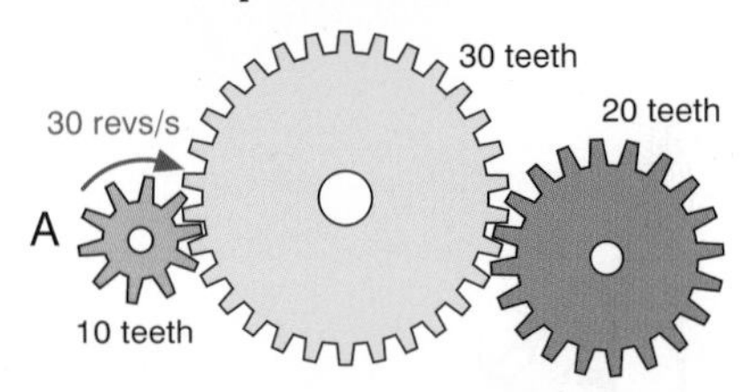

Sketch these gears and mark on each one:
a) the direction in which it is turning
b) the number of revolutions per second.

8. How many examples of simple machines can you find in the photograph? In each case name the type of machine and say whether it is a force magnifier or a distance magnifier. Why does a cyclist often zig-zag up a hill?

9. For each of these pulley systems, calculate:
a) the work done in lifting the load 1 metre
b) the work done by the effort in this case
c) the percentage (%) efficiency.

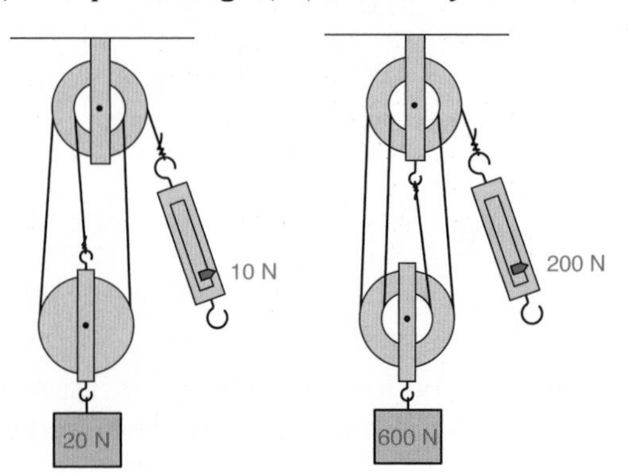

▶ Mechanics crossword

(Do not write on this book – copy the grid with tracing paper first or ask your teacher for a copy.)
For some of these questions you may need to read the next chapter.

Clues across:

1. The pull of the Earth on one ____ = 9.8 newtons.
8. Change in velocity divided by time taken.
10. A fresh weight for Sir Isaac?
13. One joule is the energy needed to move 1 N through ____ metre.
14. Speed equals distance travelled divided by ____.
15. Source of wisdom!
17. Like action or reaction?
18. Velocity and force are ____.
21. Your most valuable equipment in experiments.
23. Physics is ____ special!
25. Power is work done divided by ____.
26. An electric ____ transfers electric energy to kinetic energy.
27. Not the first unit of time?
28. A wound-up spring has ____ energy.
31. It might make you move.
32. Shortened unit of electric current.
35. What is the unit!
36. You have been, all term.
37. Newton's First Law: a stationary object, with no resultant force acting on it, remains at ____.
38. 3.141 592 653 589 793 238 462 643 383 279 502 884 197 169 399 375 105 8 . . . .
39. Speed? Not quite.
40. If a velocity–time graph is a straight line, the acceleration is ____.

Clues down:

1. and 9. A heavy lorry travelling at high speed has a lot of this.
2. Newton's Third Law: apply a force in, the reaction is ____.
3. In a little while you have the turning effect of a force!
4. Newton's Third Law: apply a force to the west; the reaction is to the ____.
5. Force is measured in ____.
6. A man walking at 2 m/s for 4 seconds covers a distance of ____ metres.
7. It keeps you going when the car crashes.
9. See 1 down.
11. The pull of gravity.
12. A joule is a unit of power – yes or no?
16. When there is a resultant force on a body, it ____.
19. Push or pull to ____ a force.
20. When you divide by 'time', you are calculating a ____.
21. An experimental inaccuracy.
22. A dynamo transfers kinetic energy to ____ energy.
24. Cold place with little friction.
28. Rate of working.
29. Take in fuel.
30. An electric fire transfers electric energy to heat and ____ energy.
33. A city where you might be inclined to experiment!
34. An engine transfers ____ energy to kinetic energy.
36. If the mass is 6 kg and the volume is 3 m^3, then the density is ____ kg/m^3.

chapter 18

VELOCITY and ACCELERATION

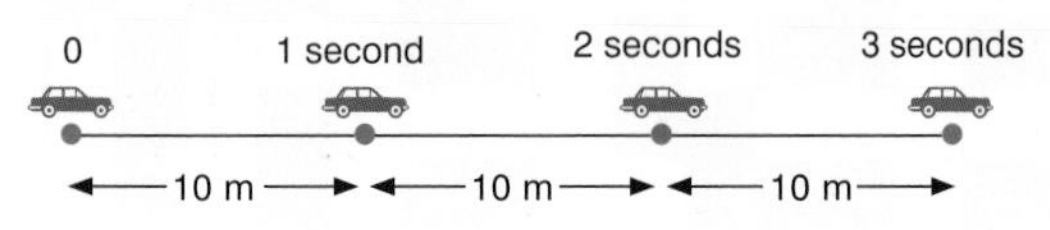

A constant speed of 10 m/s

If a car travels down the road at a ***constant speed*** of 10 metres per second (10 m/s), it means that it travels 10 metres in every second.

The speed can be found by

$$\textbf{Average speed} = \frac{\textbf{distance travelled} \text{ (metres)}}{\textbf{time taken} \text{ (seconds)}}$$

Velocity is almost the same thing as speed but it is a ***vector*** quantity (see page 86). Velocity has a ***size*** (called speed) ***and*** a ***direction.***

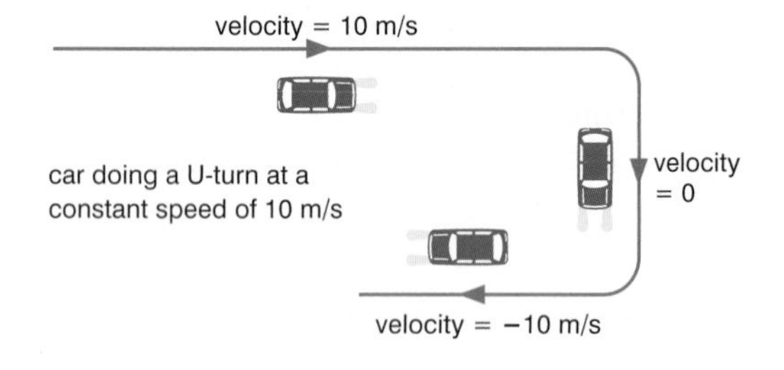

If a car is travelling at a constant speed ***in a straight line***, then it has a constant velocity. If it turned a corner then its velocity would change, even though its speed remained constant.
A car travelling in the opposite direction has a negative velocity.

▶ Acceleration

If a car-driver presses the accelerator pedal, the car goes faster – it accelerates.
The acceleration can be found by

$$\textbf{Acceleration} = \frac{\textbf{change in velocity} \text{ (m/s)}}{\textbf{time taken for the change} \text{ (s)}}$$

If the change in velocity is measured in m/s and the time is measured in seconds, then the acceleration is measured in m/s^2 (metres per second squared).

An acceleration of 2 m/s^2 means that the velocity increases by 2 m/s every second.

If a car is slowing down then it has a ***deceleration*** or a ***retardation*** or a ***negative acceleration***.

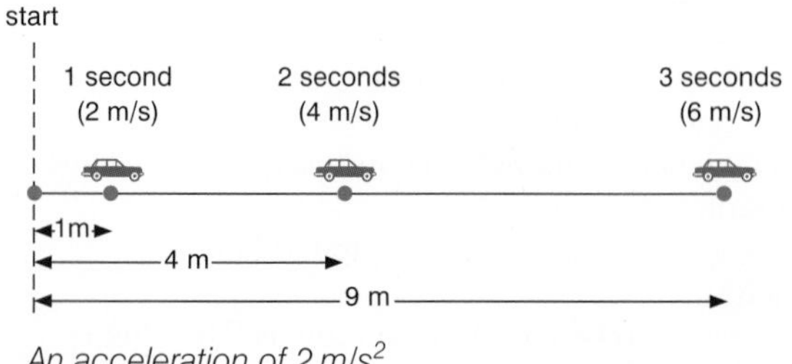

An acceleration of 2 m/s^2

▶ Ticker-timer experiments

A ticker-timer is simply a vibrator that puts little black dots on to a paper tape at the rate of 50 dots in each second.

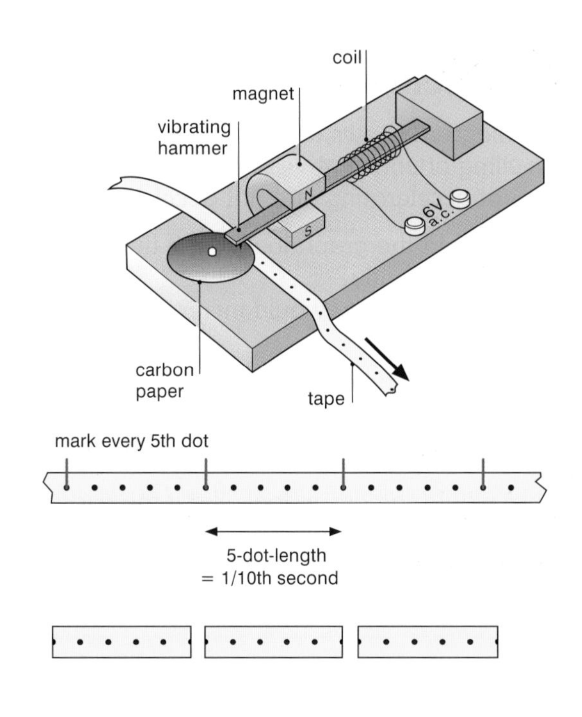

Experiment 18.1
Pull a length of tape through the ticker-timer at a ***constant velocity***.

What do you notice about the spacing of the dots?

a) Count along the tape, marking off every fifth dot. Because there are 50 dots in each second, each 5-dot length is produced in $\frac{1}{10}$th second.
b) Use scissors, at each mark, to cut the tape into several 5-dot lengths (carefully keeping them in the right order).
c) Stick the 5-dot lengths side by side in the right order to get a strip chart.
Since each strip is the distance travelled in $\frac{1}{10}$th second, you have made a strip chart of ***velocity*** against ***time***.

mark every 5th dot

5-dot-length = 1/10th second

The constant height of the strips means that the tape was moving at a constant velocity:

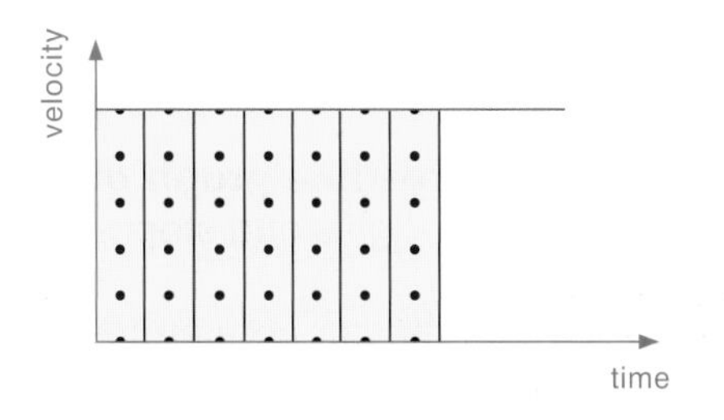

Example
Find the speed of the tape in the diagram:

Measure the tape shown. Do you find that it has travelled 1.9 cm in each 5-dot length?

$$\text{Speed} = \frac{\text{distance travelled (5 dots)}}{\text{time taken for 5 dots}}$$

$$= \frac{1.9\text{ cm}}{0.1\text{ s}} = \underline{19\text{ cm/s}}$$

Experiment 18.2
Repeat experiment 18.1 but make the tape ***accelerate*** as you pull it.

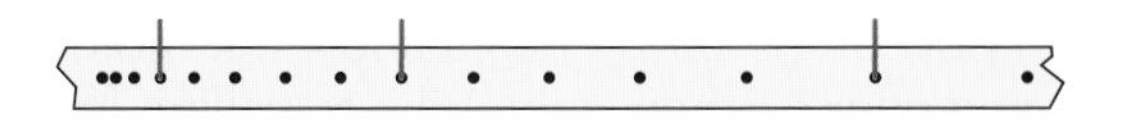

What do you notice about the spacing of the dots? Repeat steps a), b) and c) to build a velocity–time strip chart for this accelerating tape.

You can see that the velocity is increasing (because the tape was accelerating).

Another way to get this information is to use a ***motion sensor*** with a computer to draw the graph.

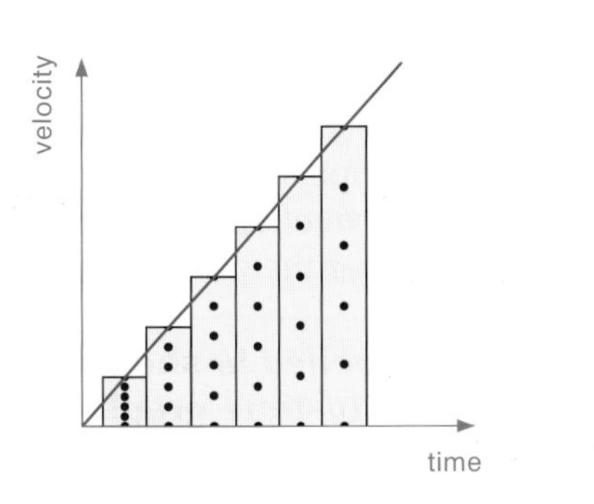

▶ Velocity–time graphs

The diagram shows a **velocity–time graph for a car** travelling at constant velocity:
It is not accelerating and not decelerating.

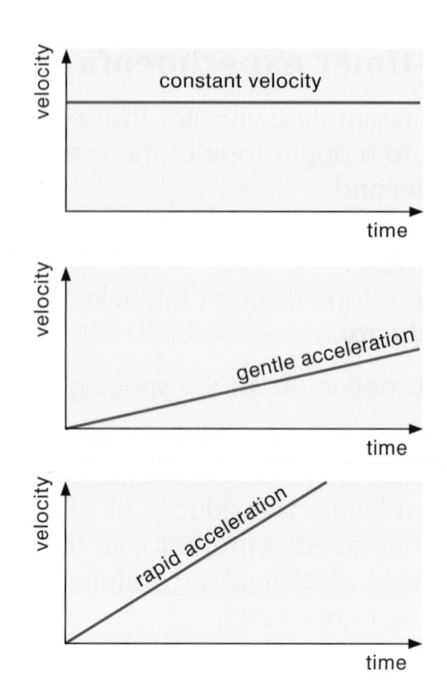

What would the graph look like if the car accelerated gently?
Because its velocity would increase, the graph would slope upwards:

In this graph, the car starts off from rest (velocity is zero) and accelerates uniformly (steadily).
A straight graph means uniform, constant acceleration.

What would the graph look like if the car accelerated more rapidly?
Since its velocity would increase more rapidly, the graph would be ***steeper***. In fact,

> **the *acceleration* is shown by the *slope* (or gradient) of the velocity–time graph.**

Page 363 explains how to find the slope of a graph.

Here is a velocity–time graph of a car starting off at one set of traffic lights and stopping at the next set of lights:
Can you see that the car accelerates from A to C, travels at constant velocity between C and D and then decelerates (brakes) rapidly to stop at E?

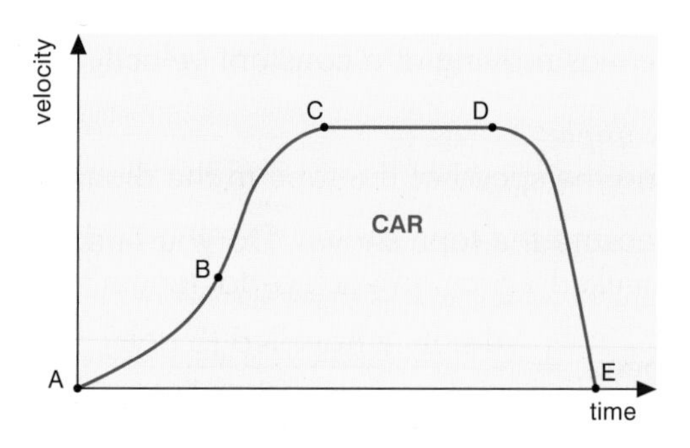

At which point (A, B or C) is it accelerating most?

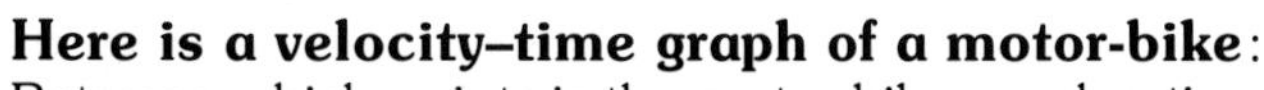

Here is a velocity–time graph of a motor-bike:
Between which points is the motor-bike accelerating most rapidly?
Which part of the graph might show the motor-bike changing gear? When is the bike travelling fastest?

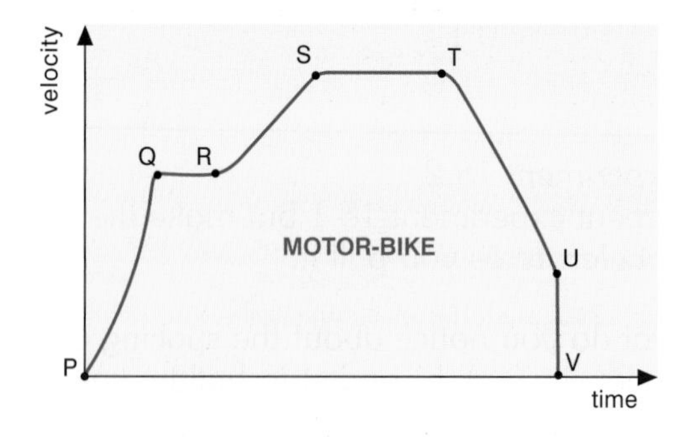

When does the rider start to brake? Which part of the graph shows the bike hitting a solid brick wall?

If this was a ticker-tape strip chart you could find the total ***distance travelled*** by measuring the total amount of tape under the graph. In fact,

> **the *distance travelled* is shown by the *area* under the velocity–time graph.**

Here is a velocity–time graph of a 'stockcar' starting a race, crashing into another car, and then reversing.

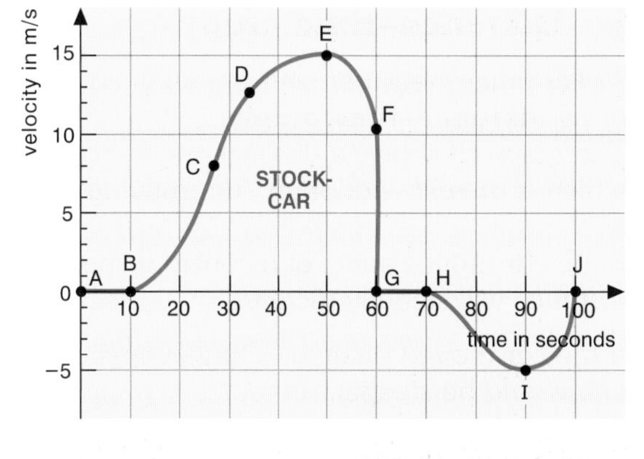

a How long do you think the driver had to wait before the starter waved his flag to start the race?

b Is his acceleration greatest at B, C, D or E?
c At which point is he travelling fastest?
d What is his velocity then?
e How much time has passed?

f When does he start to brake?
g At what time does he come to a stop after hitting the other car? How long has he been travelling then?

h For how long does he stay at rest before reversing?
i At which point (G, H, I or J) does he start to reverse his car?

When the car is reversing, it has a ***negative*** velocity (remember velocity is a ***vector*** quantity, see p. 86).

j What is his maximum velocity in reverse?
k For how many seconds does he reverse the car?
l When does he finally stop the car?

m How do the areas tell you whether the distance travelled is farther in forward gear or reverse?

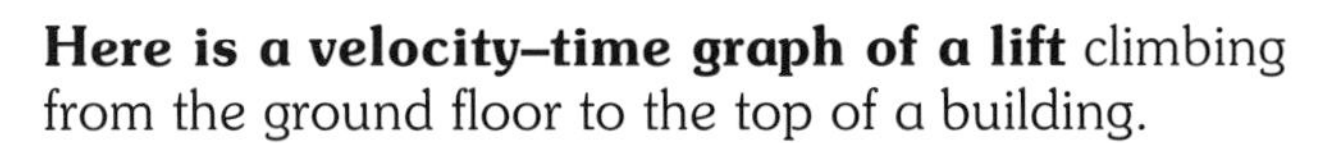

Here is a velocity–time graph of a lift climbing from the ground floor to the top of a building.

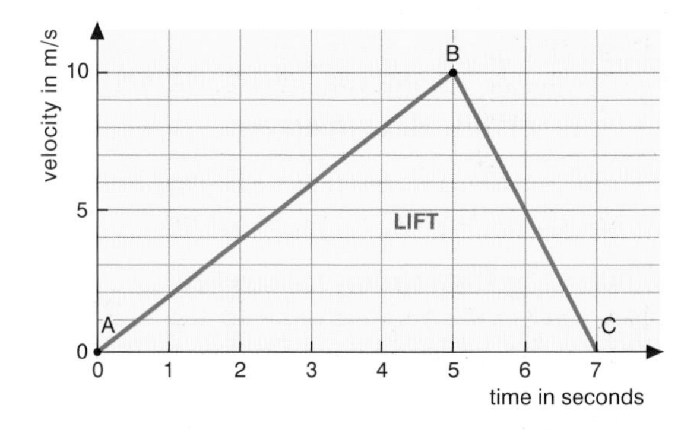

n At which point (A, B or C) is it on the ground floor?
o At which point (A, B or C) is it travelling fastest?
p What is its maximum velocity in m/s?
q How long does it take to reach this velocity?

r It then decelerates to stop at C, the top of the building. How long did the journey take?

Example 1
We can find its acceleration from the slope:

$$\textbf{Acceleration} = \frac{\textbf{change in velocity}}{\textbf{time taken for the change}} = \frac{10\text{ m/s}}{5\text{ s}} = \underline{2\text{ m/s}^2}$$

Can you see why the deceleration is 5 m/s^2?

Example 2
To find the distance travelled between A and B:

Distance travelled = **area under graph**
= area of triangle under AB
= $\frac{1}{2} \times$ base $\times$ height
= $\frac{1}{2} \times 5\text{ s} \times 10\text{ m/s}$
= $\underline{25\text{ metres}}$

Do you agree the distance from B to C is 10 m?

▶ Distance–time graphs

Ticker-timer measurements are sometimes used to draw **distance–time graphs**.
The diagram shows a distance–time graph for a car which is ***at rest*** (stationary, not moving):

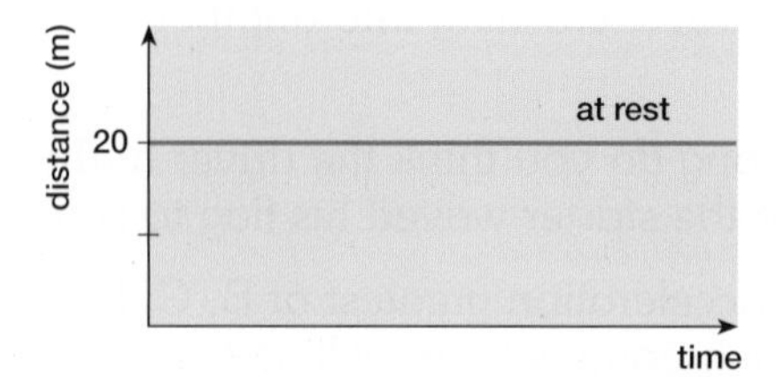

If the car moves away at a constant speed then the distance will steadily increase:

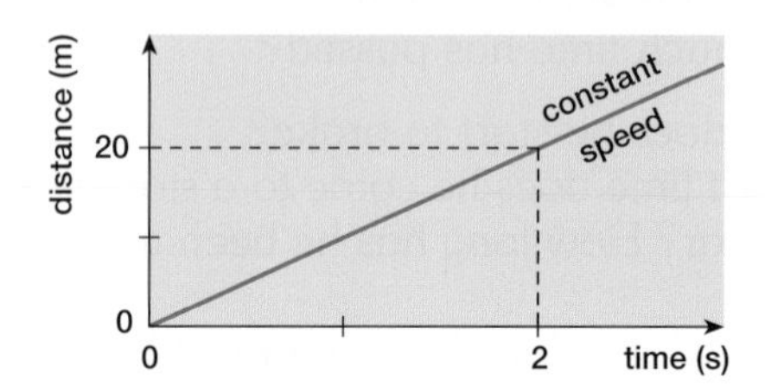

If the car moved away at a higher speed then the line would be steeper. In fact,

> **the speed is shown by the *slope* (or gradient) of the distance–time graph.**

Example From page 122 and the second graph:

$$\text{speed} = \frac{\text{distance travelled}}{\text{time taken}} = \frac{20\text{ m}}{2\text{ s}} = \underline{10\text{ m/s}}$$

If the car accelerates, then the speed increases, and so the slope of the graph increases:

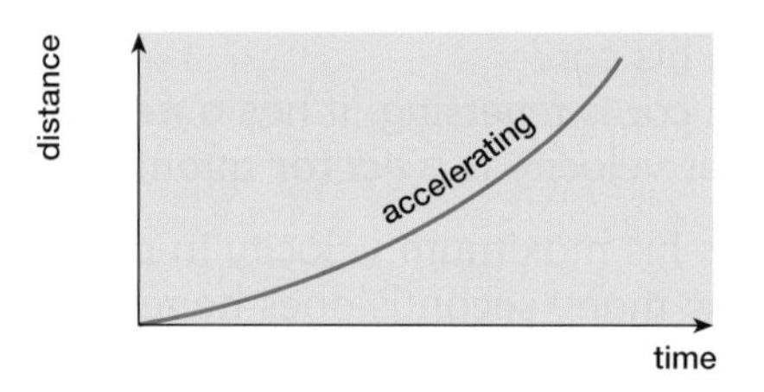

If the car decelerated, then the speed would decrease, and so the slope of the graph would decrease. The graph would curve the other way.

Here is a distance–time graph for the same **lift** that is shown on the previous page. It has the same points (A, B, C) marked:

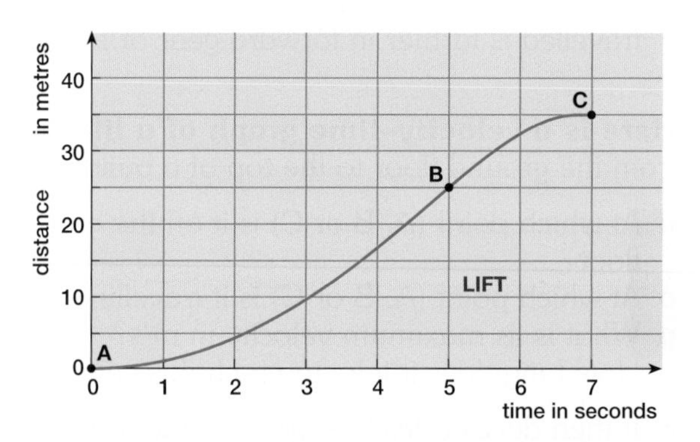

The lift accelerates from A to B, and is travelling fastest at B. Where is it decelerating?

What is the total distance travelled?
How long does it take the lift to travel 25 m?
What is the ***average*** speed between A and B?

Displacement

We have seen that velocity is almost the same thing as speed, but it is a vector quantity (see page 86).

In a similar way, ***displacement*** is almost the same thing as distance, but it is a ***vector*** quantity. Displacement has a ***size*** (called distance) ***and*** a ***direction***.

If you move to a different part of the room, you might have a displacement of 2 metres ***in an easterly direction***. The direction is important.

'The direction is important'

Equations of motion

These are 4 equations which can be used whenever an object travels with **constant, uniform acceleration** in a straight line.

We write these equations using 5 symbols: s, u, v, a, t.
Suppose an object is travelling at a velocity u and then moves with a uniform acceleration a for a time t.
Its velocity is then v and it has travelled a distance s.

s = distance travelled (metres)
u = initial velocity (m/s)
v = final velocity (m/s)
a = acceleration (m/s²)
t = time taken (seconds)

Then from page 122: **Acceleration** = $\dfrac{\textbf{change in velocity}}{\textbf{time taken}}$

or, in symbols: $a = \dfrac{v - u}{t}$ $\quad\therefore\quad \boxed{v = u + at} \quad \ldots(1)$

Also from page 122: **Average speed** = $\dfrac{\textbf{distance travelled}}{\textbf{time taken}}$

or, in symbols: $\dfrac{u + v}{2} = \dfrac{s}{t}$ $\quad\therefore\quad \boxed{s = \dfrac{(u + v)}{2}t} \quad \ldots(2)$

Using equation (1) to replace v in equation (2): $s = \left[\dfrac{u + (u + at)}{2}\right]t$ $\quad\therefore\quad \boxed{s = ut + \tfrac{1}{2}at^2} \quad \ldots(3)$

Using equation (1) to replace t in equation (2): $s = \dfrac{(u + v)}{2}\dfrac{(v - u)}{a}$ $\quad\therefore\quad \boxed{v^2 = u^2 + 2as} \quad \ldots(4)$

You can remember the 5 symbols by 'suvat'. If you know any three of 'suvat' the other two can be found.
In questions, 'initially at rest' means $u = 0$. A negative number for the acceleration means the object is slowing down (decelerating).

Example
A cheetah starts from rest, and accelerates at 2 m/s² for 10 seconds. Calculate:
a) the final velocity and b) the distance travelled.

First write 'suvat' and show what you know:

$s = ?$
$u = 0$
$v = ?$
$a = 2\text{ m/s}^2$
$t = 10\text{ s}$

a) Use equation (1): put in numbers:

$$v = u + at = 0 + 2 \times 10 = \underline{20\text{ m/s}}$$

b) Use equation (2): put in numbers:

$$s = \frac{(u + v)}{2}t = \frac{(0 + 20)}{2} \times 10 = \underline{100\text{ m}}$$

A cheetah can accelerate up to a speed of 27 m/s

▶ Acceleration due to gravity

The force of gravity pulls down on all objects here on Earth (see page 65). If objects are allowed to fall, they ***accelerate*** downwards.

If there is no air resistance or friction then all objects accelerate downwards at the ***same*** rate. You may have heard of a famous experiment that Galileo is supposed to have done from the leaning tower of Pisa: if a heavy stone and a light stone are dropped together, they accelerate at the same rate and land at the same time.

Accurate measurements show that:
the **acceleration due to gravity** = $\mathbf{9.8\ m/s^2}$.

For simple calculations we usually use $\mathbf{10\ m/s^2}$.

This means that for an object falling with no air resistance, the velocity after 1 second is 10 m/s, and after 2 seconds the velocity is 20 m/s, and so on.
If a man fell for 5 seconds with no air resistance, what would his speed be?

In practice, there is usually air resistance.
If the girl in the photo falls a long way without a parachute, then because of friction she reaches a final or ***terminal velocity*** of about 50 m/s – the speed of a fast racing car. See page 89.

There was a young man who had heard,
That a person could fly like a bird.
To prove it a lie
He jumped from the sky,
– His grave gives the date it occurred!

A parachute is designed to make the air resistance as large as possible.
With a parachute the terminal velocity is only 8 m/s.

Raindrops, snowflakes and sycamore seeds – they all fall at their own terminal velocity.

At the terminal velocity, the forces on the object are ***balanced:***

force of gravity (weight) = force of air resistance
(downwards) (upwards)

This is an example of Newton's First Law (page 69): because there is no resultant force on the object, it continues to move at constant speed in a straight line.

Experiment 18.3 Measuring the acceleration of free fall, g
An electric stop-clock (accurate to $\frac{1}{1000}$ second) is used to measure the time taken for a small steel ball to fall through a known distance, s.

When the switch is in position A, the electromagnet holds up the ball:

When the switch is moved quickly to B, the electromagnet releases the ball and the clock starts timing.
When the ball hits the trap door, the circuit is broken at contact C, and the clock stops.

To calculate g, use: $s = ut + \frac{1}{2}at^2$ (see page 127)
But $u = 0$ and $a = g$, acceleration due to gravity

$\therefore s = \frac{1}{2}gt^2$ or $g = \frac{2s}{t^2}$

How could this experiment be improved?

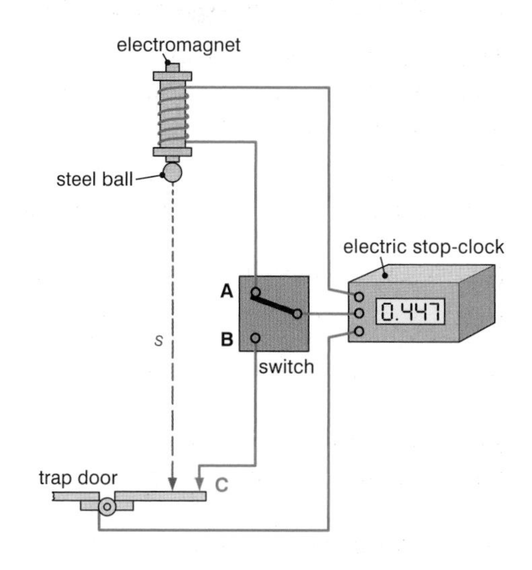

Example
A ball is thrown vertically upwards at 20 m/s.
Ignoring air resistance and taking $g = 10\text{ m/s}^2$, calculate
a) how high it goes b) the time taken to reach this height
c) the time taken to return to its starting point.

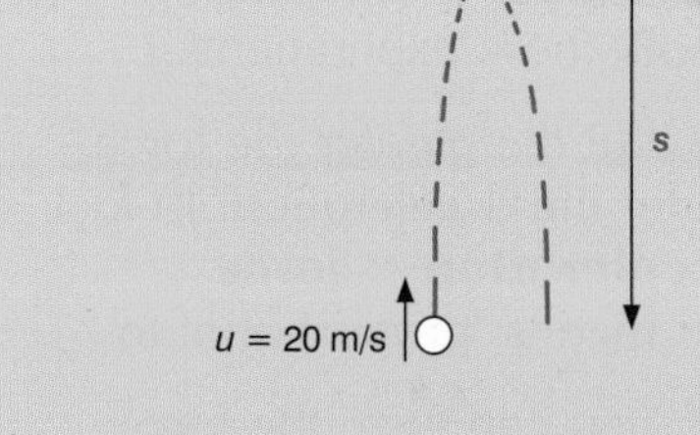

When travelling upwards it is ***de***celerating, so $a = -10\text{ m/s}^2$.
At the moment when it reaches its highest point, $v = 0$.

Write 'suvat' first:
$s = ?$
$u = 20\text{ m/s}$
$v = 0$
$a = -10\text{ m/s}^2$
$t = ?$

a) From p. 127: $v^2 = u^2 + 2as$
$0 = 20^2 + 2(-10)s$
$20s = 20^2$
$\therefore s = \underline{20\text{ m}}$
(see also p. 109)

b) From p. 127: $v = u + at$
$0 = 20 + (-10)t$
$\therefore t = \underline{2\text{ seconds}}$

c) It also takes 2 seconds to fall down again. $\therefore$ total = $\underline{4\text{ seconds}}$

Vertical *and* horizontal motion

Experiment 18.4
Set up a ruler and two coins as shown.
Press on the ruler and tap the end so that A falls vertically while B is projected sideways.

Listen to the coins hitting the ground.
Do they hit at the same time?

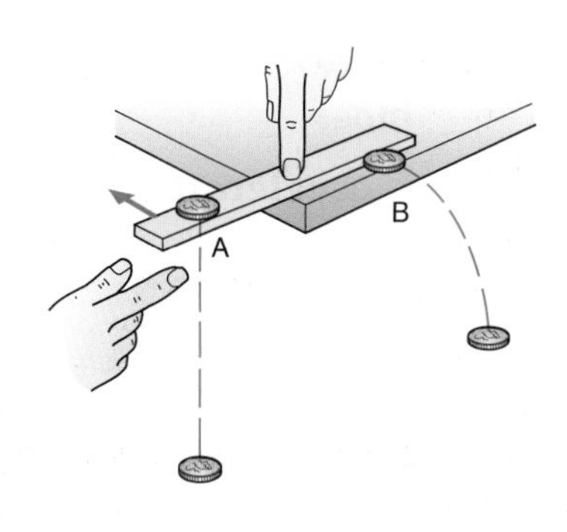

The vertical accelerations of the coins are exactly the same – even though one is moving horizontally as well! (See also page 134.)
In calculations it is better to deal with the vertical and horizontal movements separately.

The horizontal and vertical motions of a body are independent and can be treated separately.

▶ Force, Mass and Acceleration

When the forces on an object are ***un***balanced, there is a resultant force. This resultant force causes the object to change its velocity.
It accelerates or it decelerates.
In the photo, the bigger the force on the javelin, the more it accelerates. The heavier the javelin, the less it accelerates.

Experiment 18.5
Investigating acceleration
You can use a ticker-timer (see page 123) to investigate the acceleration of a trolley:

First, you can keep the mass of the trolley constant and vary the force using elastic bands as shown:
(You must keep the elastic bands stretched the same amount throughout these experiments.)

You see that the greater the force, the greater the acceleration. In fact:

$$\textbf{acceleration} \propto \textbf{force}$$

where $\propto$ means 'proportional to'.

Second, you can keep the force constant (just one elastic band) and vary the mass of the trolley.
You see that the greater the mass, the ***less*** the acceleration. In fact:

$$\textbf{acceleration} \propto \frac{\textbf{1}}{\textbf{mass}}$$

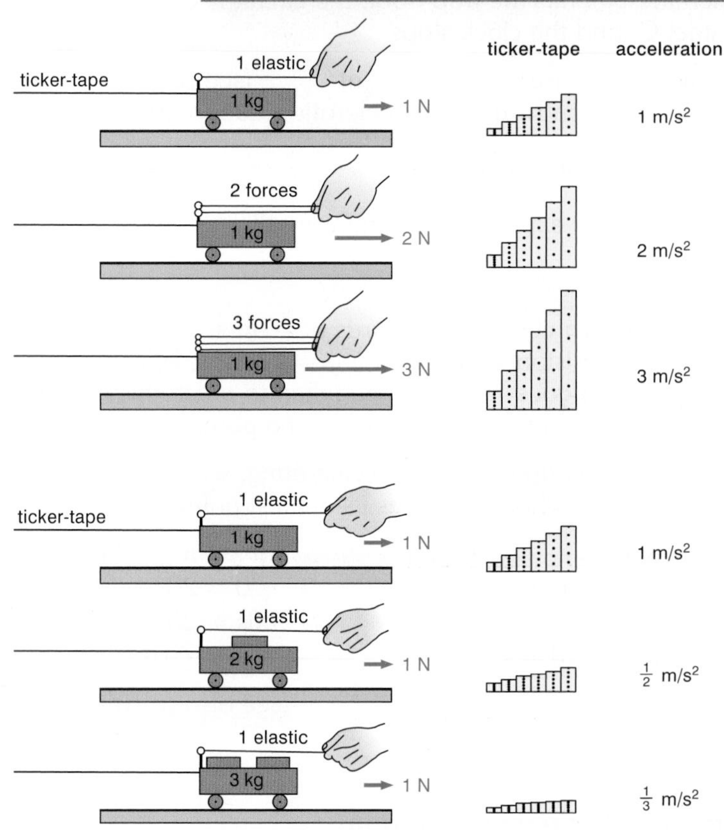

Combining these two results:

$$\textbf{acceleration} \propto \frac{\textbf{force}}{\textbf{mass}}$$

or, **force** $\propto$ **mass** $\times$ **acceleration**.

Provided we measure the force in newtons, the formula becomes:

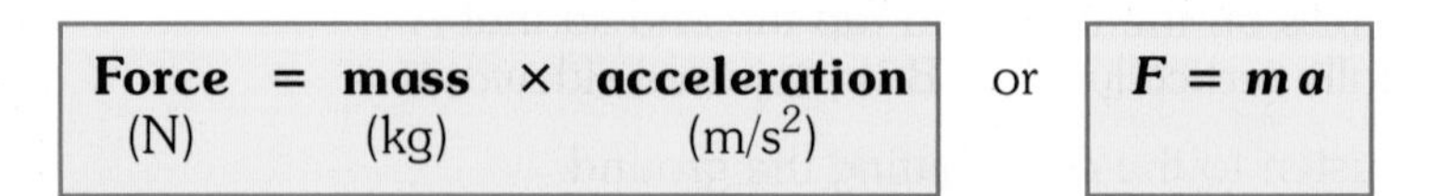

This is **Newton's Second Law of Motion**.

- F is the ***resultant*** (or unbalanced) force on the mass m.
- The force F is measured in newtons.

One newton (1 N) is defined as the force which gives to a mass of 1 kg, an acceleration of 1 m/s^2.

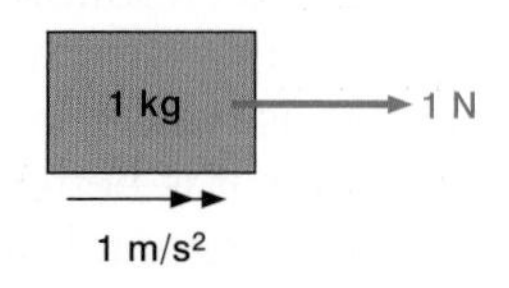

Example
When a force of 6 N is applied to a block of mass 2 kg, it moves along a table at constant velocity.
a) What is the force of friction?
When the force is increased to 10 N, what is:
b) the resultant force? c) the acceleration?
d) the velocity, if it accelerates from rest for 10 seconds?

a) When the block is moving at constant velocity (no acceleration), there is no resultant force on it (Newton's First Law, page 69). This means the forces are balanced and equal.
∴ The frictional force = 6 N

friction force — 2 kg — 6 N
constant velocity (*no acceleration*)

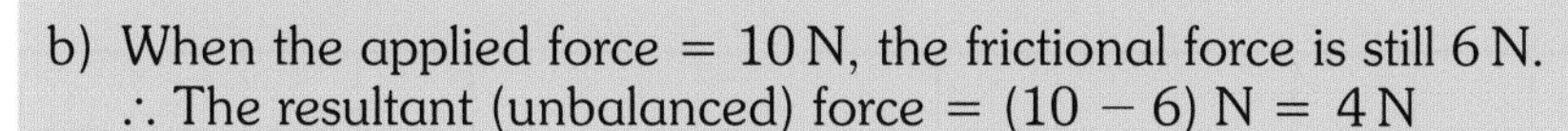

b) When the applied force = 10 N, the frictional force is still 6 N.
∴ The resultant (unbalanced) force = (10 − 6) N = 4 N

6 N — 2 kg — 10 N
2 kg — 4 N

c) Formula first: **Force** = **mass** × **acceleration**
Then numbers 4 N = 2 kg × acceleration

$$\therefore \text{acceleration} = \frac{4}{2} = 2\ \text{m/s}^2$$

d) Formula first: $$\textbf{acceleration} = \frac{\textbf{change in velocity}}{\textbf{time taken}}$$ (see p. 122)

Then numbers $$2\ \text{m/s}^2 = \frac{\text{velocity} - 0\ \text{(at rest)}}{10\ \text{s}}$$

$$\therefore \text{velocity} = 20\ \text{m/s}$$

▶ Gravitational field strength *g*

The Earth is surrounded by an invisible ***gravitational field***. This field exerts a force (called weight) on any mass which is in the field.

On the surface of the Earth, experiments show that this force is **9.8 N on every 1 kg** (see page 67).
This is the **gravitational field strength, *g* = 9.8 N/kg.**

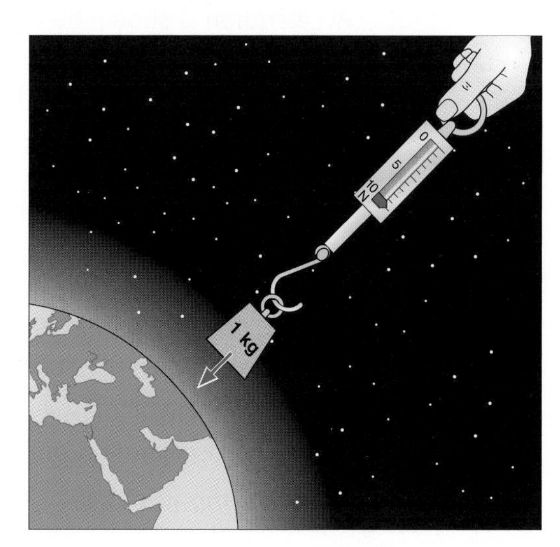

We now have 2 ways of thinking of ***g***:
1. For an object in free fall, the acceleration ***g*** = 9.8 m/s^2
2. For any object, moving or at rest, the gravitational field strength ***g*** = 9.8 N/kg.

   For ***m*** kg the force is ***m*** times as much,

   ∴ Weight = *m* × *g* newtons

We usually take ***g*** = **10 m/s^2** = **10 N/kg** (here on Earth).

On the Moon, ***g*** is only 1.6 N/kg. On Jupiter, ***g*** = 26 N/kg!

If your mass is 50 kg, what is your weight on Earth, and what would it be on the Moon, and on Jupiter?

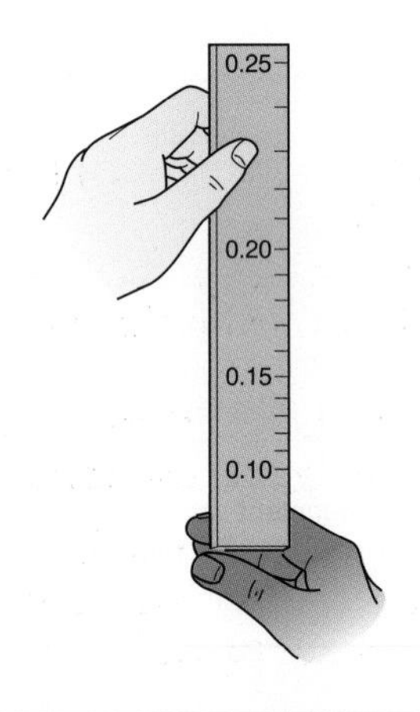

▶ Something to make – a reaction timer

You can test a friend's reaction time by holding the top of a 30 cm ruler so that it hangs vertically with your friend's fingers close to (but not touching) the zero mark at the bottom of the ruler. When you let go (with no warning), your friend has to grip the ruler as quickly as possible. The slower your friend's reaction time, the farther the ruler will fall as it accelerates downwards (at 9.8 m/s^2).

The table below shows how to mark the ruler (or a strip of card). It also shows how far you would travel in a car at 50 km/h (30 m.p.h.) (that is, how far you would travel ***before*** your foot could reach the brake pedal to ***start*** to slow down).

At this distance (in cm) from the zero mark:	4.9	5.9	7.1	8.3	9.6	11.0	12.5	14.2	15.9	17.7	19.6	21.6	23.7	25.9	28.2	30.6
Mark this time (in seconds) on your ruler:	0.10	0.11	0.12	0.13	0.14	0.15	0.16	0.17	0.18	0.19	0.20	0.21	0.22	0.23	0.24	0.25
Distance travelled (in metres) at 50 km/h (30 m.p.h.)	1.4	1.5	1.7	1.8	1.9	2.1	2.2	2.4	2.5	2.6	2.8	2.9	3.1	3.2	3.3	3.5

In a real driving situation, most people's reaction time is over ***three*** times longer than this! (See also page 83.)

Summary

Average speed (m/s) = **distance travelled** (metres) / **time taken** (seconds)

Acceleration = **change in velocity** (m/s) / **time taken for change** (s)

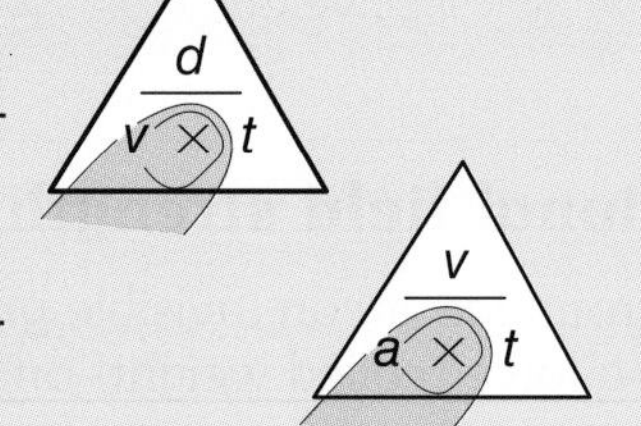

Velocity (a vector quantity) is speed in a particular direction.

Acceleration is shown by the ***slope*** of a velocity–time graph.
Distance is shown by the ***area*** under a velocity–time graph.

On a distance–time graph, the speed is shown by the ***slope***.

For constant acceleration: $\boldsymbol{v = u + at}$ $\quad \boldsymbol{s = \frac{(u + v)}{2}t}$

$\boldsymbol{s = ut + \frac{1}{2}at^2}$ $\quad \boldsymbol{v^2 = u^2 + 2as}$

The acceleration of free fall (due to gravity) = 9.8 m/s^2 (10 m/s^2)
The horizontal and vertical motions of a body are independent.

Newton's Second Law: **Force** (N) = **mass** (kg) × **acceleration** (m/s^2)

F
m × a

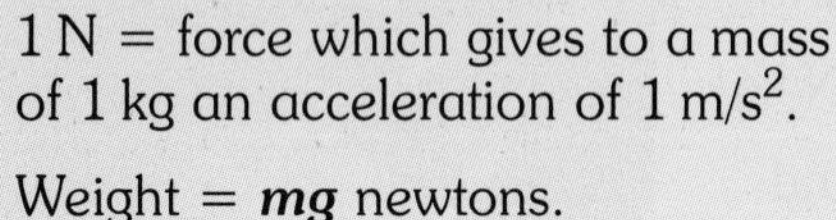

1 N = force which gives to a mass of 1 kg an acceleration of 1 m/s^2.

Weight = ***mg*** newtons.

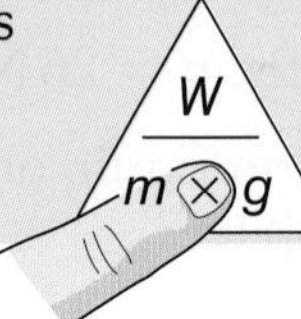

▶ Questions (Take $g = 10\,m/s^2 = 10\,N/kg$ where necessary)

1. a) A car travels 100 m in 5 seconds. What is its average speed?
b) A car accelerates from 5 m/s to 25 m/s in 10 seconds. What is its acceleration?

2. A speed-camera takes 2 photos 0.6 s apart while a car travels 12 m. What is its speed?

3. Sketch a speed–time graph of you walking to a bus-stop, catching a bus, getting off it, and walking to school.

4. The table gives some data for a Ferrari racing car at the start of a Grand Prix race:

Speed (m/s)	0	10	20	29	37	50	59	64	65	65
Time (s)	0	1	2	3	4	6	8	10	12	14

a) Plot a speed–time graph for the car.
b) What was its acceleration at:
i) 13 seconds? ii) 1 second?

5. Describe in as much detail as possible, the motion of a car which has this graph:

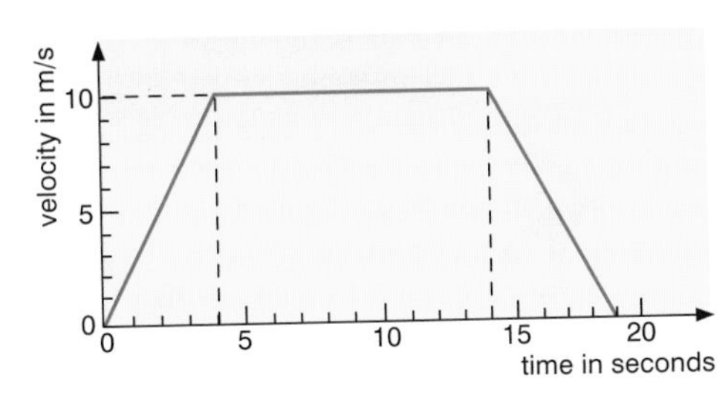

6. The Highway Code says that at 20 m/s, the reacting (thinking) distance of a driver is 14 m, and the brakes take 3 seconds to halt the car.
a) What is the reaction time in this case?
b) Plot a velocity–time graph for these data (assume the brakes are on steadily for 3 s).
c) Use the graph to estimate the speed at which the driver would hit a wall 3 seconds after first noticing it.

7. Sketch a distance–time graph for the car in question 5.

8. A sports car accelerates from rest at $4\,m/s^2$ for 10 seconds. Calculate the final velocity.

9. Explain why parachutists and snowflakes do ***not*** fall with a constantly accelerating motion.

10. A sky-diver weighing 500 N falls through the air at a steady speed of 50 m/s.
a) Draw a diagram of the forces on her.
b) What is the air resistance on her (in N)?
c) What is her mass?
d) Sketch a velocity–time graph to show what happens when her parachute opens.

11. The Eiffel Tower is 300 m high. A boy at the top slips on a banana skin and falls over the side. How long has he got to live?

12. A ball is thrown upwards with a velocity of 10 m/s. How high does it go?

13. A film stunt man dives off a cliff 20 m high, so that initially his ***vertical*** velocity is zero but his ***horizontal*** velocity is 5 m/s. Find
a) his vertical velocity as he hits the water
b) the time taken and c) the horizontal distance travelled during this time.

14. A car travelling within the town speed limit, at 13 m/s, hits a brick wall. For a passenger without a seat belt, this is like falling from the top of a house of height ***h*** metres. Find the value of ***h***. (Houses are about 7 m high.)

15. A mass of 4 kg is accelerated by a force of 20 N. What is the acceleration if there is no friction? If a frictional force of 8 N is acting, what is the acceleration?

16. A Saturn V Moon rocket has a mass at lift-off of 3.0×10^6 kg. The thrust at lift-off is 3.3×10^7 N. Find:
a) the weight of the rocket on Earth
b) the resultant (unbalanced) force at lift-off
c) the acceleration at lift-off
d) the apparent weight of the rocket in orbit.

17. A car-driver of mass 100 kg is in a crash. He decelerates, from 30 m/s to rest, in 2 seconds. Calculate:
a) his deceleration,
b) the force exerted by his seat and seat-belt.

18. An astronaut has a mass of 100 kg. What is his weight a) on Earth b) on the Moon where the gravitational field = 1.6 N/kg?

Further questions on page 144.

▶ Physics at work: Sport

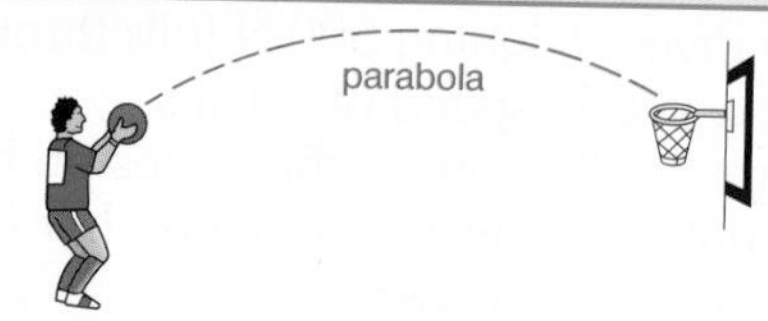

In many games, a ball is thrown or hit.
The ball does not travel in a straight line, but in a ***parabolic*** trajectory. (This is because gravity is pulling it downwards.)
To be more accurate, it is the centre of gravity of the ball that follows the parabola.

The centre of gravity (centre of mass) of a person is usually behind the navel, but it depends on the positions of the arms and legs.
When an acrobat or a diver jumps through the air, the centre of gravity moves in a smooth parabola, no matter how the person twists or spins.

The gymnast must keep her centre of mass directly over her hands, in order to balance.

In judo it's an advantage to have a low centre of mass, and stand with your legs apart, for stability.

Skiing is another sport where a low centre of mass is helpful.

Can you think of any others?

The long jump

Sprinting: The athlete accelerates because of the force *F* (the friction force, see also page 82).
His weight *W* is balanced by the reaction force *R*.
You can measure the acceleration of a runner by using ticker-tape (page 123) or a motion-sensor.

R W F

The take-off:
A final push to make *R* and *F* as large as possible.

W

R W F

The landing:
Now the forces *F* and *R* bring him to a stop.

His centre of gravity follows a parabola between take-off and landing.
The world record is about 9 m.

Using friction to turn corners

In many sports you have to turn corners.
This can only happen if there is an inwards force ***F*** provided by friction. This is a ***centripetal force***.

The faster you go, and the sharper the bend, the bigger the force ***F*** needs to be (and the bigger the angle you lean over). See also page 70.
What happens if the friction ***F*** is not big enough (for example, if the cyclist meets ice or gravel)?

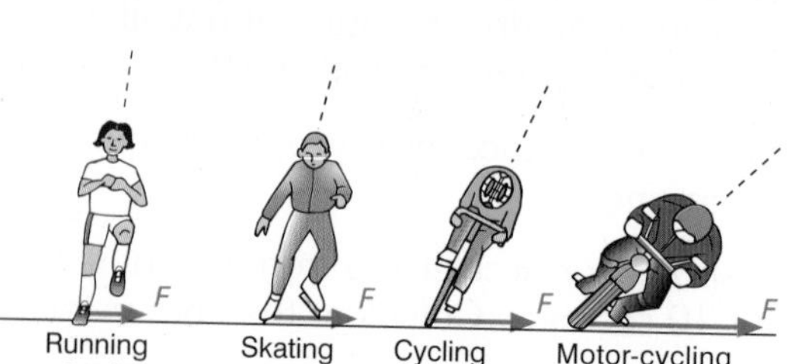

▶ Physics at work: Sport

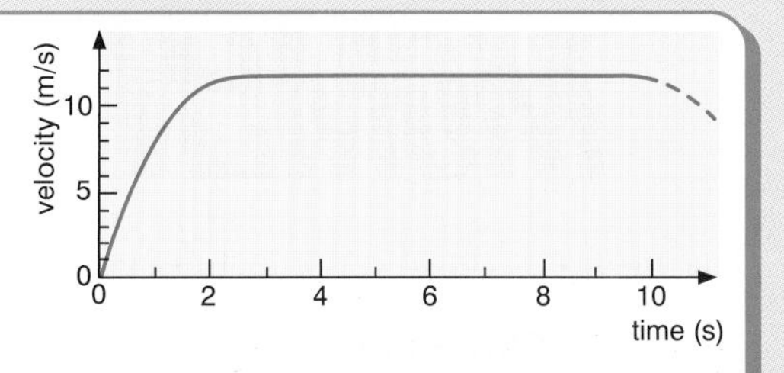

Here is a velocity–time graph for a runner in the Olympic 100-metre sprint:

He accelerates to almost 12 m/s in about 2 seconds, and then travels at almost constant speed.

If his mass is 60 kg, then:

his kinetic energy $= \frac{1}{2} \times$ mass $\times$ speed squared (p. 109)

$= \frac{1}{2} \times 60 \times 12 \times 12 =$ <u>about 4000 J</u>

$$\text{Power developed in his legs at start} = \frac{\text{energy transferred}}{\text{time taken}} = \frac{4000\text{ J}}{2\text{ s}} = \underline{2000\text{ W}}$$

In fact, to produce this much power in his legs, his whole body produces at least twice as much – more than 4 kW.
This is equivalent to four one-bar electric fires – he gets hot!

He loses the surplus energy by radiation from his skin (20%), convection (20%), evaporation of sweat from his skin (40%), and respiration (breathing out hot air, 20%).

What was his initial acceleration? Why did he stop accelerating?

Analysing movement

One way to analyse an athlete's movement is to take a video or cine film (see page 201), and then play it back slowly. Another way is to use flashing lights, like the 'strobe' lamps in a disco:

A camera is used in the dark, with a flashing lamp called a ***stroboscope***. In the photo, it was flashing 10 times in each second. How long was the shutter open? Where do his feet move fastest?

A large mass can be an advantage.

The Sumo wrestler needs a large mass and a low centre of mass so that he is not easily toppled.

When the shot-putter throws the shot, it is rather like firing a gun (page 85). The heavier she is, the less recoil, and the farther it goes.

There are examples of Physics in other sports: car-racing (page 94), snooker (p. 137), shooting (p. 138), cycling (p. 120), football (p. 136).

chapter 19

MOMENTUM MOMENTUM

The ***momentum*** of an object depends on its ***mass*** and its ***velocity***.
In fact:

Momentum = **mass** × **velocity** (kg) (m/s)	or	**Momentum** = mv

Momentum is a ***vector*** quantity. It is measured in units of **kg m/s**.
A bicycle of mass 10 kg moving at 5 m/s has a momentum of 50 kg m/s.

Consider a force F acting on a mass m for a time t so that it accelerates from velocity u to velocity v.

From page 122, acceleration $a = \dfrac{v-u}{t}$

∴ Newton's Second Law (page 130) is: $F = ma = m\left(\dfrac{v-u}{t}\right) = \dfrac{mv-mu}{t}$

or in words:

$$\text{Force (N)} = \frac{\text{change in momentum (kg m/s)}}{\text{time taken for the change (s)}} = \frac{\text{momentum after} - \text{momentum before}}{\text{time taken for the change}}$$

or, multiplying both sides by time, we get: **Force × time = change in momentum**

Example 1
Consider first a boy kicking a **stone** of mass 1 kg and accelerating it from rest to 10 m/s. Because the stone is rigid, the force of his foot acts for only $\frac{1}{100}$ second. Calculate this force.

Formula first: **Force** = $\dfrac{\textbf{momentum after} - \textbf{momentum before}}{\textbf{time taken}}$

Then numbers: $= \dfrac{(1 \times 10) - (1 \times 0)}{\frac{1}{100}} = \dfrac{10}{\frac{1}{100}}$

$= \underline{1000\text{ N}}$ (painful!)

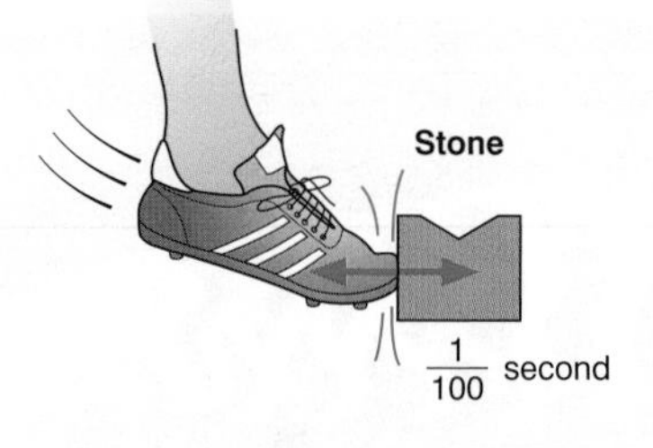

Example 2
Now consider the same boy kicking a **football** of the same mass (1 kg) to give it the same speed (10 m/s) and therefore the same momentum (10 kg m/s).
Because the ball is soft and he follows through with his foot, this force is applied for $\frac{1}{10}$ second (10 times longer than before).

Using the same formula with time = $\frac{1}{10}$, gives force = $\underline{100\text{ N}}$
Which kick hurts less? Why?

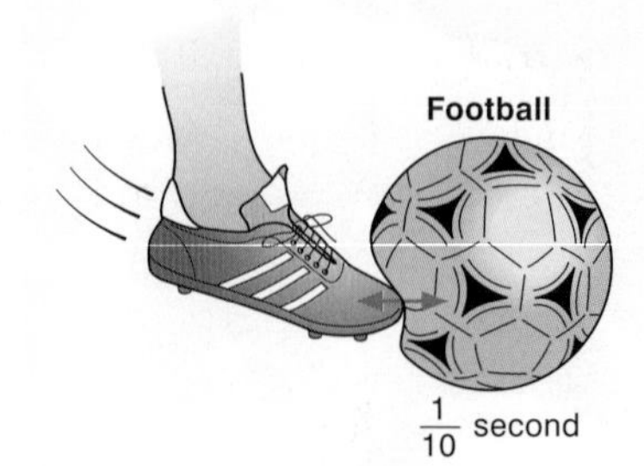

The ***longer*** the time of a collision, the ***smaller*** the force.

▶ Collisions

Here are two balls rolling towards each other so that they collide:

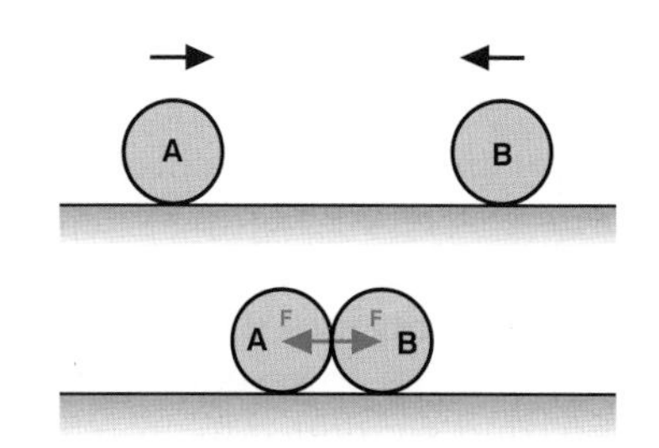

When they collide they exert a force F on each other for a short time t and so the momentum of each ball changes.
From Newton's Third Law (page 84), the forces are equal and opposite. Therefore (from the equation on the opposite page) the changes in momentum are equal and opposite.

That is, the momentum ***gained*** by one ball is equal to the momentum ***lost*** by the other ball.

This is the **Principle of Conservation of Momentum**:
When two or more bodies act on each other, their total momentum remains constant, providing there is no external force acting.

That is:

total momentum before collision	=	**total momentum after collision**

The same is true for an explosion (as an explosion is the opposite of a collision).

Example 3
A bullet, mass 10 g (0.01 kg), is fired into a block of wood, mass 390 g (0.39 kg), lying on a smooth surface. The wood then moves at a velocity of 10 m/s. a) What was the velocity of the bullet?
b) What is the kinetic energy before and after the collision?

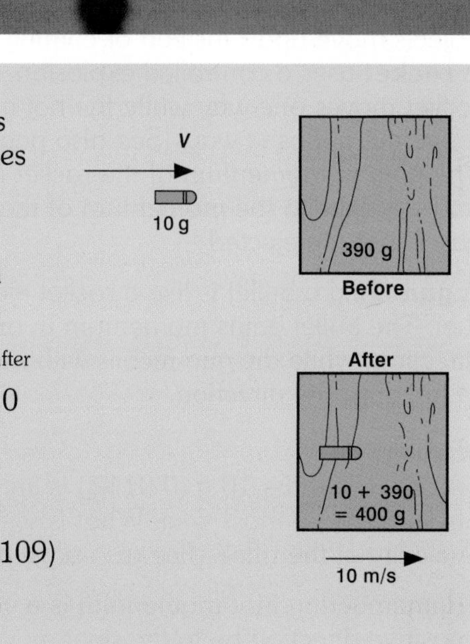

a) Formula first: **total momentum before collision** = **total momentum after collison**

$$(\text{mass} \times \text{velocity})_{\text{before}} = (\text{mass} \times \text{velocity})_{\text{after}}$$

Then numbers:

$$0.01 \times V = (0.01 + 0.39) \times 10$$
$$0.01 \times V = 0.4 \times 10$$
$$V = \underline{400\ \text{m/s}}$$

b) Total KE ***before*** collision $= \frac{1}{2} \times \text{mass} \times \text{speed}^2$ (page 109)

$$= \frac{1}{2} \times 0.01 \times 400^2 = \underline{800\ \text{J}}$$

Total KE ***after*** collision $= \frac{1}{2} \times \text{mass} \times \text{speed}^2$

$$= \frac{1}{2} \times 0.4 \times 10^2 = \underline{20\ \text{J}}$$

The difference, 780 J, is transferred to heat energy and sound energy during the collision.

Note: **momentum is *conserved* in a collision, but kinetic energy is not.**

More examples of momentum in collisions

In any collision, a force is exerted for a length of time. The examples on page 136 showed that the ***longer the time*** of a collision, the ***smaller the force*** exerted.
This idea is used to design safety into **cars**. The front and back of a car are designed to crumple, in order to spread out the time of a collision, and so reduce the force on you. (See page 69.) An **air bag** also spreads out the time. A **seat-belt** is designed to stretch slightly, to spread out the time of the crash even further, and so reduce the force on you to a safe level:
A motor-cyclist's **safety helmet** is padded inside so as to extend the time of any collision.

In an opposite way, a **hammer** is designed from hard metal. This is so that the time of the collision is as small as possible, and so the force of the hammer on a nail is as large as possible.

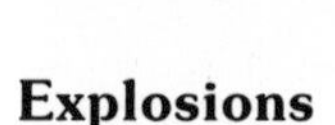

Explosions

An explosion is the opposite of a collision – objects move apart instead of coming together.
A **rocket** uses a controlled explosion. The rocket moves one way while the hot gases move the opposite way. (See also page 85.)
The gain of momentum of the rocket is equal and opposite to the momentum of the hot gases that are ejected:

A **gun** firing a bullet is like a rocket ejecting fuel. The bullet gains momentum in one direction, while the gun recoils with momentum in the opposite direction.

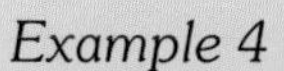

Example 4
A bullet of mass 10 g (0.01 kg) is fired at 400 m/s from a rifle of mass 4 kg. What is v, the recoil velocity of the rifle? (See also page 85.)

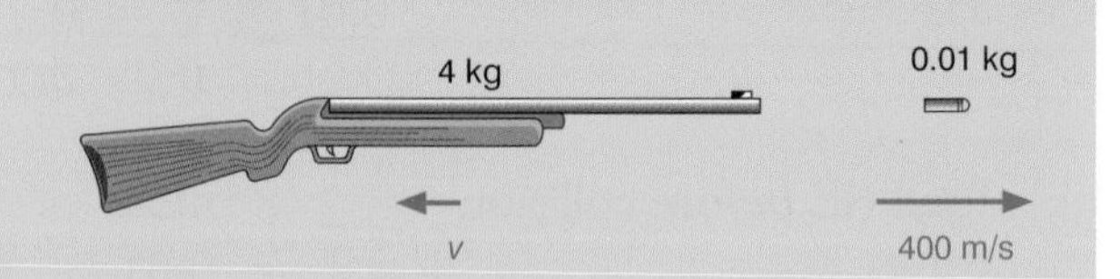

Remembering that momentum is a vector, let the positive direction be to the right.

Formula first:	**total momentum before explosion**	=	**total momentum after explosion**
Then numbers:	0	=	$(0.01 \times 400) - (4 \times v)$
	$\therefore 4v$	=	0.01×400
	$\therefore v$	=	$\underline{1\ \text{m/s}}$

Summary

Momentum = **mass** × **velocity.** It is a vector quantity.
(kg m/s) (kg) (m/s)

Newton's Second Law: **Force** = $\dfrac{\textbf{change in momentum}}{\textbf{time taken}}$ = **rate of change of momentum**

So: **Force × time** = **change in momentum** = $mv - mu$
(Compare: Force × ***distance*** = change in KE = $\frac{1}{2}mv^2 - \frac{1}{2}mu^2$ from page 109)

Principle of Conservation of Momentum: **total momentum before collision (or explosion)** = **total momentum after collision (or explosion)**
(if no external force is acting)

▶ Questions

1. Copy and complete:
a) Momentum = ... × . . . . .
b) It is a . . . . quantity.
c) Its unit is . . . . .
d) Newton's Second Law can be written as Force = mass × . . . . or Force = ... of change of . . . . .
e) The Principle of Conservation of Momentum states:
f) . . . . is conserved in any collision but kinetic . . . . is not.

2. A mass of 3 kg is moving with a velocity of 2 m/s. What is its momentum?

3. a) Explain the Physics of this limerick:

A dashing young footballer, Paul,
Scored a goal with a one-kilo ball,
But a similar kick
To a one-kilo brick
Made Paul bawl, and then fall, and then crawl.

b) When you jump down from a table, why do you bend your legs rather than keep them rigid? Why do paratroopers roll on landing? Why would you prefer to fall on to a bed than on to concrete?
c) Why are seat-belts designed to stretch in a collision? Why is the front of a car designed to collapse in a serious collision?
d) Why does a cricket batsman, a tennis player or a golf player 'follow through'?
e) Professor Messer invents a foam-rubber hammer ('to make less noise'). Will it work? Why?

4. A truck of mass 2 kg travels at 8 m/s towards a stationary truck of mass 6 kg. After colliding, the trucks link and move off together.
a) What is their common velocity?
b) What is the KE before and after the collision?
c) Explain the apparent loss in energy.

5. In a sea battle, a cannonball of mass 30 kg was fired at 200 m/s from a cannon of mass 3000 kg. What was the recoil velocity of the gun?

6. A heavy car A, of mass 2000 kg, travelling at 10 m/s, has a head-on collision with a sports car B, of mass 500 kg. If both cars stop dead on colliding, what was the velocity of B?

7. A man wearing a bullet-proof vest stands still on roller skates. The total mass is 80 kg. A bullet of mass 20 grams is fired at 400 m/s. It is stopped by the vest and falls to the ground. What is then the velocity of the man? How does this compare with what you see in TV films?

8. A Saturn V Moon rocket burns fuel at the rate of 13 000 kg in each second. The exhaust gases rush out at 2500 m/s.
a) What is the change in momentum of the fuel in each second?
b) What is the thrust? (See also question 16, page 133.)

Further questions on page 141.

Hooke's Law

1. Some students carry out an experiment to find out how a spring stretched when small loads were added to it. The results are shown in the table. One reading is incorrect.

Load (N)	0	2	4	6	8	10	12	14
Extension (mm)	0	16	32	58	64	80	96	112

a) Use these results to plot a graph. [3 marks]
b) Use your graph to find:
i) the extension when the load is 3 N
ii) the load for an extension of 40 mm. [2]
c) Label the incorrect point on the graph with the letter 'E'. [1]

Moments

2. A person pushes a door open by applying a force of 10 N at a point 0.8 m from the hinge of the door. The turning effect or moment of this force, in N m, is:
A 0.08 **B** 8.0 **C** 10.8 **D** 12.5 **E** 80.0
(NI)

3. A uniform metre rule of mass 100 g balances at the 40-cm mark when a mass ***X*** is placed at the 10-cm mark. What is the value of ***X***?

4. The diagram shows a simple machine for lifting water from a river.

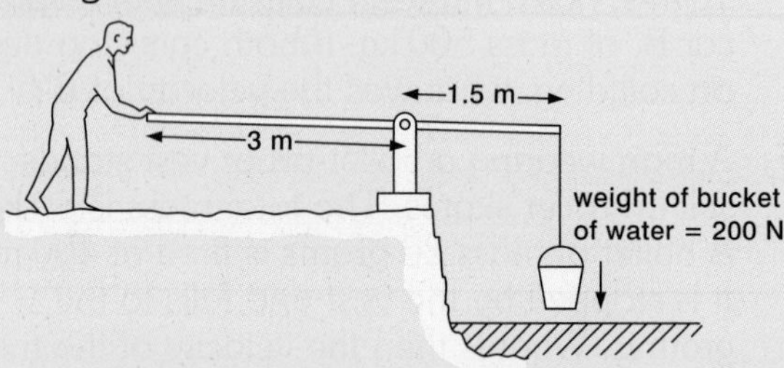

a) Calculate the turning force (moment) of the bucket of water. [2]
b) Calculate the downwards force the operator must use to balance the moment of the bucket of water. [4] (AQA)

5. A spanner was used to undo a nut. A force of 20 newtons (N) was applied at a distance of 20 centimetres (cm) from the nut.
Calculate the moment of the force being used to undo the nut. [3] (AQA)

Pressure

6. The diagram shows side and front views of a car tyre in contact with the road.

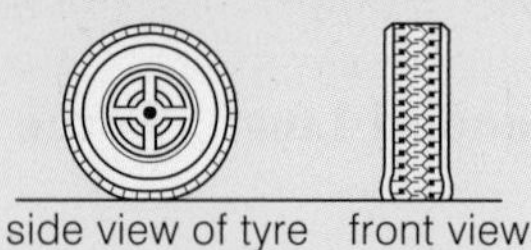

A tyre company stated that the area of the tyre in contact with the road was about the same as the area of the sole of one of your shoes.
a) Describe (with a diagram) how you would estimate the area of sole of your shoe. [3]
b) The car weighs 12 000 N. What is the force acting on one tyre if the weight of the car is evenly distributed amongst the tyres? [1]
c) If the area of contact of the tyre is 80 cm^2 ($0.008\ m^2$), calculate the pressure of the air in the tyre. [2]

7.

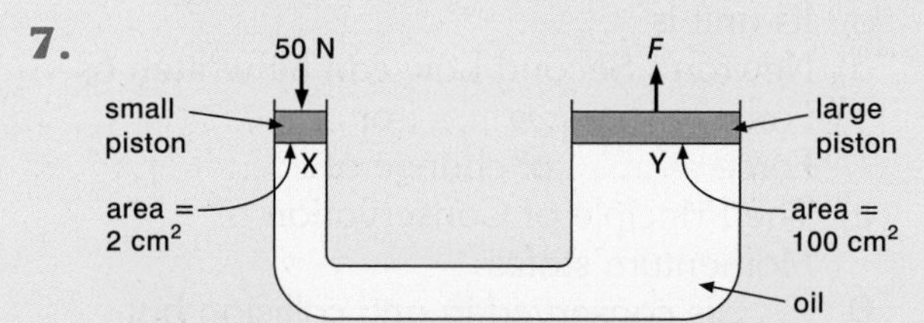

The diagram above shows the principle of the hydraulic car jack.
a) If a force of 50 N is applied to the small piston, calculate the pressure produced in the oil at X. [2]
b) What is the pressure exerted by the oil at Y? [1]
c) Calculate the upward force, *F*, acting on the large piston. [1]

Forces

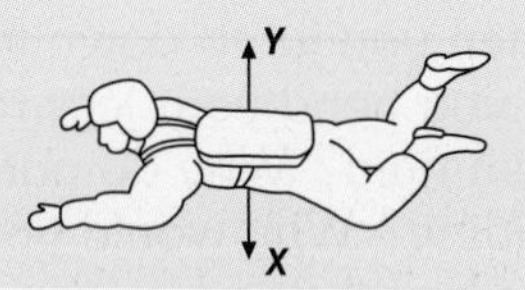

8. A skydiver is falling from an aeroplane.
a) Name i) force *X*, ii) force *Y*. [2]
b) State how **each** force changes as the sky diver speeds up. [2]
c) Why does the skydiver reach a steady speed (terminal velocity)? [2]
d) Describe and explain what happens when the sky diver opens the parachute. [3] (WJEC)

 Note: although a question may be marked (AQA) etc, it will often be relevant to other examinations as well.

9. Susan is riding her motorbike at 100 km/h around a bend, which is part of a circle.
a) Although her speed stays the same she is accelerating. Explain why. [2]
b) Draw a diagram to show the horizontal forces on Susan and the motorbike. [2]

10. The table gives data about a falling skydiver.

Speed (m/s)	Weight (N)	Air resistance (N)	Resultant force (N)
0	700	0	700
10		280	420
20		490	210
30		630	70
40	700		0

a) Copy and complete the table. [2]
b) Use the data to draw a graph of resultant force against speed. [3]
c) What is meant by ***terminal velocity***? [1]
d) State the terminal velocity of this skydiver. [1]

11. The table shows the annual risk of being in a road accident where somebody is injured.

Country	Annual risk per 100,000 people	Annual risk per person (%)	Annual risk per car (%)
Ireland	150	0.150	0.009
UK	385	0.385	0.026
USA	675	0.675	0.023

a) Which of these countries appears to have the safest roads? [1]
b) Explain why all the figures in the first column are larger than all the figures in the second column. [1]
c) The risk per car in the USA and the UK is about the same but the risk per person is much higher for the USA. Suggest reasons for this. [2]
d) Despite the risks, people still use the roads. Suggest reasons for this. [2]

12. Andrew is designing a lift to carry people in an office block. He knows that the tension in the cable depends on
- the weight of the lift and the people in it,
- the forces from the lift motor and the brakes.

He uses a computer to calculate the cable tension in hundreds of different situations.
a) Explain why it is important for Andrew to know the cable tension. [2]
b) Explain why Andrew uses a computer instead of doing lots of experiments. [2]

▶ Momentum

13. A trolley having a mass of 2 kg and velocity 4 m/s collides with another trolley having a mass of 1 kg which is at rest. If the trolleys stick together after the impact, calculate:
a) the total momentum before the collision,
b) the common velocity after the collision,
c) the kinetic energy before the collision,
d) the kinetic energy after the collision.
Why are the answers to (c) and (d) different?

14. A golfer strikes a stationary golf ball with his golf club. The mass of the ball is 0.045 kg.
a) What is the momentum of the ball before it is struck? [2]
b) When the club strikes the ball it is in contact for 0.001 s and exerts a force of 3600 N on the ball. Calculate the velocity at which the ball leaves the club. [3] (AQA)

15. Use the idea of momentum to explain why 'crumple zones' should reduce passenger injury in a car crash. [3] (AQA)

▶ Energy sources (see also page 60)

16. a) Choose an example of a renewable energy source and suggest what you would say to persuade people to support the development of this energy source. [2]
b) Many people might disagree with your ideas. Suggest what arguments they might use. [2] (OCR)

17. a) Give one reason why solar power would not be a suitable replacement for power stations burning fossil fuels. [1]
b) Explain why some people believe that nuclear power would not be an acceptable replacement. [3] (WJEC)

18. A village uses solar panels and windmills as alternative energy sources.
a) The panels and windmills are expensive to install. Eventually all the money spent on them will be recovered. Explain why. [2]
b) The energy of the air hitting the blades of a windmill is 20 000 J each second. The energy transferred to the power lines is 5000 J each second. Calculate the efficiency of the windmill. [2] (OCR)

▶ Energy (see also page 60)

19. a) For each of these 4 appliances state the main energy transfers which take place in them:
Battery *Lamp*
Loudspeaker *Microphone* [4]

b) When a driver applies the brakes of a car, its kinetic energy is transferred to heat energy.
i) Where is this heat energy produced? [1]
ii) Why does frequent speeding up and slowing down increase the amount of petrol used by a car? [3]
iii) Explain why petrol is used up even when a car is travelling at a constant speed. [1]
iv) Describe how cars can be designed to reduce the amount of petrol used when travelling at a steady speed. [2]

c) The primary source of energy for most cars is crude oil which is a fossil fuel.
i) Give **one** other example of a fossil fuel. [1]
ii) Explain why car designers are concerned with fuel economy when designing a new car. [3] (Edex)

20. A coal-fired power station generates electrical power for the national grid. Explain how the energy from the coal is used to generate electricity. [3] (WJEC)

21. a) i) State **two** environmental problems caused by burning coal to generate electricity. [2]
ii) How may these environmental problems be reduced? [2]

b) Some data for Didcot coal-fired power station are given below.

Number of generators	4
Maximum continuous power rating of a generator	500 MW
Energy content of coal used	2.66×10^{10} J per tonne
Total quantity of coal used per day	18 289 tonnes

Use the given data to calculate:
i) the total electrical energy output per day [3]
ii) the total input of coal energy per day [2]
iii) the efficiency of the power station. [3]

c) Energy is conserved. Explain what happens to all wasted energy during energy transfers. [2] (AQA)

22. The diagram shows what happens to each 100 joules of energy from crude oil when it is used as petrol in a car.

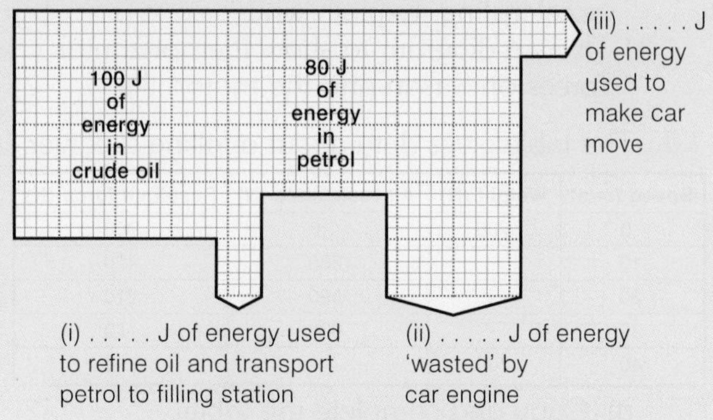

a) What are the missing values (i), (ii), (iii)? [3]
b) Calculate how efficient the car engine is at transferring the energy **from petrol** into useful movement. [2]
c) Two students are discussing the diagram. The first says that **none** of the energy released from the crude oil is really lost. The other says that **all** of the energy released from the crude oil is really lost.
What do you think? Explain your answer as fully as you can. [4] (AQA)

23. The diagram shows a coal-fired power station.

a) Explain where the energy in coal originally came from. [2]
b) Calculate how efficient the power station is at changing the energy in the coal into electrical energy. [3]
c) In what form is most energy lost from this power station? [1]
d) What eventually happens to this 'lost' energy? [1] (AQA)

24. Energy cannot be created and it cannot be destroyed, but energy can be wasted. Discuss this statement. [7] (AQA)

25. Give **three** advantages of a hydro-electric power station over a coal-fired power station. [3] (AQA)

▶ Energy, work and power

26. A worker on a building site raises a bucket full of cement at a slow steady speed, using a pulley as shown:

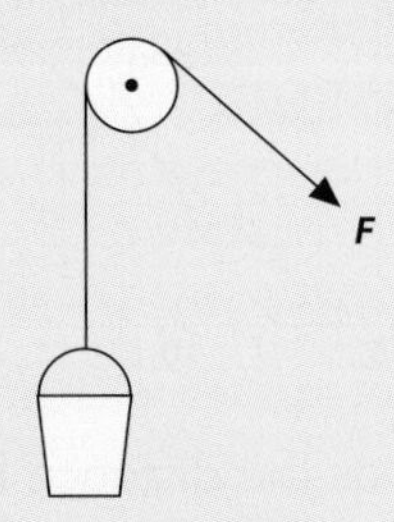

The weight of the bucket and cement is 200 N. The force ***F*** exerted by the worker is 210 N.

a) Why is ***F*** bigger than the weight of the bucket and cement? [1]

The bucket is raised through a height of 4 m.

b) Through what distance does the worker pull the rope? [1]

c) How much work is done on the bucket and cement? [3]

d) What kind of energy is gained by the bucket? [1]

e) How much work is done by the worker? [2]

f) Where does the energy used by the worker come from? [1] (AQA)

27. A car is travelling at 30 m/s on a level road. At this speed the car has to overcome a total force opposing motion of 600 N.

a) How far does the car go in 10 seconds? [1]

b) How much work does the car engine do in this time? [2]

c) What is the power? [2]

d) Burning the petrol supplies 60 kJ (60 000 J) of energy to the engine every second. Why is your answer to (c) smaller than this value? [2]

e) What is the efficiency? [2] (OCR)

28. A car is travelling at a steady speed of 15 m/s.

a) Calculate the distance moved in 10 s. [1]

b) The total resisting force (friction and air resistance) is 800 N. Calculate the work done by the car in 10 s in overcoming this resisting force. [4]

c) Calculate the power developed at the driving wheels. [4] (AQA)

29. In a downhill race a man skis a total distance of 3000 m. The vertical height the man drops is 800 m.

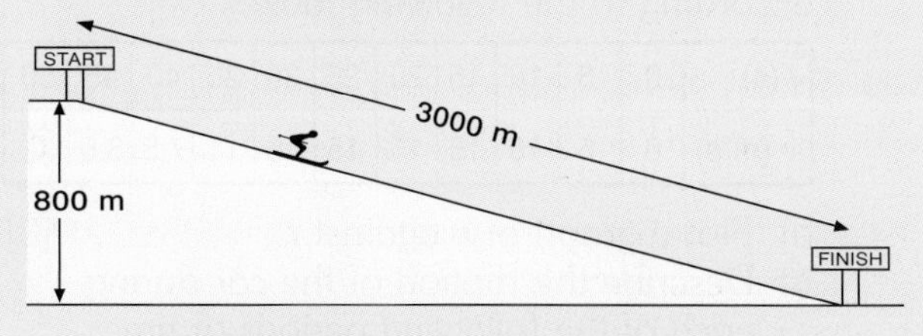

a) The man and his ski equipment have a total mass of 90 kg. What is their weight? [2]

b) Copy the diagram and draw and label three forces acting on the man as he skis down the slope. [3]

c) The average frictional force pushing against the man is 210 N. Calculate the total work done against friction as he skis down the slope. [2]

d) Calculate the loss in gravitational potential energy of the skier during the race. [3]

e) Calculate the kinetic energy of the skier as he passes the finish. (Ignore any work done by the skier himself.) [1]

f) Calculate the speed of the skier at the finish. [3] (OCR)

30. A motor is used on a building site to raise a block of stone. The weight of the block is 720 N and it is raised 20 m in 24 s. Calculate:

a) the work done on the block, [3]

b) the useful power supplied by the motor. [3]

31. A boy of mass 50 kg races up a flight of 40 steps, each of height 15 cm, in 5 s. Calculate:

a) the work done [3]

b) his average power. [3]

c) Why is the energy he transfers greater than the calculated work done? [2] (WJEC)

32. An electric pump whose efficiency is 60% raises water to a height of 15 m. If water is delivered at the rate of 360 kg per minute, what is the power rating of the pump? [5] What is the energy lost by the pump? [1] (WJEC)

▶ Velocity and acceleration

33. The velocity, v, of a car varies with time, t, according to the following table:

t (s)	0	5	10	15	20	25	30	35	40	45	50
v (m/s)	0	5	10	15	15	15	15	11	7.5	3.5	0

a) Plot a graph of v against t. [3]
b) Describe the motion of the car during each of the following periods of time:
0–15 s 15–30 s 30–50 s. [3]
c) Calculate the acceleration of the car during the period 0–15 s. [3]
d) How far did the car travel in the following periods of time?
0–15 s 15–30 s 0–30 s [7]
e) State what forces would be acting on the car when it is travelling at constant velocity. What is the resultant of these forces equal to? [4]

34. When a car driver has to react and apply the brakes quickly, the car travels some distance before stopping.
Part of this distance is called the 'thinking distance'. This is how far the car travels while the driver reacts to a dangerous situation.
The table below shows the thinking distance (m) for various speeds (km/h).

Thinking distance (m)	0	9	12	15
Speed (km/h)	0	48	64	80

a) On graph paper, draw a graph of thinking distance against speed. [3]
b) Describe how thinking distance changes with speed. [1]
c) A driver drank two pints of lager. Some time later the thinking time of the driver was measured as 1.0 second.
i) Calculate the thinking distance for this driver when driving at 9 m/s. [1]
ii) A speed of 9 m/s is the same as 32 km/h. Use your graph to find the thinking distance at 32 km/h for a driver who has not had a drink. [1]
iii) What has been the effect of the drink on the thinking distance of the driver? [1] (AQA)

35. A man runs a race against a dog. Here is a graph showing how they moved:

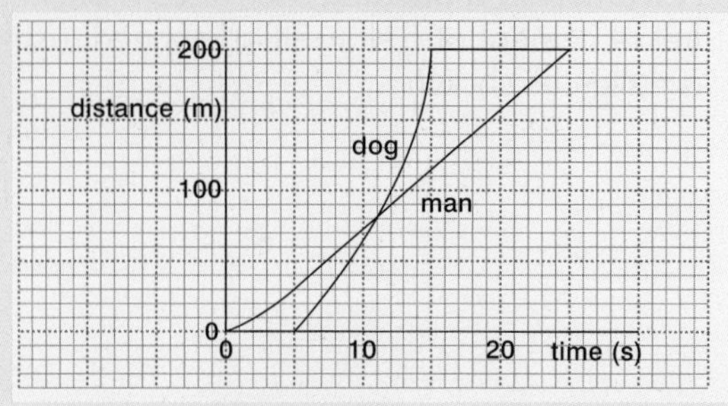

a) What was the distance for the race?
b) After how many seconds did the dog overtake the man?
c) How far from the start did the dog overtake the man?
d) What was the dog's time for the race?
e) Use the equation $v = \frac{d}{t}$ to calculate the average speed of the man.
f) After 8 seconds, is the speed of the man increasing, decreasing, staying the same?
g) What is the dog's speed after 18 s? (AQA)

36. An underground train takes 1 minute to travel from one station to the next.
The train accelerates from rest to a speed of 25 m/s in 20 s, travels at this constant speed for 30 s before coming to rest under a uniform braking force.
a) Draw a graph to represent the motion. [3]
b) From the graph, calculate:
i) the time that the train is travelling with a speed greater than 20 m/s,
ii) the distance travelled by the train when it is moving at constant speed,
iii) the retardation of the train,
iv) the braking force applied to stop the train, given that the mass of the train is 100 000 kg. [6] (WJEC)

37. A roller-coaster carriage is released at the top of a steep slope. Then it
- accelerates to the bottom of the slope
- is slowed by a water splash
- continues at constant speed
- is stopped by brakes at the end of the ride.

a) Sketch a graph to show how the speed of the roller-coaster carriage changes during this time. [4]
b) Describe the energy changes involved. [4]

▶ Force and acceleration

38. A car of mass 900 kg is travelling at 30 m/s. A small child runs in front. The driver's reaction time is 0.6 s and it takes a further 2 s to brake the car to a halt. Calculate:

a) the 'thinking distance' travelled by the car during the reaction time, [2]
b) the braking deceleration, [3]
c) the average braking force, [3]
d) the change in kinetic energy, [3]
e) the work done by the brakes. [1]
f) Draw a speed : time graph for the car. [2]
g) Explain how the graph will change if:
   i) the driver drinks alcohol,
   ii) the road is wet. [2]

39. Ian has a mass of 70 kg. He dives from a high diving board. His vertical velocity at different times is shown in the graph below. Gravitational field strength is 10 N/kg.

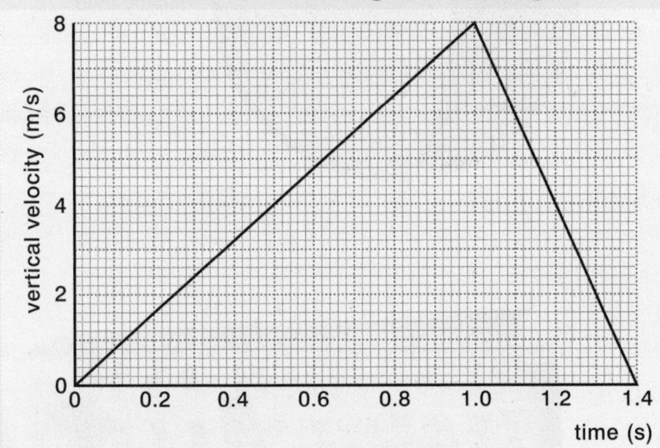

a) From the graph calculate
   i) the time that he took to reach the water,
   ii) the height of the diving board,
   iii) Ian's deceleration in the water,
   iv) the retarding force on Ian in the water,
   v) the depth in the water that Ian reached. [13]

b) i) How far did Ian travel in the first 0.5 s of the dive?
   ii) What is the loss of Ian's potential energy after he has fallen this distance?
   iii) Use the graph to calculate Ian's kinetic energy 0.5 s after he has started to dive.
   iv) Give an explanation for the difference between the answers to part (ii) and (iii). [10] (AQA)

40.

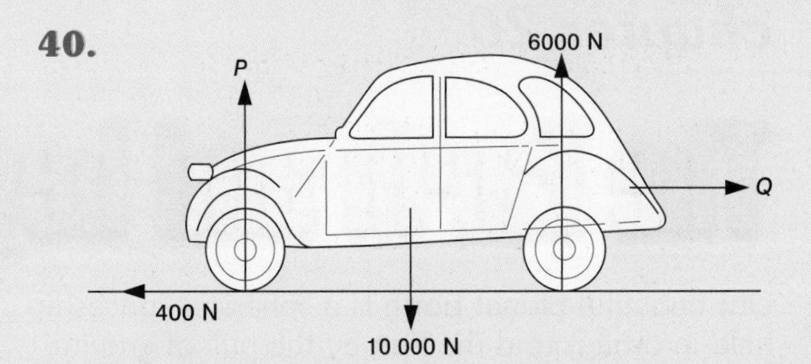

A front-wheel drive car is travelling at **constant velocity**. The forces acting on the car are shown in the diagram above. $\boldsymbol{Q}$ is the force of the air on the moving car. $\boldsymbol{P}$ is the total upward force on both front wheels.

a) Explain why i) $\boldsymbol{P} = 4000$ N,
   ii) $\boldsymbol{Q} = 400$ N. [2]
b) Calculate the mass of the car. [1]
c) The 400 N driving force to the left is suddenly doubled.
   i) Calculate the resultant force driving the car forward.
   ii) Calculate the acceleration of the car.
   iii) Draw a sketch graph showing how the velocity of the car changes with time. (Start your graph just before the driving force is doubled.) [5]
d) i) Passengers in a car are advised to wear a safety belt. Explain, in terms of Newton's Laws, how a safety belt can reduce injuries. [2]
   ii) What other design feature in a car can offer protection in a crash? [4] (WJEC)

41. A car is travelling at 20 m/s. The driver brakes constantly until the car stops. This takes 2 s, once the brakes come on.

a) Calculate the deceleration of the car. [3]
b) The car and driver have a mass of 1000 kg. Calculate the braking force on the car. [3] (OCR)

42. A mass of 5 kg changes velocity steadily from 4 m/s to 24 m/s in 10 s. Calculate:

a) the average velocity, [3]
b) the distance travelled in 10 s, [3]
c) the acceleration, [3]
d) the force needed for this acceleration, [3]
e) the momentum after 10 s, [3]
f) the kinetic energy at the start, [3]
g) the kinetic energy after 2 s. [3]

chapter 20

The Earth and beyond

Our beautiful planet Earth is a spherical spaceship held in orbit round the Sun by the pull of gravity. The atmosphere and everything on Earth is held to it by gravity. This gravitational force also crushes the Earth itself so that the rocks inside are very dense.

Inside the Earth there are 3 main layers, as shown:

- The **core** is very hot metal, mostly iron and nickel. Most of it is liquid, but the centre is under such high pressure that it is solid. The Earth's magnetic field is probably due to currents in the core. The core is hot because it is radioactive (see page 338).
- The **mantle** is mostly solid rock but some of it is 'slushy' – a mixture of solid and hot molten (liquid) rock, rather like lava from a volcano. The mantle can move very slowly in giant convection currents.
- The **crust** is a very thin solid layer, like the shell on an egg.

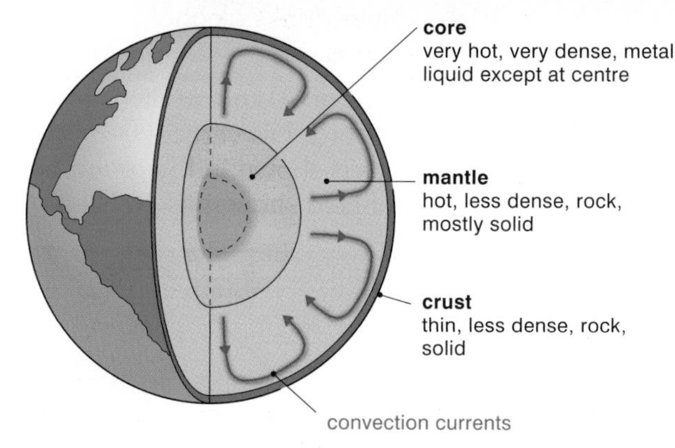

How do we know this? We make deductions by observing the effects of ***earthquakes***.
Whenever there is an earthquake, shock waves called **seismic waves** travel into the Earth.

There are 3 kinds of seismic wave: **P**, **S** and **L**.
They travel through the Earth in different ways.

The **L** waves are waves that travel along the Earth's crust. These are the waves that damage buildings:

The **P** (**p**rimary) waves:

- These are ***longitudinal*** waves, like sound waves. Think of them as '**p**ush-and-**p**ull' waves that you can make on a 'slinky' spring (see also page 166, bottom):
- They travel quickly (30 times as fast as sound in air).
- They can travel through liquids as well as solid rock.

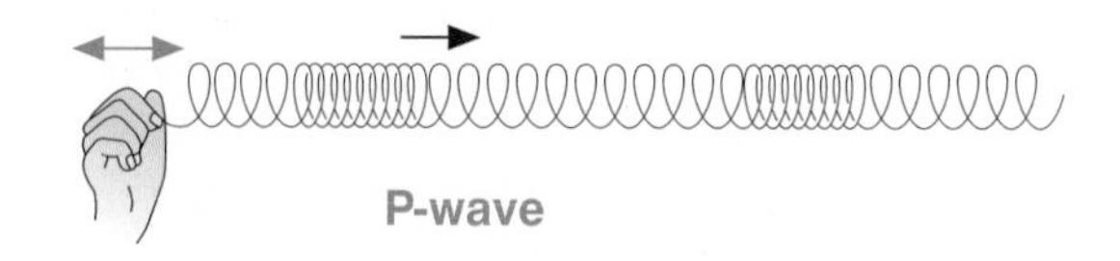

The **S** (**s**econdary) waves:

- These are ***transverse*** waves. Think of them as '**s**hake-**s**ideways' waves, like the transverse wave you can make on a rope or a 'slinky' spring (see page 166, top):
- They travel more **s**lowly than the P-waves.
- They can travel through **s**olid rock but not through liquids, because they are transverse waves.

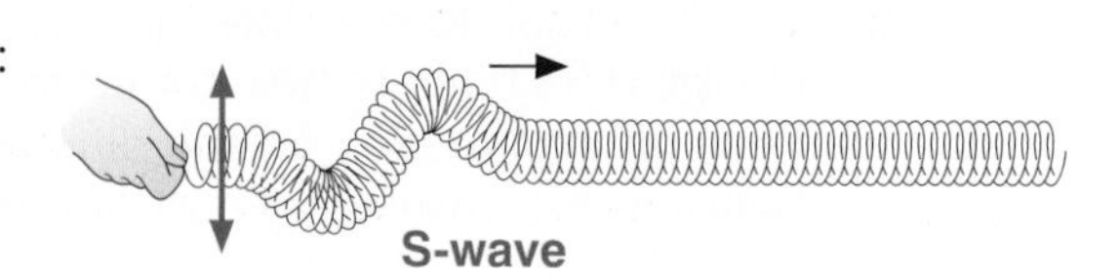

Travelling through the Earth

The P-waves and the S-waves both travel through the Earth, but they behave differently:

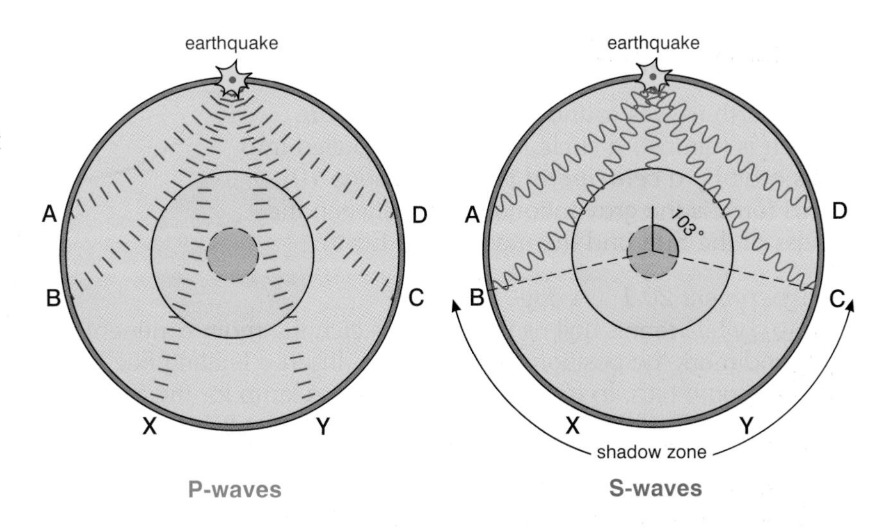

Because the core is liquid, the S-waves cannot travel through it. This creates a ***shadow-zone***:

Seismograph stations all over the world detect the waves. In the diagrams, stations A, B, C and D all receive S-waves as well as P-waves. But stations X and Y only receive P-waves. From this we know that the core is liquid, and its size can be calculated.

The 2 diagrams can be combined into one:

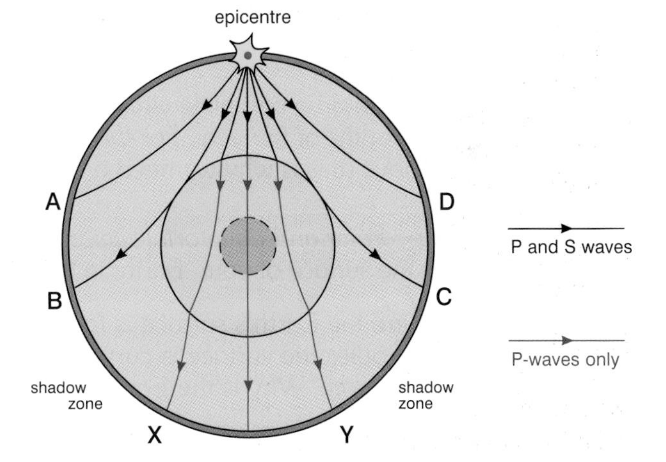

Notice that the paths are curved.
This happens because of ***refraction***.
Both the waves travel faster if the rock is denser. The deeper they go into the Earth, the denser the rock and so they speed up. Because they are changing speed, the waves are ***refracted*** (see page 168 and page 184).
If the density of the rock changes gradually, the waves are bent in curves. If the density changes suddenly, the waves are bent sharply.

Example
An earthquake occurs 9000 km from seismic station D (see diagram above).
The ***average*** speeds for P and S waves are P : 10 km/s S : 6 km/s
What is the gap between the times that the 2 waves arrive at D?

a simple seismometer

From page 122: **time taken =** $\dfrac{\textbf{distance travelled}}{\textbf{speed}}$

$$\therefore \text{ Time taken by P-waves } = \frac{9000\text{ km}}{10\text{ km/s}} = 900\text{ s}$$

$$\text{Time taken by S-waves } = \frac{9000\text{ km}}{6\text{ km/s}} = 1500\text{ s}$$

$$\therefore \text{ Time interval } = 1500\text{ s} - 900\text{ s} = 600\text{ s} = \underline{10\text{ min}}$$

In practice this calculation is done in reverse, to find the distance to the earthquake (or nuclear bomb test).

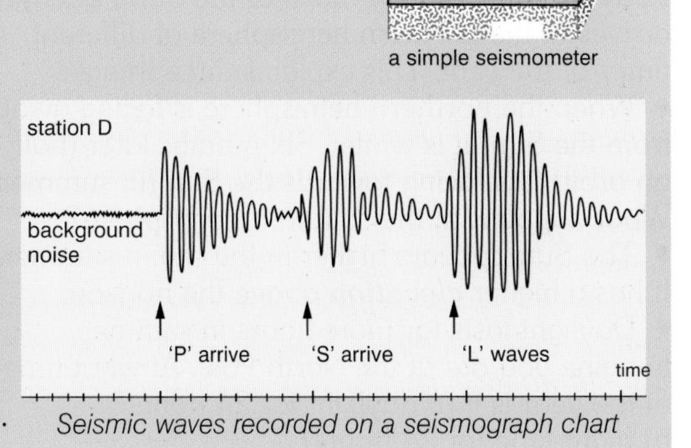

Seismic waves recorded on a seismograph chart

▶ Earth and Sun

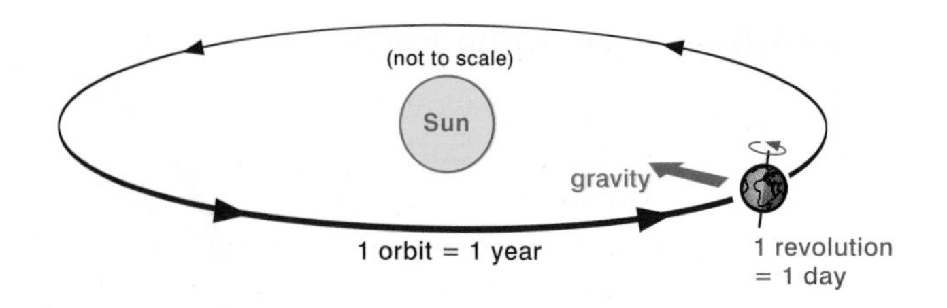

The Earth moves round the Sun, in an ***orbit***, which is (almost) a circle. The Earth is pulled into this orbit by a ***centripetal force*** (see page 70). This force is the gravitational force between the mass of the Sun and the mass of the Earth.

Experiment 20.1 A day
Use a white tennis ball as the Earth. Sketch the main continents on it, and mark the position of your school. Insert a knitting needle as its north-south axis. In a darkened room, use a lamp for the Sun.
Turn your 'Earth' slowly, so your 'school' is in daylight and then night.
For the real Earth, one revolution takes 24 hours, a complete day.
For other planets it is different.

The Sun 'rises' in the east. Which way should you rotate your 'Earth'?

Experiment 20.2 A year
Walk with your 'Earth' in a complete circle round the 'Sun'.
Mark out the 12 months of the year. For the real Earth, 1 complete orbit takes $365\frac{1}{4}$ days (this is why we need a leap year every 4 years).

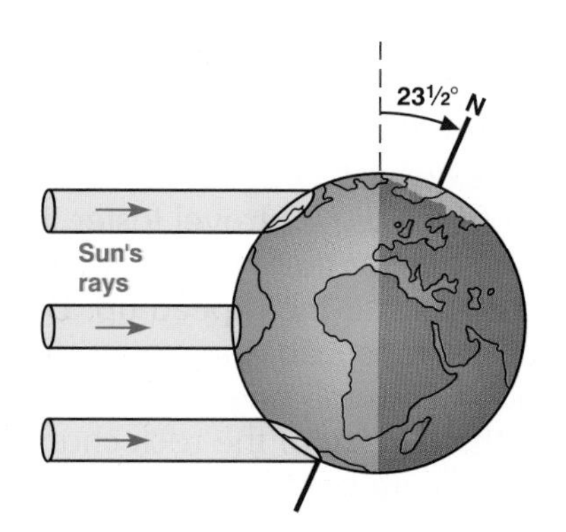

Experiment 20.3 Polar and equatorial regions
Look carefully at the surface of your 'Earth' in the light from the 'Sun'.

At the equator, where the Earth's surface is facing the Sun, it looks bright. But near the poles the surface is curved away and so the energy is spread out more. Why is the Earth cold at the poles?

The seasons of the year

In fact, the Earth's axis is not at right-angles to the orbit as you might have assumed. ***It is tilted*** at an angle of $23\frac{1}{2}^{\circ}$.

Experiment 20.4 Summer and winter
Repeat experiment 20.2 but with the axis ***always tilted towards the same end of the classroom.***

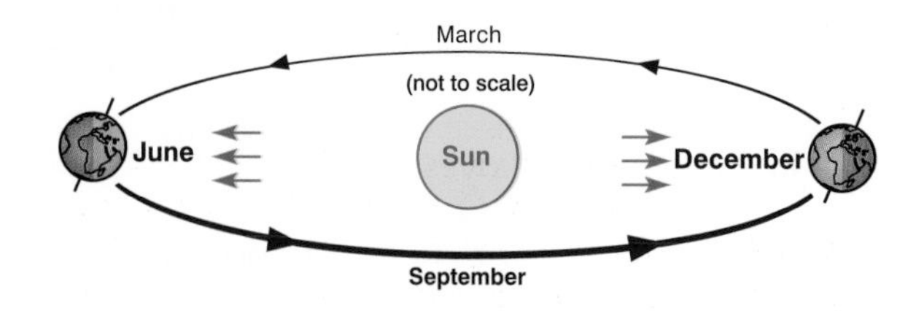

Look carefully at how much of the Sun's energy arrives at the northern hemisphere at different times of the year. This explains three things:

- When the northern hemisphere is facing away from the Sun, it is winter. Six months later (half an orbit) it is facing towards the Sun (in summer). What happens in the southern hemisphere?
- The Sun appears higher in the sky in summer. It has a higher ***elevation*** above the horizon.
- Daylight lasts for more hours in summer. Imagine you are at the North Pole. At what time of the year is it a) daylight for 24 hours? b) night-time for 24 hours?

▶ Earth and Moon

The Moon is a stark and hostile world.
Its mass is smaller than the Earth's, so its gravitational field is less. It is only one-sixth of the Earth's field (g = 1.6 kg/N, see page 131). How high could you jump on the Moon?

Because its gravity is less, all the molecules of gases have escaped, so it has no atmosphere.

An astronaut on the Moon

The Moon is a satellite in circular orbit round the Earth, trapped by gravity. One orbit of the Moon takes 1 month (1 'moonth').

It orbits the Earth once every 27 days. It also spins on its axis once every 27 days. This is why it always keeps the same side towards us.

It shines by reflecting the light from the Sun. The Moon looks different to us at different times of the month. It has ***phases***.
The diagram shows the Moon at 8 different positions round the Earth with the 8 phases:

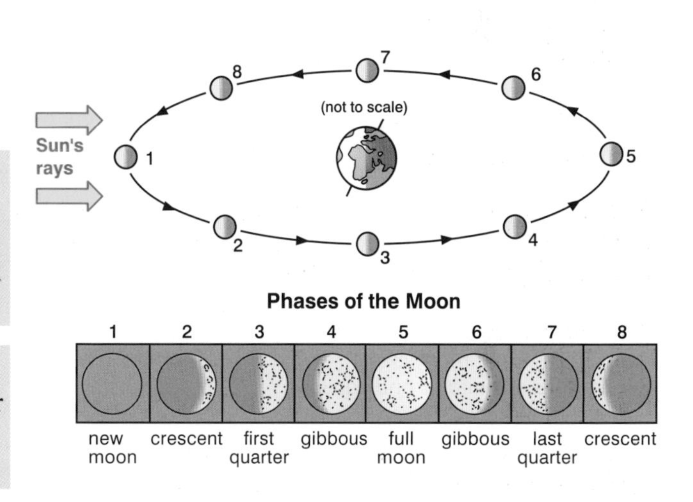

Experiment 20.5 Phases of the Moon
Use a lamp or an overhead projector for the Sun's rays, with a white ball for the Moon. Then stand in the centre (as the Earth) and look at the 'Moon' as it is moved round in orbit.

Experiment 20.6 Phases of the Moon
Observe the real Moon for a full month, weather permitting. Keep a diary of its appearance each evening, a) in writing and b) in sketches.

Try to find out when the next **lunar eclipse** will occur (see the diagrams on page 173).

Tides

Just as the Earth pulls on the Moon to keep it in orbit, so the Moon pulls on the Earth (Newton's Third Law, see page 84).
This gravitational force also pulls on the Earth's seas so that the water piles up. For complex reasons, there are ***two*** bulges as shown:

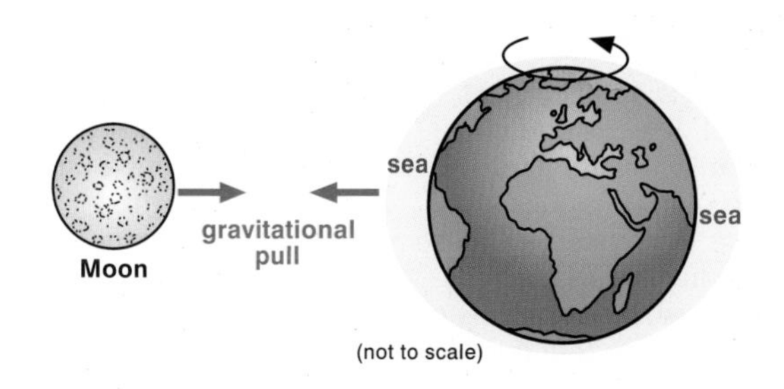

As the Earth rotates, each part of the world gets two high tides each day, and two low tides.
The energy stored in these tides can be used to produce electricity (see page 15).

▶ The solar system

Our solar system consists of a central star (the Sun), surrounded by planets, including Earth.

The first six planets are visible to the naked eye, and have been known since ancient times. The others were discovered through telescopes.

The planets are held in their orbits by the gravitational pull of the Sun (see page 71).

The planets vary enormously in their size (see below, with the Sun in yellow on the same scale).

They also vary in their distance from the Sun (see the diagram on the opposite page). This means that the conditions vary from planet to planet:

Saturn, photographed by the Voyager-2 space probe

1 Mercury

Mercury is closest to the Sun, and small for a planet (about the size of our Moon).
It has no atmosphere and is covered in craters.
The side facing the Sun is very hot, about 430 °C (that is hot enough to melt lead).

2 Venus

Venus is almost as big as the Earth, but very unpleasant.
It is covered in clouds of sulphuric acid, with an atmosphere of carbon dioxide at very high pressure.
Because of the CO_2 and the Greenhouse Effect (see page 48), it is even hotter than Mercury.

3 Earth

From space, Earth is a blue planet with swirls of cloud.
It is the only planet with water and oxygen and living things.
It is at the right distance from the Sun, with the right chemicals, to support life. Other stars may have planets with the same conditions.

4 Mars

Mars – the red planet – is a cold desert of red rocks, with huge mountains and canyons.
There is no life on Mars.
It has a thin atmosphere of carbon dioxide, and two small moons.
Between Mars and Jupiter there are thousands of rocks, called asteroids (see the opposite page).

5 Jupiter

Jupiter is the cold giant of the planets.
It has no solid surface, being mainly liquid hydrogen and helium, surrounded by gases and clouds.
The Great Red Spot is a giant storm, three times the size of Earth. Some of the 16 moons have volcanoes.

6 Saturn

Saturn is another 'gas giant', very like Jupiter.
The beautiful rings are not solid. They are made of billions of tiny chunks of ice, held in orbit by the pull of Saturn's gravity. As well as the rings, Saturn has more than 20 moons.

7 Uranus

Uranus is another giant planet. It looks pale green, with very faint rings and 15 moons.
It was discovered by William Herschel in 1781.
It is unusual because its axis is tilted right over, so that it is 'lying on its side' as it goes round the Sun.

8 Neptune

Neptune is the twin of Uranus. They are both about 4 times the size of Earth.
Neptune is bluish, due to a thick atmosphere of cold methane.
It has 8 moons, one with volcanoes.
The Great Dark Spot is a storm about the size of Earth.

Pluto

Pluto used to be called a planet but now it is called a dwalf planet.
It was discovered in 1930.

We don't know much about it.
Pluto's orbit is not as circular as the others. Most of the time its orbit is outside Neptune's, but between 1979 and 1999 its orbit was inside Neptune's.
Astronomers have been searching for a 'planet 10' beyond Pluto.

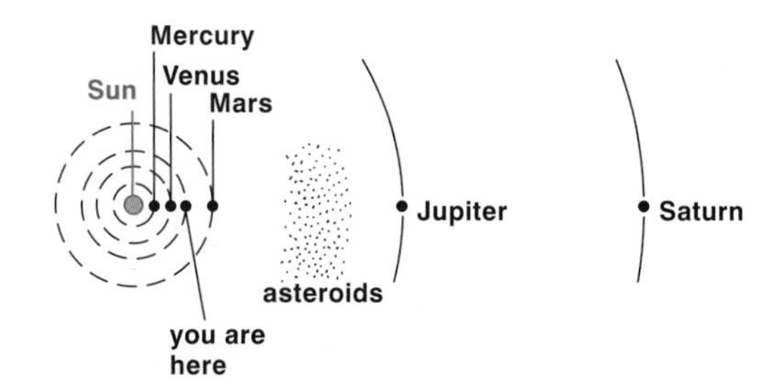

The distances to the planets, drawn to scale. On this scale, the nearest star would be 1 km away!

Experiment 20.7 The solar system

Make a scale model of the solar system:

a) For the Sun use a grapefruit or cardboard disc of diameter 11 cm.

b) For the Earth, make a ball of plasticine 1 mm in diameter. From column **C** in the table below, make all the other planets to the same scale (eg. Jupiter = 11 × Earth = 11 mm).

c) Hold your 'Earth' at distance of 12 metres from the 'Sun'. From column **B** below, hold the other planets at the correct scale distances (eg. Jupiter = 5 × Earth = 5 × 12 = 60 m).

You'll need to go on the playing field! On this scale, the nearest star would be another grapefruit about 3000 km away!

The moving planets

The orbit of each planet is not quite a circle. It is a slightly-squashed circle called an ***ellipse***. They are not all travelling at the same speed. The planets nearer the Sun travel more quickly. They have a shorter 'year' (column **H** below).

Planets do not give out their own light. They just reflect the Sun's light. Because they are much nearer than stars, they appear to move slowly against the background of the stars:

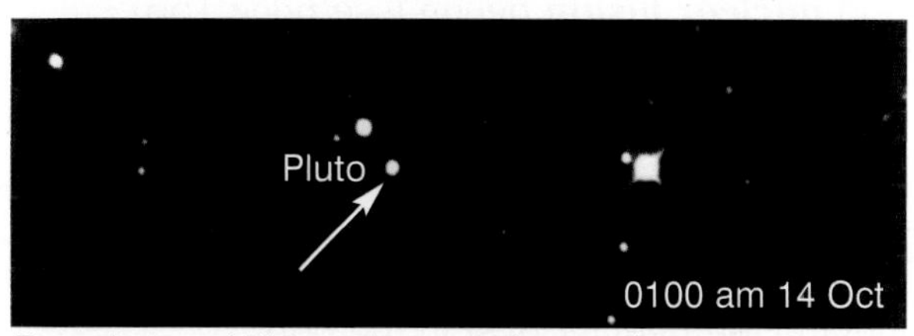

Two photos of Pluto, taken exactly one day apart. It has moved against the constellation of stars.

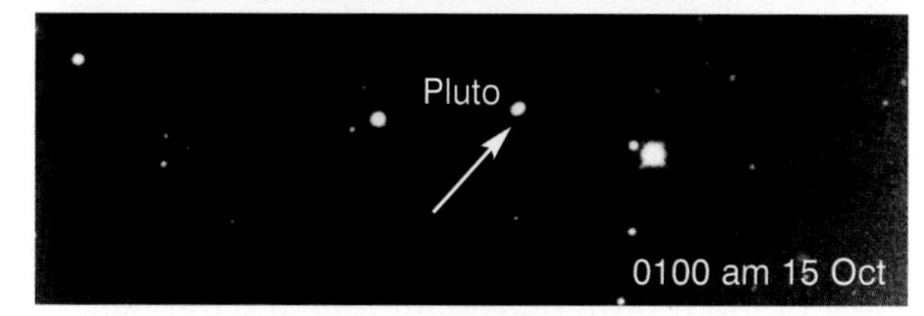

Here is a table of data about the solar system. Look down each column to see what patterns you can find (see also question 8, on page 162).

The planets	A Average distance from the Sun (millions of km)	B (relative to Earth = 1)	C Diameter (Earth = 1)	D Density (kg/m³)	E Average temperature (°C)	F Mass (Earth = 1)	G Surface gravity (N/kg)	H Time for 1 orbit (years)	I Number of moons
1 Mercury	58	0.4	0.4	5500	+430 to −180	0.1	4	0.2	0
2 Venus	108	0.7	0.95	5200	+470	0.8	9	0.6	0
3 Earth	150	1.0	1.0	5500	+15	1.0	10	1.0	1
4 Mars	228	1.5	0.5	4000	−30	0.1	4	1.9	2
Asteroids									
5 Jupiter	778	5	11	1300	−150	320	26	12	16
6 Saturn	1427	9.5	9	700	−180	95	11	30	20 + rings
7 Uranus	2870	19	4	1300	−210	15	11	84	15 + rings
8 Neptune	4497	30	4	1700	−220	17	12	165	8
9 Pluto	5900	39	0.2	2000	−230	0.002	0.6	248	1

▶ The birth of our solar system

Most astronomers believe that all the planets were formed at the same time as the Sun, about 4500 million years ago.

It began with a huge cloud of dust, ice and gas, called a **nebula**.
The gas was mostly hydrogen, with some helium.
These gases make up most of the Universe.

Because of its own gravity, the cloud began to shrink. As it got smaller and smaller, this 'protostar' got hotter and hotter. (As the particles fell inwards, they lost potential energy but gained kinetic energy, so the gas got hotter.)

Eventually the centre of the cloud was white hot and ***nuclear fusion*** began (see page 156).
It became the **star** we call the Sun.

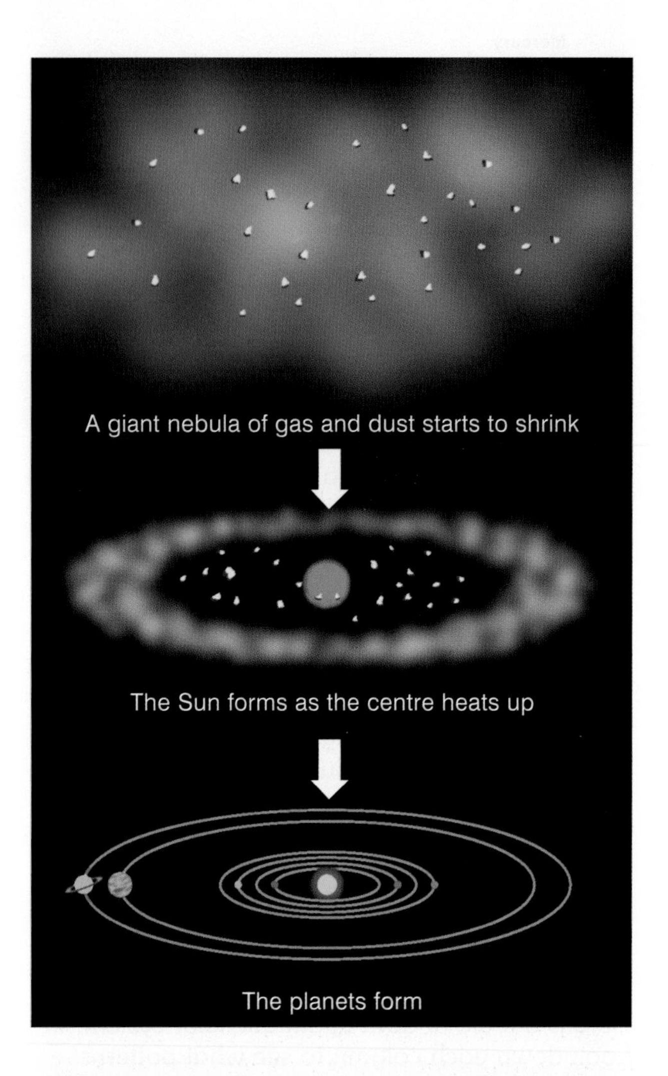

A giant nebula of gas and dust starts to shrink
The Sun forms as the centre heats up
The planets form

Meanwhile the planets began to form.
The heat of the Sun pushed some of the lighter gas and ice into a doughnut-shaped cloud, leaving the heavier dust in the gap.

Because of gravity the dust began to stick together, eventually making the 4 rocky inner planets. In the same way, the outer planets formed from the ice, and then collected the gas to become 'gas giants' like Jupiter.

Asteroids are some of the rocky left-overs. Most of them orbit between Mars and Jupiter. Those that might collide with the Earth are called Near Earth Objects (NEOs).

Meteors ('shooting stars') are tiny fragments that burn up in the Earth's atmosphere.

The famous Halley's comet, which returns every 76 years

Comets may be left-overs from when the solar system was formed. They are lumps of ice, dust and gas.
They have very elliptical orbits, which bring them close to the Sun and then far out in the solar system (often beyond Pluto):

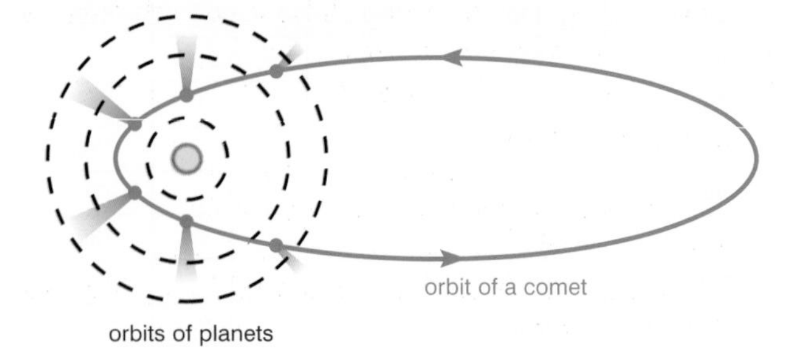

When they fall near to the Sun they speed up as the pull of gravity increases. The dust and gas are blown away from the Sun and shine in the sunlight, to form a long tail.

▶ Escaping from our planet

Imagine firing a gun from the top of a very high mountain. The shell would fall back to the Earth, just like a football. See A in the diagram:
If the shell is fired faster, it falls farther away (see B and C).
If the shell could be fired even faster (at 8 km/s, 25 times the speed of sound) then it would still fall towards Earth, but because of the curving of the Earth, the shell would stay the same height above the ground (see D):

A
B
C
D it is still falling, but does not hit the ground

It is then a ***satellite***, in orbit.
Gravity keeps pulling it into orbit (see also p. 71).

Satellite orbits

To stay in orbit at a certain height, the satellite must have the right speed to balance the pull of gravity. Low satellites have to travel faster to stay up.
A satellite in high orbit takes longer to complete 1 orbit, just like the outer planets round the Sun.

A satellite moves very fast, but it can seem to be standing still! If the satellite is put at just the right height and speed, it will take 24 hours to go round the Earth. This is the same time as the Earth takes to spin once, so the satellite appears to stay over one place! See also page 154.

An astronaut in orbit ***feels*** weightless, because the astronaut ***and*** his clothes ***and*** the spaceship are all falling at the same rate together.
However to become truly weightless the astronaut would have to travel far away from the gravitational force of the Earth and the Sun.

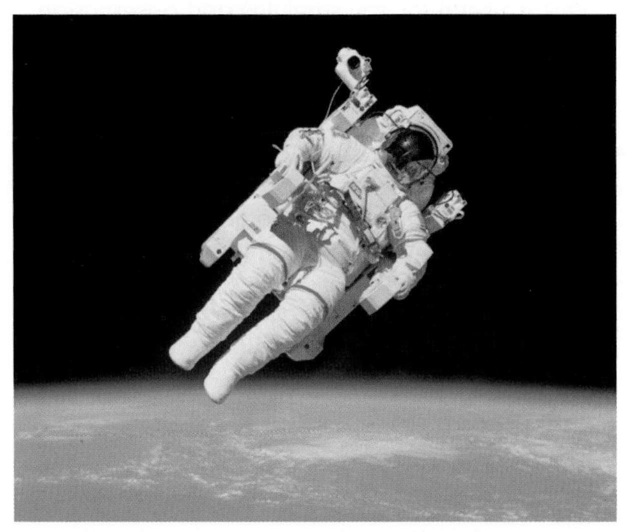

Bruce McCandless in 'weightless' orbit near his spacecraft. He is using small jets of gas to manoeuvre.

Gravitational force

Experiments show that:

- There is a gravitational pull between ***all*** masses.
- Larger masses give a stronger pull.
- The further the distance away, the smaller the gravitational pull.
  In fact, at twice the distance the pull is $\frac{1}{4}$.
  At 3 times the distance, the pull is $\frac{1}{9}$.
  This is called an 'inverse square' law.
  What do you think the pull would be at 4 times the distance?

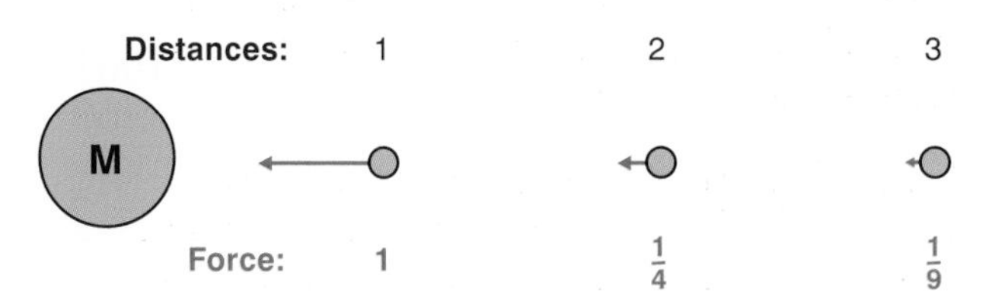

▶ Physics at work: Satellites

Satellites are objects that orbit round the Earth. They are launched by rockets (see page 160). Like the Moon, they are held in orbit by the gravitational pull of the Earth (see page 153).

Launching the space shuttle

To stay in a particular orbit, they have to travel at the right speed.
The higher the satellite, the ***slower*** the speed and the longer it takes for 1 complete orbit (see the example in the box below).

Satellites can be launched into ***polar*** orbits or ***equatorial*** orbits:

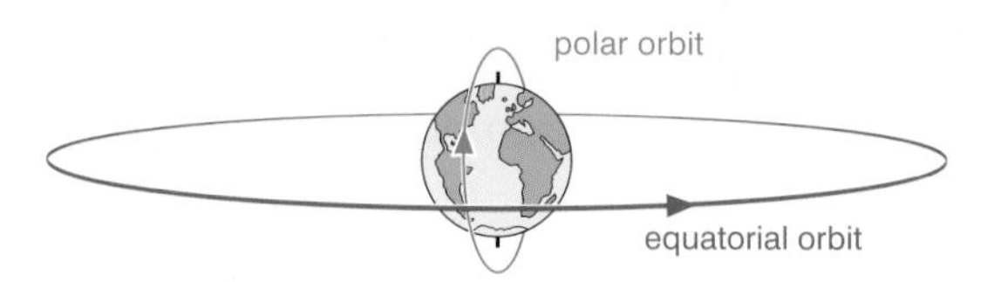

Polar orbits are usually lower. As the Earth turns on its axis, the satellite passes over a different part of the Earth on each orbit.
This is useful for spy satellites and observation satellites (see opposite page).

Equatorial orbits are better for some uses. Although a satellite is moving very fast, it can appear to be stationary. If it is placed in the right orbit, at a height of 36 000 km above the equator, it takes 24 hours to go round its orbit. This is the same period as the Earth, so the satellite appears to hover over one place.
This is a ***geo-stationary*** or ***synchronous*** orbit.
This is ideal for communications satellites.

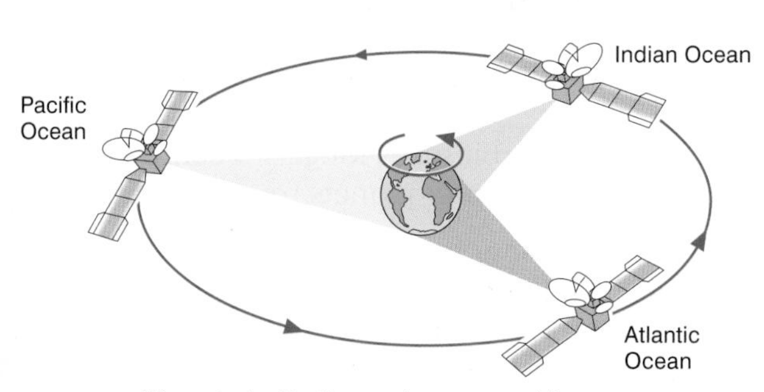

The whole Earth can be covered by just three geo-stationary satellites

Calculating the speed of a satellite

If the radius of the orbit is $\boldsymbol{R}$, then the circumference of the circle is $\mathbf{2}\,\pi\,\boldsymbol{R}$.

If the time for 1 orbit (the 'period') is $\boldsymbol{T}$.

$$\text{Speed} = \frac{\text{distance travelled}}{\text{time taken}} = \frac{\mathbf{2}\,\pi \boldsymbol{R}}{\boldsymbol{T}}$$

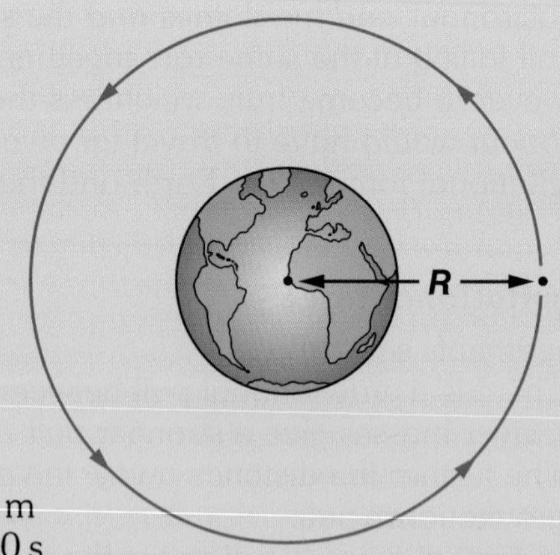

Example
A satellite at a height of 700 km above the Earth's surface, orbits with a period of 100 minutes. What is its speed?
The radius of the Earth is 6400 km.

Radius of orbit, $\boldsymbol{R}$ = 6400 + 700 km = 7100 km = 7 100 000 m
Time of orbit, $\boldsymbol{T}$ = 100 minutes = 100 × 60 seconds = 6000 s

$$\therefore \text{Speed} = \frac{2\,\pi R}{T} = \frac{2 \times \pi \times 7\,100\,000}{6000} = \underline{7435\text{ m/s}} \quad \text{(about 16 000 mph!)}$$

- Can you calculate the speed of a satellite in a geo-stationary orbit?

▶ Physics at work: Satellites

Communications satellites like 'Intelsat' use a geo-stationary orbit (see opposite page).

If you telephone to America, a microwave radio signal is transmitted from a 'dish' aerial up to the satellite. The satellite then transmits it down to another dish aerial in America. See the diagram on page 211.
Microwaves are used because they travel in a narrow beam, in a straight line, and pass easily through the Earth's atmosphere.

An 'Intelsat' communications satellite

In the same way, a TV satellite placed in a geo-stationary orbit over the equator, can broadcast programmes to all of Europe:

The waves spread out (by diffraction, see page 169) as they leave the dish.
You need a dish aerial on your house to pick up the weak signal (see page 182).

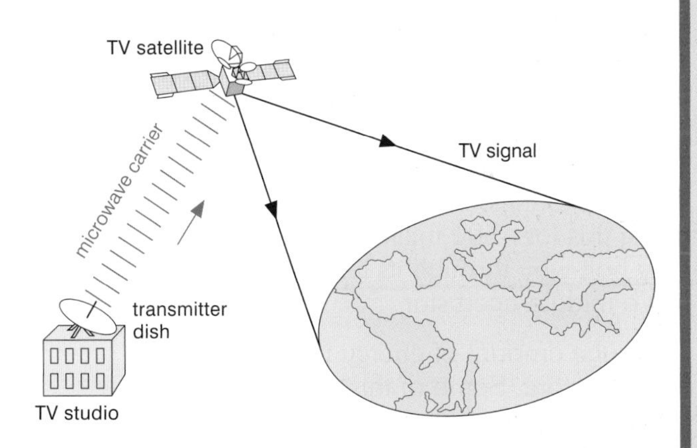

Weather satellites like 'Meteosat' are often put in geo-stationary orbits (see photo below).

Earth observation satellites like 'Landsat' can give us a detailed picture of the Earth's surface. They can warn of forest fires and locust swarms, show up diseased crops, hurricanes and oil leaks (see below).

Military satellites are used for spying.

Navigation satellites are used by ships, cars, planes to locate their position very accurately. They use a GPS (Global Positioning System) to tell their position on Earth within 10 metres!

Astronomical satellites can take sharper photographs outside the Earth's atmosphere, see page 215.
They can also travel to other planets.

From a weather satellite. A clear day except snow and cloud in Scotland.

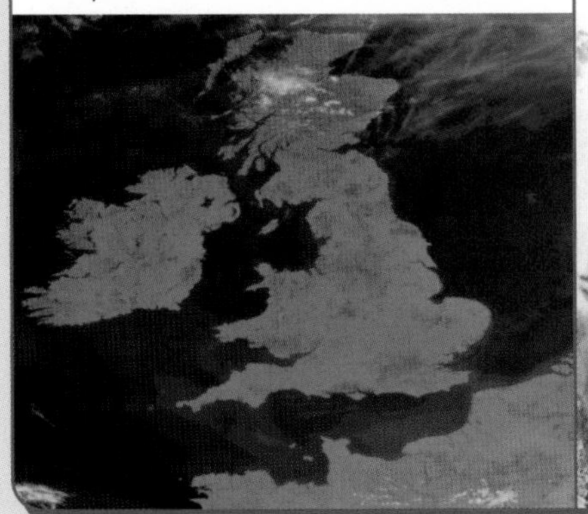

An oil-slick in the Gulf (detected by infra-red, and given artificial colour)

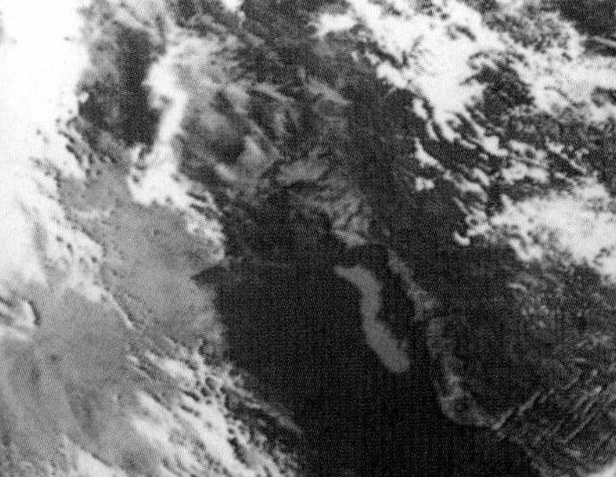

A Landsat image: a volcano (Vesuvius). Artificial colour (blue=houses in Naples).

▶ Our star: the Sun

The Sun is a star like the others that we see at night. Compared to other stars, the Sun is very ordinary and average.

The Sun is not burning like a fire – it is a huge controlled hydrogen bomb.
The core is at a temperature of 14 million °C, and at a very high pressure. The hydrogen atoms are broken into pieces, which smash into each other at high speed.
This can make the pieces of hydrogen atoms join together and become helium.
Each time this happens, some mass is ***lost***.
It reappears as ***energy*** (see page 348).
This is **nuclear fusion**.

Hydrogen $\xrightarrow{\text{nuclear fusion}}$ Helium + energy

The Sun is using up its mass at the rate of 4 million tonnes each second! It has been doing this for 4500 million years, but is still only half-way through its life-time.
It is a middle-aged star.

The huge amount of energy is radiated out from the core. Some of this energy eventually arrives on Earth, keeping us alive (page 101).

Inside the Sun there is a constant battle of forces: the pull of gravity is trying to crush the Sun, while the force of radiation from the very hot core is trying to make it expand.
As long as these forces are ***balanced***, the star is stable.

Eventually it will run out of fuel and change to another kind of star (see opposite page).

Sun-spots are slightly cooler areas on the Sun's surface – they are only at 4000 °C!
Sun-spots come and go, being common every 11 years (the 'sun-spot' cycle).
It is possible that the Sun's power may vary slightly from time to time, affecting the Earth's weather – perhaps even causing the Ice Ages.

In the night sky, the other stars appear to form patterns, called **constellations**.
Some famous ones are Cassiopeia and the Plough:

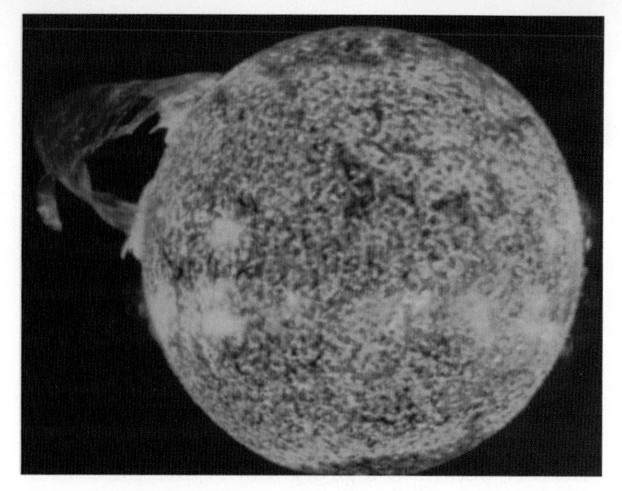

An ultra-violet photograph of the Sun, showing a huge prominence curving in the Sun's magnetic field

Twinkle, twinkle sunny star,
Now we know just what you are.
Nuclear fusion, burning brightly,
– Nice of you to switch off nightly!

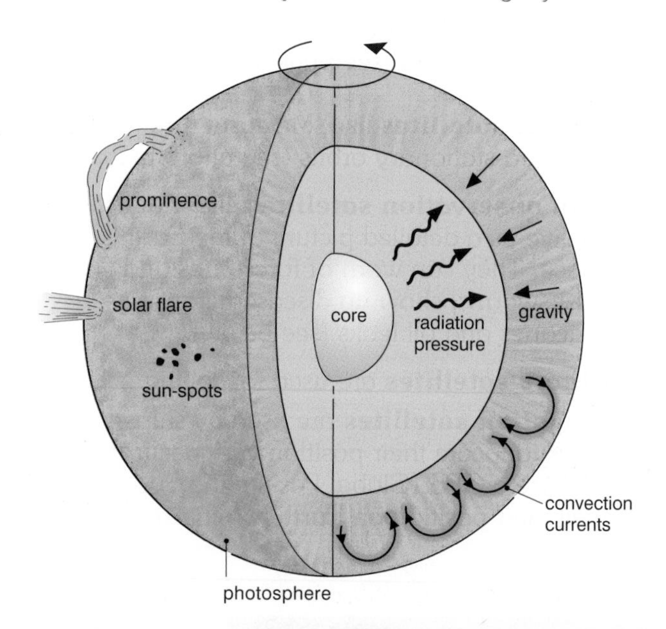

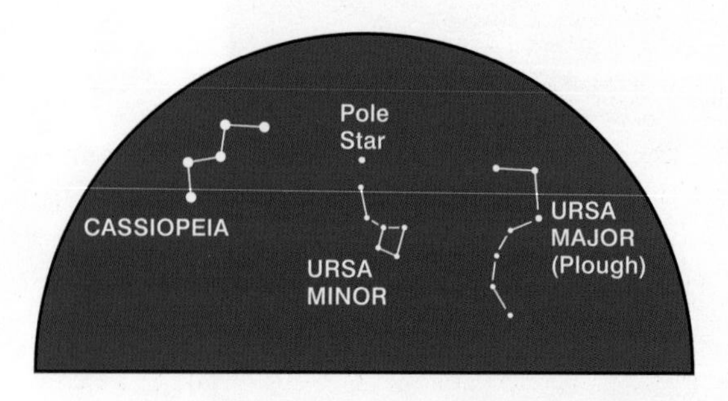

The north sky in Winter

▶ The life and death of a star

Stars eventually use up their fuel and start to change. What happens then depends on the mass of the star.

1 Small stars like our Sun (yellow dwarfs)

After about 10 000 million years, the hydrogen is all used up and the star starts to use helium as a fuel. This makes heavier elements (up to element 26, iron). While doing this it expands and cools to become a **red giant**. When this happens to the Sun, it will swallow up Mercury and Venus, and kill all life on Earth.

Later, the star uses up all its fuel and collapses under gravity to form a very hot and dense **white dwarf**. Its density is a million times the density of water!

Finally it cools like a dying fire to become a **black dwarf**.

2 Larger stars

A larger star burns hotter (a **blue giant**) and runs out of fuel sooner. It expands and cools to a **red supergiant**.

Later it runs out of fuel, becomes unstable, collapses, and blows off its outer layer in an enormous explosion called a **supernova**. During this explosion heavy elements (heavier than iron, up to element 92, uranium) are produced and thrown out across space, to become dust for making new stars (see page 152). Our planet Earth and all life on it are made from elements produced earlier by stars. You are star material!

The core of a supernova collapses to a very dense **neutron** star (also called a pulsar). If this still has a big mass it continues to collapse under its own gravity. The pull of gravity becomes so strong that nothing can escape it, even light. It is a **black hole**.

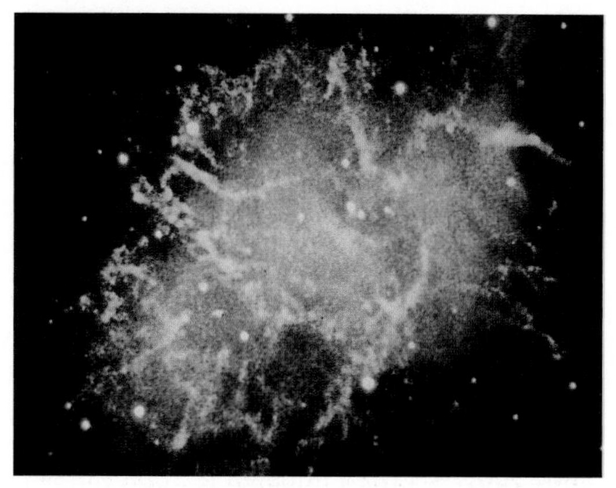

Death of a big star – a supernova

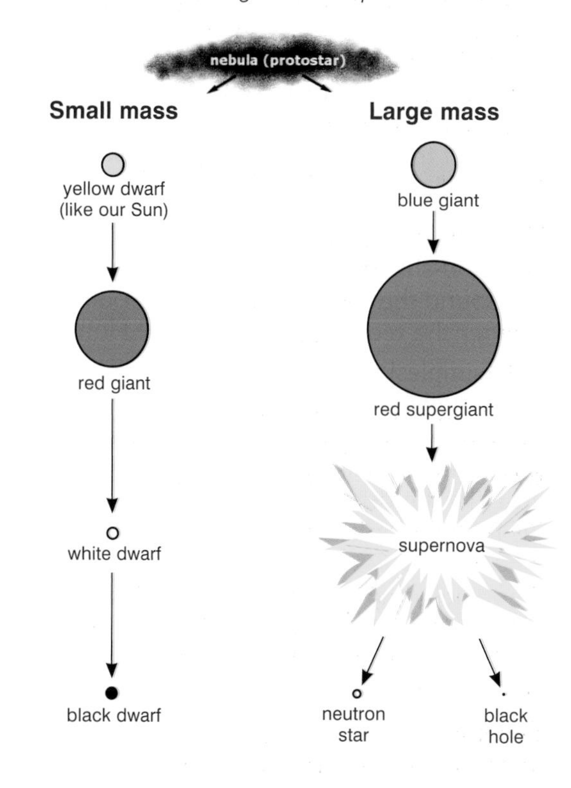

▶ Our galaxy: the Milky Way

If you look carefully at the night sky, you can see a faint band of stars running across it. You are looking edge-on at our Galaxy.

It is a collection of about 100 000 million stars! Our Sun is just one of them, placed somewhere near the edge, in a ***spiral arm***.

The Galaxy is huge. It takes 8 minutes for light to travel from the Sun to Earth; 4 years for light to travel from the nearest star; but 100 000 years for light to travel across our Galaxy!

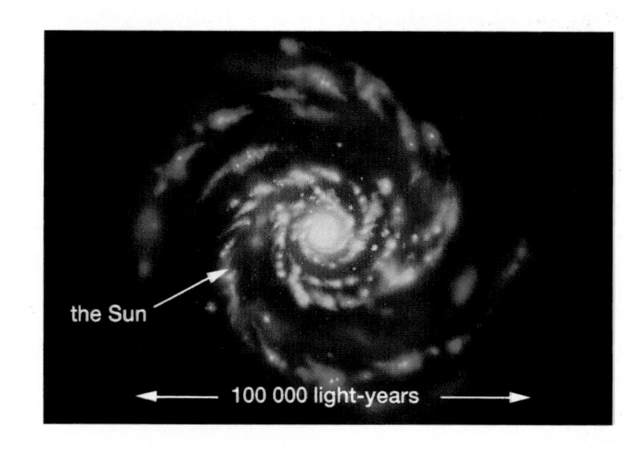

▶ The Universe

The Milky Way is our galaxy, but it is not the only galaxy. Using telescopes and radio-telescopes, astronomers have found billions of other galaxies in the Universe! See also page 215.

When we look at stars, we see them as they ***were***, when the light left them. We are looking into the past. Some stars are 10 000 million light-years away, and so we see them as they were, 10 000 million years ago! So the Universe is even older than this.

The 'whirlpool' galaxy, 37 million light-years away. It is a 'spiral' galaxy, like ours, but others are different shapes.

The expanding Universe

Astronomers can analyse the spectrum of light from a star (see page 207). They have found that the light from other galaxies ***always has a red-shift***. This is like the pitch of a police-car siren going lower as it races past. It is called a ***Doppler*** shift.

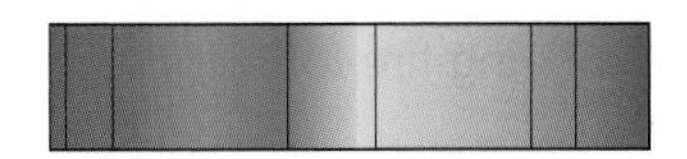

Spectrum of light from our Sun.
Light from a distant galaxy is red-shifted:

This red-shift was discovered by Edwin Hubble. He found that the ***farther*** away a galaxy, the ***bigger*** the red-shift. This is called Hubble's law.

The simplest explanation for this is:
– ***all*** the galaxies are moving ***away*** from us,
– the farthest galaxies are moving away fastest.
This means that the whole Universe is expanding, just like the dots on a balloon move farther apart as the balloon expands:

Imagine going backwards in time, with the Universe (or the balloon) shrinking. It means that at some time in the past, all the material in the Universe was lumped together. This was when it was created (and time began) in the **Big Bang**.
By measuring the rate of expansion of the Universe as it is now, astronomers can calculate that the Big Bang was almost 14 000 million years ago!

The Universe has been expanding ever since then, against the pull of gravity between the galaxies.

If the total amount of matter in the Universe was big enough, the pull of gravity would stop the expansion, and the Universe would contract to a 'Big Crunch'. At present the evidence is not certain, but it looks as though the Universe will keep on expanding forever.

Astronomers are still searching for better ways to measure the total mass of the Universe, including the 'dark matter' that cannot be seen.

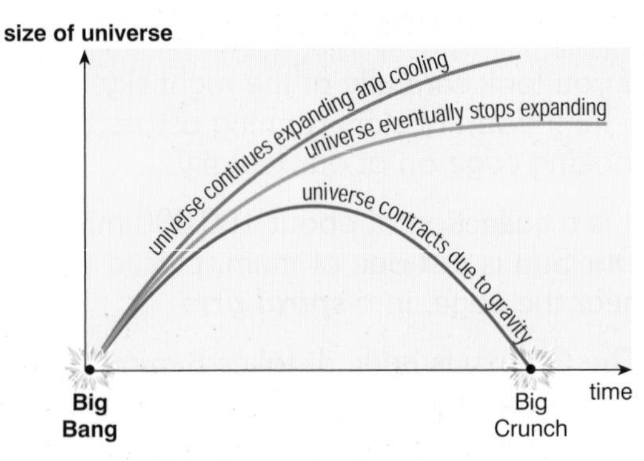

3 possible futures for the Universe

▶ The search for life in the Universe

Are we alone, or is there other life in the Universe? Scientists are looking for evidence in 3 ways:

- ***By analysing meteorites*** that fall to Earth from space. They are looking for fossils of bacteria, but no convincing evidence has been found yet.
- ***By sending robots to explore our Solar System.*** A key condition for life (as we know it) is liquid water. The most promising places for this are:
  - Mars which may have had water in the past,
  - Europa, one of Jupiter's moons. It is covered in ice and it may have liquid water underneath.

  The robots can look for life (eg. bacteria) or the chemical changes that are caused by life.
- ***By listening with radio telescopes*** (page 215) for any radio signals from advanced civilisations like ours. This is the SETI project (Search for Extra Terrestrial Intelligence). This search is made more difficult because there is a constant background 'noise' due to the microwave radiation that was left over from the Big Bang explosion.

A Sojourner lander robot working on Mars

Radio-telescopes listening for signals from space

Your place in the universe:

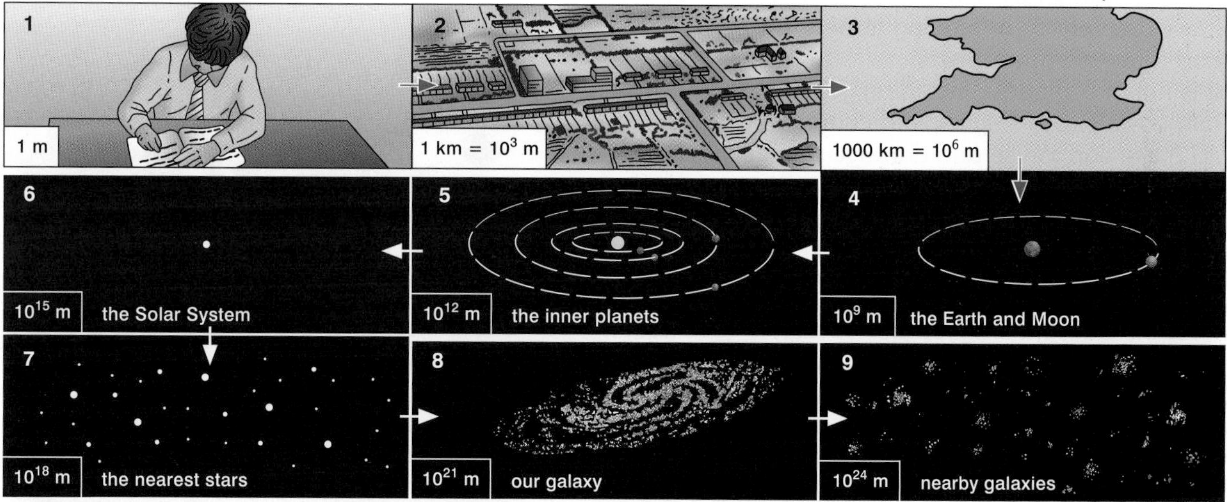

Summary

Seismic waves tell us about the Earth's structure.
The movement of the Earth and Moon explains: a day, a year, the seasons, the Moon's phases.
The Solar System is the Sun and 9 planets.
The Sun is a star in our galaxy, the Milky Way.

The Sun's energy comes from nuclear fusion.
A star is formed from a nebula, and changes as it gets older.
The Universe is expanding, as shown by the red shift, which suggests the Big Bang theory.

▶ Physics at work: Space travel

Rockets

To escape the gravitational field of the Earth, we use a rocket (see also page 85 and page 138):

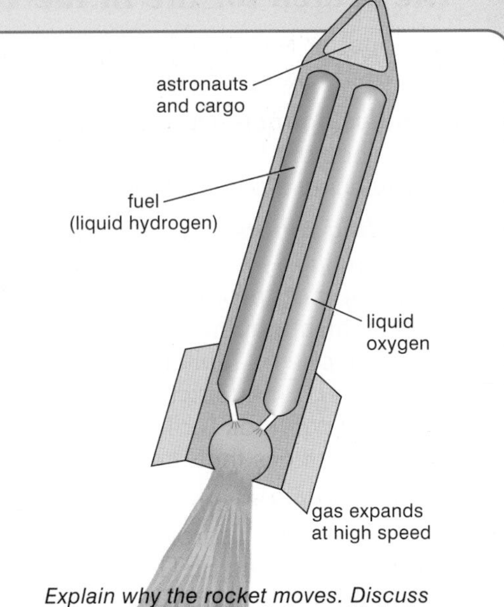

Use Newton's 3rd Law (page 84) to explain why the balloon moves:

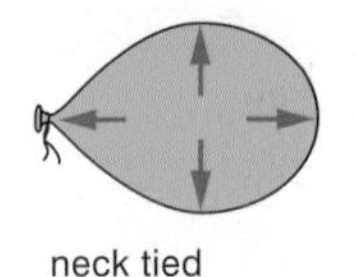

neck tied
Balanced forces

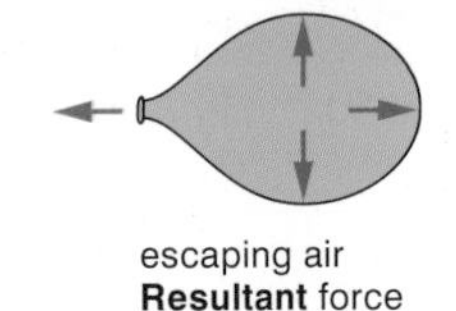

escaping air
Resultant force

Which way does Professor Messer's machine-gun boat move?

Explain why the rocket moves. Discuss how to make it move (i) faster (ii) farther.

Is it safe?

A rocket is dangerous – it is a controlled explosion!

And interplanetary space is a hostile environment. It is extremely cold (tending to near absolute zero) and it has no atmosphere.
It is full of radiation, including gamma-rays, X-rays and cosmic rays (high-speed charged particles). All these can ionise and damage the human body (see page 214).

When astronauts are in 'free fall' orbit round the Earth (see page 153), or far out in space, they feel weightless. Lack of gravity allows the body to weaken during a long journey. Muscles get smaller, the heart becomes weaker, and bones get thinner.

An astronaut working in 'micro-gravity'

Artificial gravity

In a space station astronauts can use exercise machines for their muscles.
A better solution might be to create 'artificial gravity' with a rotating spaceship:

In the rooms near the rim, an astronaut will feel a force 'upwards' from the floor. This force is called a centripetal force (see page 70). It gives the impression of weight.

▶ Physics at work: Journey to Mars

How long would it take to travel to Mars?

To get to Mars you would have to travel about 24×10^7 km.
Suppose your average speed was 40 000 km/h (about 25 000 mph).

Formula first: $\text{average speed} = \dfrac{\text{distance travelled}}{\text{time taken}}$ (see page 122)

Then numbers: $40\,000 \text{ km/h} = \dfrac{24 \times 10^7 \text{ km}}{\text{time taken}} = \dfrac{240\,000\,000 \text{ km}}{\text{time taken}}$

$$\therefore \text{ time taken} = \frac{240\,000\,000 \text{ km}}{40\,000 \text{ km/h}} = \frac{24\,000}{4} = 6000 \text{ hours} = \underline{250 \text{ days}}$$

(8 months)

The distance to Europa (one of Jupiter's moons) is 2 000 000 000 km (2×10^9 km).

a How long would it take to get there?

The nearest star is 4 light-years away (about 150 000 times farther than Mars).

b How long would it take to get there?

c How long would it take a radio signal?

d Discuss what these calculations suggest to you.

Designing the spacecraft

Discuss the design of a spaceship to take 3 humans to Mars.

You will need to consider:

- The length of time of the voyage.
- How to take enough fuel, food and water.
- How they will keep warm.
- The need to shield them from cosmic rays.
- How to maintain a stable atmosphere.
- The effect of low gravity on their health.
- How they will land on Mars.
- How they will return to Earth.

e Sketch your design ideas and add notes.

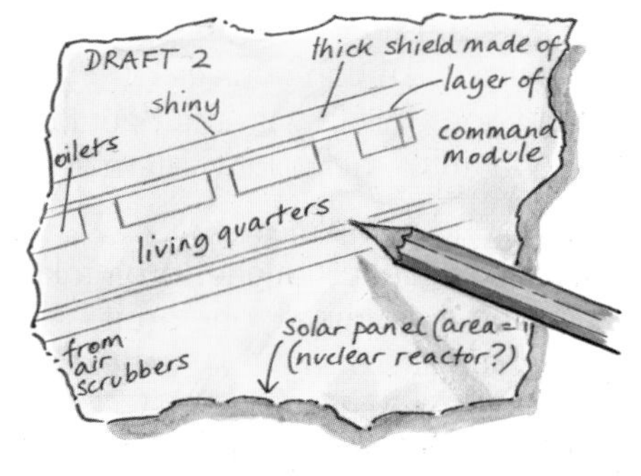

Visit: www.nasa.gov or try Google with 'spaceship Mars'

Living on Mars

Look at the data on page 151, and research about Mars using the internet.

Discuss the design of a base for living on Mars.

f Sketch your ideas and add notes.

g Use your scientific knowledge to write about a day in the life of an astronaut living on Mars.

Unmanned spacecraft (probes)

The only visitors on Mars so far have been robots (see page 159).

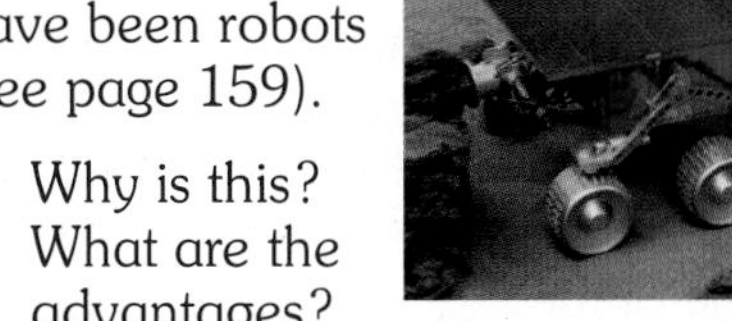

h Why is this? What are the advantages?

i Robots can send back data to us by radio. Discuss whether human beings need to go, at enormous cost?

▶ Questions

1. Copy and complete:
   a) The Earth has . . layers: . . . . , ... , ...
   b) Seismic P-waves are . . . . waves and can go through . . . . and . . . .
      S-waves are . . . . waves and can go through . . . . but not . . . .

2. Copy and complete:
   a) A day is the time for:
   b) A month is the time for:
   c) A year is the time for:
   d) The Earth's axis is tilted at an angle of . . . and this causes our . . . .
   e) The names of the planets, in order, are:
   f) The shape of a planet's orbit is an . . . .

3. Look at page 150 and the table on page 151. Write two paragraphs about each planet to describe what you would see, and feel, if you visited each one in a spaceship.

4. On a sheet of graph paper, draw bar-charts of:
   a) the diameter of each planet (see p.151),
   b) the density of each planet.
   What pattern can you see?
   How does this fit with the nebula theory of the origin of the Solar System?

5. Copy and complete:
   a) A satellite in a higher orbit takes . . . . time to complete one orbit.
   b) A satellite or a planet is held in . . . . by a . . . . force.
   c) The Solar System was formed by:
   d) Comets travel in very . . . . orbits.
   e) The Sun is an ordinary . . . . , part of our . . . . called the Milky Way.
   f) Solar energy comes from nuclear . . . .
   g) Our Sun is a . . . . dwarf star.
      Later it will change to a red . . . . ,
      then a . . . . dwarf, then a . . . . dwarf.
   h) A larger star (a blue . . . .) will change to a red . . . . and later explode in a . . . . leaving a . . . . star, or a . . . . hole.
   i) The universe has been . . . . since the time of the . . . .
   j) The light from distant galaxies shows a . . . . -shift due to the . . . . effect.
      The farthest galaxies move . . . .

6. Explain how the Solar System was formed.

7. Look at the table on page 151.
   a) Looking down column **B** and column **E**, what pattern do you notice? Why is this?
   b) Looking down columns **B** and **H**, what pattern do you see? Why is this?
   c) Looking down column **D**, what do you notice? List the names of the planets, divided into two clear groups.
   d) What do you notice from column **C**? Why do astronomers consider Pluto to be odd?
   e) From column **G**, how much would you weigh on i) Mercury ii) Jupiter?
   f) What is the connection between column **D** and columns **C** and **F**?
   g) For an asteroid, estimate i) the distance from the Sun, ii) the average temperature, iii) the length of its 'year'.

8. a) What is a geo-stationary orbit?
   b) Communication satellites and Earth observation satellites are placed in different orbits. Explain why.

9. Describe what will happen to our star, the Sun, in the future.

10. Use the internet, CD-ROMs or books to write a paragraph about each of the following:
   a) red giants, b) white dwarfs, c) supernovas, d) neutron stars (pulsars), e) black holes, f) quasars, g) cosmic microwaves, h) SETI.

11. Describe in your own words the evidence for the Big Bang theory for the origin of the Universe.

12. Describe in your own words how scientists are searching for extra-terrestrial life.

13. Estimate the possibility of ET (extra-terrestrial) life elsewhere in our galaxy, if there are:
   – 100 000 000 000 stars in our galaxy,
   – 1 in 100 are suitable stars with planets,
   – 1 in 10 planets are at a suitable distance from the star.

14. Explain why the journey for a human to travel to Mars will be difficult.

15. Professor Messer dreamed he was living on an asteroid. Why was his dream a nightmare?

Further questions on page 163.

▶ The Earth and beyond

1. Two types of shock waves are P and S waves. Study the diagram below.

Cross Section of the Earth

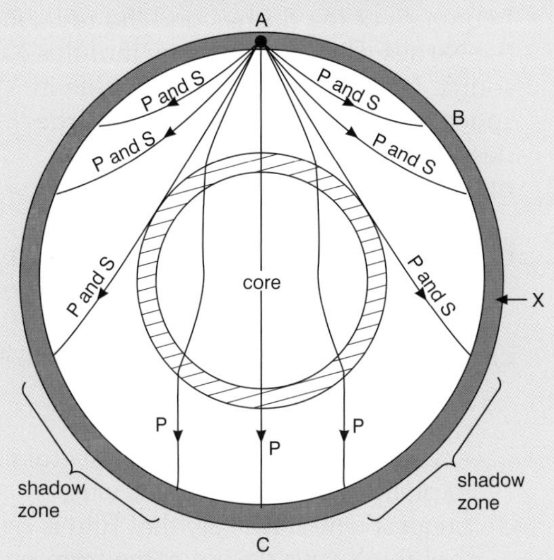

a) Name the layer marked X. [1 mark]
b) What is the exact point of origin of an earthquake called? [1]
c) What type of wave is an S wave? [1]
d) Explain why the wave AB is curved. [2]
e) No S waves travel through the core. What does this tell us about the structure of the Earth's core? [1] (NI)

2. The diagrams show the paths of seismic waves through the Earth's mantle and core.

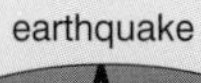

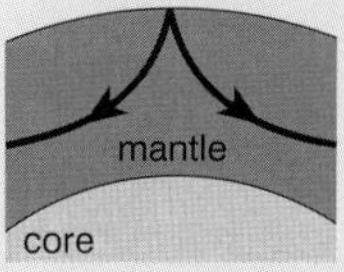

diagram X

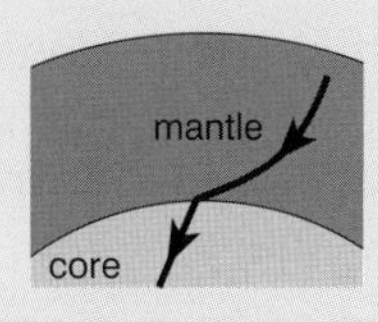

diagram Y

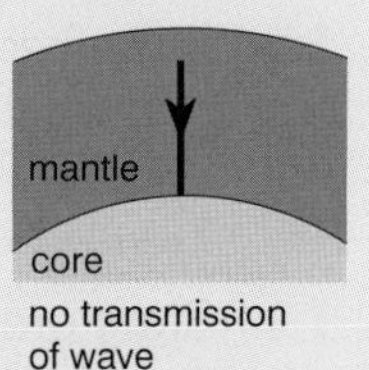

diagram Z

a) Describe and explain what happens to the waves travelling through the mantle in diagram X. [3]
b) Describe and explain what happens to the wave in diagram Y. State which type of wave is shown. [3]
c) Explain why the wave shown in diagram Z does not travel through the core. State which type of wave it is. [2]

3. a)

12 hours	24 hours	1 week	4 weeks	6 months	1 year

Complete the following sentences by choosing a time **from the above list**.
i) The Moon moves around (orbits) the Earth once every . . . . . [1]
ii) The Earth spins on its axis once every . . . . . [1]
iii) The Earth moves around (orbits) the Sun once every . . . . . [1]

b) Use the data on page 151 to answer the following.
i) Name the planet that takes 44 times as long as Mars to orbit the Sun. [1]
ii) Name the object whose diameter is about half that of Mercury. [1]
iii) What is the distance in kilometres between Mars and the Sun? [1]
(WJEC)

4. In 1609, Galileo was one of the first scientists to use a telescope. He used it to look at the planet Jupiter. The 4 diagrams show what he observed during one night:

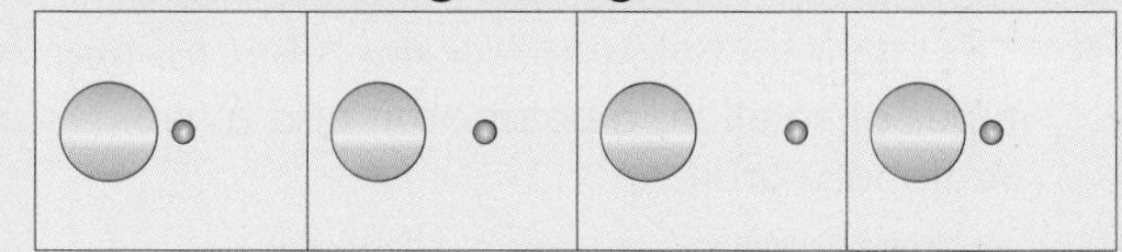

The small object close to Jupiter had not been seen before. It was later named Io.
a) Suggest a conclusion that Galileo could draw from his observations. [1]
b) Explain how Galileo's observations went against the belief that all heavenly bodies revolve around the Earth. [3]
c) Galileo published his findings in a book called *The Starry Messenger*. Why did Galileo publish his findings? [1] (Edex)

5. Describe how the orbit of a comet is different from the orbit of a planet. [2] (OCR)

6. This question is about asteroids.
a) How were asteroids formed? [2]
b) What is the evidence for past collisions of asteroids with the Earth and the Moon? [2]
c) i) Explain what is meant by Near Earth Objects (NEOs). [2]
ii) How can they be detected? [4]
iii) Why may they be a problem? [2]

7. This question is about satellites and space exploration.

a) Communication satellites are usually put into an orbit high above the equator called a geostationary orbit.

i) Explain what is meant by a *geostationary orbit*. [1]

ii) What is the advantage of such an orbit for communicating from one part of the world to another? [1]

iii) What is the time for a complete orbit of this satellite? [1]

iv) Why must it be placed at a particular height above the Earth's surface? [1]

v) Why is the orbit above the equator? [1]

b) Monitoring satellites scan the whole Earth each day. By means of a sketch and a brief explanation show how this is possible. [3]

c) Astronauts have landed on the Moon and it is hoped they will be able to visit Mars in the future.
Describe, briefly, **two** factors which make this difficult and expensive. [4] (WJEC)

8. The diagram below shows the orbits for two types of satellite, a polar orbit and a geostationary orbit.

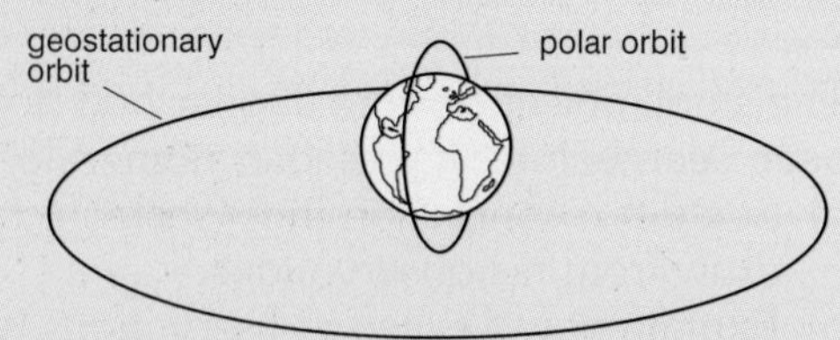

A satellite in stable Earth orbit moves at a constant speed in a circular orbit because there is a single force acting on it.

a) What is the direction of this force? [1]

b) What is the cause of this force? [1]

c) What is the effect of this force on the **velocity** of the satellite? [1]

d) In which of the orbits shown above would this force be bigger? Explain the reason for your answer. [2]

e) Explain why the kinetic energy of the satellite remains constant. [2]

f) Suggest, with reasons, one use of a satellite in a geostationary orbit. [2]

g) Suggest, with reasons, one use of a satellite in a polar orbit. [2] (AQA)

9. Rita works for a scientific institute called 'Search for Extraterrestrial Intelligence' or SETI. Using a radio telescope, she has discovered a distant source of X-rays that is flashing on and off. She measures the frequency of the flashes and the red-shift of the X-rays. Rita has a theory that this X-ray source is not proof of intelligent life in space, but a new kind of star that she describes with some new equations.

a) Explain the terms ***frequency*** and ***red-shift***. [3]

b) Explain how the amount of red-shift helps Rita to decide how far the X-ray source is from the Earth. [1]

c) When Rita makes her observations, she
– uses a computer
– takes lots of readings
– repeats the measurements several times.
Explain why she works this way. [3]

d) Suggest why it is likely that Rita is right that the X-rays do not come from an intelligent source. [2]

e) Rita decides to publish her theory. Explain how and why she would do this. [2]

f) Rahul works in an observatory in India. He discovers another X-ray source that flashes just like the one that Rita found. Explain why this neither proves nor disproves Rita's theory about the new star. [2]

g) What else must happen before other scientists will be able to accept Rita's theory? [2]

10. Stars begin their life in dense clouds of gas and dust.
Describe how they are formed and how they produce energy. [5] (OCR)

11. Stars are formed from massive clouds of dust and gases in space.

a) What force pulls the clouds of dust and gas together to form stars? [1]

b) Once formed, a star can have a stable life for billions of years. Describe the **two** main forces at work in the star during this period of stability. [2]

c) What happens to this star once this stable period is over? [4]

d) Suggest what might then happen to a planet close to this star. [1] (AQA)

12. The table gives information about 3 satellites.

Satellite name	Purpose	Orbit height above surface	Approximate orbit time
Hubble space telescope	to observe stars	612 km	1.5 h
NOAA-17	to monitor the Earth and oceans	833 km	1.7 h
Intelsat-906	communications	36 000 km	24 h

a) How does the orbit time for a satellite change with distance from the Earth? [1]
b) i) How many times does Intelsat-906 orbit the Earth in one day? [1]
ii) Why is this important for a geo-stationary satellite? [1]
iii) State two uses for a communications satellite like Intelsat-906. [2]
c) NOAA-17 takes 100 minutes to orbit.
i) How many orbits does it make in one day? [1]
ii) Explain how its polar orbit allows it to monitor ***all*** the Earth's oceans. [2]
d) The Hubble telescope was used to study the faint light from distant stars. Suggest why it was placed in orbit just above the Earth's atmosphere. [2] (Edex)

13. Telescopes can be used on Earth or in space (see page 215).
a) What are the advantages and disadvantages of using telescopes in space? [4]
b) How do different telescopes help us to make deductions about the Universe? [4]

14. This is a question about space travel.
a) Discuss and explain the problems involved in putting an astronaut on Mars. [4]
b) How are scientists devising ways to overcome these problems? [4]

15. a) What is meant by red-shift? [1]
b) What is a light year? [1]
c) A galaxy is 1200 million light years away from us. It is moving away from us at 30 000 km/s, which is one-tenth (0.1) of the speed of light. Assuming that its speed has been constant and that all the matter in the Universe was originally in one place, how long has it taken for the galaxy and us to be this far apart? [2]
d) What is the significance of this value? [1]
(OCR)

16. Stars do not stay the same forever.
a) Over billions of years the amount of hydrogen in a star decreases. Why? [1]
b) Describe how a massive star (at least five times bigger than the Sun) will change at the end of the main stable period. [4]
To gain full marks you should write your ideas in good English. Put them in a sensible order and use the correct scientific words.
c) The inner planets of the Solar System contain atoms of the heaviest elements.
i) Where did these atoms come from? [1]
ii) What does this tell us about the age of the Solar System compared with many of the stars in the Universe? [1]
(AQA)

17. a) What is meant by the term **red-shift**, as used in astronomy? [2]
b) Explain how the red-shift helps us to find out more about
i) galaxies [2]
ii) the universe. [2] (Edex)

18. a) How does the 'big bang' theory of the Universe account for its creation? [2]
b) Outline the evidence which supports the 'big bang' theory. [3] (AQA)

19. a) Explain how stars are thought to have formed from clouds of dust and gas. [3]
b) During the stable period of a star, high temperatures create forces which act outwards. These forces are balanced by its inward gravitational forces. Briefly describe the circumstances whereby each of the following will occur.
i) Our Sun becomes a red giant. [2]
ii) A red giant becomes a white dwarf. [2]
c) Briefly describe the process of nuclear fusion. [2]
d) What evidence is there to suggest that the Earth and the other planets in the Solar System were formed from material produced when earlier stars exploded? [3]
e) What evidence is there that the Universe is expanding? [4] (Edex)

chapter 21

Waves waves waves

Energy is often moved from one place to another by waves.

You can see this if you throw a stone into a pond – waves spread out from the splash and soon the water at the edge of the pond moves as it gets some of the energy of the splash.

There are ***two*** kinds of waves:

▷ Transverse waves

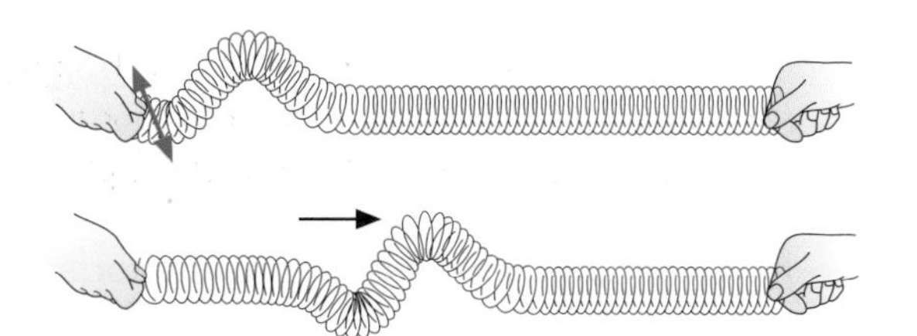

Experiment 21.1
To see a transverse wave, stretch a long 'slinky' spring along a smooth bench or floor and then give one end a quick wiggle ***at right angles*** to the spring. What happens?

You can see that energy is moving down the spring although each bit of the spring is only vibrating. It is vibrating at right angles (***transversely***) to the spring.

In this chapter we will be investigating transverse water waves. Light waves are also transverse waves.

▷ Longitudinal waves

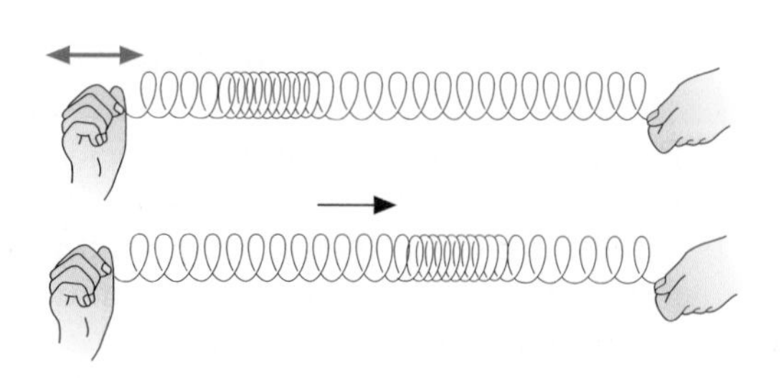

Experiment 21.2
Stretch the same 'slinky' spring along the floor and this time give it a quick wiggle to and fro ***along the length*** of the spring. What happens?

If you look closely you can see that again, energy is passing down the spring although each bit of the spring is only vibrating (or 'oscillating'). It is vibrating lengthways or ***longitudinally***.

We will investigate longitudinal waves in chapter 29 (page 224). Sound waves are longitudinal waves.

▷ The ripple tank

We can study transverse waves by using water waves in a ripple tank.

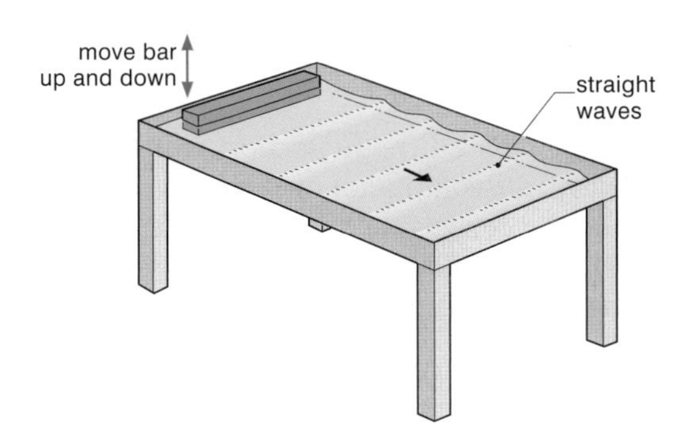

Experiment 21.3
Put some small pieces of cork on the water in a ripple tank (or your bath) and move the wooden bar up and down to make straight waves.

As the waves move down the ripple tank, which way do the pieces of cork move?
Are these transverse or longitudinal waves?

Here is a side view of a transverse water wave:

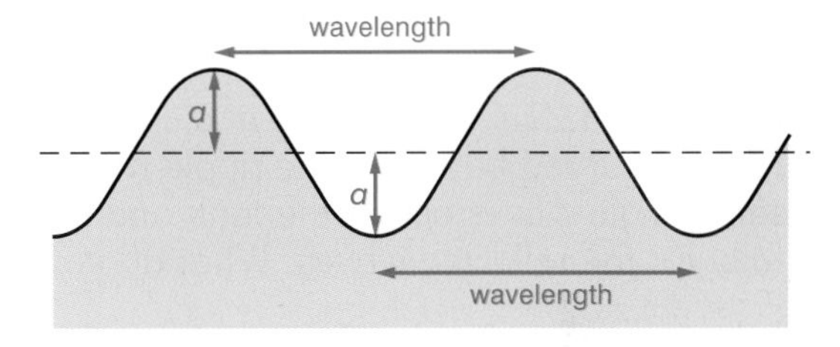

The high parts are called ***crests*** and the low parts are called ***troughs***.
The distance marked ***a*** is called the **amplitude**.
It is the height of a crest above the average water level (or the depth of a trough below it).

The distance between two successive crests (or troughs) is called the **wavelength**.

If the wooden bar (or a piece of cork) in the experiment moves up and down through two complete vibrations in each second, then the **frequency** of the vibrations is two cycles per second. This is also written as 2 hertz or 2 Hz.

A wave which vibrates only up and down like this is said to be ***polarised*** vertically.

The time **period** of a wave, ***T***, is the time taken to produce one complete wave.
It is connected to the **frequency** by:

$$\textbf{Time period, } T = \frac{1}{\textbf{frequency, } f}$$

Experiment 21.4
Move the wooden bar up and down at a low frequency first and then at a higher frequency.
What do you notice about the speed of the waves and their wavelength?

You should be able to see that:
a) the speed stays the same and
b) when the frequency increases, the wavelength decreases:

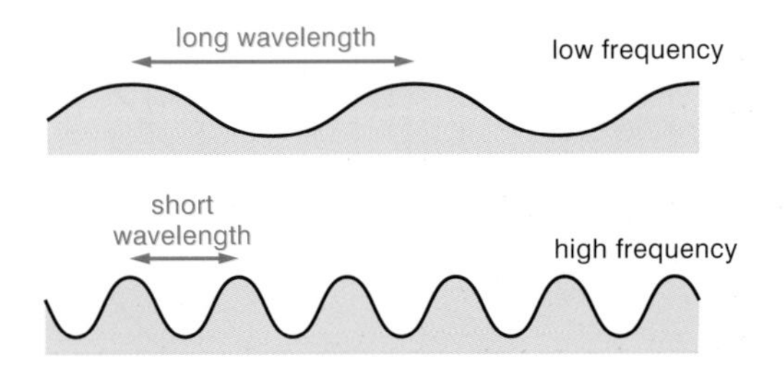

Example
A wave has a frequency of 2 hertz (2 cycles per second) and a wavelength of 10 cm.
What is the speed of the wave?

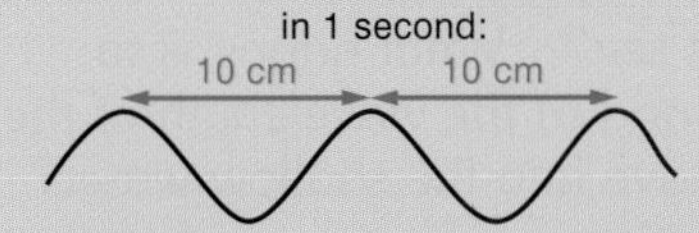

In each second, two waves are produced, each of length 10 cm.
Therefore the wave travels forward 20 cm in each second. That is, the speed of the wave is 20 cm/s.

In fact, for any wave:

wave speed (m/s)	=	**frequency** (Hz)	×	**wavelength** (m)

▷ Reflection and Refraction

Experiment 21.5   Reflection at a straight mirror
Place a straight metal or plastic barrier in the tank so that it reflects the waves (like a mirror).
Send single waves across the ripple tank and look carefully for the waves reflected off the 'mirror'.

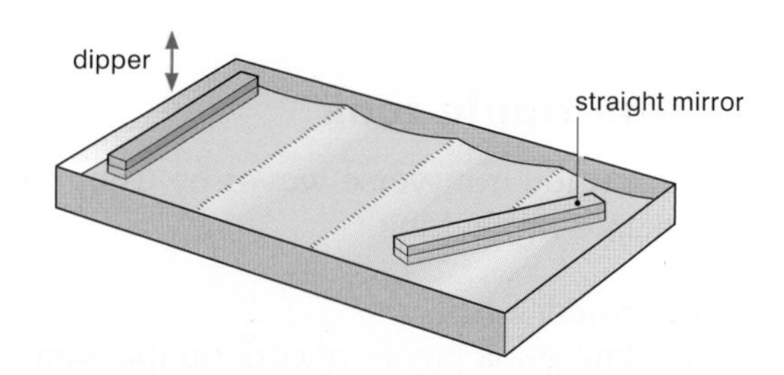

What do you notice about the angle at which the wavefront hits the mirror and the angle at which the reflected wave leaves the mirror?
What do you know about the angles when light is reflected from a mirror? (See page 177.)

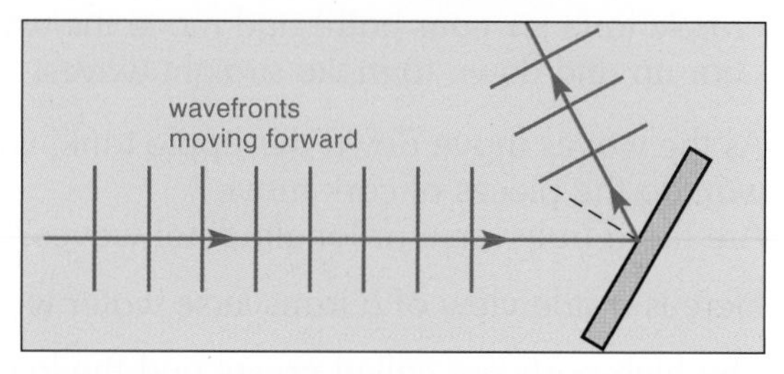

Experiment 21.6   Reflection at a curved mirror
Place a curved barrier in the ripple tank. Again send single waves across the tank and look carefully for the reflected waves. What do you see?

The waves are ***converged*** to a point called the ***focus***: (See also page 181).

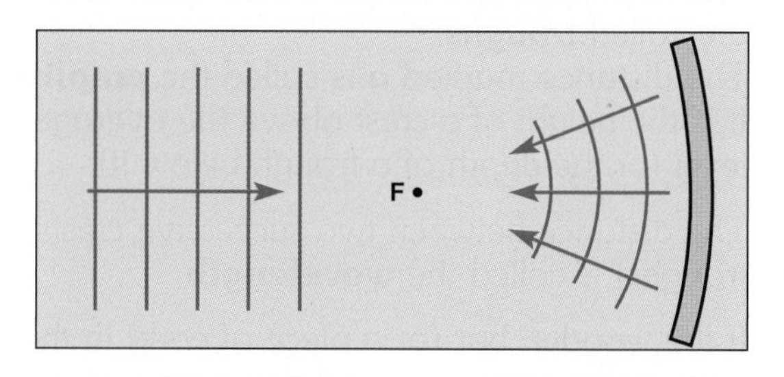

Experiment 21.7
Repeat experiment 21.6 with the curved mirror turned round the other way.

Can you see that the waves are spread out or ***diverged*** by the mirror? (See also page 181.)

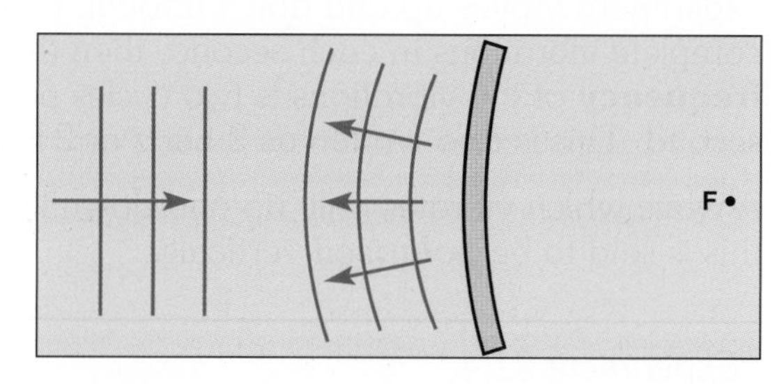

Experiment 21.8   Refraction
Place a thick sheet of glass in the tank and adjust the depth so that the water is very shallow over the glass sheet.
Again send waves across the tank and look at the waves when they travel through the shallow water.

What do you see?   What happens to:
a) the speed and b) the wavelength of the waves when they move into the shallow water?

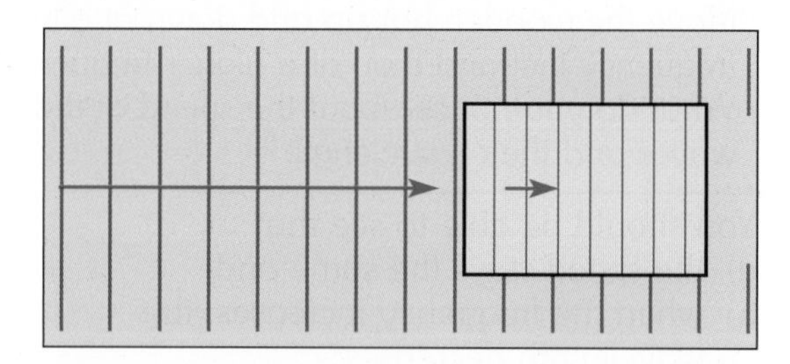

Experiment 21.9   Refraction
Now place the sheet of glass at an angle and repeat the experiment.

What happens to the ***direction*** of the waves as they slow down in the shallow water?

When waves change their speed and direction like this, we say they have been ***refracted***. Sound and light waves can also be refracted (see page 184).

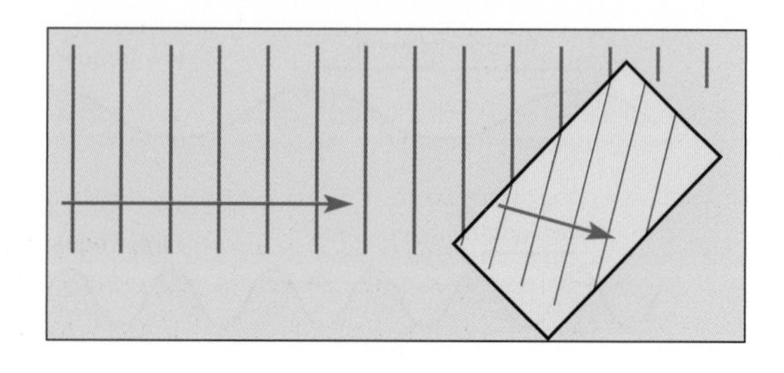

▷ Diffraction

Do waves change shape as they go through a gap in a wall?

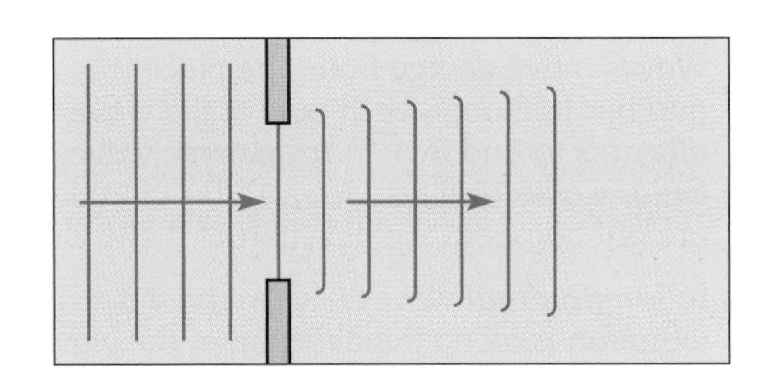

Experiment 21.10 Wide gap
Use a ripple tank to send straight waves across water to a wide gap between two barriers.

Look at the waves after they have passed through the gap. Are they still perfectly straight?

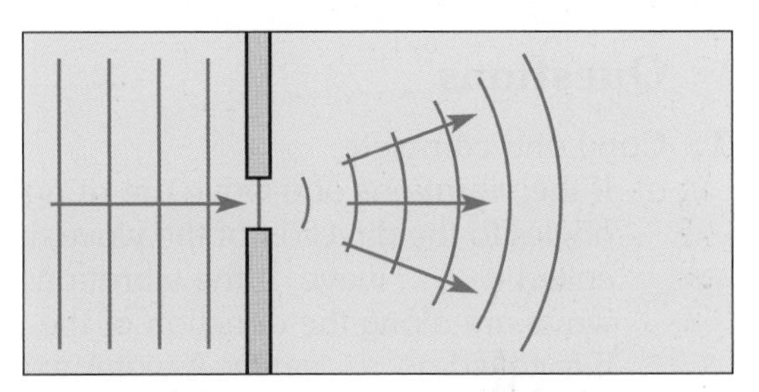

Experiment 21.11 Narrow gap
Now send straight waves through a narrow gap.

What happens to the shape of the waves?

In the first diagram, with a wide gap, the waves travel almost straight on ('rectilinear propagation'). There is a clear shadow behind each barrier.

In the second diagram, the waves spread out after passing through the gap. This is called ***diffraction***.

Diffraction is most obvious when the ***width*** of the gap is about the same size as the ***wavelength*** of the waves.

The photograph shows some sea waves passing through into a harbour. If you look carefully you can see the waves spreading out in the harbour:

Sound waves are diffracted through doorways, because the wavelengths of sounds are about the same size as doorways. This is one reason why you can hear people without seeing them.

Light is seen to be diffracted only if it passes through a very narrow slit. This shows us that ***light is a wave motion***, with a very small wavelength.

Radio and TV waves can also be diffracted. (Your teacher may be able to show you this with 3 cm radar waves.)

This affects the lives of people living near hills:

Radio waves (longer wavelength) can diffract round the hill to be received by the house, but TV waves (short wavelength) do not bend as much. The house will have poor TV reception.

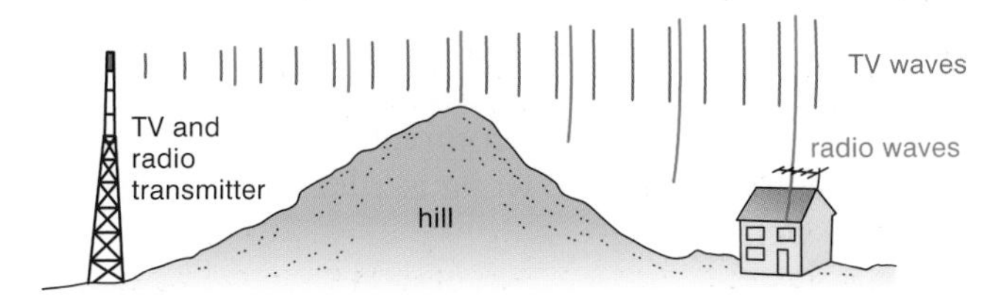

Summary

Waves move energy from one place to another (although each part of the wave only ***vibrates*** to and fro). In ***transverse*** waves (like water waves and light waves) this vibration is at right angles to the direction of the wave.

In ***longitudinal*** waves (like sound waves) the vibration is along the direction of the waves.

The shorter the wavelength, the higher the frequency.

For ***all*** waves: speed = frequency × wavelength

Waves can be reflected and refracted.

Waves can be diffracted through a narrow opening.

▷ Questions

1. Copy and complete:

a) If the vibrations of a wave are at right angles to the direction of the wave, it is called a . . . . wave. If the vibrations of a wave are along the direction of the wave, it is called a . . . . wave. A water wave is a . . . . wave and a sound wave is a . . . . wave.

b) The wavelength of a wave is the . . . . between two successive . . . . (or . . . .). The frequency of a vibration is measured in . . . . or . . . . .

2. a) The speed of a wave (measured in . . . .) is equal to the frequency (measured in . . . .) multiplied by the . . . . (measured in . . . .).

b) When waves travel through a narrow gap in a wall, they are . . . . .
This effect is most obvious when the . . . . of the gap is similar to the . . . . .

3. A certain sound wave has a frequency of 170 hertz (cycles per second) and a wavelength of 2 metres. What is the speed of sound?

4. A radio station produces radio waves of frequency 200 000 hertz (cycles per second) and a wavelength of 1500 metres.

a) What is the speed of radio waves?

b) Another station produces waves at 500 kHz. What is their wavelength?

5. The ***time period*** of a wave, T, is the time taken to produce one complete wave.
If the frequency of a wave is 2 Hz, what is its time period?

6.

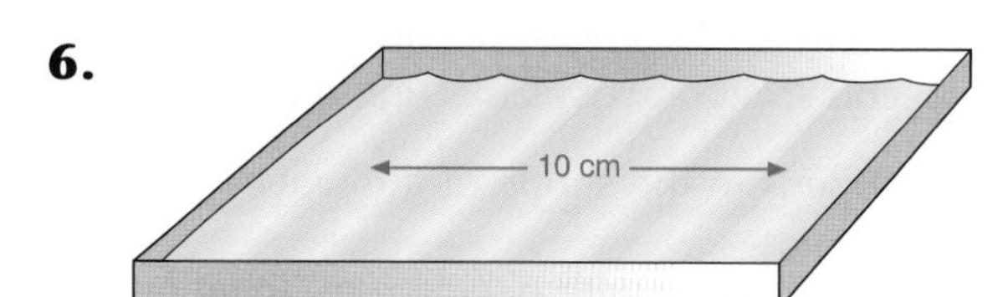

The diagram shows some waves in a ripple tank travelling at a speed of 6 cm/s.
What is a) the wavelength b) the frequency?

7. The diagram shows straight waves approaching a straight barrier at an angle of 45°. Draw a diagram to show how the waves are reflected from the 'mirror'.

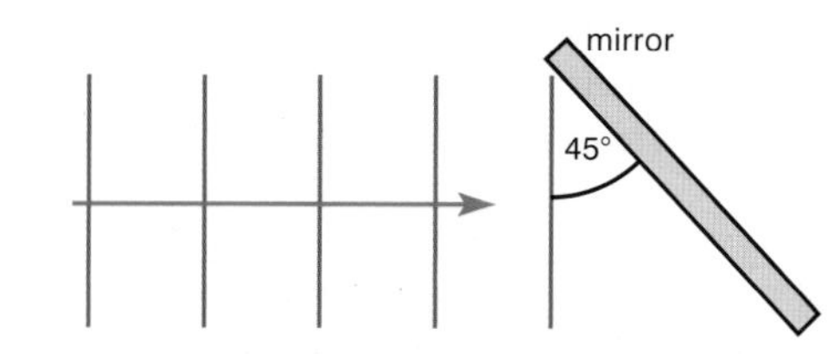

8. A vibrating source **S** produces circular water waves near a straight reflector:

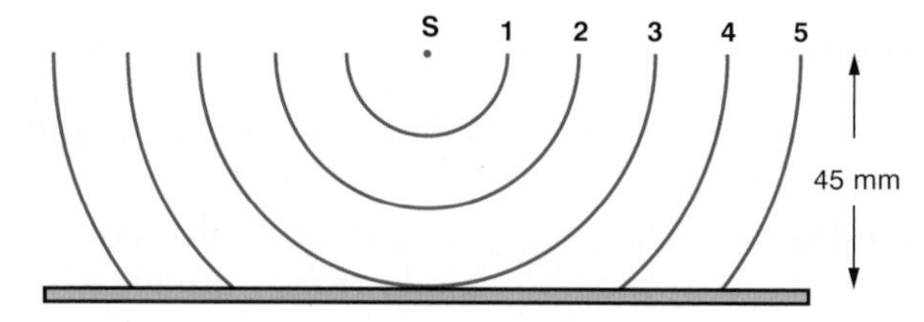

a) Copy and complete the diagram to show how crests 4 and 5 are reflected.

b) What is the wavelength of these waves?

c) If their speed is 60 mm/s, what is the frequency?

Further questions on page 236.

chapter 22

Visible light is a form of energy that we can detect with our eyes.

Objects that produce their own visible light are called ***luminous sources*** (for example, the Sun, other stars, lamps, televisions, glow-worms).

Other objects are ***illuminated*** by this light and reflect it into our eyes (for example, this page and most objects in this room, and the Moon).

Light waves

Light is a wave motion, rather like the water waves that you can see on a pond or in a ripple tank:

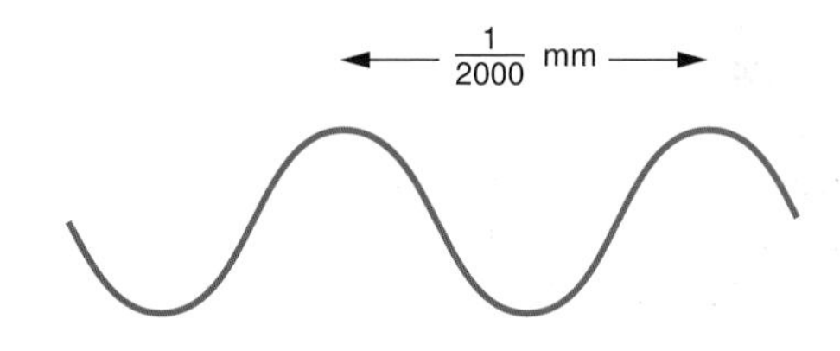

The light that we can see has a ***wavelength*** of only about $\frac{1}{2000}$ of a millimetre!

The speed of light

Light travels at a very high speed – about a million times faster than the speed of sound.
Light takes only 8 minutes to travel from the Sun, – a fast car would take about 150 years!
The speed of light in air is 300 million metres ***per second***.

A ***light-year*** is a distance – how far is it?

Rays of light

For experiments with visible light, we often use a very narrow beam of light called a ***ray***.

> *Experiment 22.1*
> Use a ***raybox*** to shine a narrow beam of light across a piece of paper.

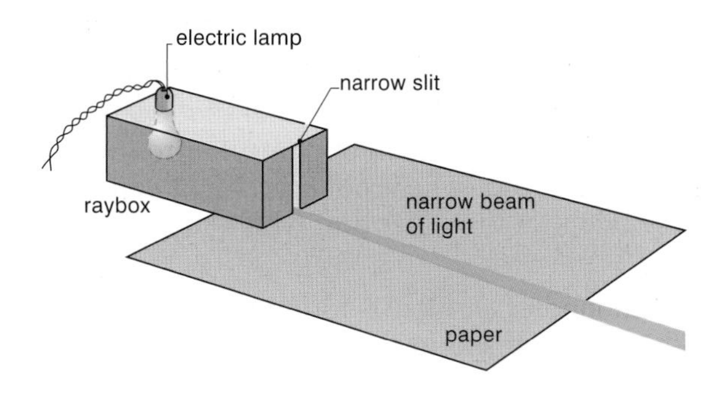

What do you notice about the edges of the beam?
Why can't you see round corners?

Light travels in straight lines.
It is 'rectilinear'.

▷ Shadows

Shadows are formed when some rays of light continue to travel in straight lines, while other rays are stopped by an object.

The kind of shadow that is formed depends on the size of the source of light:

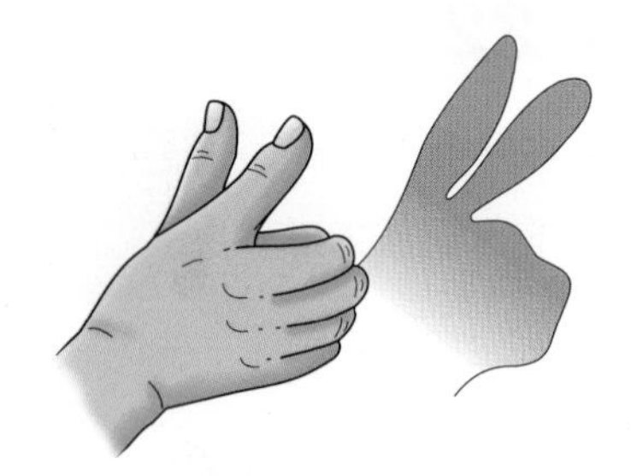

1 Shadow due to a small source of light

> *Experiment 22.2*
> Hold a table-tennis ball between a ***small*** electric lamp and a white screen.

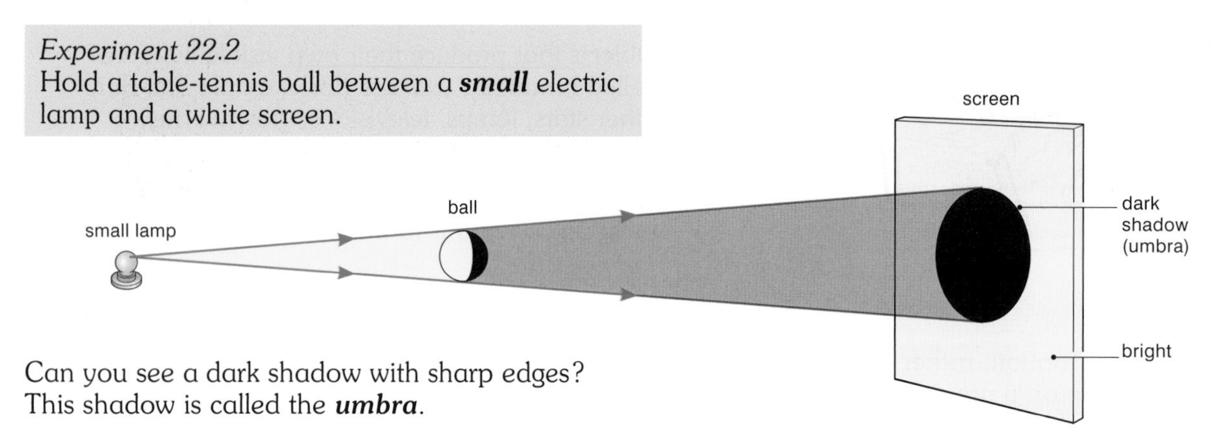

Can you see a dark shadow with sharp edges?
This shadow is called the ***umbra***.

2 Shadow due to a large source of light

> *Experiment 22.3*
> Change the small lamp for a **large** pearl lamp.

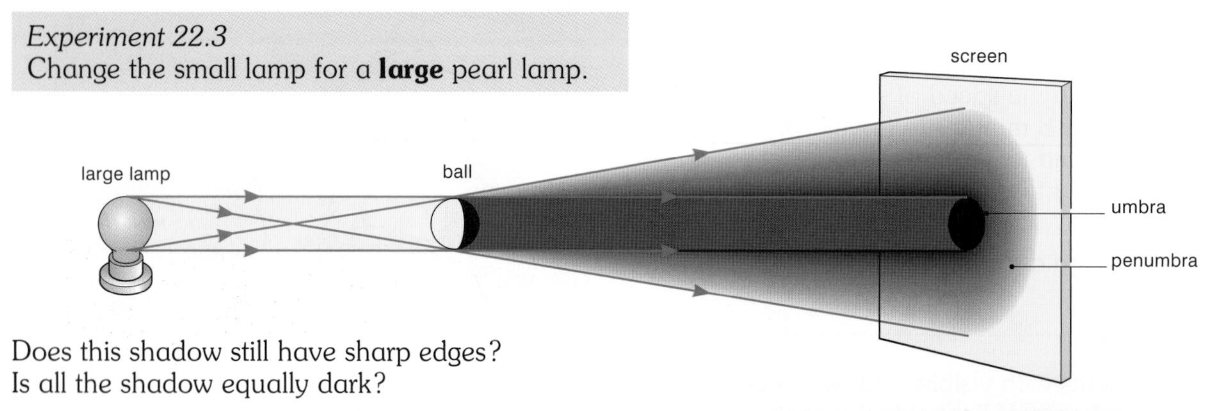

Does this shadow still have sharp edges?
Is all the shadow equally dark?

The dark centre of the shadow is still called the umbra.
No light from the lamp reaches the umbra.
It is surrounded by a fuzzy partial shadow or ***penumbra***.
The penumbra varies in brightness – from very dark near the umbra (where not much light can reach it) to very bright at the outer part where more light can reach.

Which kind of shadow is formed on a clear sunny day?
Which kind of shadow is formed on a cloudy day?

How many things can you deduce from this photo?

▷ Eclipses

Eclipses occur because the Moon and the Earth cast large shadows.

solar eclipse

1 Eclipse of the Sun (solar eclipse)

This happens when the Moon comes between the Sun and the Earth, so that the Earth (in the shadow) darkens during the day.
At the time of total eclipse, only the flames of the outer edge of the Sun can be seen.

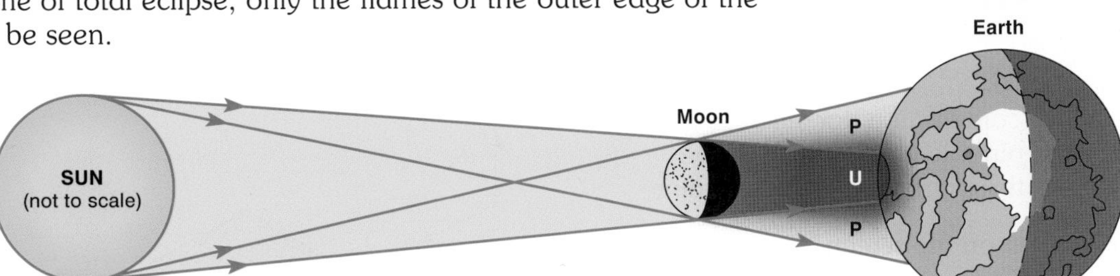

> *Experiment 22.4*
> Set up a model of an eclipse of the Sun (in a darkened room) using a large pearl lamp for the Sun, a table-tennis ball for the Moon and a white football for the Earth.

Total eclipses of the Sun do not occur often. After the eclipse in August 1999, the next total eclipse in Britain is on 7 October 2135.

2 Eclipse of the Moon (lunar eclipse)

This happens when the Moon moves into the Earth's shadow. When the Moon is in the Earth's umbra, no sunlight can reach the Moon and it is seen to go dark.

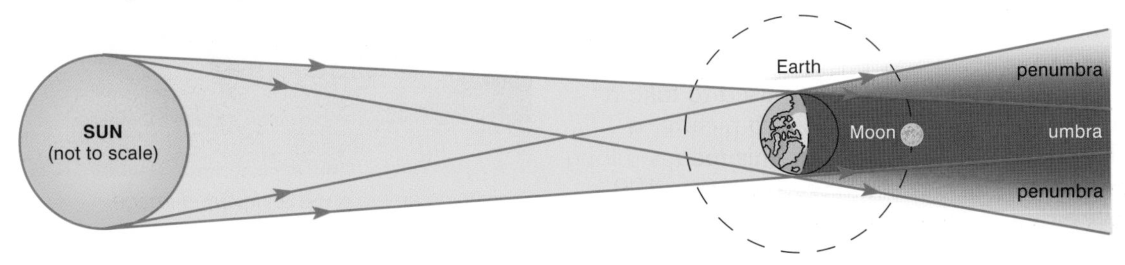

> *Experiment 22.5*
> Use the same apparatus as in experiment 22.4 to show an eclipse of the Moon.

Eclipses of the Moon can be seen much more often than eclipses of the Sun. Try to find out the date of the next eclipse of the Moon to be seen from where you live.
How long does it take for the Moon to move once round the Earth?

partial lunar eclipse

▷ The pinhole camera

A simple pinhole camera consists of a closed box with a screen at one end and a small pinhole at the other end.
It uses the fact that the light going in through the pinhole travels in straight lines to the screen.

The diagram shows one way to make a pinhole camera.

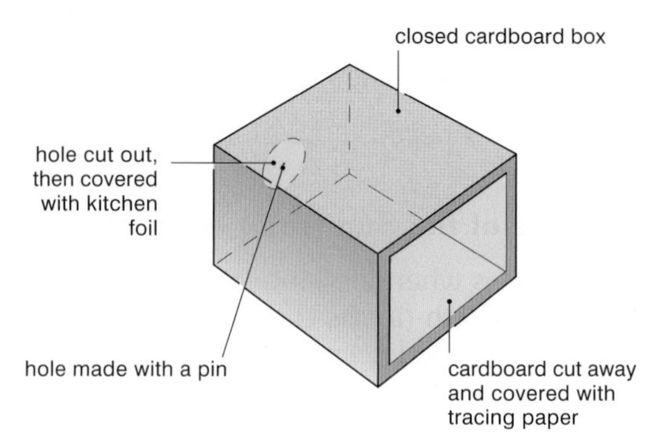

> *Experiment 22.6*
> Use a pinhole camera in a darkened room and point the pinhole at a lamp or at the outside world through a gap in the curtains.

What do you see on the screen?

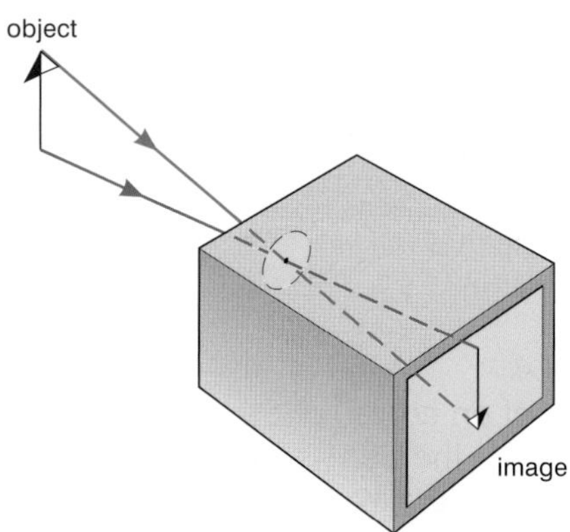

The picture that you see is called an ***image***. In this case, because the picture is formed on a screen, it is called a ***real*** image.
In the pinhole camera the image is ***inverted*** – it is turned upside-down and turned left-to-right.
The diagram shows just two of the many rays of light going through the pinhole.

> *Experiment 22.7*
> Look carefully at the image on your screen and notice how sharp and dim it is.
> Then use a pencil to make the pinhole larger.

Is the image sharper or more blurred?
Is the image dimmer or brighter?
Why does this happen?

This kind of camera is ***not*** often used to take a photograph because to get a sharp picture, a small pinhole must be used and this means a long exposure time for the film. (See also page 199.)

A pinhole camera gives an inverted image

Improve your eye-sight without spectacles

If you normally wear glasses you can also make things look sharper by looking through a small pinhole in a piece of card.
In this way you are converting your eye (which is like a lens camera, see page 198) to a pinhole camera.

Summary

Shadows and eclipses occur because light travels in straight lines.

A small source of light causes a sharp shadow, called the umbra.
A large source of light causes a blurred shadow with two parts, the umbra and the penumbra.

In a pinhole camera, the image is real and inverted.
A small hole gives a sharp but dim image.
A large hole gives a bright but blurred image.

A light-year is the distance travelled by light in one year.

▷ Questions

1. Copy out and complete:
   a) Light travels in . . . . lines.
   b) A shadow formed by a small source of light has . . . . edges and is called an . . . . .
   A large source of light causes a shadow with two parts called the . . . . and the . . . . .
   c) In an eclipse of the Sun, the . . . . comes between the Sun and the . . . . .
   In an eclipse of the Moon, the . . . . comes between the . . . . and the . . . . .
   d) In a pinhole camera, the image is . . . . and . . . . .
   If the pinhole is made larger, the image becomes . . . . and . . . . .

2. How could a detective tell from a photograph, without seeing the sky, whether it was a bright clear day or cloudy?

3. *Professor Messer thinks he's clever,*
Put him right or he's gone for ever:

4. A laser beam of light can be bounced off the Moon (from a special mirror left there by astronauts). The light travels there and back in 2.6 seconds. What is the distance to the Moon?
(Speed of light = 300 000 000 m/s = 3×10^8 m/s)

5. Draw a diagram to explain an eclipse of the Sun. Label all parts of the diagram, including the umbra and penumbra.
Why is your diagram not drawn to scale?

6. Draw a diagram of a pinhole camera with two rays passing from the object through the pin-hole. How can you make the image
a) sharper b) brighter c) larger?

7. Professor Messer fell asleep and dreamed that light no longer travelled in straight lines.
Why was his dream a nightmare?

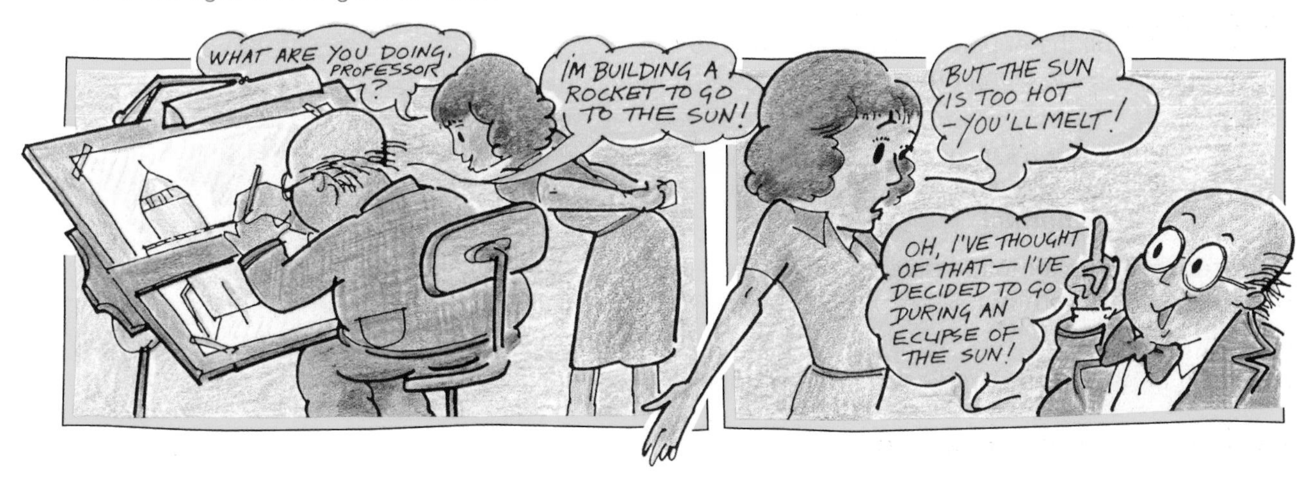

chapter 23

REFLECTION

Have you ever reflected sunlight from a mirror to dazzle someone or to flash messages to a friend?

Experiment 23.1 Investigating reflection

Get hold of a ***plane*** (flat) glass mirror and look at it carefully.
Where is the light reflected from – the glass at the front of the mirror or the 'silvering' at the back?

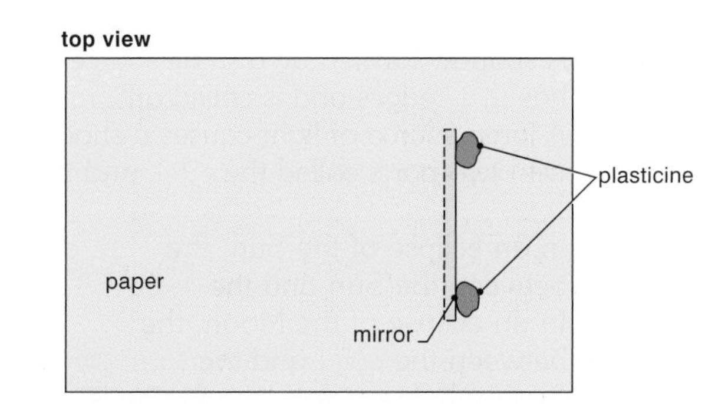

Place the plane mirror on a sheet of white paper, holding it vertical with plasticine or a block of wood.

Draw a line along the ***back*** of the mirror (where the light is reflected).

Use a raybox to shine a ray of light on to the mirror. This is called the ***incident ray***.

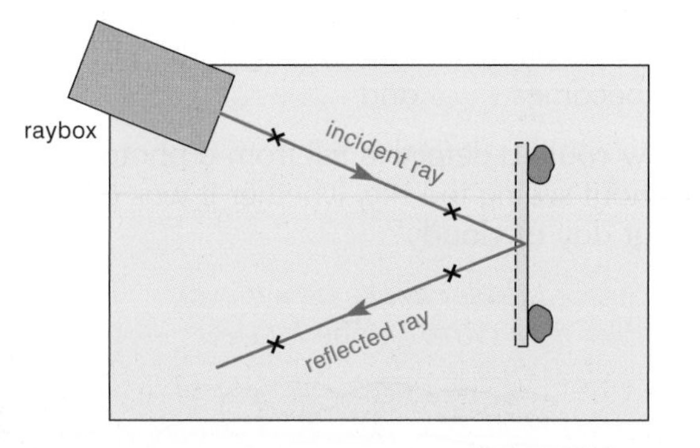

Can you see the reflected ray of light?

Mark the position of each ray with crosses as shown in the diagram.

Now take off all the apparatus and use a ruler to join up the crosses, along the lines of the rays of light.

Use a protractor or a set-square to draw a line at right angles (90°) to the mirror, as shown. This line is called the ***normal***.

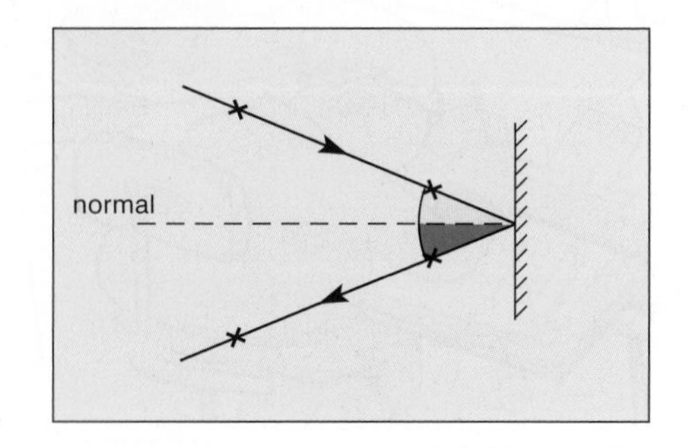

Now use a protractor to measure the angles on each side of the normal.
What do you find?

Now try the experiment again with different angles.

What do you find each time?

The diagram shows the correct scientific words:

The two rays and the normal line are ***all in the same plane*** (the plane of this piece of paper).

The **law of reflection**:

the angle of incidence	=	**the angle of reflection**

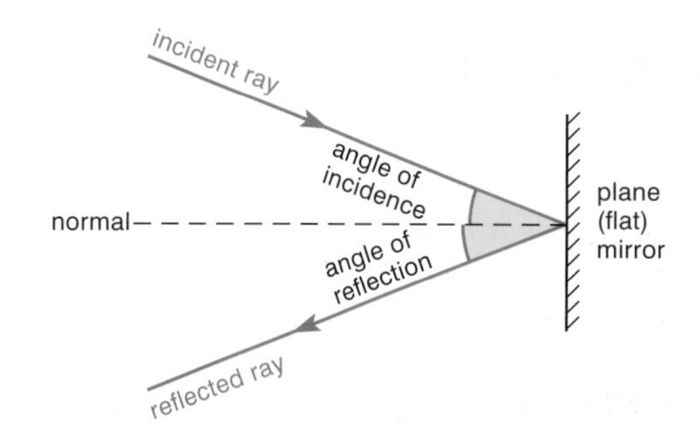

Water waves are reflected in the same way as light waves (see the ripple tank experiments on page 168).

If the angle of incidence is 45°, what is the angle of reflection and what is the angle turned by the ray of light?

The periscope

Periscopes are used in submarines so that people below the surface can see what is happening above the surface.

The diagram shows how light is reflected by the two mirrors.

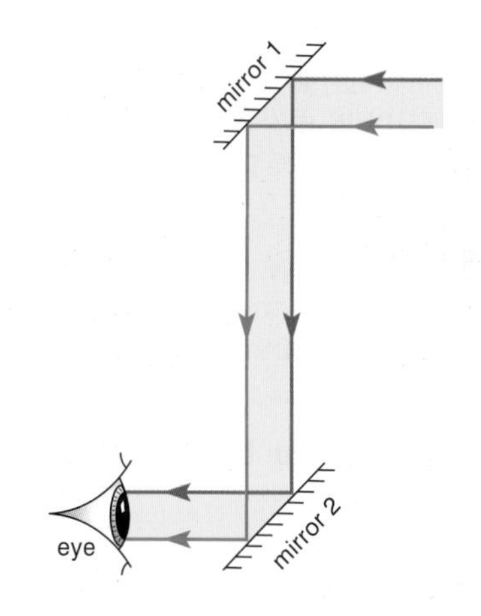

Remember that the angle of incidence = the angle of reflection and the rays should be turned through 90° each time. At what angle should each mirror be placed?

Building a periscope

You can use two handbag mirrors and a plastic bottle (like a washing-up-liquid bottle) to build a periscope which you can use to see over people's heads or to spy round corners.

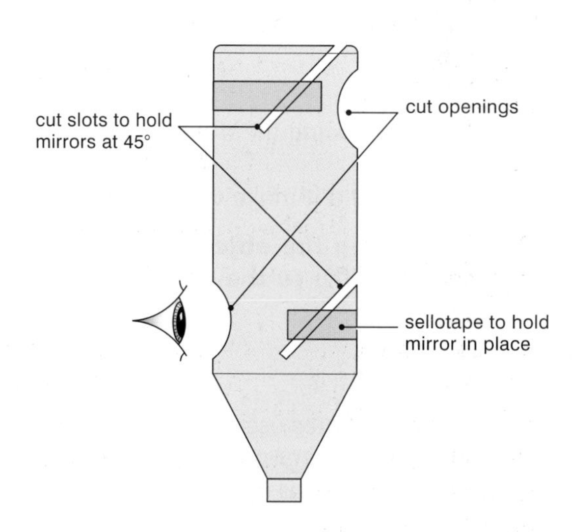

▷ The image in a plane mirror

When you look into a plane (flat) mirror, you can see a picture or ***image*** of yourself. Whereabouts do you think this image is – on the mirror or behind the mirror?

Experiment 23.2 To find out where the image is
Use a vertical sheet of clear glass as a mirror and place a lighted candle (or Bunsen burner) on the bench in front of it. (Take care.) ⚠

Then move an unlighted candle or Bunsen burner (of the same height) into a position behind the mirror so that it ***appears*** to be burning.

When the image of the flame appears to be on the unlighted candle, we know where the image is – it must be in the same position as the unlighted candle.

Measure the distance of the image ***from the mirror*** and the distance of the lighted candle (called the object) ***from the mirror***. What do you notice?

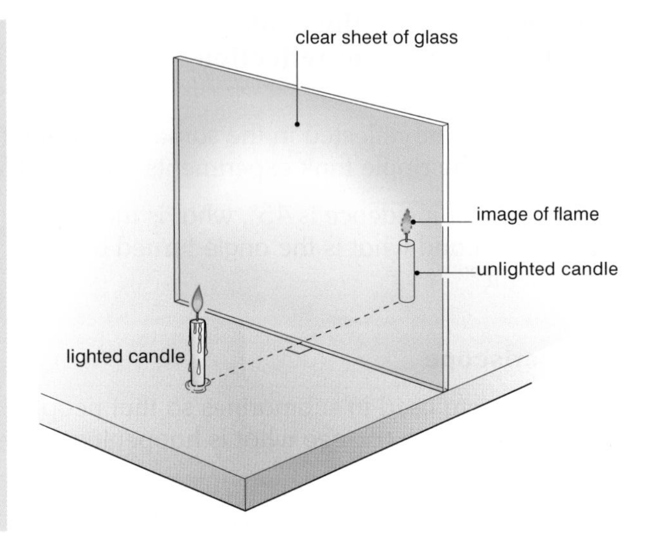

The image is as far behind the mirror as the object is in front.

This means that if your face is 30 cm in front of a mirror, then your image is 30 cm behind the mirror.

The diagram shows how the reflected rays appear to come from the image:

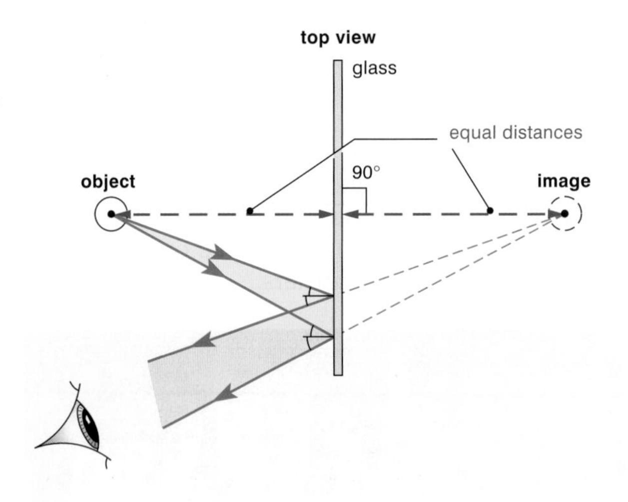

Imagine a line joining the object to the image in this experiment.
What angle would it make with the mirror?

The line joining the object to the image is at right angles (90°) to the mirror.

'Imagic' tricks

Magicians sometimes use images in their tricks.
If you put your finger on the <u>un</u>lighted candle in this experiment, it looks as if your finger is burning!

Another trick is to fix the unlighted candle in a tall beaker which you then fill with water so that it appears to burn underwater.
Can you think of some other tricks like these?

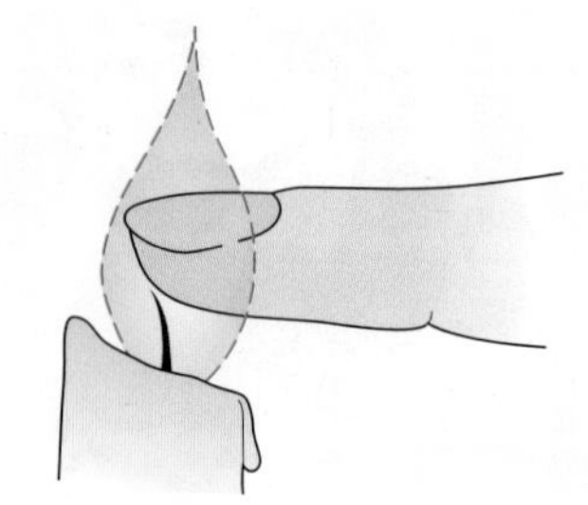

Virtual images

If you go behind the mirror to look for the image, you cannot see it.
This ***virtual*** image only ***appears*** to be behind the mirror (because the rays of light are reflected to your eye as though they came from behind the mirror). This virtual image cannot be formed on a screen (as a real image can, see page 174).

Look at the ***size*** of the real flame compared with its virtual image.
What do you notice?

The image in a plane mirror is virtual and is the same size as the object.

Experiment 23.3
Look into a mirror and hold your ***right*** ear.
Which ear is your image holding?

Write your name on a piece of paper and pin it to your chest as a label. What do you notice when you look in the mirror?
This is an example of ***lateral inversion***.

The image in a plane mirror is laterally inverted.

A tattooist once wrote on my throat
A short but mysterious note
And so to inspect it
I tried to reflect it
And found I'd got ƎTAOЯHT *on my throat!*

Regular and diffuse reflection

We have been looking at ***regular*** reflection from mirrors and other shiny flat surfaces.
Other surfaces, like this sheet of paper, also reflect light but because the surface is rough, the light is reflected off in all directions.
This is *diffuse* reflection.

If a surface does not reflect any light at all, it looks dull black, like this black ink.

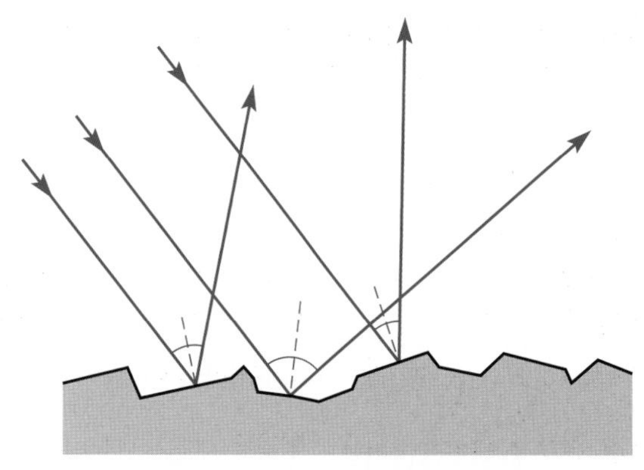

Diffuse reflection from a rough surface

Summary

The law of reflection:

the angle of incidence = the angle of reflection

The image in a plane mirror is:
a) as far behind the mirror as the object is in front, with the line joining the object and the image at right angles (90°) to the mirror
b) the same size as the object
c) virtual
d) laterally inverted.

▷ Questions

1. The angle between an incident ray and the mirror is 30°.
a) What is the angle of incidence?
b) What is the angle of reflection?
c) What is the total angle turned by the ray?

2. Complete this verse:

A periscope's a useful thing,
As everyone agrees.
It has . . mirrors fixed inside,
At . . degrees.

3. A boy with a mouth 5 cm wide stands 2 m from a plane mirror. Where is his image and how wide is the image of his mouth?
He walks towards the mirror at 1 m/s. At what speed does his image approach him?

4. AMBULANCE (printed mirror-reversed) Explain where you would see this sign and explain why it is written like this.
Write STOP as it should be written on the front of a police car.

5. In experiment 23.2 you can often see ***two*** images, close together. Why is this?

6. Explain how to read the following message which was found on some blotting paper:

Meet me at midnight (handwritten, mirror-reversed)

7. Explain what is happening in the photo.

8. Draw a labelled ray diagram to show this:

A periscope builder called Fred,
Fixed two mirrors to the head of his bed.
The light was reflected,
And so he inspected,
The hair on the top of his head.

9. Professor Messer fell asleep and dreamed that all objects stopped reflecting light. Why was his dream a nightmare?

10. Explain, with a diagram, how 2 plane mirrors could be fixed at a dangerous T-junction so that motorists approaching the junction could see all other traffic.

11. An object O is placed near a plane mirror:

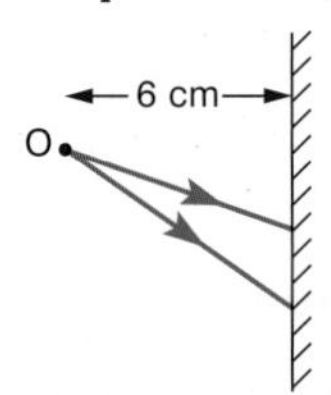

Draw it full size, with two rays from O to the mirror. Working carefully, draw in the two reflected rays and the position of the image.

12. A room with only one window is dark even at mid-day. Explain how you could use wall-paper, furnishings and mirrors to brighten the room.

Further questions on page 236.

chapter 24

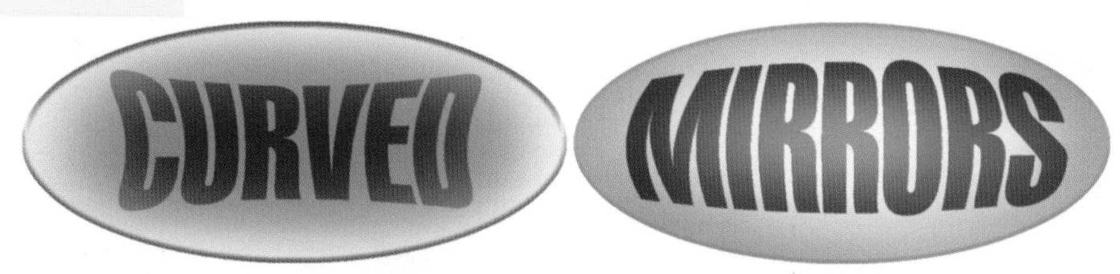

Have you ever looked into a spoon to see an image of your face? Have you noticed that the front and back of a spoon give different images?

concave

convex

Two kinds of curved mirror:
A mirror that curves in (like a cave) is called a ***concave*** mirror.
A mirror that curves outwards is called a ***convex*** mirror.

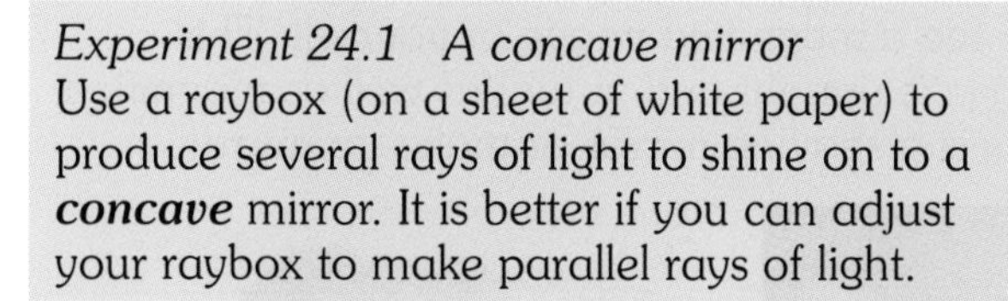

Experiment 24.1 A concave mirror
Use a raybox (on a sheet of white paper) to produce several rays of light to shine on to a ***concave*** mirror. It is better if you can adjust your raybox to make parallel rays of light.

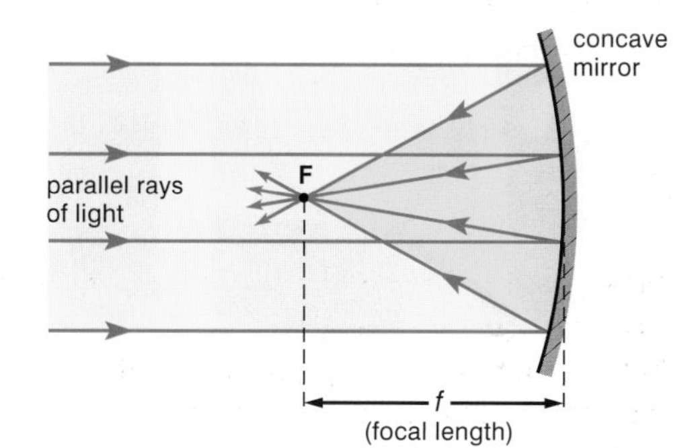

Can you see that the rays coming off the concave mirror are getting closer together – we say they are ***converging***. (See also page 168.)
A concave mirror is a converging mirror.

If the rays of light from the raybox are ***parallel*** rays, the rays reflected off the mirror converge to one point called the ***principal focus*** (marked F on the diagram).

Parallel rays of light are reflected through the principal focus of a concave mirror.

The distance from F to the mirror is called the ***focal length*** of the mirror.

Experiment 24.2 A convex mirror
Repeat the experiment, with a ***convex*** mirror.

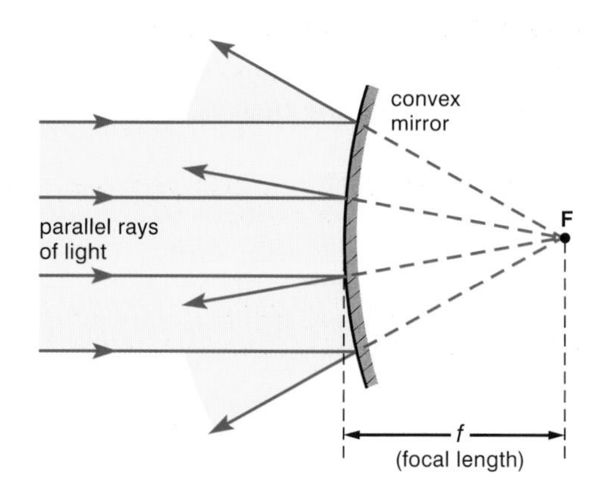

Can you see that the light rays are spread out or ***diverged*** by this mirror?

A convex mirror is a diverging mirror.
Parallel rays of light are reflected so that they appear to come from the principal focus (F) of a convex mirror.

Where have you seen curved mirrors used?

▷ Uses of concave reflectors

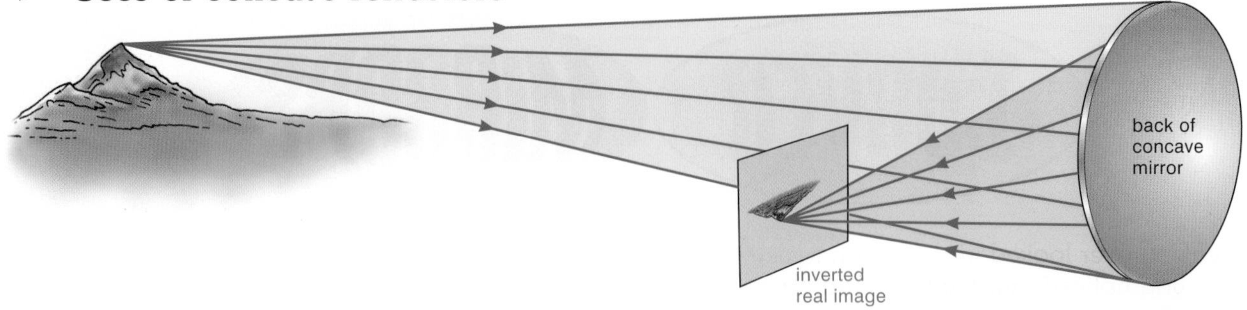

Concave mirrors are used to collect light energy. They are also used to collect sound, heat radiation, radar, and TV signals.

This idea is used in TV satellite dishes, solar ovens (see page 48), telescopes (see page 215), and radar receivers (see page 211).

In an opposite way, energy can be sent out in a beam from a concave mirror. This is used in car headlamps (see Q 4 opposite), film projectors (see page 204), and electric fires (see page 264).

If you move ***close*** to a concave mirror, then you will see a magnified image of yourself. This image is upright and virtual as well as magnified. This is a shaving mirror or make-up mirror:

Satellite 'dishes' collect energy for TV sets

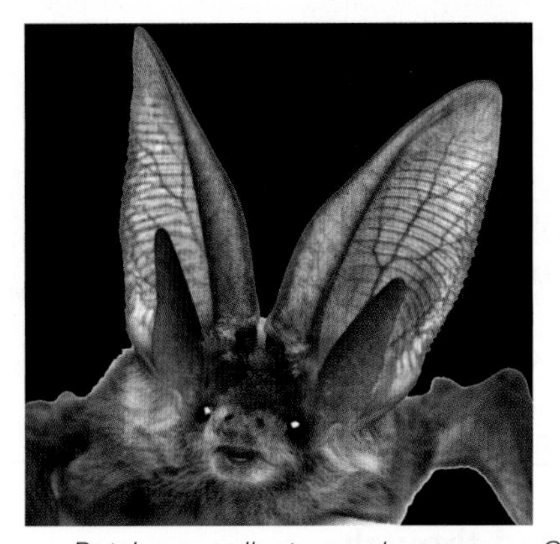

Bats' ears collect sound energy

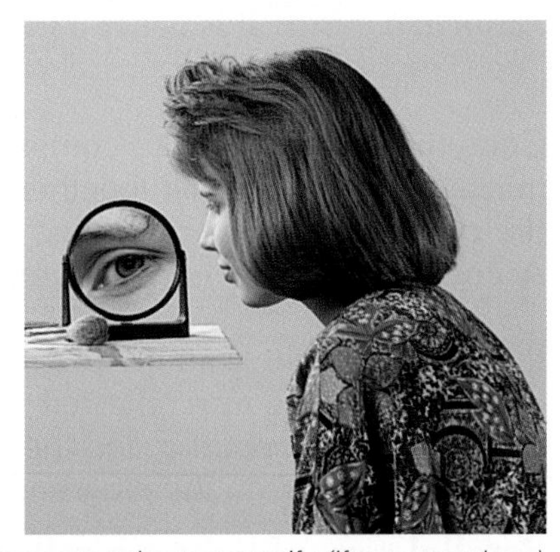

Concave mirrors magnify (if you are close)

▷ Uses of convex mirrors

Convex mirrors always produce virtual, upright images. The image is always 'diminished' (it is always smaller than the object).

Convex mirrors are useful if you want a wide field of view, such as in a car driving mirror or a shop security mirror:

A hairy young student called Raven
Used a diverging mirror for shaving,
His image was diminished,
And when he had finished,
His shaving looked more like engraving!

Summary

Concave mirrors **converge** parallel rays of light towards a principal focus in front of the mirror.

For distant objects, concave mirrors produce inverted real images. For a near object, the image is upright and magnified and virtual.

Convex mirrors **diverge** parallel rays of light away from the principal focus which is behind the mirror.
The image in a convex mirror is always erect, virtual and diminished.
Convex mirrors have a wide field of view.

▷ Questions

1. Copy and complete:
   a) A concave mirror . . . . rays of light, whereas a convex mirror . . . . rays of light.
   b) Parallel rays of light are reflected by a concave mirror to a point called the . . . . The focal length is the distance from the . . . . to the mirror.
   c) For a convex mirror, parallel rays of light are reflected . . . . a point called the . . . .

2. The diagram shows a dish-aerial which is used to receive television signals from a satellite.

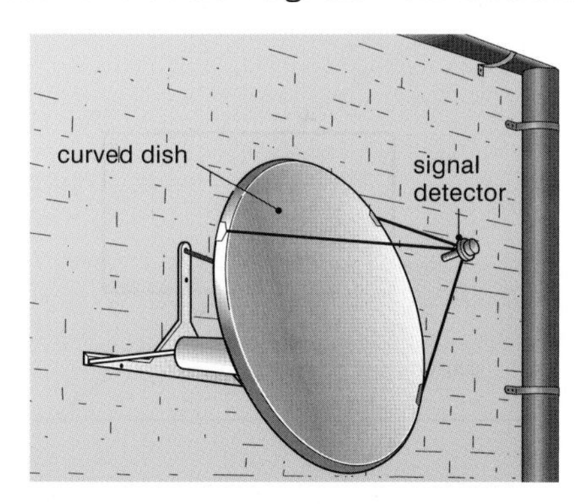

The signal detector (aerial) is fixed in front of the curved dish.
   a) What is the purpose of the dish?
   b) Should it be concave or convex?
   c) Where should the aerial be positioned to receive the strongest possible signal?
   d) Explain what change you would expect in the signal if a larger dish was used.

Further questions on page 236.

3. Draw up two lists headed 'Concave mirrors' and 'Convex mirrors' from the following: shaving mirror, car headlight mirror, searchlight mirror, driving mirror, dentist's inspection mirror, torch mirror, projector lamp mirror, staircase mirror on a double-decker bus, make-up mirror, reflecting telescope mirror, solar furnace mirror, satellite TV dish, shop security mirror.

4. A car headlight bulb contains two filaments labelled A and B in the diagram. Filament B is at the principal focus of the concave mirror. Copy the diagram and draw in the rays of light from each filament. Why is this arrangement used in car headlights?

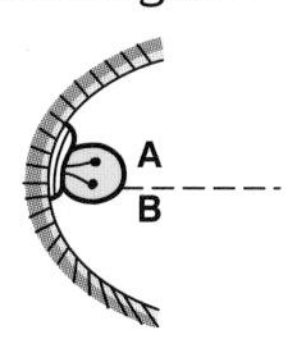

5. *Professor Messer's in a dither,*
He's seeing things in his driving mirror:

chapter 25

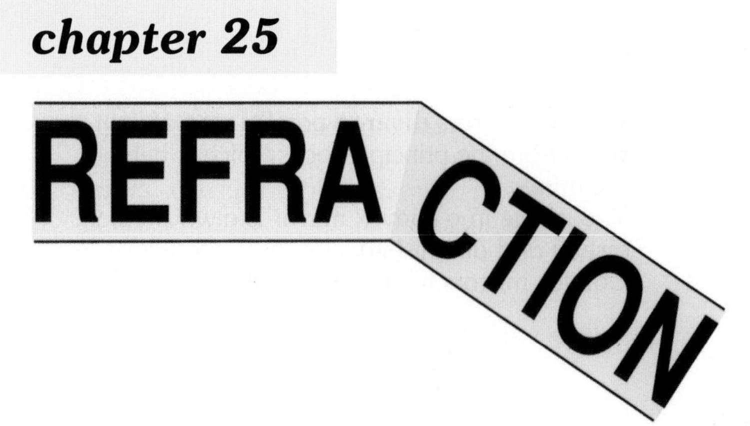

Why do swimming pools look shallower than they really are? Is this pencil really broken?

Before we can answer questions like these, we must find out what happens when light travels from one substance into another.

Experiment 25.1
Place a glass block on a sheet of white paper and draw a line all round it. Use a raybox to shine a ray of light from the air into the glass.

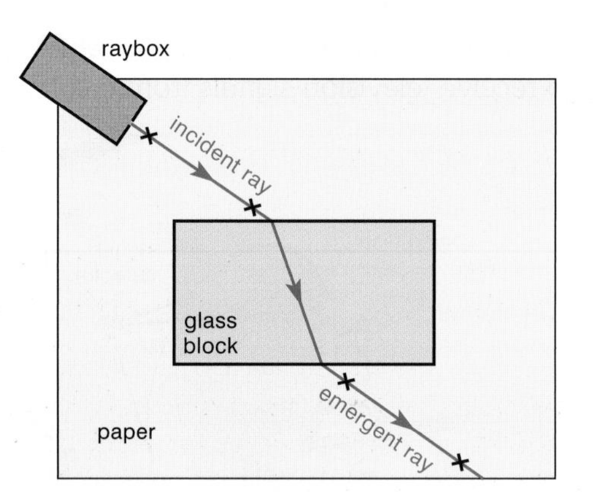

Look carefully for the ray inside the glass and where it comes out again.
Has the ray of light travelled straight into the glass or has it been bent?

Mark the incident ray and the emergent ray with crosses as shown in the diagram.
Then take off all the apparatus and use a ruler to join the crosses and re-construct the path of the ray of light.

You can see that the ray was bent or ***refracted*** where it entered the glass and where it left the glass.

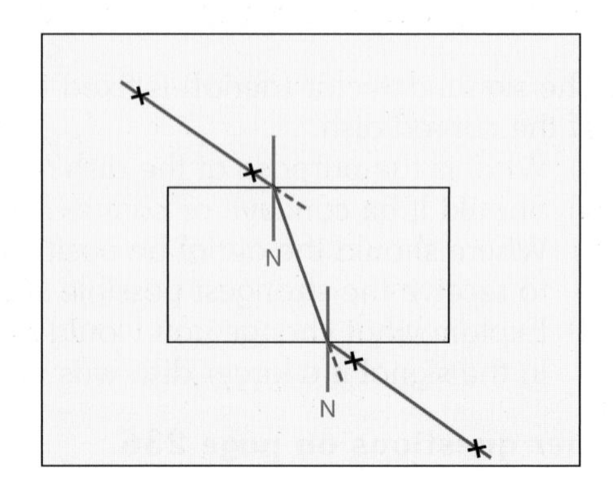

Use a protractor or a set-square to draw lines at right angles (90°) to the glass surface, as shown. Each of these lines is called a ***normal*** (N).

Where is the light bent towards a normal and where is it bent away from a normal?

Rays of light travelling from air into glass are bent or refracted *towards* the normal.

Rays of light travelling from glass into air are refracted *away from* the normal.

Here are some of the scientific words used to describe experiment 25.1.
Angle ***i*** is called the ***angle of incidence***.
Angle ***r*** is called the ***angle of refraction***.

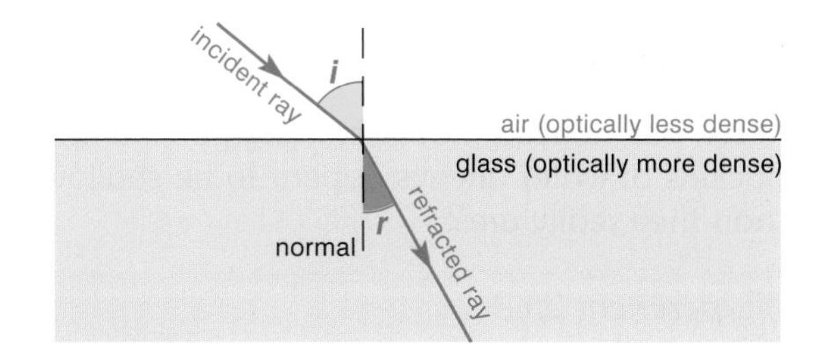

Changing speed

We saw on page 171 that the speed of light in air is about 300 000 000 metres/second.
But light travels more ***slowly*** in glass – in fact only two-thirds as fast.
This is the reason why light is refracted – in the same way as a fast car travelling on the road is 'refracted' if it moves at an angle on to thick mud.
The mud slows down one side of the car, and the path of the car bends:
The more it is slowed, the more it bends.

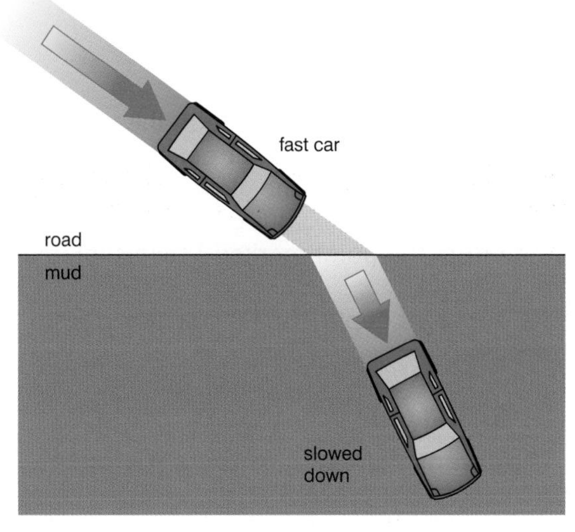

A motorist drove down a straight country road,
When the kerb gave the steering a twitch,
One wheel was slowed by the edge of the road,
And refracted him into a ditch.

Look again at the ripple tank experiment 21.9 on page 168. This shows how waves are refracted when they slow down.
When the waves slow down, the wavelength gets shorter, but the frequency stays the same.

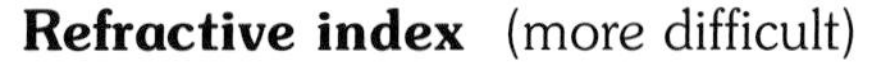

Refractive index (more difficult)

A number called the ***refractive index*** of a substance is the ratio of the two speeds:

$$\textbf{Refractive index of a substance} = \frac{\textbf{speed of light in air}}{\textbf{speed of light in substance}}$$

The refractive index of most kinds of glass is about 1.5.
The refractive index of water is less (1.33) which means that the light is not bent as much when it enters water, because it is not slowed as much.

Experiment 25.2 (more difficult)
In experiment 25.1, measure ***i***, the angle of incidence, and ***r***, the angle of refraction.
Then use your calculator or a book of mathematical tables to find 'sine ***i***' and 'sine ***r***'.
The refractive index of the glass can be found from another formula called ***Snell's Law***:

$$\textbf{Refractive index} = \frac{\textbf{sine } i}{\textbf{sine } r}$$

Substance	Refractive index	Speed of light (m/s)
Air	1.0	300 000 000
Water	1.33	225 000 000
Perspex	1.5	200 000 000
Glass	About 1.5	200 000 000
Diamond	2.4	120 000 000

▷ Real and apparent depth

Have you noticed that swimming pools and buckets of water always appear to be shallower than they really are?

Experiment 25.3
Place a thick glass block or a beaker of water on top of some writing.

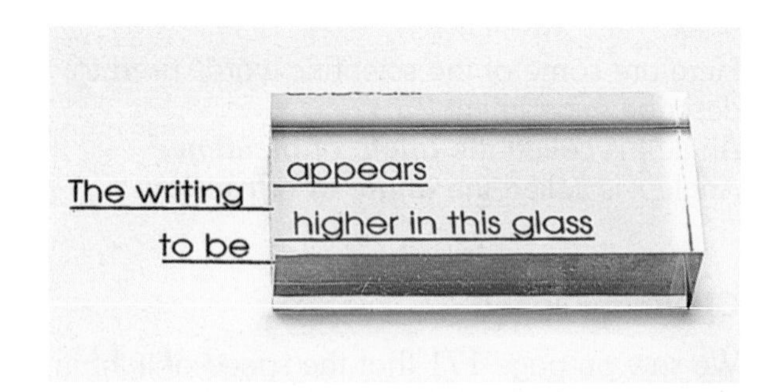

What do you notice?

The apparent depth is ***less*** than the real depth because rays of light are refracted away from the normal as they leave the glass or the water. When they are received by your eye, they ***appear*** to come from the point marked I (a virtual image):

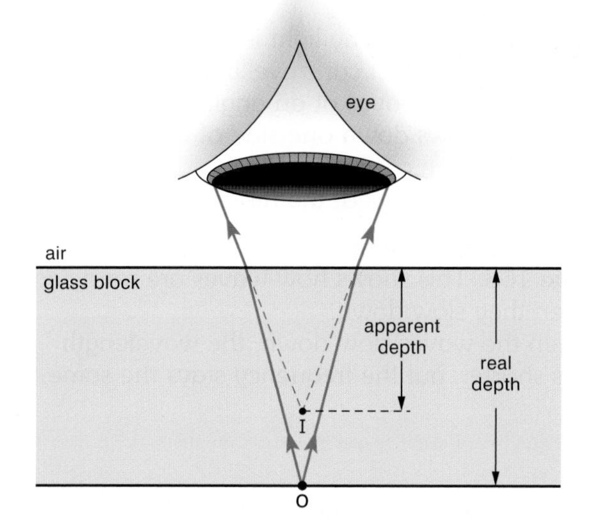

The apparent depth can be calculated from the formula:

$$\frac{\textbf{Real depth}}{\textbf{Apparent depth}} = \textbf{refractive index of the substance}$$

This means that water is really 1.33 times as deep as it appears to be; and glass is about 1.5 times as deep as it appears to be.

Experiment 25.4 An 'imagic' trick
Put a coin in the bottom of an empty cup and position your head so that the coin is just out of sight.

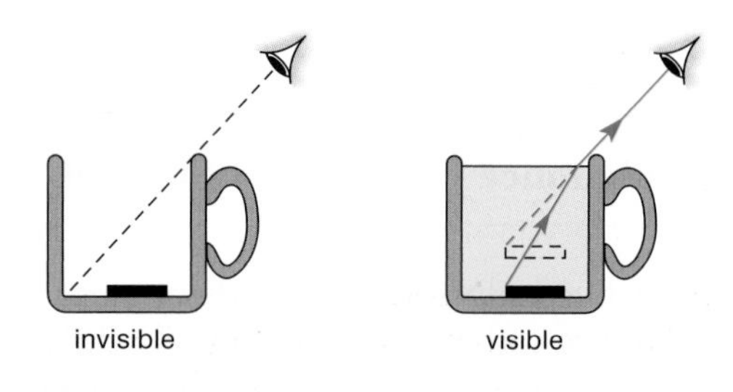

How can you bring it into view without moving your head or the cup or the coin (and without using a mirror)?

Simply pour in water. Rays of light from the coin are refracted away from the normal and into your eye so that you can see it. Its apparent position is higher than its real position.

Experiment 25.5 The 'broken' pencil
Hold a pencil at an angle in a beaker of water.

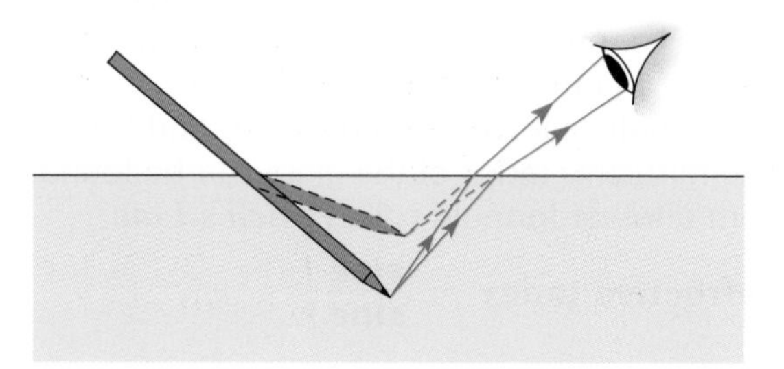

What do you see? (See the photo on page 184.)

Use the diagram to explain why the pencil appears to be broken.

▷ Total internal reflection

Experiment 25.6
Place a semi-circular glass block on a sheet of white paper. Draw a line round it and mark the mid-point of the straight side.
Use a raybox to shine a ray of light at right angles to the curved surface so that it goes straight through to the mid-point of the straight side.

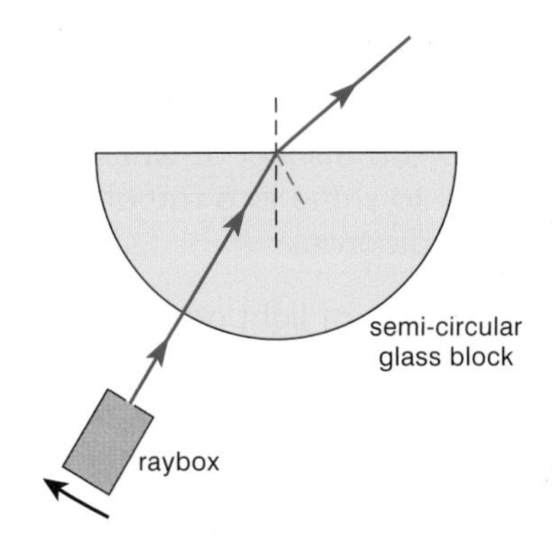

Can you see the ray of light is refracted away from the normal as it comes out of the glass block?
Can you also see that only a small amount of light is reflected back?

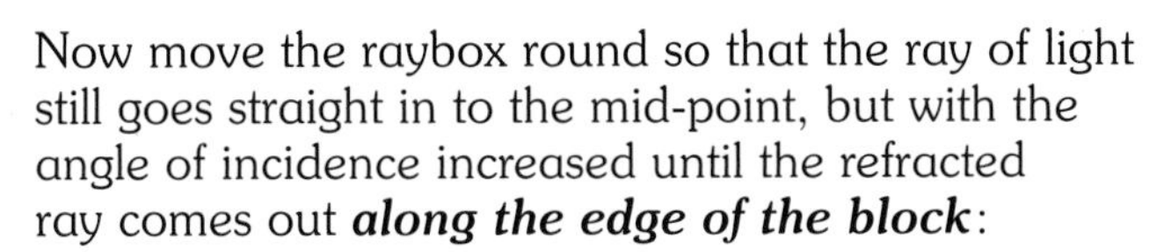

Now move the raybox round so that the ray of light still goes straight in to the mid-point, but with the angle of incidence increased until the refracted ray comes out ***along the edge of the block***:

When this happens, the angle of incidence in the glass is called the ***critical angle*** (***C***).

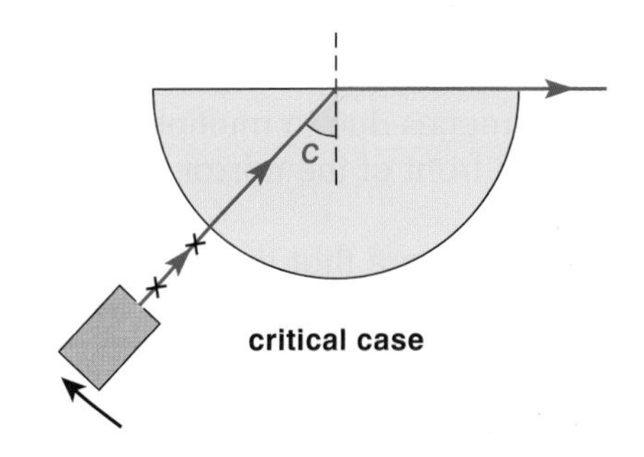

critical case

Mark the ray with crosses so that you can take the apparatus off and re-construct the diagram.
Use a protractor to measure the critical angle (***C***).
Is it the same size as the angle in this diagram?

Now move the raybox to increase the angle of incidence so that it is ***greater*** than the critical angle.
What happens?

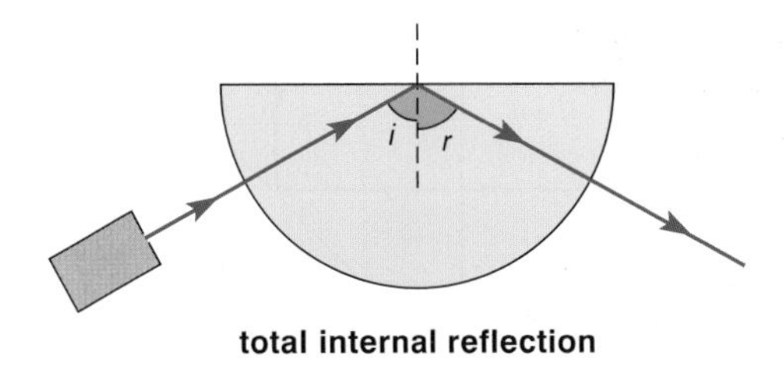

total internal reflection

We call this ***total internal reflection***: 'total' because ***all*** the light is reflected and 'internal' because it only takes place ***inside*** the more dense substance.

As with ordinary reflection, angle $\boldsymbol{i}$ = angle $\boldsymbol{r}$.

Total internal reflection takes place only
when i) the rays are travelling in a dense medium towards a less-dense medium

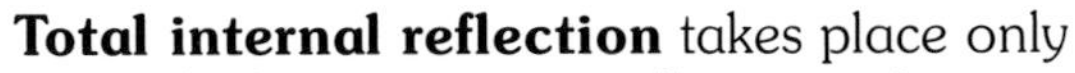

and ii) the angle of incidence is greater than the critical angle.

The critical angle is about 42° for glass, and 49° for water.

More difficult: using the formula on page 185 and referring to the middle diagram above:

$$\text{refractive index} = \frac{\text{sine } 90°}{\text{sine } \boldsymbol{C}}$$

$$\therefore \quad \textbf{refractive index} = \frac{\mathbf{1}}{\textbf{sine } \boldsymbol{C}}$$

▷ Uses of total internal reflection: totally reflecting prisms

Experiment 25.7
Put a prism of glass with angles measuring 45°–45°–90° on a sheet of white paper.
Use a raybox to shine in a ray of light as shown in the diagram:

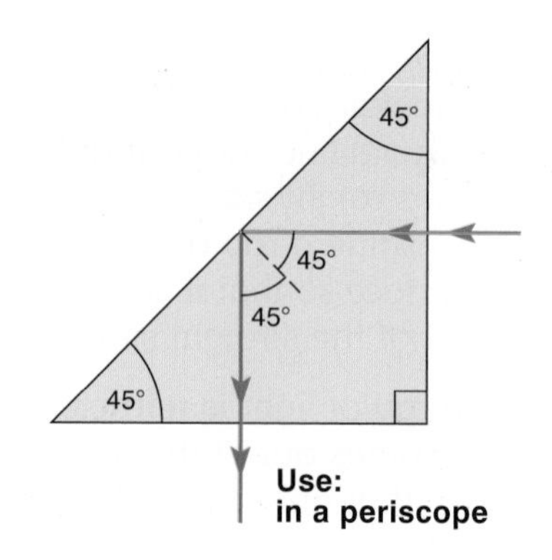

Notice that the ray of light goes straight through the first surface. Inside the glass it meets the second surface at an angle of incidence of 45°. This is ***greater*** than the critical angle of 42° and so total internal reflection takes place.
The long side of the prism acts as a mirror and turns the ray of light through 90°.

In periscopes, prisms are better than ordinary plane mirrors (page 177). This is because ordinary mirrors give multiple images due to multiple reflection in the glass at the front of the mirror. This is shown in the diagram:
Two other reasons are that the silvering on a mirror is easily damaged and that a mirror reflects less light than a totally reflecting prism.

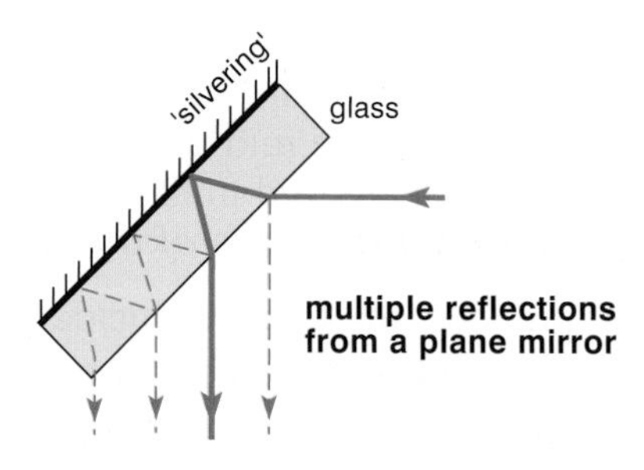

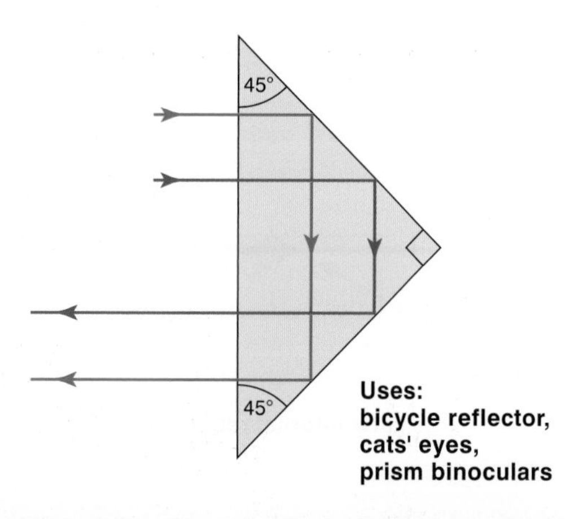

Experiment 25.8
Use a 45°–45°–90° prism in a different way, as shown in the diagram.

Use your raybox to show that the rays are totally reflected. This is because the angle of incidence (45°) is greater than the critical angle (42°).

The rays of light are turned through 180° so that they return to the direction they came from.

Bicycle reflectors and "cats' eyes" reflectors on roads are built like this so that they reflect back the light from car headlights.

Diamonds are cut in a similar way to make use of total internal reflection so they sparkle.

Notice that the two rays of light are inverted when they come out of this prism.
This is used in prism binoculars (telescopes) where prisms are used to invert the image and to shorten the binoculars.

▷ More examples of total reflection

The light-guide (or optical fibre)

Experiment 25.9
Shine a lamp into one end of a curved glass or perspex rod.

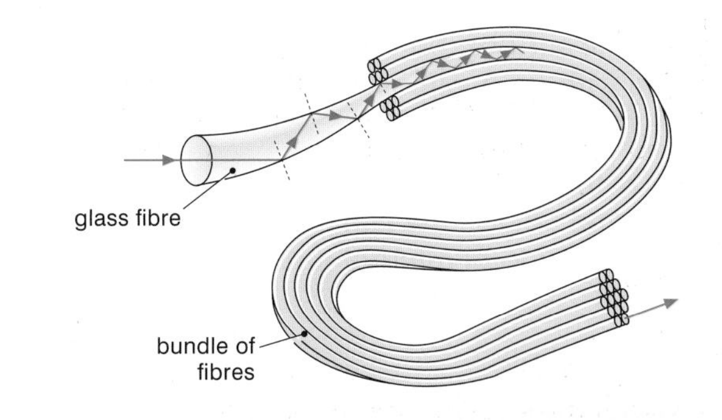

What do you see at the other end?

This is possible because of total internal reflection of the rays of light inside the rod (as long as the angle of incidence is greater than the critical angle).

Light guides are used to illuminate the dials on radios and in cars. The streams of water in illuminated fountains act as light guides.

If the glass is a thin flexible fibre, it can even be tied in knots or pushed down your throat to take photos of the inside of your lungs!

Optical fibres are sometimes used to make lamps and illuminated Christmas trees. They have important uses in medicine and in communications (see pages 192, 218).

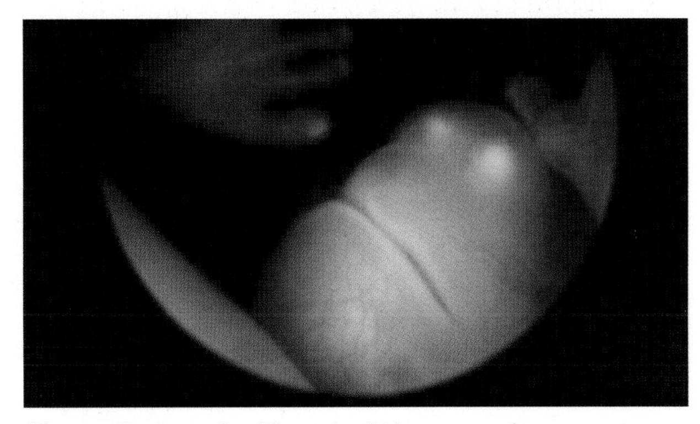

Fibre optic view of a 10-week-old human embryo

Mirages

Sometimes, on a hot day, you might see what look like pools of water on a long stretch of road.

This happens because there are hot layers of air near the hot road and cooler (denser) layers of air higher up.

A ray of light is gradually refracted more and more towards the horizontal. Eventually it meets a hot layer near the ground at an angle greater than the critical angle, and total internal reflection takes place:

A mirage: Concorde aeroplane and its reflection

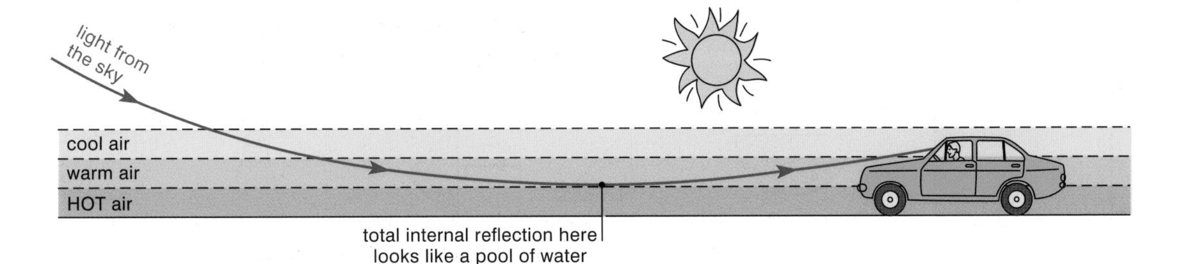

A fish's-eye view

A fish or a diver under water can see everything above the surface, but their view is squeezed into a cone with an angle of 98° (twice the critical angle for water).

Outside this cone, the surface looks like a silvered mirror and reflects the light from objects inside the pond (by total internal reflection).

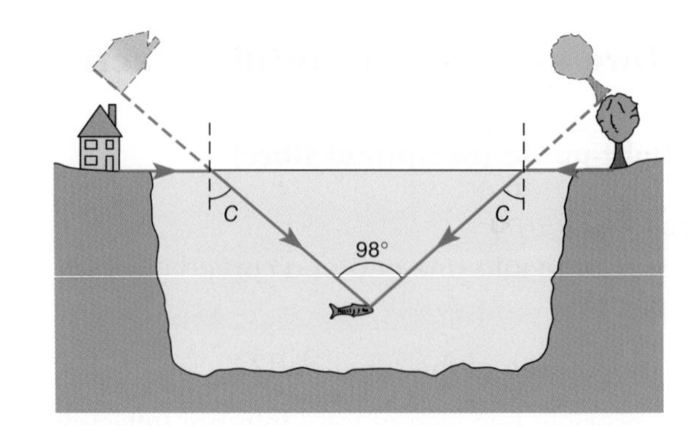

Summary

Rays of light travelling into a 'denser' medium (e.g. from air into glass) are refracted ***towards*** the normal. When leaving the denser medium (e.g. glass to air) the rays are refracted ***away from*** the normal.
This happens because light travels slower in glass than in air.

$$\text{Refractive index of a substance} = \frac{\text{speed of light in air}}{\text{speed of light in substance}}$$

The apparent depth of an object is less than its real depth. Total internal reflection takes place if
i) the rays are travelling in a dense medium towards a less-dense medium ***and***
ii) the angle of incidence is greater than the critical angle ***C***.
($\boldsymbol{C}$ = 42° for glass, $\boldsymbol{C}$ = 49° for water.)

Optical fibres have important uses in medicine and in communications.

▷ Questions

1. Copy and complete:
a) A ray of light travelling from air into glass is . . . . or bent . . . . the normal. A ray of light travelling from water to air is . . . . or bent . . . . the normal.
b) The speed of light in water is . . . . than the speed of light in air.
c) A swimming pool looks . . . . than it really is, because light from the bottom is refracted . . . . the normal on passing into air.
d) Total internal reflection takes place in a glass prism if the angle of incidence in the . . . . is . . . . than the . . . . angle.
The angles of a totally reflecting prism are . . degrees . . degrees . . degrees.

2. Copy and complete the diagrams:

3. Explain with the aid of diagrams:
a) A swimming pool appears to be shallower than its real depth.
b) Diving for a coin is more difficult than it seems to be from the side of the pool.
c) A hunter who is spear-fishing does not aim his spear at where the fish appears to be.
d) In a photograph of a man rowing a boat, the oars seem to be broken.

4. Explain with the aid of diagrams:
a) To a diver under water, most of the surface looks silvery. Bubbles of air rising from the diver look silvery.
b) Light can be shone in inaccessible places by means of a 'light guide'.

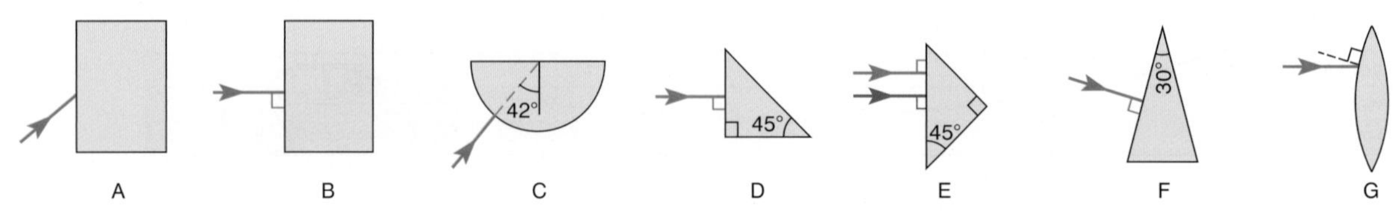

5. Explain with diagrams, how 45°–45°–90° prisms can be used to reflect light through
a) 90° b) 180°.

6. The diagram shows a bicycle reflector made of red transparent plastic:

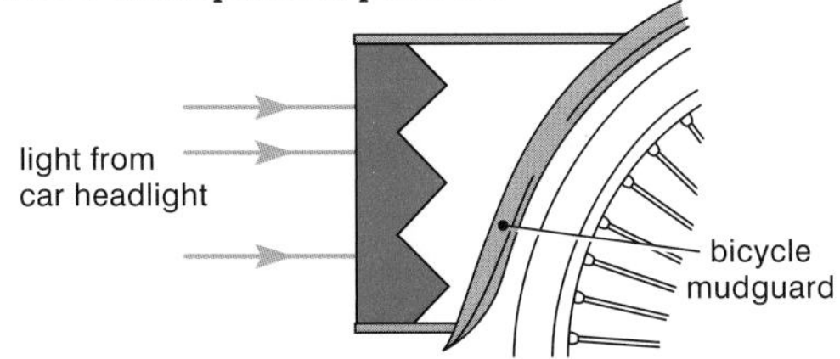

a) Copy and complete the diagram.
b) What does the car driver see?
c) Why is the reflected light red?

7. Draw a diagram showing a periscope made with 45°–45°–90° prisms. What are the advantages over ordinary plane mirrors?

8. Attic windows, roof-lights on caravans and the covers over fluorescent lights are sometimes made of ribbed plastic or glass:

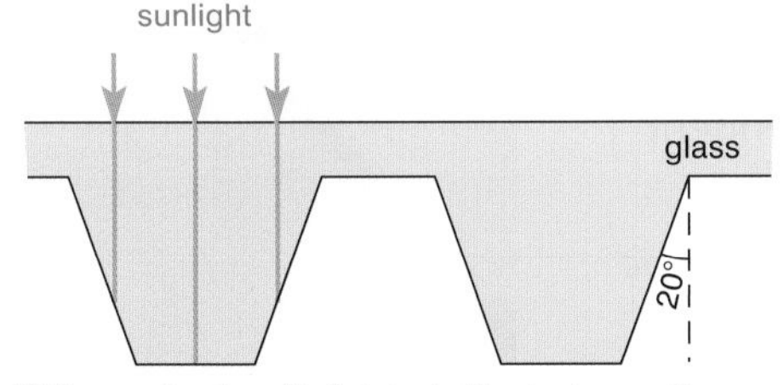

a) Where is the light totally internally reflected and where is it refracted?
b) Copy and complete the diagram carefully.
c) Why is the glass made like this?
d) What would happen if the glass was reversed?

9. Professor Messer goes out of his depth again:

10. Write an essay about the use of optical fibres in medicine. Use your scientific knowledge to explain how they work.
Why is the use of fibre optics an advantage to the patient?

11. Write an essay about the use of optical fibres in modern communications.
Use your scientific knowledge to explain how they work.
Explain the several advantages of using fibre optics.

12. Look at the table on page 185 showing the speeds of light.
Which substance refracts light:
a) the most?
b) the least?
c) If light travels from perspex to water, will it bend towards or away from the normal?

13. Here are some angles of incidence ***i*** with the corresponding angles of refraction ***r*** for glass:

i	0	10	20	30	40	50	60	70	80	90
r	0	7	13	19	25	30	35	39	41	42

a) Plot a graph of angle ***i*** against angle ***r***.
b) What angle of incidence will give an angle of refraction of 10°?
c) What angle of refraction would you get for an angle of incidence of 36°?

14. Professor Messer fell asleep and dreamed that light could not be refracted by any substance. Why was his dream a nightmare?

Further questions on page 237.

THE WATER LOOKS SHALLOW AND I'VE GOT MY FISHING BOOTS ON!
WHAT WENT WRONG?

▷ Physics at work: Fibre optics

An optical fibre is a very narrow ***light-pipe*** (page 189).
It is as thin as a human hair and as flexible.
It has a very narrow core made of very pure glass.

The core is surrounded by a 'cladding', also made of a very pure glass. However, this glass has a slightly lower refractive index (see page 185).
Because the cladding (like air) has a lower refractive index than the core, the light rays inside the core are ***totally internally reflected***.

The edges of the core act like a perfect mirror and the light rays can travel a long way inside the glass. Optical fibres can be as long as 200 km.

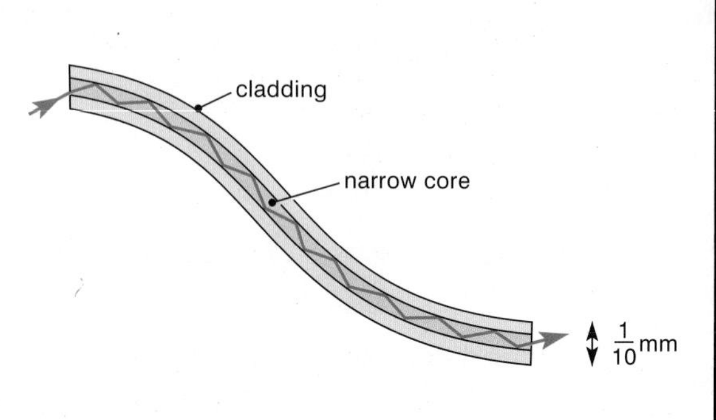

Communications

Optical fibres are being used to replace the copper wires in the telephone system, see page 314.
The telephone conversations are sent down the optical fibre by switching on and off a LED (Light-Emitting Diode, page 318) or a laser (see opposite page). Infra-red rays are often used.
At the other end, the signals are converted back to electricity by a ***photo-diode*** which acts rather like an LDR (Light-Dependent Resistor, page 319). The signals are ***digital*** (see page 218).

Over 90 000 telephone conversations can be carried at the same time by one fibre! Fibre optics can also be used to send data from one computer to another, or to carry up to 90 TV channels.

Optical fibres have many advantages over copper cables. They are thinner, cheaper and carry more signals. They have no cross-talk between two telephone conversations and are almost impossible to 'bug'. The signal needs to be boosted over long distances (100 km) but not as often as copper cables (8 km).

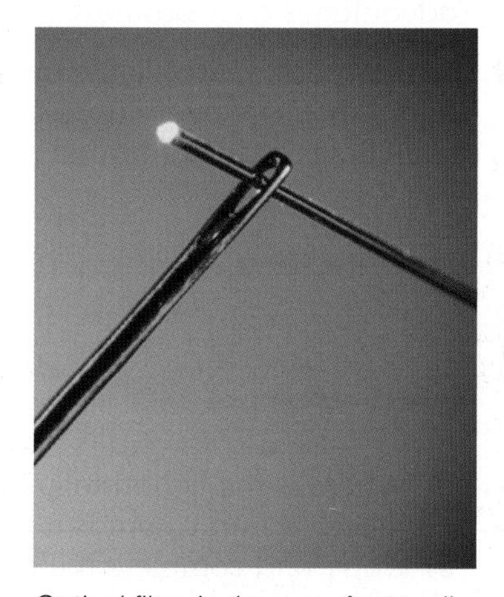

Optical fibre in the eye of a needle

Medicine

Doctors use optical fibres (in an **endoscope**) to look at the inside of people's lungs and stomachs:

A large number of optical fibres are held together in a bundle which goes down the patient's throat.
Light is sent down some of the fibres to illuminate the stomach.
Reflected light comes back up the bundle of fibres to form an image of bright and dark spots of light.

If an ulcer is found, a laser beam can be sent down the fibres to burn and seal the ulcer neatly.

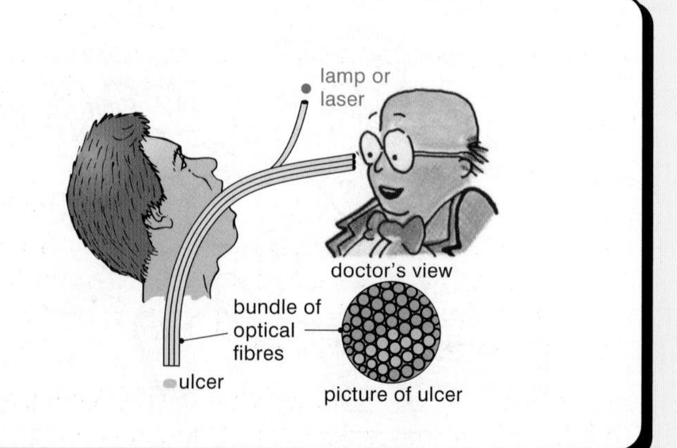

▷ Physics at work: Lasers

A gas laser

100% mirror
(1) unexcited atoms
(2) excited atoms have absorbed energy
(3) excited atoms emit light waves
(4) light waves hit an excited atom and the light is amplified
99% mirror
laser beam

A laser is a device for producing a very intense beam of light.
It is an ***amplifier*** of light.
The word LASER stands for Light Amplification by Stimulated Emission of Radiation.

It consists of a narrow tube containing a gas:
Electricity is passed through the gas so that it starts to glow (just like the fluorescent lamp on your classroom ceiling).
This is because the electricity 'excites' the atoms of the gas by giving them energy which they absorb.
Later they give out this energy, but in the form of a light wave.
If this light wave hits an atom which is already 'excited' then this atom ***also*** gives out light (this is 'stimulated emission'). The light has been ***amplified*** and is brighter.

If most of the atoms are excited by the electricity, then the light waves build up like an avalanche to make a very bright light. The mirrors help by reflecting the light up and down the tube many times. One of the mirrors is not a perfect reflector so some of the light escapes. This is the laser beam.

Laser light is a special kind of light. Ordinary light is rather like the people in a football crowd: it moves in different directions with different wavelengths. Laser light is more like a column of soldiers: it all moves in the same direction, with exactly the same wavelength and with all the waves exactly in step or '***in phase***'.

Lasers can be made from solids and liquids as well as gases. Some are smaller than a pea, others are as big as a desk.

Uses of lasers

Because laser beams are perfectly straight, they are used by surveyors and engineers to ensure that roads, pipes, railway lines, oil tankers and buildings are built accurately.

Laser beams are very intense (***never*** look up the beam) and they can melt or even vaporise substances. Laser beams are used to 'drill' holes and 'cut' sheets of metal. They do this more quickly and accurately than metal drills and saws.
The cloth for your jeans was probably cut by a laser.

Drilling a hole with a laser beam. The metal is melted then vaporised.

Lasers are used to weld pieces of metal together. The laser melts the metal, which mixes and later solidifies.
Lasers are used by doctors to 'weld' skin. For example, if the retina in a person's eye becomes loose, a laser is used to 'weld' it back on. Optical fibres can be used to guide the laser beam (see opposite page).

Lasers are used in 'compact disc' CD players (see page 306) and laser-TVs can project big TV pictures on to a wall.

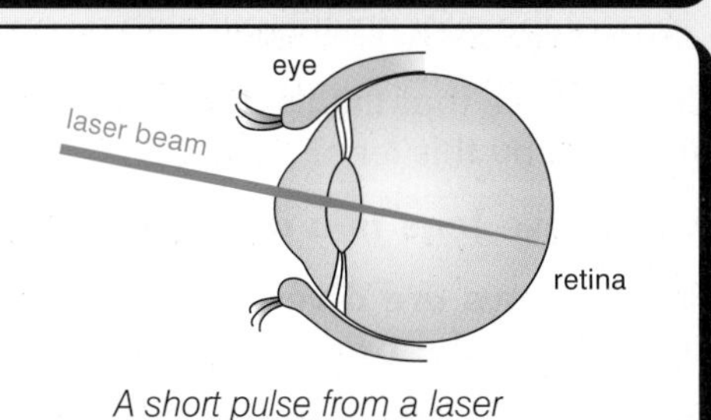

A short pulse from a laser can weld a loose retina

chapter 26

Lenses are very useful – in fact, you are using the lens in your eye to focus on these words now.

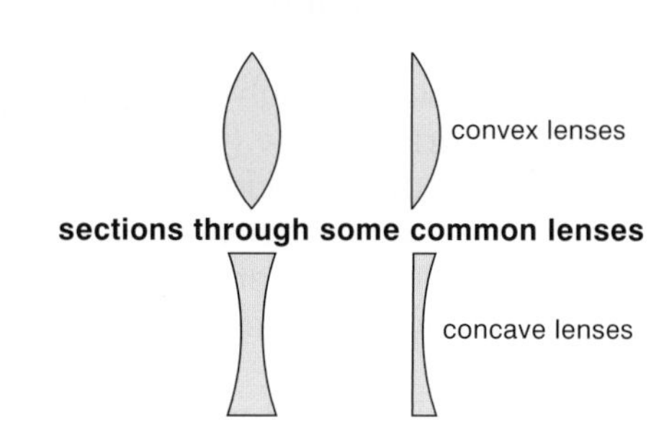

Two kinds of lenses

A lens which is thicker at the centre than at the edges is called a ***convex*** lens. A lens which is thinner at its centre is called a ***concave*** lens.

Experiment 26.1 A convex lens
Use a raybox (on a sheet of white paper) to produce several rays of light to shine on to a convex lens. It is better if you can adjust your raybox to make ***parallel*** rays of light.

Can you see that the rays after the convex lens are getting closer together? We say they are ***converging***. **A convex lens is a converging lens**.

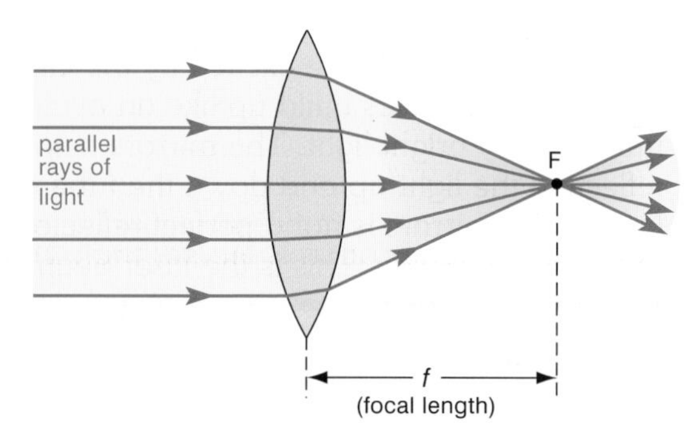

Convex (converging) lens
(used like this as a burning glass or the objective lens of a telescope)

If the rays of light from the raybox are ***parallel*** rays, the rays refracted through the lens converge to one point called the ***principal focus*** (marked **F** on the diagram).

Parallel rays of light are refracted through the principal focus of a convex lens.
The distance from F to the centre of the lens is called the ***focal length*** of the lens.

Experiment 26.2 A concave lens
Repeat the experiment, but with a concave lens.

Can you see that the light rays are spread out or ***diverged*** by this lens?

A concave lens is a diverging lens.

Parallel rays are diverged so that they appear to come from the principal focus (F) of a concave lens.

Objects always look ***smaller*** through a concave lens.

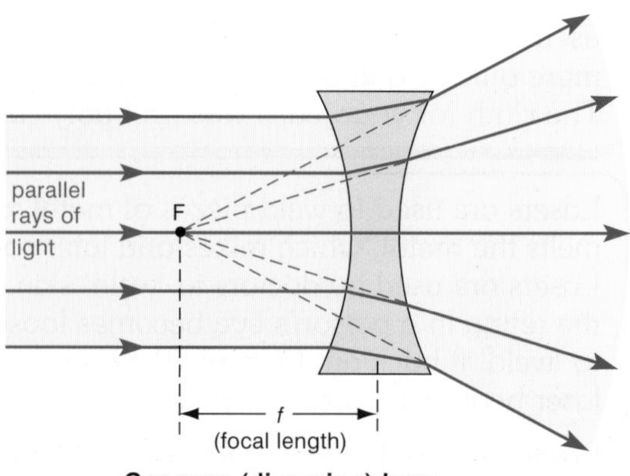

Concave (diverging) lens

▷ Forming images

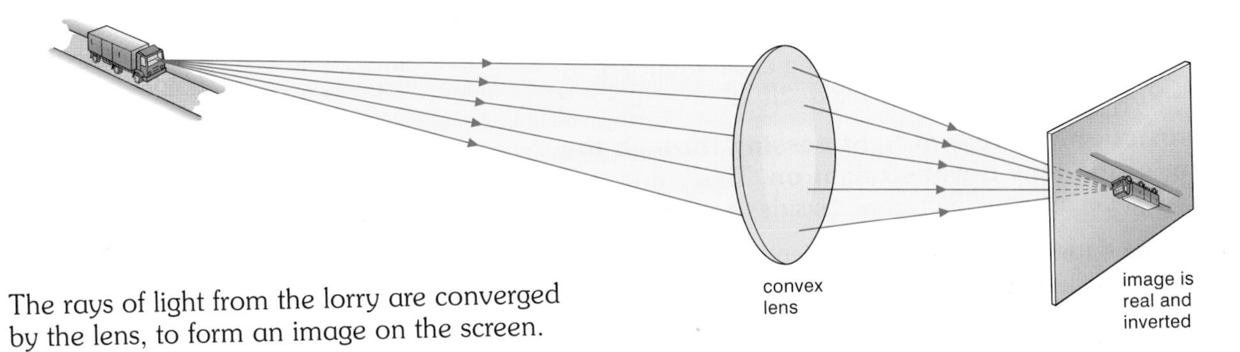

The rays of light from the lorry are converged by the lens, to form an image on the screen.

This image is:
- **inverted** (upside-down)
- **real** (because rays of light go through it, and the image is shown on the screen).

A camera uses a lens like this (see page 198):

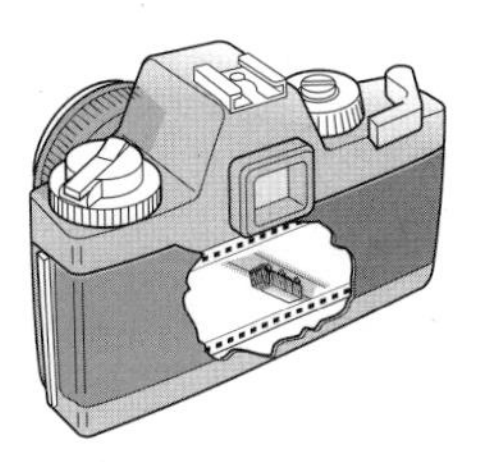

As the lorry moves nearer to the lens, the image moves ***away from*** the lens.
A camera would have to be re-focussed by moving the film away from the lens.

Images can be **magnified** larger than the object (that is, the magnification is greater than 1) or **diminished** smaller than the object (the magnification is less than 1).
What does 'magnification = 1' mean?

$$\textbf{Magnification} = \frac{\textbf{height of image}}{\textbf{height of object}}$$

Experiment 26.3 Measuring the focal length
Hold a convex (converging) lens so that the parallel rays of light from a distant house or tree are focussed on a piece of paper.

The piece of paper must then be at the principal focus F. The distance from the (inverted) image to the centre of the lens is the focal length f.

Do this experiment first with a fat lens, and then with a thin lens. What do you find?

A fat lens is a strong lens, with a short focal length.

A thin lens is a weak lens, with a long focal length.

▷ Ray diagrams for convex (converging) lenses

To draw accurate ray diagrams you can use two constructions:

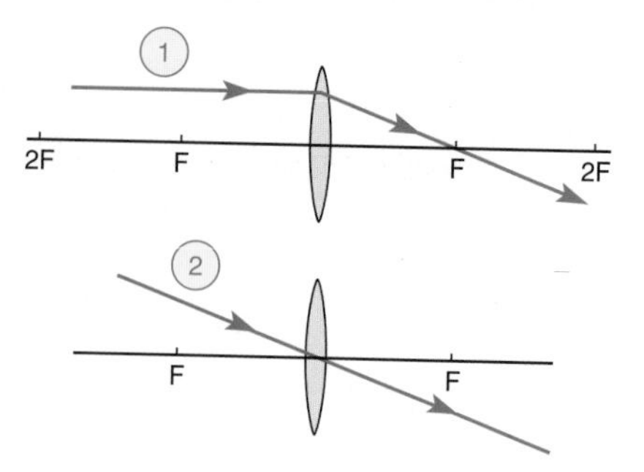

Construction ①: parallel rays of light are refracted through the principal focus F. (See experiment 26.1.) It is also useful to mark points called '2F' at twice the distance.

Construction ②: rays of light passing through the centre of the lens travel straight on. This is true for a thin lens because, at the centre, its sides are parallel.

The image depends on where the object is placed:

Distant Object
outside 2F

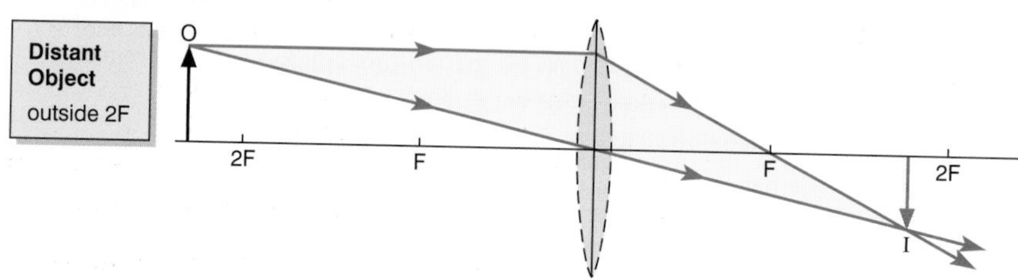

Image
between F and 2F
inverted, diminished, real

Uses:
in a camera, in your eye at this moment

Object
between F and 2F

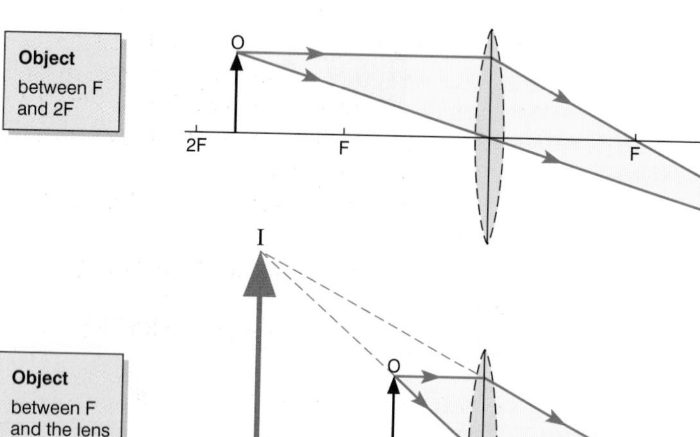

Image
outside 2F
inverted, magnified, real

Uses:
slide projector, film projector

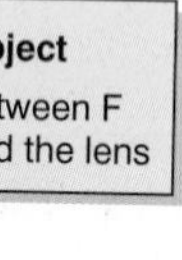

Object
between F and the lens

Image
appears to be the same side of the lens
upright, magnified, virtual

Uses:
magnifying glass, eye lens in a telescope, spectacles for long sight (see next chapter)

The last diagram is particularly important – it shows how a magnified ***virtual*** image is seen through a magnifying glass (as in the photograph):

In a **di**verging lens, the image is always erect, virtual, diminished.
Use: spectacles for short sight

▷ Questions

1. Copy out and complete:

a) A convex lens . . . . rays of light, whereas a concave lens . . . . rays of light.

b) Parallel rays of light are refracted by a convex lens to a point called the . . . . The focal length is the distance from the . . . . to the lens.

c) A fat convex lens is a . . . . lens, with a . . . . focal length.

d) The image in a convex lens depends upon the distance of the . . . . from the lens.

e) For a concave lens, parallel rays of light are refracted . . . . a point called the . . . . The image in a concave lens is always . . . . than the object.

2. Copy and complete these ray diagrams:

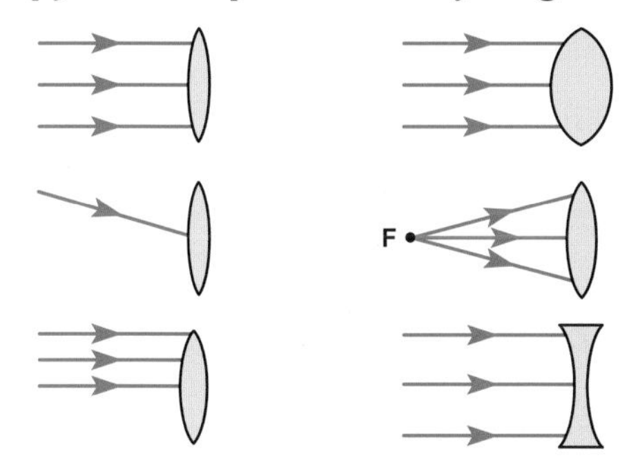

3. a) Describe an experiment to measure the focal length of a convex lens.

b) Your eye contains a convex lens – why is it unwise to look at the Sun?

c) Why is it unwise to leave glass bottles in a forest?

4. *Detective Messer looks for clues*
But hasn't a clue which lens to choose:

5. We can imagine that a lens is made up of many prisms:

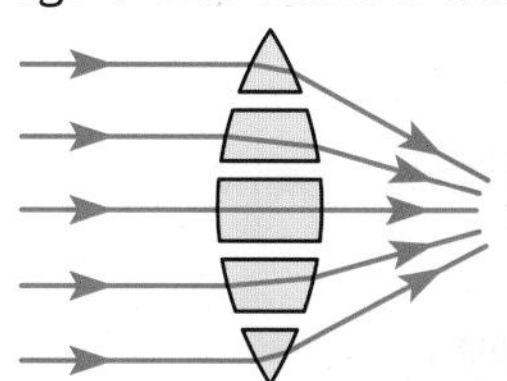

Use the diagram to explain why a convex lens converges rays of light.
Draw a similar diagram for a diverging lens, showing clearly what happens at each surface.

6. An object 4 cm high is placed 15 cm from a convex lens of focal length 5 cm. Draw a ray diagram on graph paper (full size or $\frac{1}{2}$ scale) to find the position, size and nature of the image.

7. Professor Messer did an experiment with a convex lens. He put an object at different distances (25 cm, 30 cm, 40 cm, 60 cm, 120 cm) from the lens. In each case he measured the distance of the image from the lens. His results were: 100 cm, 24 cm, 60 cm, 30 cm, 40 cm. Unfortunately his results are written in the wrong order.

a) Rewrite the image distances in the correct order.

b) Plot a graph of object distance : image distance.

c) What would be the image distance if the object distance was 90 cm?

d) Which of his object distances gives the biggest image?

e) When the object distance equals the image distance, they are each at twice the focal length from the lens. What is the focal length of this lens?

chapter 27

In this chapter we look at optical instruments that use ***converging*** lenses to form ***real*** images:
in a camera, in your eye, and in a film projector.

▷ The lens camera

A camera consists of a light-tight box with a convex (converging) lens at one end and the film at the other end:

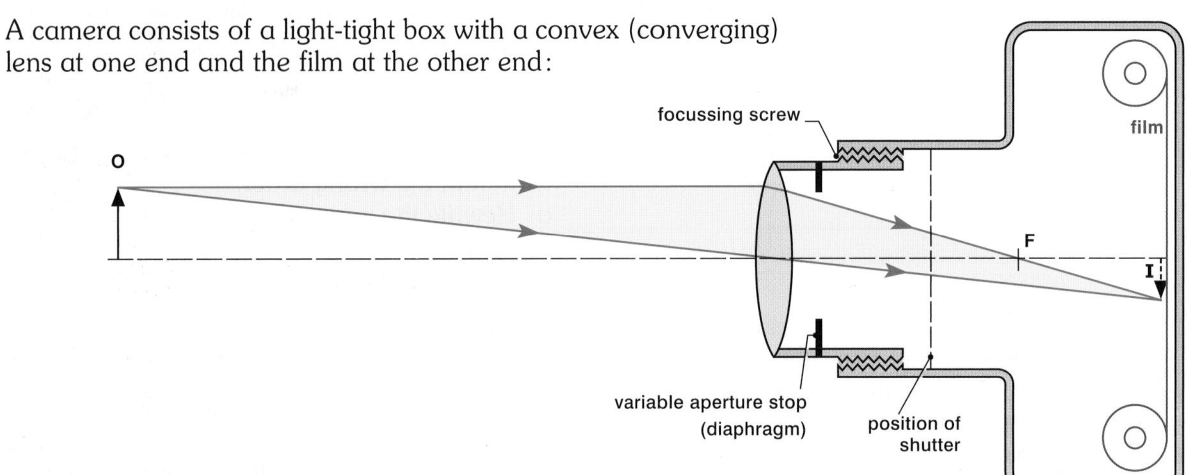

The image on the film is small, and upside-down.
It is a ***real*** image, because the light-rays actually pass through it.

> *Experiment 27.1*
> Convert the pinhole camera (described on page 174) into a lens camera by enlarging the hole at the front of the box and holding a lens over the hole (choose a lens with a focal length about as long as the box).
> Adjust the position of the lens for either near or far objects to make a sharp image on the screen.

I knew a famous photographer chap,
He snapped everyone in the town,
But his clients weren't happy,
– their tempers would snap,
When he made them all stand
upside down.

Is the image erect (upright) or inverted (upside-down)?
If the object is coloured, is the image coloured?
Why is the image brighter than in a pinhole camera?

▷ Operating a camera

1 Focussing

A camera is focussed by moving the lens. If the object moves ***nearer*** the lens, its image moves farther away (see page 195). So to keep the image in focus on the film, the lens must be moved farther ***out***.

Night scene
1 second at f/2, fast film

2 The shutter

The amount of light entering the camera can be controlled by the length of time that the shutter is open. Fast moving objects will appear blurred unless the exposure time is very short (perhaps only $\frac{1}{500}$ second).

Moving object
1/500th second at f/4, fast film

3 The aperture stop or diaphragm

The amount of light entering the camera can also be controlled by varying the size of the hole in the diaphragm which is just behind the lens. To take a photograph in dim light, a large hole is needed.

The control which varies the size of the hole is marked in ***f***-numbers. A value of ***f***/8 means that the diameter of the hole is $\frac{1}{8}$th of the focal length. A value of ***f***/11 is a ***smaller*** opening.

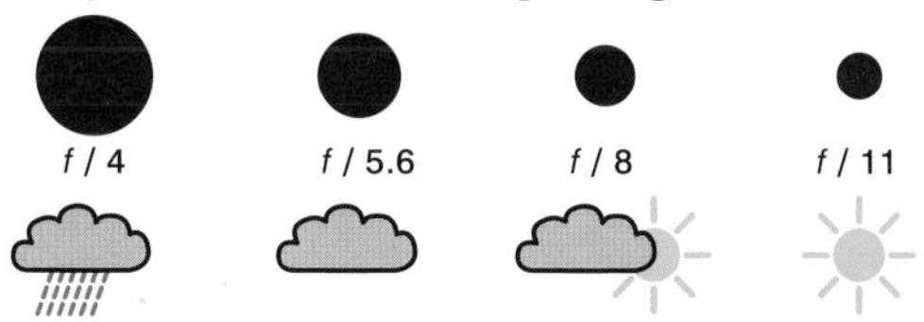

Still object; detail important
1/15th second at f/8, slow film

The size of this hole also controls the ***depth of focus***. A large depth of focus means that both near and far objects will appear to be in focus at the same time. This is obtained by having a ***small*** hole in the diaphragm.

Large depth of field
1/15th second at f/16, slow film

4 The film

The film is coated with chemicals that are sensitive to light. The chemicals are changed by the light of the image. A 'fast' film is more sensitive to light than a 'slow' film. Fast film can be used in dim light or when a fast shutter speed is needed to 'freeze' a moving object. However, a fast film may not give the quality of a slow film.

In **digital cameras**, the film is replaced by a CCD (charge-coupled device), an array of tiny solar cells converting light to electricity.

Setting the controls

The shutter time and aperture that you should use depend upon:
a) the brightness of the object,
b) the sensitivity or 'speed' of the film,
c) the kind of effect that you want.

Look at the values given with each photograph:
Can you decide why the photographer chose the values in each case?

Small depth of field
1/1000th second at f/2, slow film

▷ The human eye

Study the different parts of the diagram of your eye:

The light enters your eye through the transparent ***cornea***, passes through the ***lens*** and is focussed on the ***retina***.
The retina is sensitive to light and sends messages to your brain by way of the ***optic nerve***.

The ***iris*** changes in size to vary the amount of light that enters through the ***pupil***.
When is your pupil small and when is it large?

Notice that the image on your retina is inverted (see also the ray diagram on page 196). Although the image is inverted, your brain has learned to interpret this correctly.

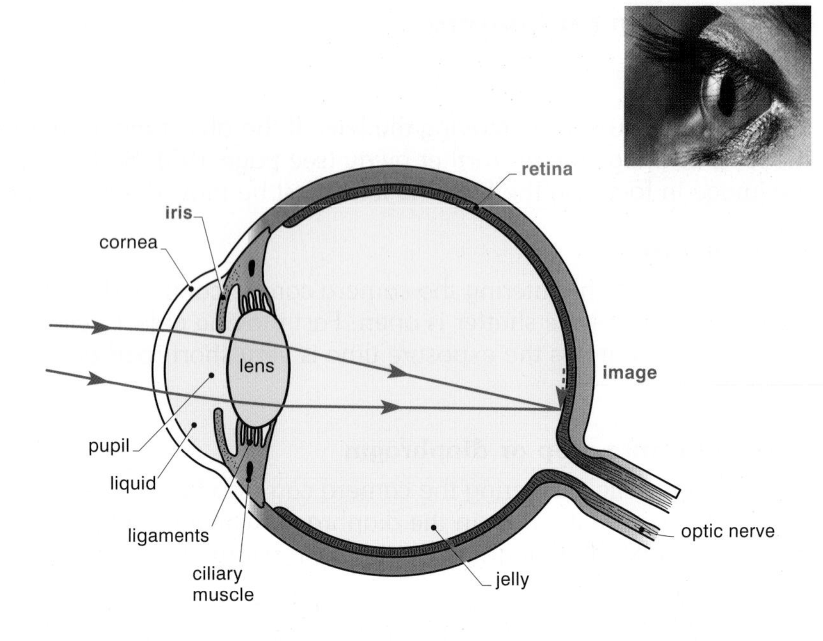

top view of your right eye

Focussing your eye

Most of the bending of the light rays is done by the curved cornea, but your lens can change its shape to alter its focal length slightly.
To focus distant objects the lens is ***thin*** (with the ***ciliary muscles*** relaxed).

To focus on a near object (like this book) the ciliary muscles tighten (rather like the draw-strings on a duffel bag). This allows the lens to become fatter (with a shorter focal length) and so it can focus the light from near objects.

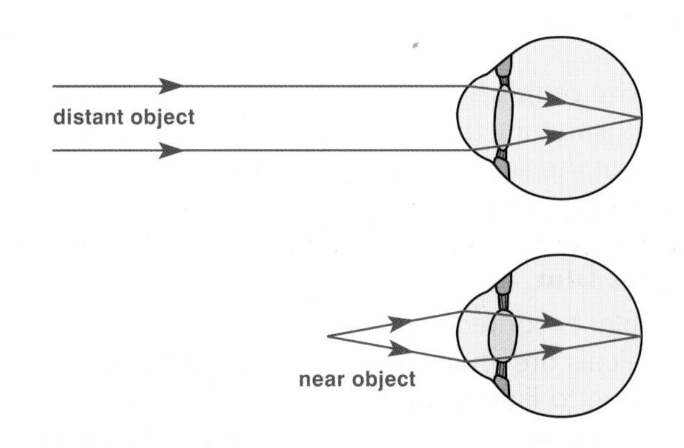

Experiment 27.2 Find your eye's blind spot
Cover your ***left*** eye, hold the page at arm's length and keep staring at Professor Messer's magic wand. Then move the page slowly towards your eye until the rabbit disappears.

This happens because the light rays from the rabbit are then arriving at a place where there is a gap in your retina. This gap is where the optic nerve leaves your eye.

Experiment 27.3 Binocular vision
Hold a pencil in one hand and close one eye. Then, with another pencil in your other hand, try to make the two pencil points touch.

Now try this with both eyes open.
Which is easier? Why is this?

Having two eyes open allows you to see an object from two slightly different angles and so you can judge distances more accurately.

Experiment 27.4 Persistence of vision
Look at the little man at the bottom corner of this page and the next few pages.
What appears to happen if you flick over the corners of the pages quickly?

The effect of the image on your retina lasts for about $\frac{1}{10}$th of a second – so that you can appear to see an image after it has disappeared.
This effect is used in television and in the cinema where you are shown about 25 pictures every second, each picture changing slightly from the one before.

Experiment 27.5 Optical illusions
Judge each of the diagrams with your eyes and then answer each of the questions by measuring with a ruler.

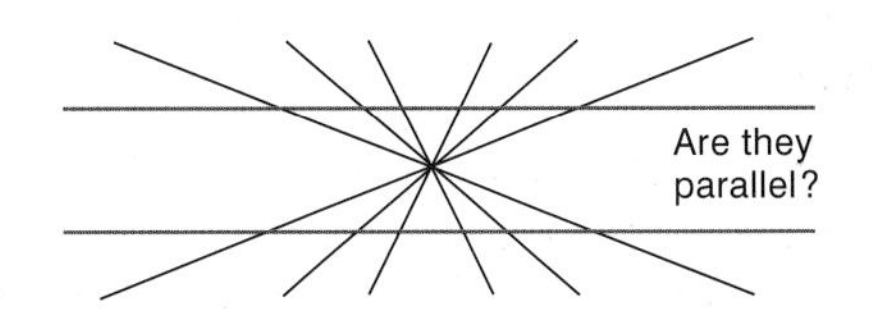

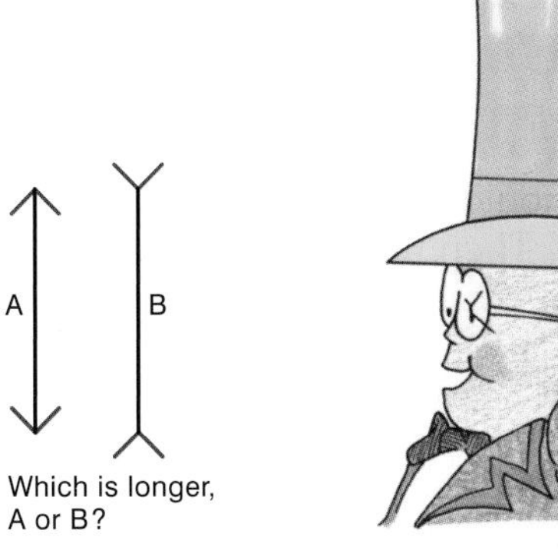

Which is longer, A or B?

Which is greater, the height of the hat or the width of the brim?

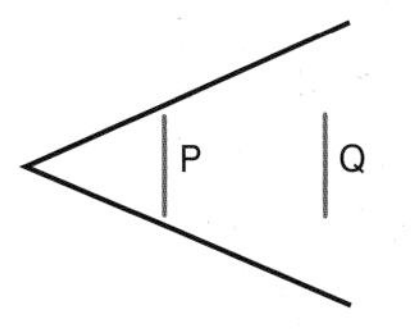

Which is longer, P or Q?

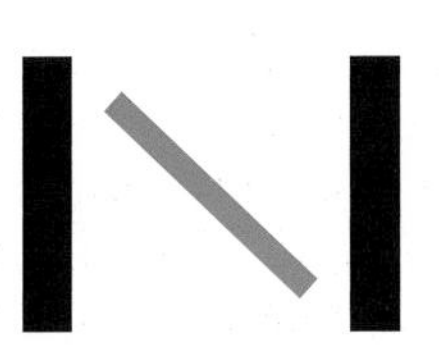

All equal in length or one shorter?

▷ Defects of vision

People with normal vision can focus clearly on very distant objects at 'infinity'. We say their ***far point*** is at infinity.

People with normal vision can also focus clearly on near objects such as this book. The closest point at which they can see an object clearly is called the ***near point***. The near point for adults is often about 25 cm from the eye (less for teenagers).

Experiment 27.6
Hold a book at arm's length and move it closer to find the nearest distance that you can focus it clearly without straining your eyes.
What is the distance to your near point?

A long-sighted student called Menzies,
Wore her spectacles only on Wednesdays.
She said, "It's not bliss,
But the question is this:
Are these glasses for keeps or just lenses?"

Long sight

A person who can see distant objects clearly but cannot focus near objects is said to be ***long-sighted***.

This is because his eye-ball is too short or his eye-lens is too thin even though his ciliary muscles are fully squeezed.

This means that rays of light from a close object are focussed towards a point ***behind*** his retina.

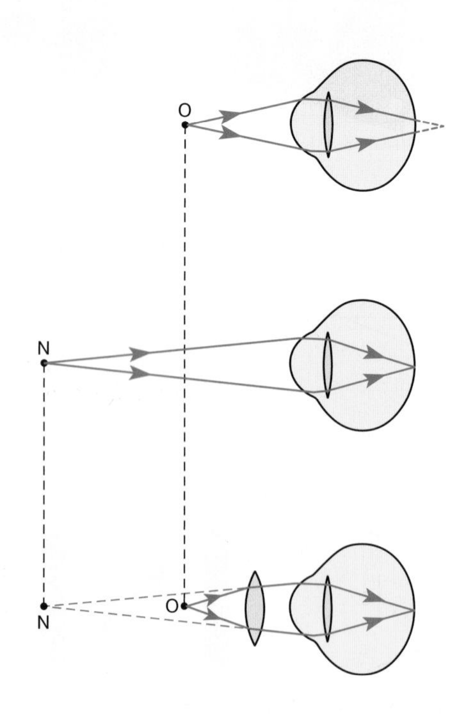

Rays of light from this person's near point N can just be focussed by his eye.

This person should wear a ***convex*** (***converging***) spectacle lens. Then rays of light from a close object are converged so that, as far as his eye is concerned, the rays appear to come from his near point N. Then they can be focussed by his eye.

Did you hear about the optician who fell into his lens-grinding machine and made a spectacle of himself?

Short sight

A person who can see near objects clearly but cannot focus distant objects is ***short-sighted***.

This is because her eye-ball is too long or her eye-lens is too strong. (The lens is too thick even though the ciliary muscles are relaxed.)

This means that rays of light from a distant object (at 'infinity') are focussed ***in front of*** her retina:

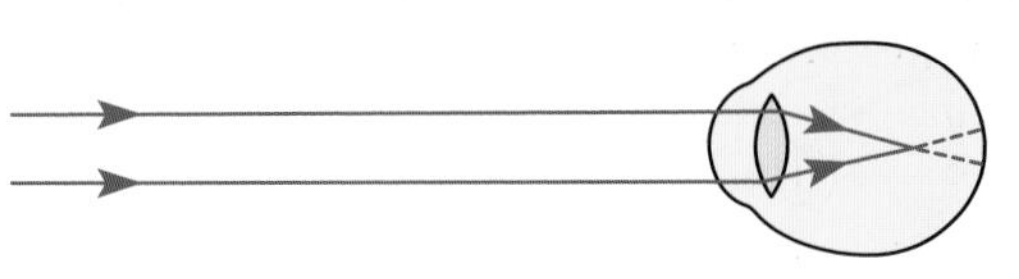

Rays of light from this person's far point F can just be focussed:

This person should wear a ***concave (diverging)*** spectacle lens.
Then rays of light from a distant object are diverged so that they appear to come from her far point F and so they can be focussed by her eye:

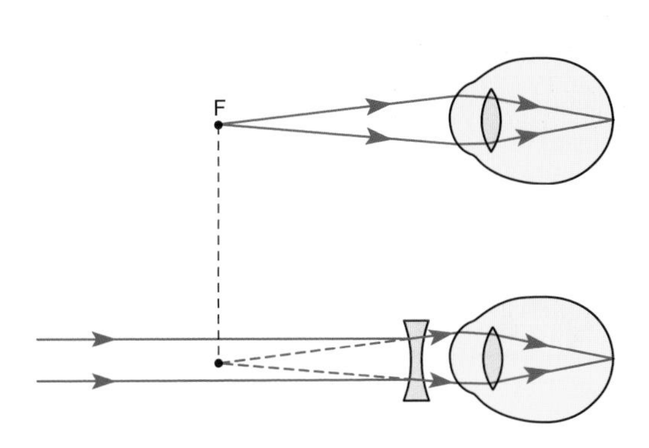

Other defects

In old people, the eye lens may become cloudy (a cataract). A surgeon can replace it. The pressure in the eye may build up too high (glaucoma). The retina may become detached (see page 193).

A short-sighted teacher called Rose,
Confused the optician she chose,
And so in her specs,
Each lens was con<u>vex,</u>
And her far point was the end of her nose.

> *Experiment 27.7 Colour-blindness*
> If some specially coloured charts are available, your teacher may be able to test you for colour-blindness (the commonest kind is when red and green look the same to the colour-blind person).

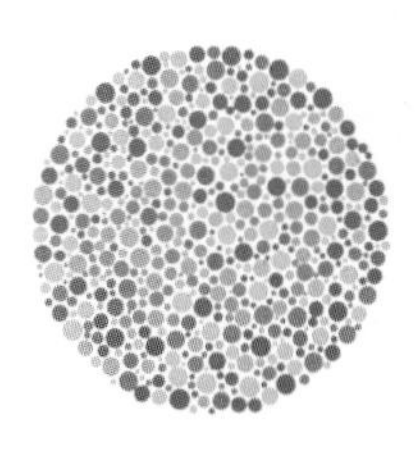

Can you see a number here?

Things aren't always what they seem, Is the Professor in a dream?

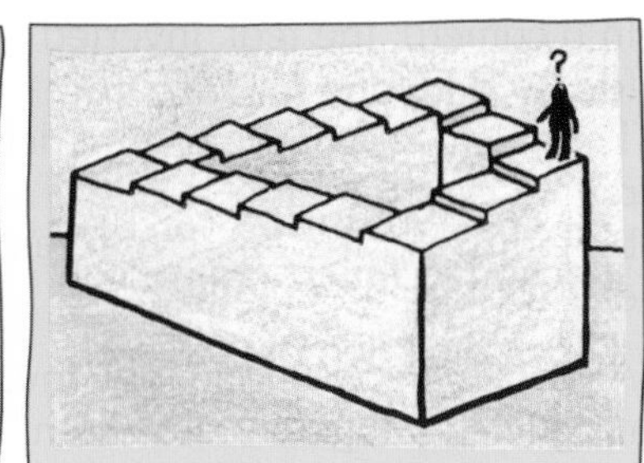

▷ The projector (for slides, movies or a data-projector)

A projector contains a ***lamp*** and a ***concave mirror*** to make the image brighter. The lamp is placed so that the rays of light are reflected back along their own path.

concave mirror
lamp

To give a brighter picture, a ***condenser*** is included. It is usually made of two plano-convex lenses, as shown.

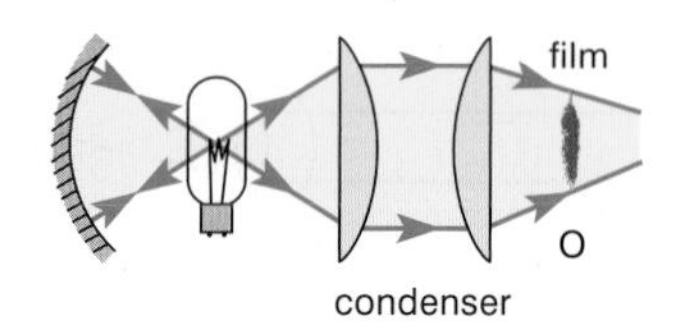

Now the light is converging towards the screen and the film is illuminated both brightly and evenly.

The light is then scattered by the film and focussed by a convex (converging) ***projection lens*** on to the screen:

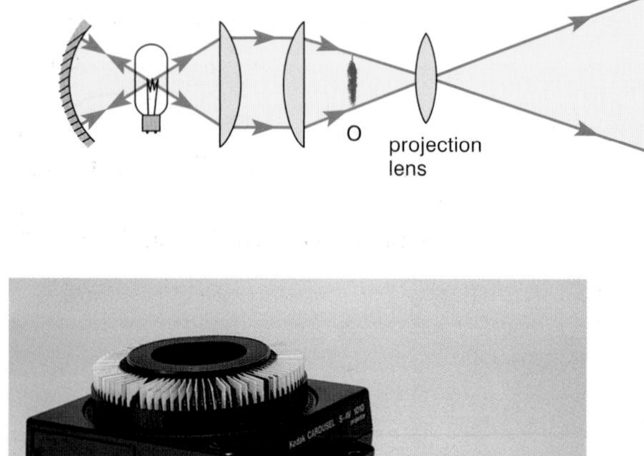

The film O is placed between F and 2F of the projection lens (see the middle diagram on page 196), so that the image I is real, inverted and magnified. The film is put in the projector upside-down so that the picture is seen the right way up.

How is the projector focussed?
Why does a projector usually contain a fan?

Summary

In a camera, the real, inverted image is focussed by moving the lens.
The brightness of the picture depends on:
a) the exposure time
b) the size of the aperture (*f*-number)
c) the 'speed' of the film.

A projector produces a real, inverted, magnified image on the screen.

In your eye, the real, inverted image is focussed by changing the shape of the lens.

With long sight, a person cannot see near objects. It is corrected by a convex (converging) spectacle lens. With short sight, a person cannot see distant objects. It is corrected by a concave (diverging) spectacle lens.

For a reflecting telescope, see page 215.

▷ Questions

1. Comparing the eye and a camera, make lists of a) similarities and b) differences.

2. Draw a large labelled diagram of a human eye. Explain the function of a) ciliary muscles b) the iris c) the retina. Where is the blind spot? Why are two eyes better than one?

3. Use a ruler to check the illusions on page 201. In order to look slimmer, should you wear clothes with vertical or horizontal stripes? What optical illusions are used by artists?

4. Among animals, the 'hunters' usually have their eyes facing forward at the front of their heads, whereas the 'hunted' usually have eyes at the sides of their head. Why is this?

5. In a certain murder investigation, it was important to discover whether the victim was long-sighted or short-sighted. How could a detective decide by examining the spectacles?

6. With the aid of ray diagrams, explain what is wrong with the eyes of a person with a) long sight b) short sight. Use ray diagrams to explain how each fault can be corrected.

7. The diagram shows possible settings for a camera using a fast film. Analyse the data.

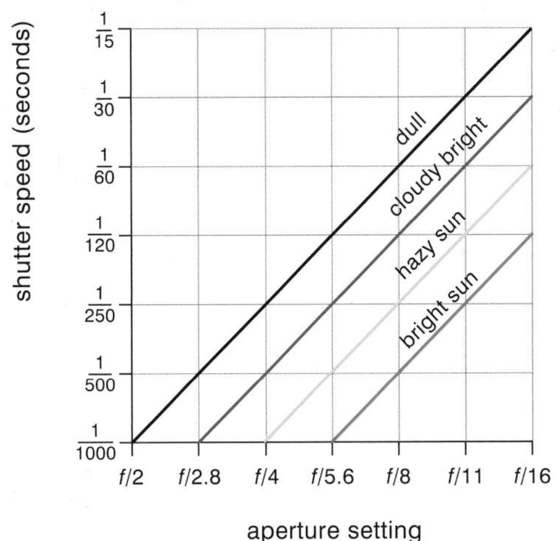

Explain, with reasons, the settings of shutter and aperture you would use for these photos:

a) a speeding car in bright sunlight
b) a rabbit in a wood
c) a friend in hazy sunlight with a mountain behind her
d) the friend's head without the background.

8. Professor Messer fell asleep and dreamed that light could not be refracted. Why was his dream a nightmare?

9. Another night, Professor Messer dreamed that no optical instrument (camera, projector, microscope, telescope) had been invented. In what ways would our lives be different if this was so?

10. Discuss the Physics of this cartoon:

chapter 28

What causes the colours in a rainbow or in a diamond? To answer this, we must investigate the ***spectrum*** of white light.

Experiment 28.1 Newton's experiment
Use a raybox (or a slide projector) with a narrow slit. Shine a ray of white light through a glass ***prism*** and on to a white screen, as shown:

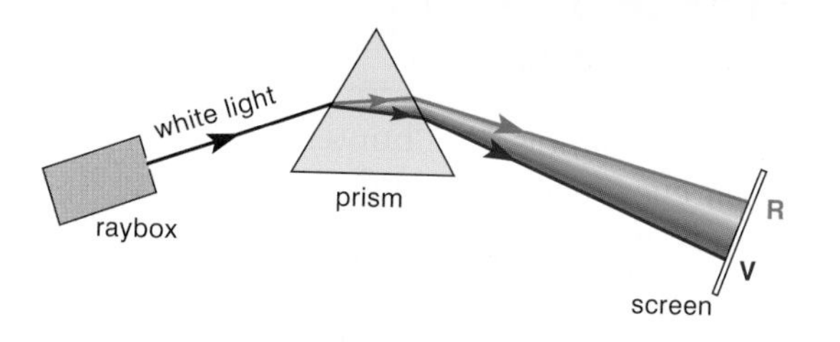

What do you see on the screen?
How many different colours can you count?

It seems that white light is really a mixture of several colours and can be split up by a prism. We say that the white light from the raybox has been ***dispersed*** by the prism to form a ***visible spectrum***.
The colours, in order, are **R**ed, **O**range, **Y**ellow, **G**reen, **B**lue, **I**ndigo, **V**iolet. (Remember them by forming them into a boy's name: **ROY G. BIV.**)

A name you should remember
As long as you may live.
The colours of the spectrum,
Are known by ***ROY G. BIV****.*

The colour that is ***deviated*** (bent) least by the prism is red. Violet is deviated through a larger angle, as shown in the diagram.

Experiment 28.2 An almost pure spectrum
In that experiment, the spectrum was 'impure' because the colours overlapped each other on the screen. To form a 'pure' spectrum, a lens must be used to focus the rays.
Move the screen (or lens) until the colours do not overlap and the spectrum is (almost) pure.

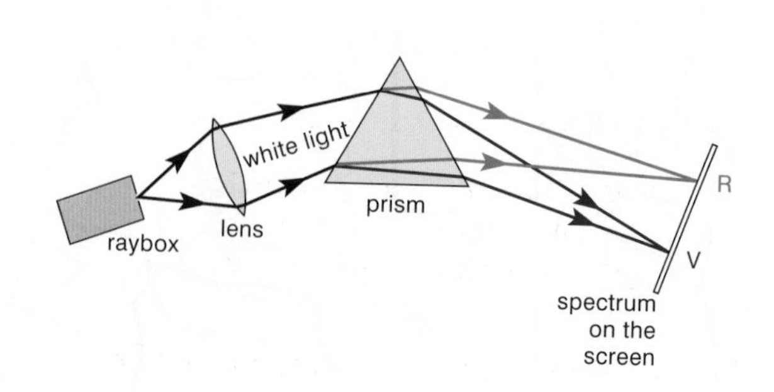

The colours of a rainbow or a diamond or a cut-glass necklace are caused by the refraction and dispersion of white light into a spectrum.

Experiment 28.3 Newton's colour wheel
If white light is composed of the colours **ROY G. BIV**, then we should be able to get white light by adding these colours together.

Spin a colour wheel quickly so that the colours appear to be mixed together.

Why does it appear to be white?
What do we really mean by white light?

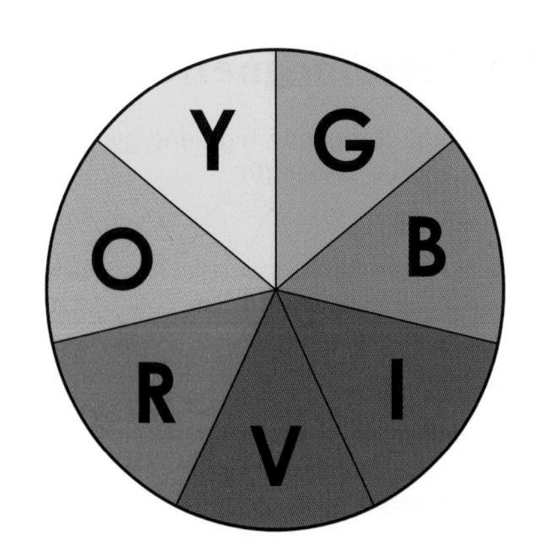

How does a prism disperse white light into a spectrum?

Different colours of light have different wavelengths (rather like waves at sea have a different wavelength from ripples on a pond – see page 167).

All colours of light travel at the same speed in a vacuum. When they enter a transparent substance like glass, they all slow down but by ***different*** amounts.
Because they slow down, they are refracted (see page 185) but because they slow down by ***different*** amounts, different colours are refracted through ***different*** angles.

Violet (the shortest wavelength) is slowed down the most and so is refracted through the largest angle. Red (a longer wavelength) slows down less and so is deviated through a smaller angle.

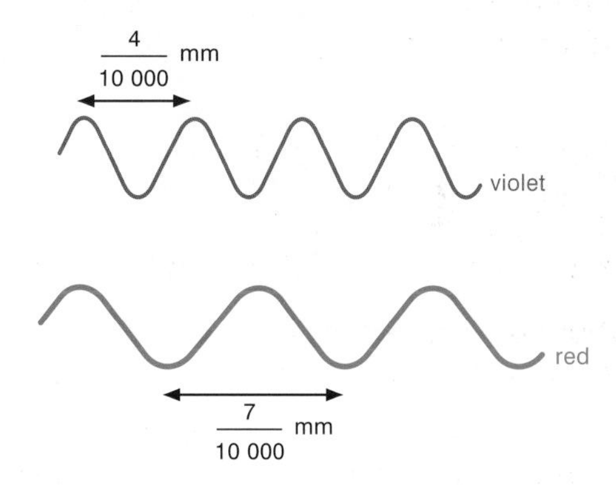

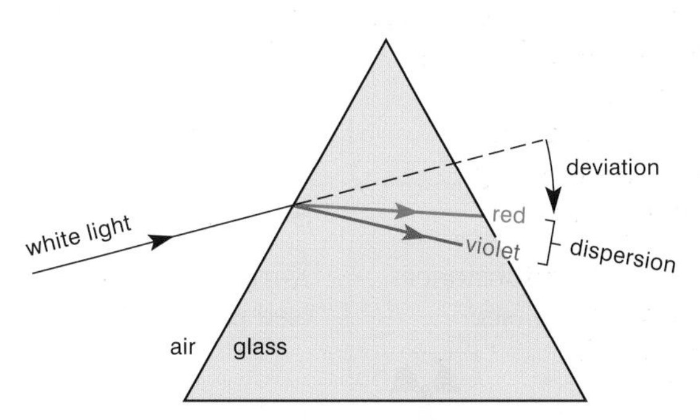

The full electro-magnetic spectrum

The visible spectrum is only a small part of a much larger spectrum containing many other wavelengths which we cannot see with our eyes.

The full electromagnetic spectrum is shown in more detail on the next six pages.
Study each part of it carefully.

The visible spectrum

▷ The electromagnetic spectrum

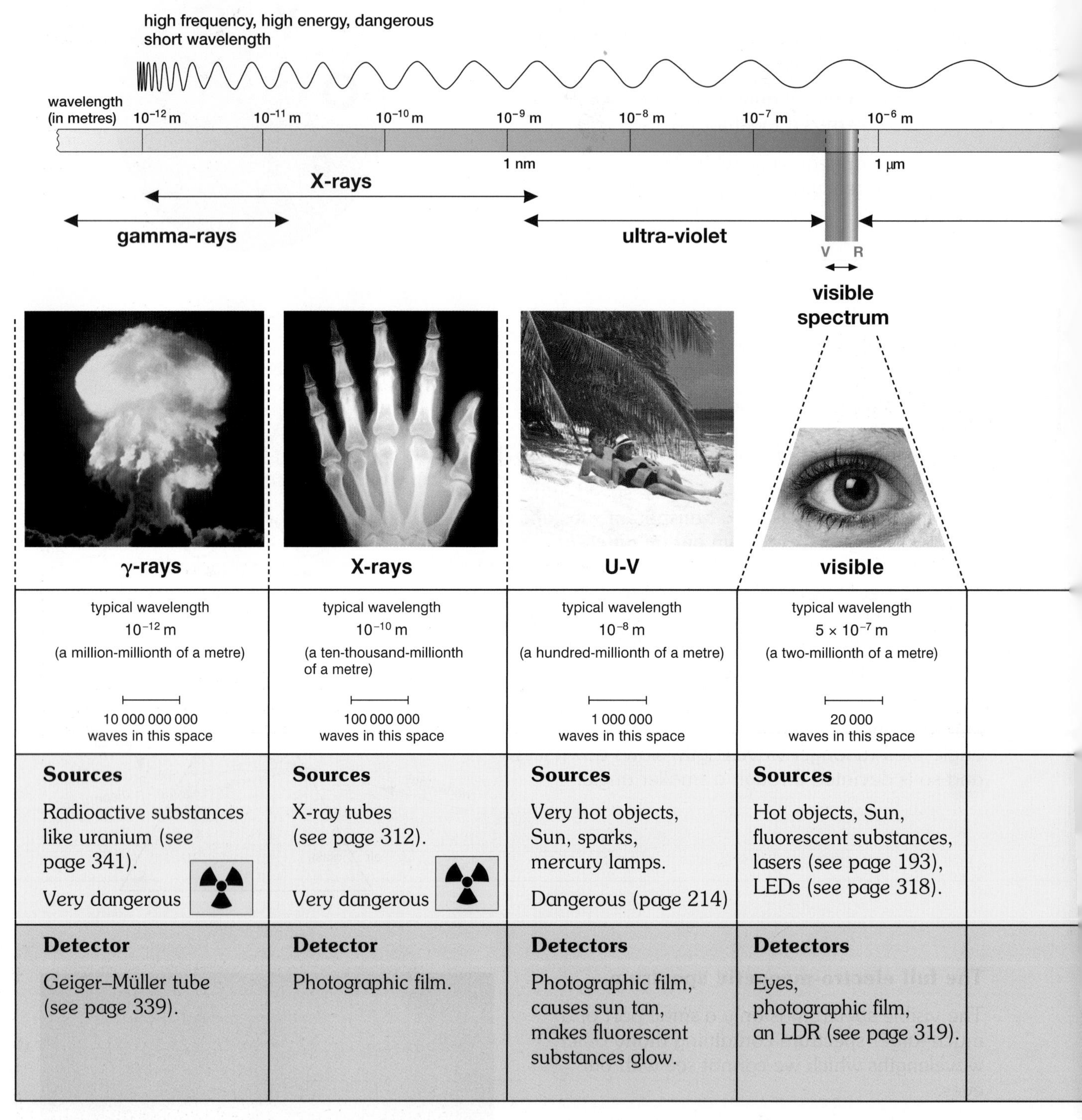

γ-rays	X-rays	U-V	visible
typical wavelength 10^{-12} m (a million-millionth of a metre)	typical wavelength 10^{-10} m (a ten-thousand-millionth of a metre)	typical wavelength 10^{-8} m (a hundred-millionth of a metre)	typical wavelength 5×10^{-7} m (a two-millionth of a metre)
10 000 000 000 waves in this space	100 000 000 waves in this space	1 000 000 waves in this space	20 000 waves in this space
Sources Radioactive substances like uranium (see page 341). Very dangerous	**Sources** X-ray tubes (see page 312). Very dangerous	**Sources** Very hot objects, Sun, sparks, mercury lamps. Dangerous (page 214)	**Sources** Hot objects, Sun, fluorescent substances, lasers (see page 193), LEDs (see page 318).
Detector Geiger–Müller tube (see page 339).	**Detector** Photographic film.	**Detectors** Photographic film, causes sun tan, makes fluorescent substances glow.	**Detectors** Eyes, photographic film, an LDR (see page 319).

All travel at the same speed, in a vacuum (space), at 3×10^8 m/s (300 million metres per second).

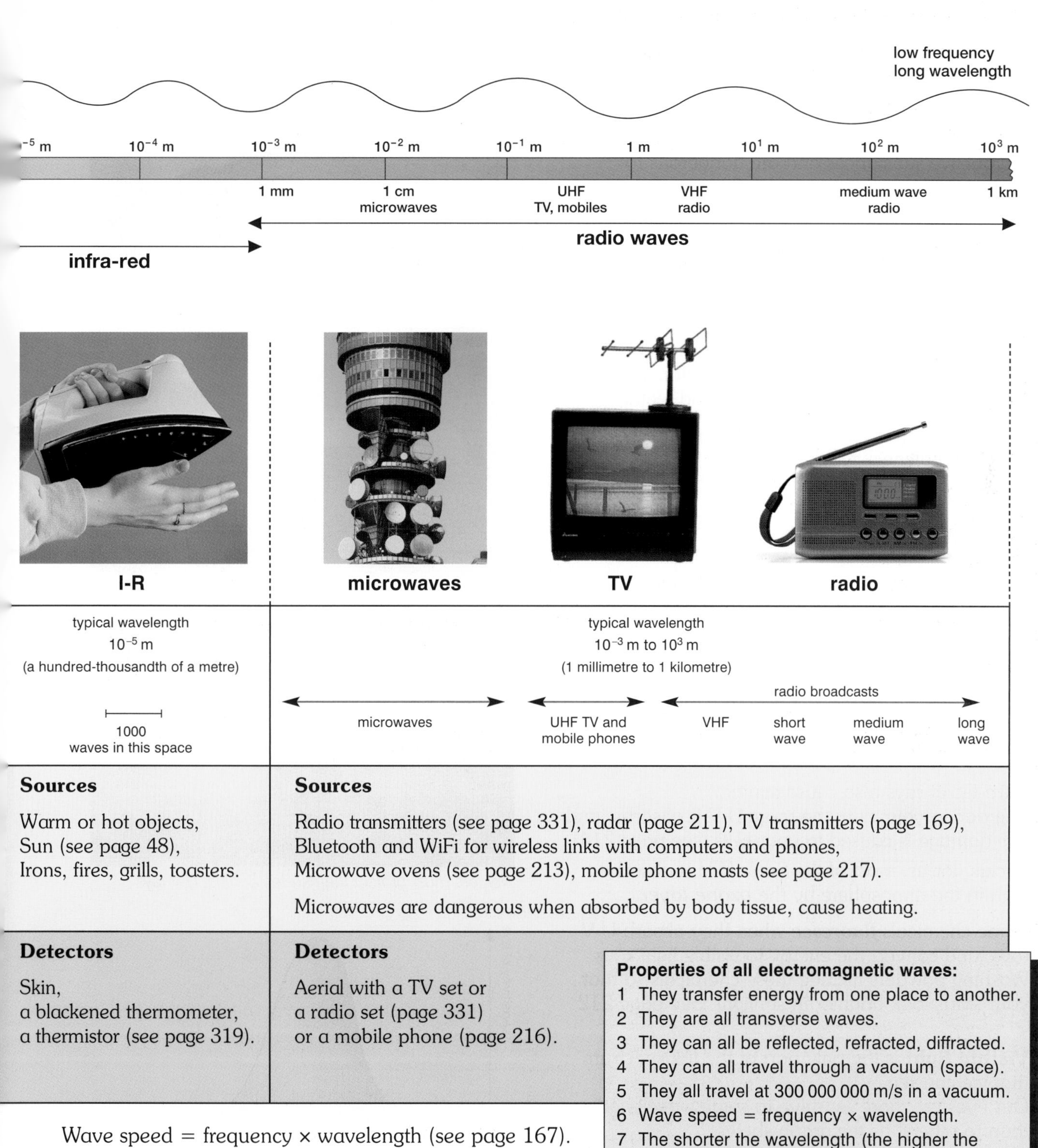

Sources	**Sources**
Warm or hot objects, Sun (see page 48), Irons, fires, grills, toasters.	Radio transmitters (see page 331), radar (page 211), TV transmitters (page 169), Bluetooth and WiFi for wireless links with computers and phones, Microwave ovens (see page 213), mobile phone masts (see page 217). Microwaves are dangerous when absorbed by body tissue, cause heating.
Detectors	**Detectors**
Skin, a blackened thermometer, a thermistor (see page 319).	Aerial with a TV set or a radio set (page 331) or a mobile phone (page 216).

Properties of all electromagnetic waves:

1. They transfer energy from one place to another.
2. They are all transverse waves.
3. They can all be reflected, refracted, diffracted.
4. They can all travel through a vacuum (space).
5. They all travel at 300 000 000 m/s in a vacuum.
6. Wave speed = frequency × wavelength.
7. The shorter the wavelength (the higher the frequency), the more dangerous they are.

Wave speed = frequency × wavelength (see page 167).
Each section is discussed on the next pages:

▷ The electromagnetic spectrum

Each part of the full electromagnetic spectrum (see the previous page) has its own properties and uses.

Gamma-rays have a very short wavelength and are very penetrating. They are produced by radioactive substances and are dangerous to humans unless used very carefully.
Gamma-rays can be used to sterilise food so that it does not rot so quickly (see page 347).
They can be used by doctors to get important medical information (see pages 346 and 357).

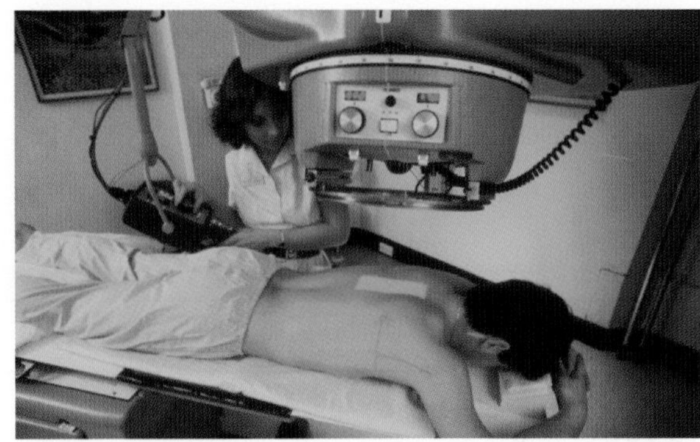

A carefully-controlled beam of gamma-rays is used to kill cancer cells. This is radiotherapy.

X-rays are very like gamma-rays, but are produced by an X-ray tube (see page 312). Doctors and dentists use X-rays to check bones and teeth.

As with gamma-rays, X-ray photos are used by engineers to check welds and metal joints.
In factories, X-rays can be used to check that food does not have metal or stones in it.
Airport security uses X-rays to scan your bags.

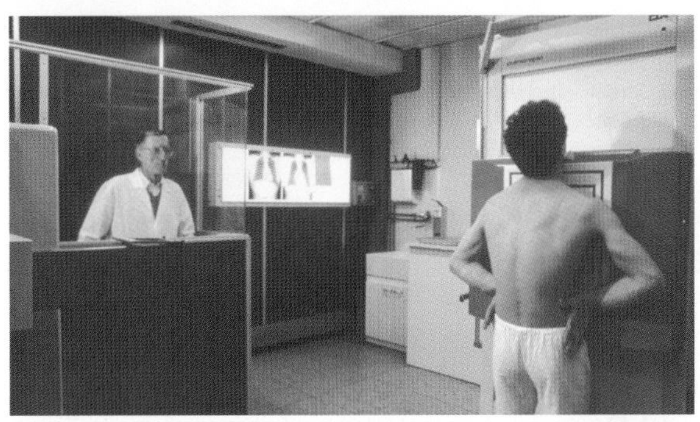

Taking an X-ray (operator screened at a safe distance)

Ultra-violet rays can also be dangerous to us. Hot objects like the Sun produce ultra-violet rays, and so do the electric arcs used in electric welding:

Small amounts of ultra-violet rays are good for us, producing vitamin D in our skin.
Large amounts of ultra-violet are bad for our eyes and cause skin cancer!

Ultra-violet rays also cause suntan.
The darker your skin, the more U-V it absorbs, and less damage is caused to your cells (see page 214).

Luckily for us, most of the Sun's U-V is absorbed high in the atmosphere by the ***ozone layer***.

Some chemicals ***fluoresce*** when they absorb U-V rays, and convert the energy to visible light rays. Washing powders include these chemicals so that your shirts look brighter in sunlight. See page 212.

Welders must protect themselves against U-V rays

Visible light is the only part of the full spectrum that we can see.
Our eyes are more sensitive to some wavelengths than to others, as the graph shows:

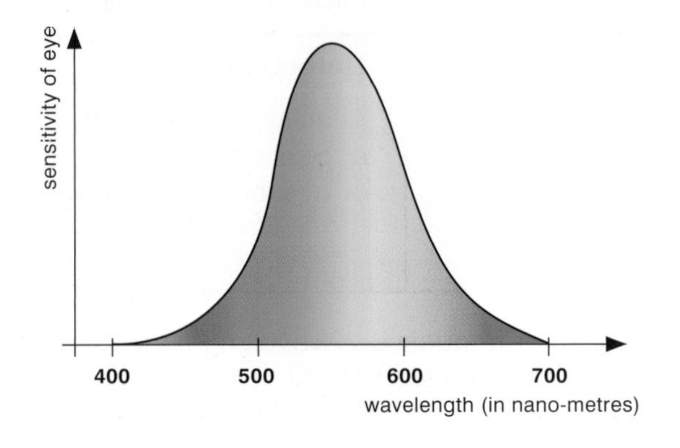

Your eye is most sensitive to green-yellow light.

Infra-red rays are given out by warm objects.
In fact, every object that has a temperature above absolute zero gives out infra-red waves – including you!
Because of this, fire-fighters can use infra-red viewers to search for unconscious people in smoke-filled buildings, and to search for survivors trapped alive beneath buildings by earthquakes.

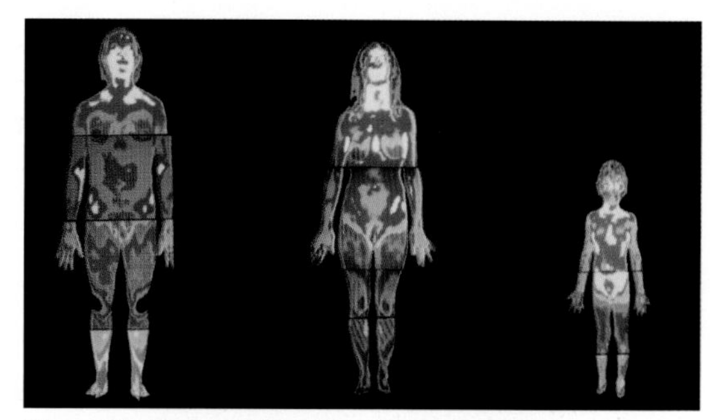
These thermographs are coloured so that white/red = most radiation; blue = least

Special photographs taken with infra-red rays are called ***thermographs***. They can help doctors to detect circulation problems (where the skin is cooler, so the thermograph is bluer) and arthritis and cancer (where the skin is warmer).

Astronomers take infra-red photos to get data about the temperatures of planets and stars.

Burglar alarms are designed to detect the infra-red rays from an intruder.
See also pages 48–50 and 213–215.

Radio waves have a longer wavelength.
They are divided into several types.

Microwaves are the shortest of the radio waves.
They are used in microwave ovens (see page 213).
They are used for communicating with satellites (see page 155 and the diagram below) and for detecting the echoes from objects for **radar**:

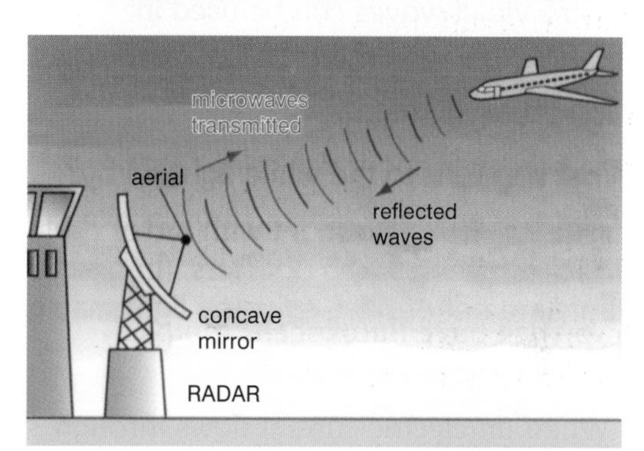

Reflected microwaves detect planes or rainclouds

UHF waves are used for TV and mobile phones.
VHF waves are used by local radio and the police.

Medium waves can be reflected by the charged ***ionosphere*** and the Earth, to travel long distances (as 'sky' waves), as shown below:

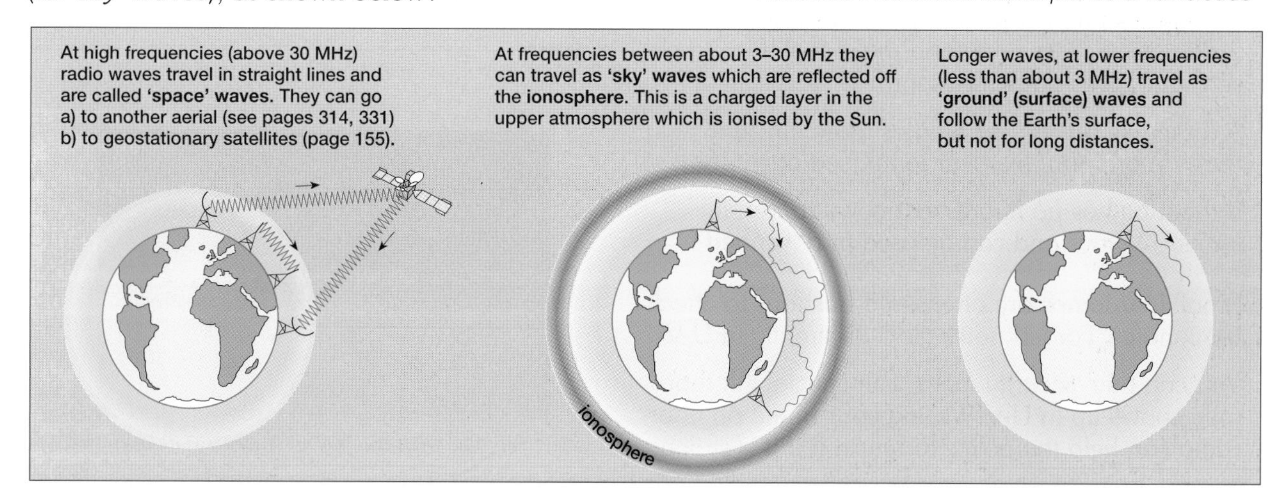

At high frequencies (above 30 MHz) radio waves travel in straight lines and are called **'space' waves**. They can go
a) to another aerial (see pages 314, 331)
b) to geostationary satellites (page 155).
At frequencies between about 3–30 MHz they can travel as **'sky' waves** which are reflected off the **ionosphere**. This is a charged layer in the upper atmosphere which is ionised by the Sun.
Longer waves, at lower frequencies (less than about 3 MHz) travel as **'ground' (surface) waves** and follow the Earth's surface, but not for long distances.
ionosphere

▷ Physics at work: Electromagnetic waves

Electromagnetic waves can have very short or very long wavelengths (see page 208). Different wavelengths have different properties and different uses.

Different wavelengths give us different information about a hand:

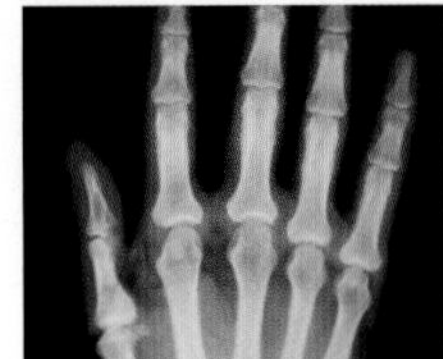

X-rays transmitted

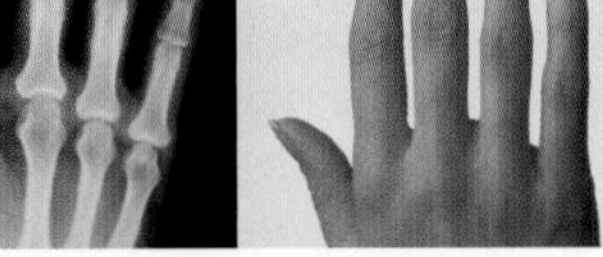

Visible light reflected

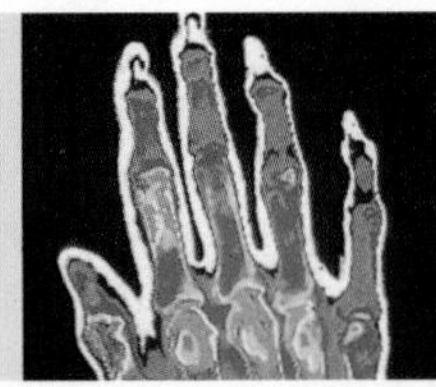

Infra-red rays emitted

X-rays

X-rays are used in medicine (see page 312).

They can also be used in detective work. This old painting has been photographed in visible light and then using X-rays:

You can see that the artist changed his mind and then painted the head in a new position.

Visible light

X-rays

Ultra-violet

Ultra-violet waves can be used in detective work, in ***forensic science***. The photograph shows a cheque photographed in ultra-violet – you can see it is a forgery.

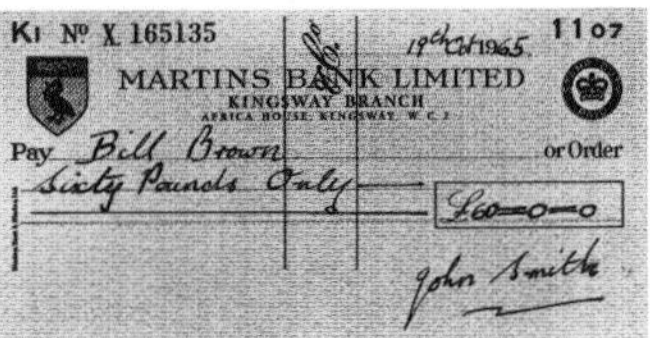

KI Nº X 165135
19th Oct 1965
11 07
MARTINS BANK LIMITED
KINGSWAY BRANCH
Pay Bill Brown or Order
Sixty Pounds Only
£60—0—0
John Smith

In visible light,

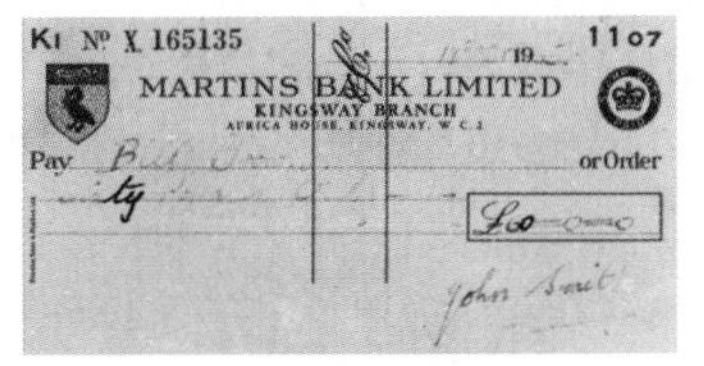

KI Nº X 165135
19
11 07
MARTINS BANK LIMITED
KINGSWAY BRANCH
Pay or Order

and in ultra-violet.

Some chemicals glow or ***fluoresce*** in U-V rays. 'Day-Glo' paints do this.

You will have ***fluorescent lamps*** in some of your classrooms:

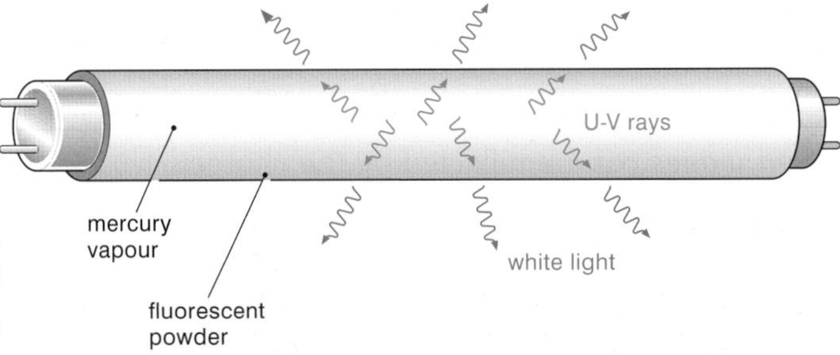

When an electric current flows through the mercury vapour, it gives out U-V rays. These hit some fluorescent powder on the inside of the glass tube, and make it glow, so that it gives out white light.

A lamp like this is 4 or 5 times more efficient than an ordinary filament lamp (see page 265).

If the tube has no fluorescent powder then it emits U-V rays and can be used in a 'sunbed'.

Small U-V lamps are often used in shops to check bank-notes. Forged notes glow differently in U-V.

You can buy 'security' pens with special ink that only shows up in U-V. Writing your name on your radio can help the police to return it if it is stolen.

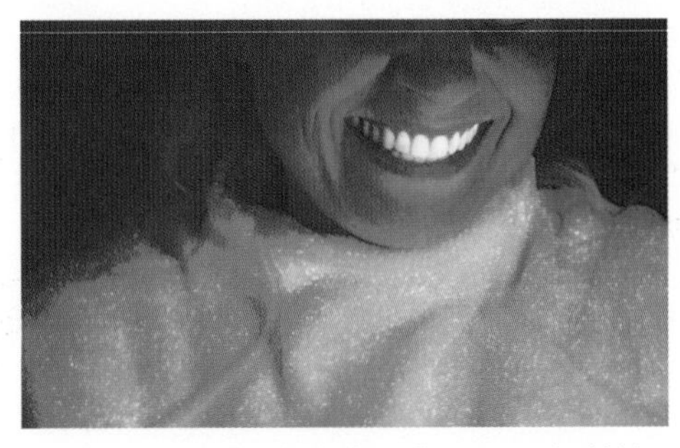

Teeth and washed clothes fluoresce in U-V.

▷ Physics at work: Electromagnetic waves

Infra-red

All objects emit infra-red rays (sometimes called 'heat radiation'). The hotter the object, the more energy it emits (see the thermograms on page 50 and page 211).

Your body is emitting and absorbing I-R all the time. If you absorb it, it heats you up. This is why you feel warm in sunlight.

Grills, toasters and radiant fires (see page 264) use I-R to heat up objects.
Radiant heaters are used to dry the paint quickly on newly-sprayed cars.
They can also be used in hospitals to treat muscular problems.

The Earth radiates infra-red rays. If there are no clouds to reflect them back, it can cause a frosty night.

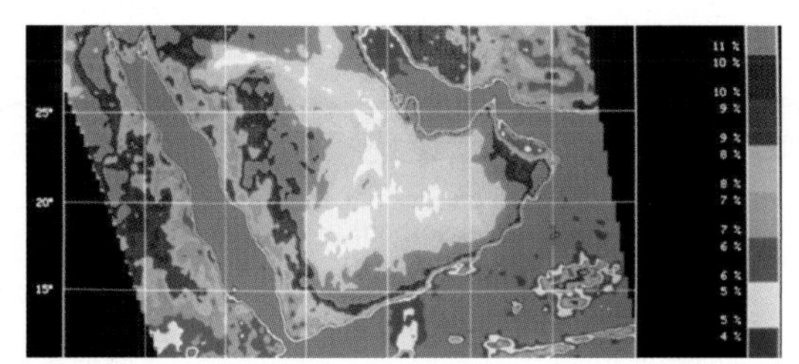

An infra-red satellite photo of Arabia

A special camera detects I-R emitted by the warm animal

Firefighters use I-R imagers to find people in thick smoke

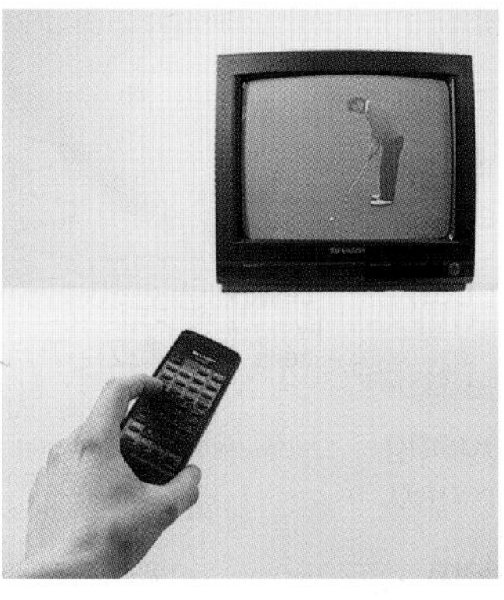

A remote control transmits I-R, with a receiver in the TV

Burglar-alarm sensors detect the I-R of an intruder's body

Microwaves

A microwave oven uses radio waves to cook food very quickly. The microwaves are produced by a 'magnetron' and guided to a metal stirrer which reflects the waves into different parts of the oven. Microwaves are reflected by metal but absorbed by food.

The waves have a wavelength of 12 cm at a frequency of 2500 MHz. At this frequency the electromagnetic waves are absorbed by water molecules in the food, heating them up and so cooking the food. The food turns on a turntable so it is cooked evenly.

The door has a wire mesh over the window (to reflect the microwaves back inside). The door must have a safety switch to turn off the microwaves if you open the door – otherwise you could cook your fingers!

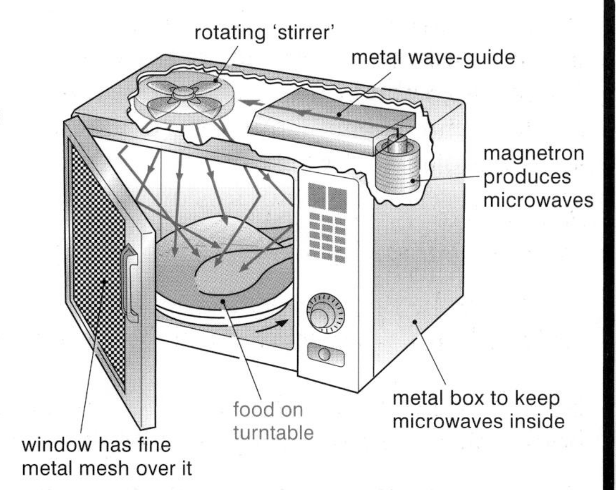

A microwave oven

Microwaves are also used for communicating with satellites and mobile phones (pages 155, 217).

▷ Dangers of electromagnetic radiation

Radiation danger symbol

Gamma-rays (γ) and X-rays

These are the most penetrating and dangerous, because of their high frequency (short wavelength). They have enough energy to ***ionise*** atoms in your body (by stripping electrons off the atoms). This ionisation can cause mutations and cancer.

High-energy gamma-rays can be used to kill cancer cells in a patient's body. But is is important to focus the beam on the cancer cells and to reduce absorption by healthy cells. To ensure this, the γ-ray source is rotated round the patient as shown here:

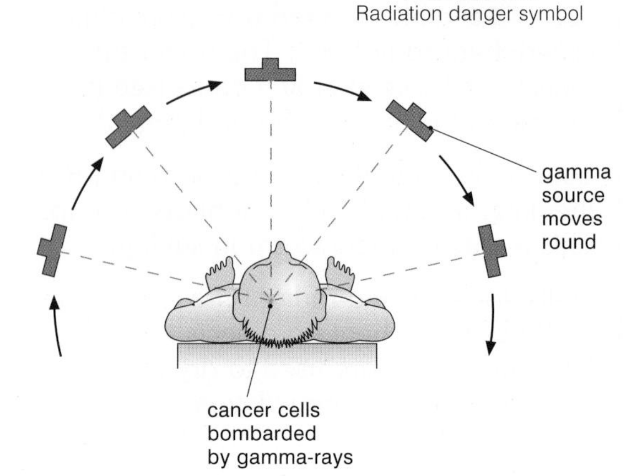

See also the first photo on page 210.

The operators should wear dosemeter badges (see page 350) to measure how much radiation they receive.

Ultra-violet (UV)

UV waves can also ionise atoms and cause mutations and skin cancer.

Dark skin and sunscreen cream can help to protect you:

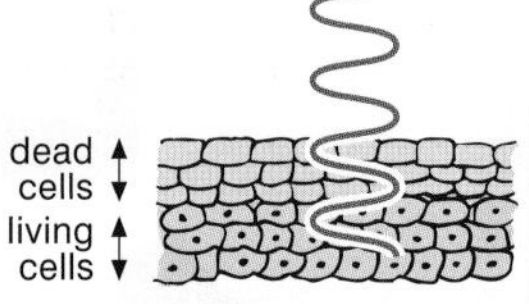

If you have pale skin, UV waves can get through to damage the living cells.

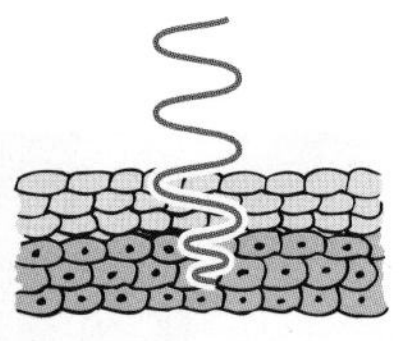

If you have dark skin, more UV is absorbed, by melanin in the skin.

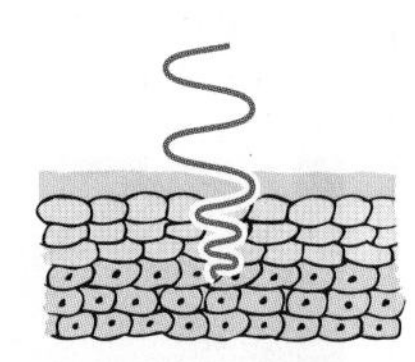

If you use sunscreen cream, it absorbs some UV, so less damage is caused.

Cream with a sun protection factor (SPF) of 15 reduces the UV to $^1/_{15}$th of the intensity. A cream helps, but a hat and clothes are safer.

UV can penetrate several cm of water and damage you even while you are swimming.

UV also damages your eyes, causing cataracts and damage to your retina.

This has become a bigger problem because the ozone layer, which helps to protect us from the sun's radiation, has been damaged by CFC pollution.

Infra-red radiation and microwave radiation

These are absorbed by skin cells, causing surface or internal heating which may damage the cells. Microwave ovens (page 213) may have some leakage at the door, so it is wise not to stay too close. Mobile phones radiate microwaves (see page 216):

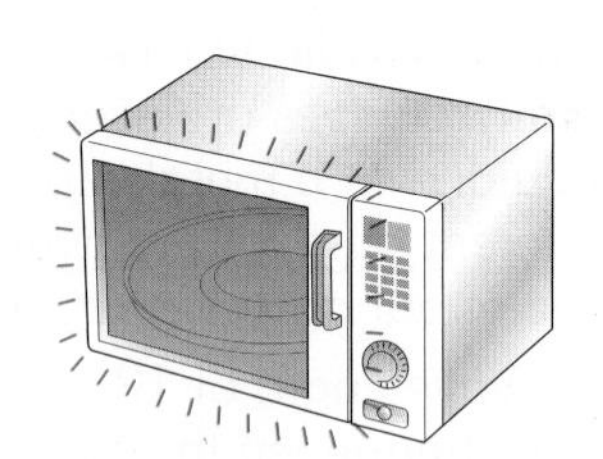

Inverse square law

The best way to be safe from electromagnetic radiation is to increase your distance from the source.

At ***twice*** the distance, the intensity reduces to just a ***quarter*** (because it is spread over 4 times the area). At 3× distance, the intensity is only $\frac{1}{9}$th.

This is called an ***inverse square law*** (see also the gravitational force diagram on page 152).

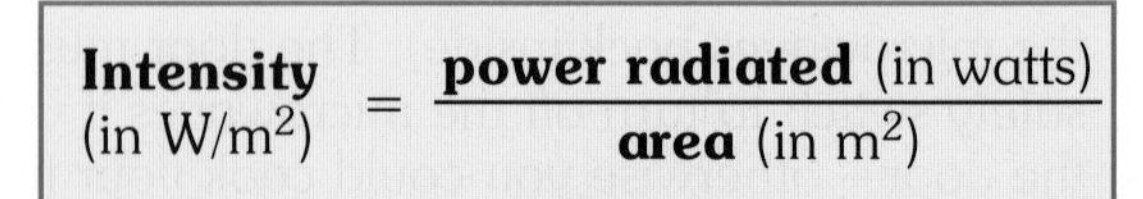

$$\textbf{Intensity} \text{ (in W/m}^2\text{)} = \frac{\textbf{power radiated} \text{ (in watts)}}{\textbf{area} \text{ (in m}^2\text{)}}$$

▷ Physics at work: Telescopes

To find out about the Universe and what happened after the Big Bang (see page 158), scientists use telescopes.
Stars, galaxies and black holes emit electro-magnetic waves across the full spectrum.

Each wavelength gives us different information about the Universe.
However, most wavelengths are filtered out by the Earth's atmosphere, so some telescopes have to be launched into space.

Optical telescopes

Optical telescopes look at the visible light from stars, galaxies (and light reflected from planets).
They use a large concave mirror to focus the light to an image (see also page 182).
Then the image is magnified by a converging lens (a magnifying glass, see page 196).

The bigger the mirror, the more light is collected, so we can see even very faint objects.
To get a clearer picture, the telescope can be launched into space, so it is outside the Earth's (dirty) atmosphere.

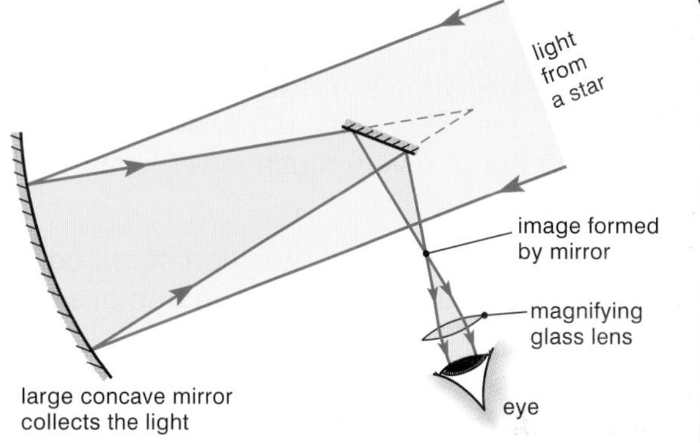

Reflecting telescope, invented by Sir Isaac Newton, 1669

Radio telescopes

Radio telescopes collect radio waves emitted by galaxies, and the waves left over by the Big Bang (see page 159).

They can be used on Earth or in space.

The Jodrell Bank telescope:

Infra-red telescopes

These are used to study the parts of galaxies where stars are born.

They can be used on mountain-tops (above most of the Earth's atmosphere) or, for better images they are launched into space.

To 'see' the radiation they have to be cooled to almost absolute zero.

Ultraviolet telescopes

These work best outside the Earth's atmosphere.
They are used to study very hot (young) stars, because they emit a lot of UV.

X-ray telescopes

They can only be used in space as, fortunately for us, X-rays are absorbed by the atmosphere.
They are used to study black-holes, supernovae and quasars.

Gamma-ray telescopes

These also can only be used in space.
They are used to study the emissions from the hottest objects, like quasars.

The Crab nebula, viewed by telescopes using different wavelengths (with colour added):

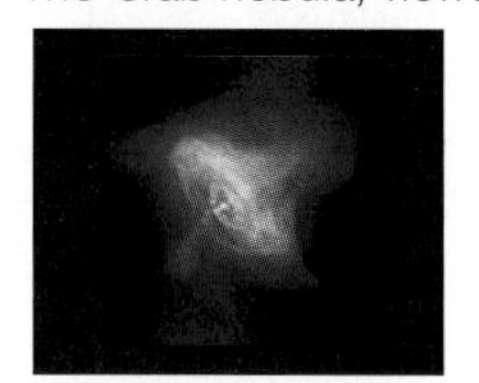

X-ray image of the nebula

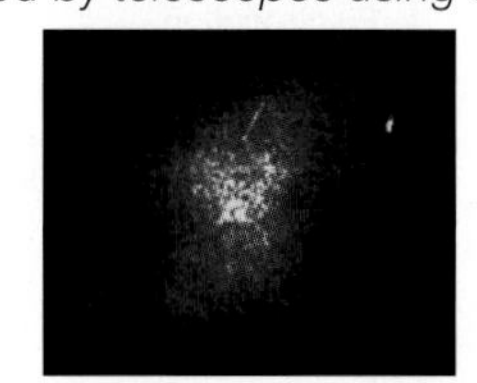

U-V image of the nebula

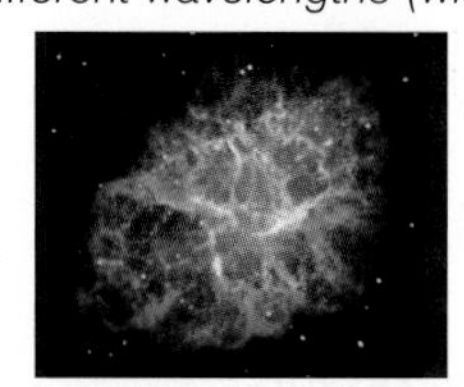

Optical (visible) image of it

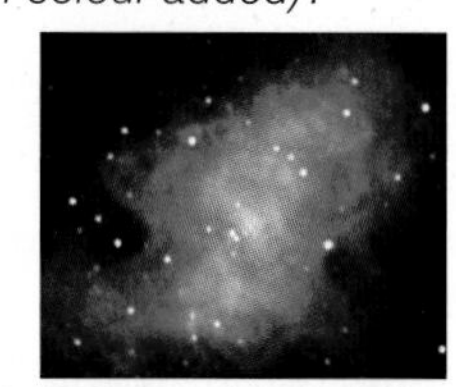

I-R image of the nebula

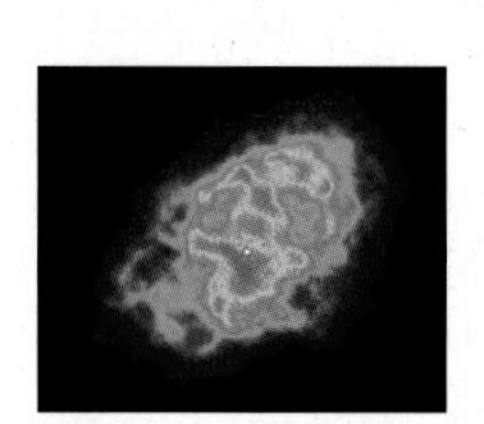

Radio image of the nebula

▷ Physics at work: Mobile phones

'ear! 'ear! for mobile phones, it's said, But do the waves affect your head?

A mobile phone is a small radio (see also page 331). To send a message it radiates energy from its aerial, as **microwaves**. The waves typically have a frequency of 900 MHz and a wavelength of 33 cm.

Modern mobile phones use **digital** signals (see page 218).

There are concerns about their effects on health.

Possible danger 1
Microwave radiation is absorbed by your body and causes heating. This is like a microwave oven (see page 213) but at a very much lower level.

Possible danger 2
High levels of microwave radiation cause cancer in rats. The effect of weak microwaves on humans is not clear.

a Discuss why it is hard to find this out.

SAR (Specific Absorption Rate)

All phones are tested to find their SAR value, in W/kg. This is the number of watts absorbed (of the energy radiated per second) per kilogram (of your body or brain). The legal limit in the UK is 2 W/kg. In USA it is 1.6 W/kg.

The table shows details of 4 mobile phone models:

Model	Weight g	Battery talktime min	Camera resolution MP	SAR W/kg
SonyEric T610	95	540	1.0	1.21
Motorola A1000	160	200	1.3	0.74
Siemens SF65	97	240	1.3	0.38
Nokia 9300	167	240	1.0	0.24

Source: Manufacturers' web-sites

b Analyse and discuss the table. Explain which model you would want to purchase, and ***why***.

c Discuss whether or not the evidence in the table is (i) reliable, (ii) valid (see pages 7, 359).

Risks

There is a risk in many parts of your life. But the risks vary enormously, as shown in the table:

Lifetime risk of dying by:	
Heart disease	1 in 3
Cancer	1 in 4
Traffic accident	1 in 180
Accidental poisoning	1 in 600
Fall on stairs	1 in 700
Fire	1 in 1600
Electrocution	1 in 20 000
Bitten by dog	1 in 100 000

Source: Office of National Statistics

The risk of getting an ***acoustic neuroma*** (a benign tumour in the head) is about 1 in 100 000.
Swedish research shows the risk of getting this tumour is up to 4 times greater in regular users of mobile phones.

d Discuss your feelings about the risks involved.
Will the risk of an acoustic neuroma affect how often you use your phone?
Would it affect your usage if the risk was 10× as great? Or 100×? What is an acceptable level of risk for you?

Emma says, *"The most dangerous thing about mobile phones is when drivers use them."*

e Discuss this statement. What is the law on this?

▷ Physics at work: Mobile phone aerial masts

Mobile phone aerial on a mast

Each time you talk on a mobile phone, you are connected to the aerial for your local area or ***cell***.

This aerial then connects to other aerials, in sequence, to connect you with the other person.

An aerial (like your phone) radiates energy, as microwaves. The vibrating waves from one aerial induce an alternating current in the next aerial, with the same frequency.

The diagram shows a typical beam of microwaves from a mobile phone mast.

Some people are concerned about the radiation from these aerials. There are about 40 000 aerials in the UK. Microwaves are reflected by metal (as in a microwave oven, see page 213).

a Explain why the person at **A** does not receive much radiation even though they are close to the mast.

b Discuss why the person at **B** has some protection from the microwave radiation.

c Give 2 reasons why **C** receives the most radiation.

d The radiation follows an 'inverse square law' (like the gravitational force, pages 152, 214). If the radiation at 100 m is 1.0 mW/m^2, what is it at (i) 200 m (ii) 300 m?

You can measure the level of radiation with a meter like this

The table gives information about levels of microwave radiation (in milli-watts per square metre). Study and discuss the table.

e Which recommended limit is the safest? Why do you think the limits change over time? Discuss why there is no absolute or definite limit.

f Compare the maximum level measured at a school with the current UK limit. Which is higher? What is the ratio? Is it safe?

g When you use a mobile phone the level is higher (as it is closer than the mast). Discuss why there is usually more local anger against masts than mobile phones.

Radiation from masts	Power density in mW/m^2
Typical levels measured in UK national survey sample	0.1 – 1.0
Maximum level measured in a UK school playground	8.3
Recommended limits:	
UK before year 2000	35 000
USA current level	5700
EU and UK current level	4500
Swiss current level	450

Source: ICNIRP international commission

▷ Physics at work: Analogue and Digital Communications

Analogue or Digital?

Electromagnetic waves are used for communications, in radio and telephones. The signals can be ***analogue*** or ***digital***.

The sound wave you may have seen on an oscilloscope is an **analogue** signal:
The voltage can vary smoothly, and have any voltage between the lowest and the highest.

Some examples are: a dimmer switch, the hands on a clock, an ammeter with a pointer.

Another kind of signal is **digital**. Digital signals can have only certain definite values, usually just ***on*** (value = 1) or ***off*** (value = 0):

Some examples are: an on/off switch, a digital clock, digital voltmeter, bar-code stripes. Mobile phones, CD and DVD discs (page 306) all use digital signals and store data digitally.

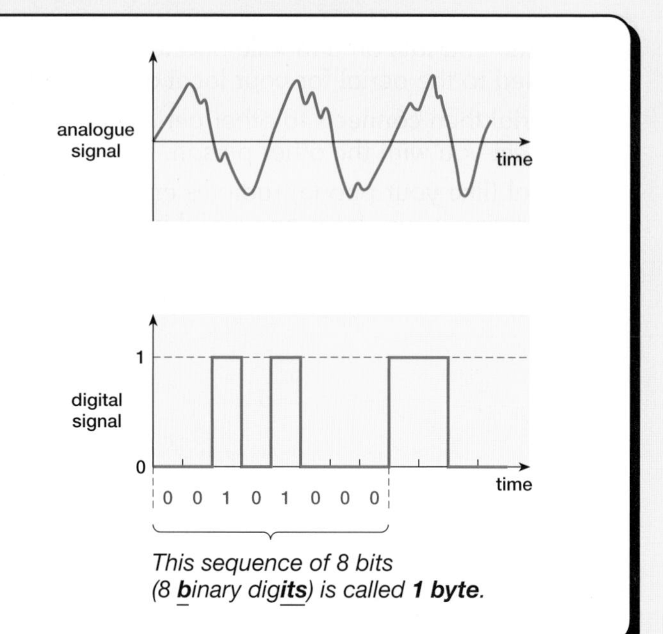

*This sequence of 8 bits (8 binary digits) is called **1 byte**.*

Converting signals

Analogue signals can be converted to digital signals by an **ADC** (Analogue to Digital Converter).
This ***encodes*** the signal into a digital format.

Computers use only digital signals:

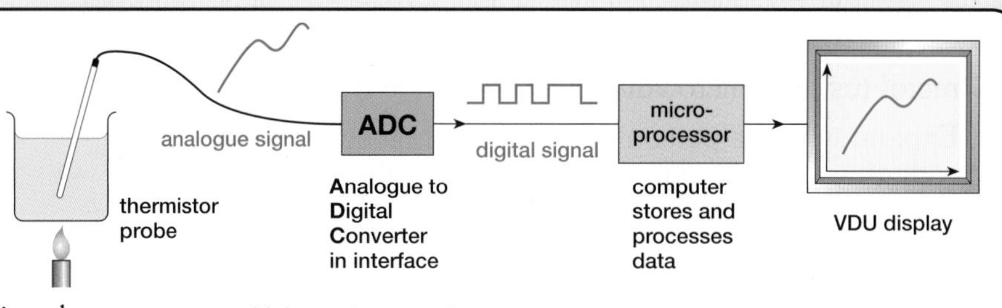

Using a temperature sensor

Digital transmission

Telephone systems (see page 314) are linked by optical fibres (page 192).

This block diagram shows the idea:

Digital signals are used in the optical fibre, because:

- the final signal has a better quality,
- more signals can be sent in each second.

See the opposite page for more details.

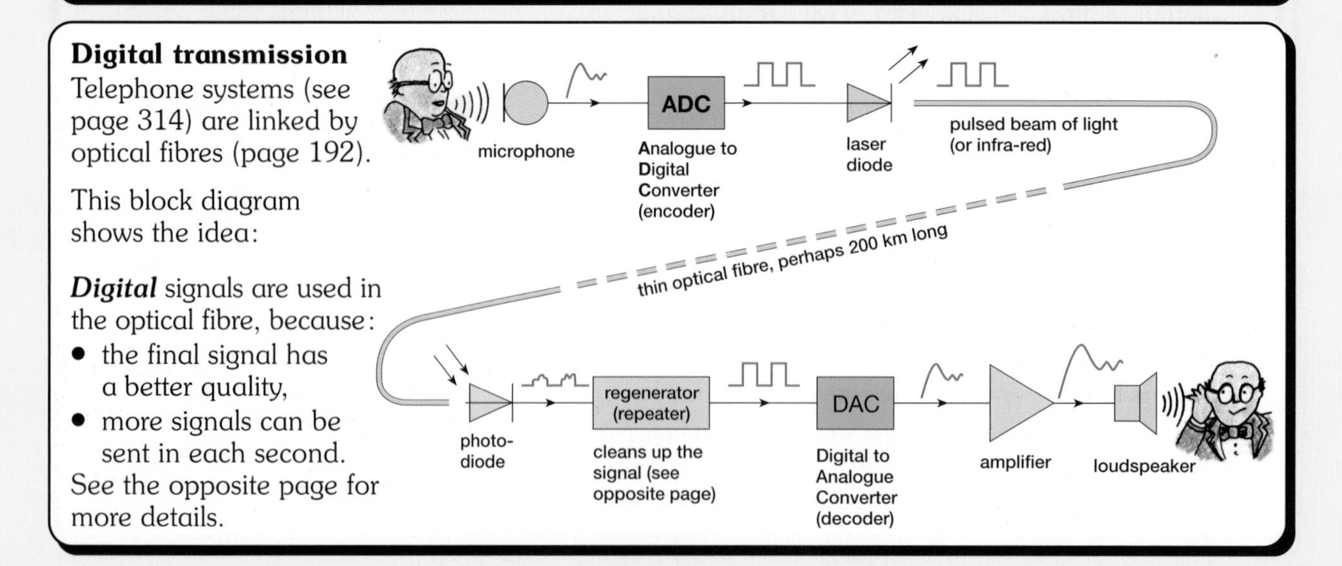

▷ Physics at work: Analogue and Digital Communications

An advantage of using digital signals

When a signal is transmitted, it gradually becomes altered or damaged in ***two*** ways:

- The signal strength is reduced or ***attenuated*** after travelling a long distance.
- The signal picks up '***noise***' (random extra signals).

The diagrams show why a digital signal gives a better quality result:

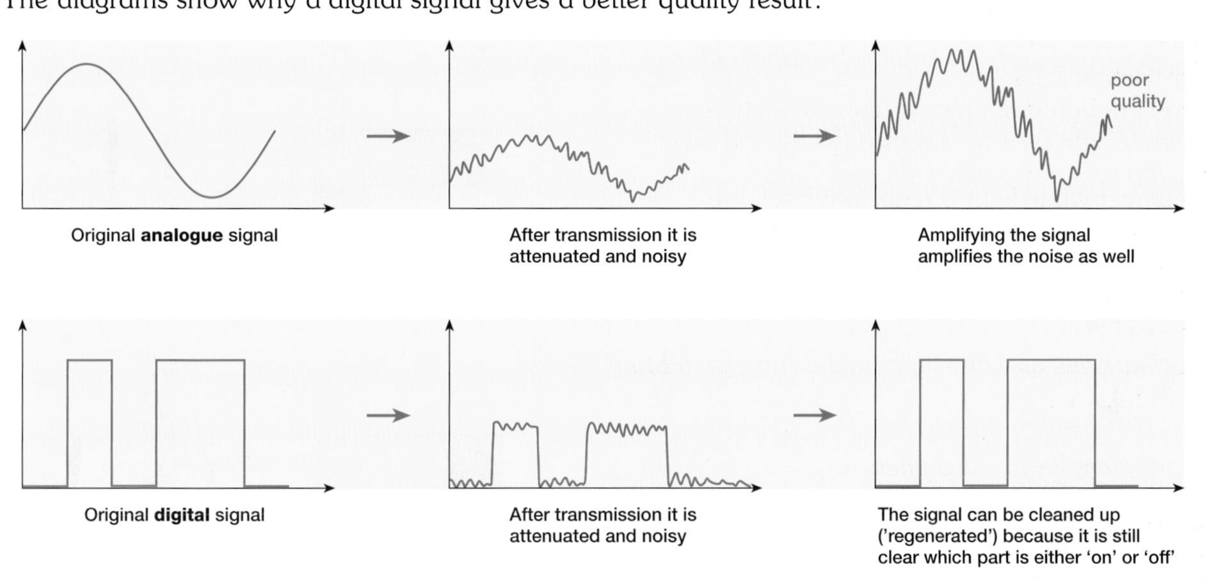

Original **analogue** signal

After transmission it is attenuated and noisy

Amplifying the signal amplifies the noise as well

Original **digital** signal

After transmission it is attenuated and noisy

The signal can be cleaned up ('regenerated') because it is still clear which part is either 'on' or 'off'

Another advantage of digital signals

Using digital signals in an optical fibre, more information can be transmitted in each second.

A high-frequency on/off signal is used, switching 100 million times in each second (100 MHz) or even more.

Two or more signals can also be interleaved or ***multiplexed*** as shown:

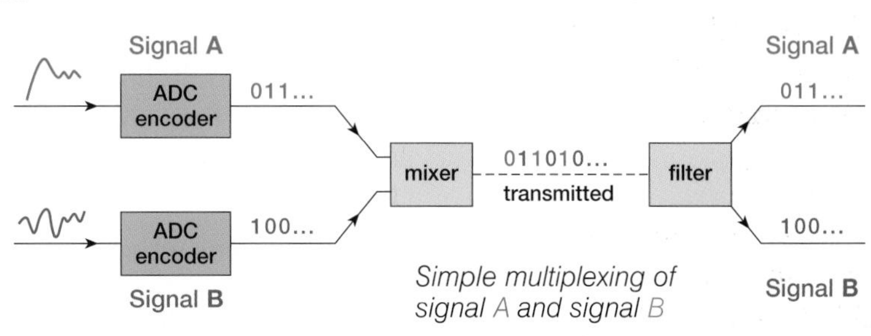

Simple multiplexing of signal A and signal B

Analogue to Digital Conversion (ADC)

This is done by ***sampling*** the signal regularly to find its height. The ***binary*** value is then sent as a digital signal:

To make a CD-ROM the sound is sampled 44 000 times a second (44 kHz) with 65 536 voltage levels (16 bits), to keep good quality reproduction.

DAC is the reverse process.

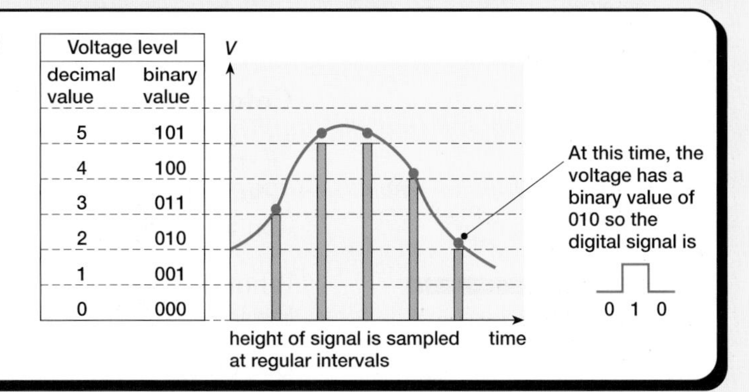

Voltage level	
decimal value	binary value
5	101
4	100
3	011
2	010
1	001
0	000

height of signal is sampled at regular intervals

▷ Subtraction of colour (absorption)

Within the visible part of the electromagnetic spectrum, colour plays an important part in our lives and in our environment. Colour mixing is important to artists and to textile designers.

Filters

Experiment 28.4
Look through a red plastic filter. What happens to white light as it passes through the red filter?

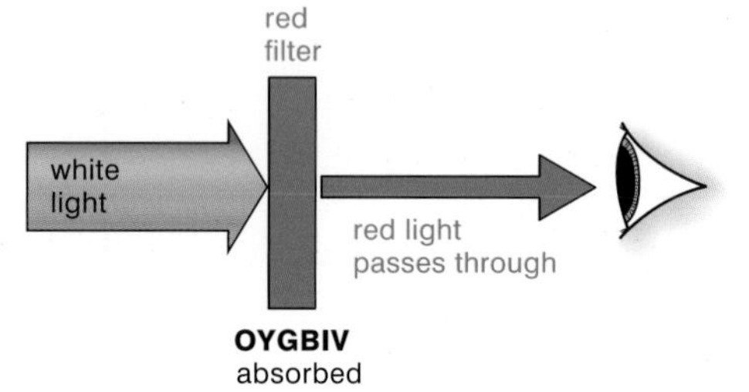

Remember white light is really a mixture of several colours (ROY G. BIV).
Which one of these colours passes through a red filter? Which colours are ***absorbed*** (and subtracted) by a red filter?

What happens as white light passes through a ***blue*** filter?

Experiment 28.5 Two filters
Place a red filter and a green filter together and look through them.

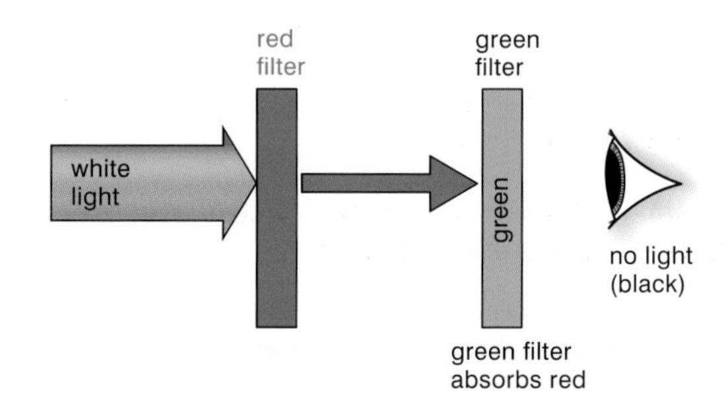

What happens to the white light? Why?

Coloured objects in white light

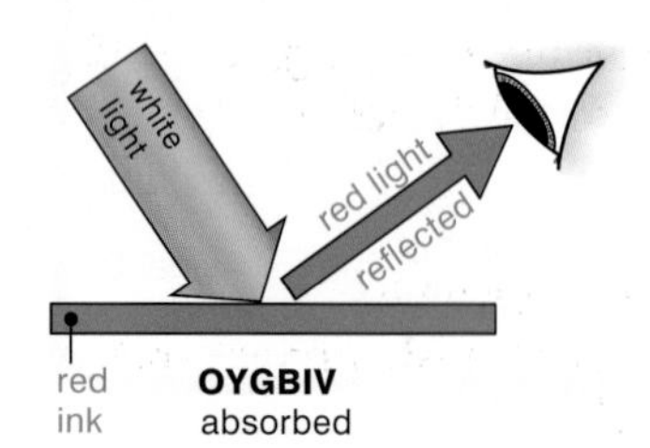

When white light shines on this white paper, the white light is reflected to your eye. When white light shines on this black ink, all the light is absorbed and none is reflected.

What happens when white light shines on this red ink (or red paint or red cloth)?
White light (really ROY G. BIV) shines on the red ink which absorbs OYGBIV leaving only red to be reflected to your eye and so the ink appears red.

red light
no light
reflected (black)
blue
object
blue absorbs red light

Coloured objects in coloured light

Experiment 28.6
Place a red filter over a raybox or a torch and shine the red light on to a bright blue object (in a dark room).

What do you see?
Explain why this is so.

Use different colours and fill in a table to show your results.

Coloured objects viewed through filters

Experiment 28.7
Look at white paper through a green filter.
Look at a green book through a green filter.
What do you see?

Look at a red book through a green filter.

What do you see? Explain why this is so.

Use other filters and brightly coloured objects to fill in a copy of this table:

Colour of object	Filter	Appearance
White	Green	Green
Green	Green	Green
Red	Green	Black
Blue	Red	

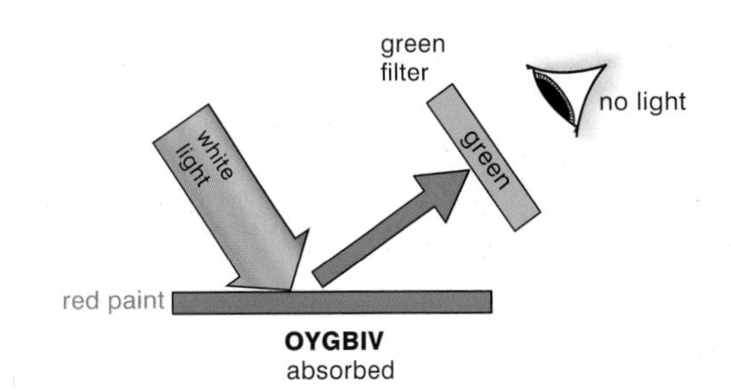

Look at the cartoon through a red filter and then through a green filter. Explain what you see.

Look at the cartoon in red light and in green light and explain what you see.

▷ Addition of colours

Experiment 28.8 Primary colours
Use 3 rayboxes (or slide projectors) fitted with a red, a green and a blue filter and shine them on to the same white screen so that the colours overlap and add together.

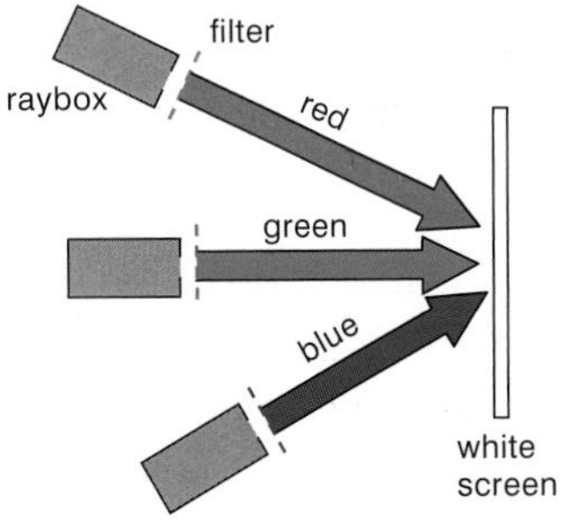

Where all 3 colours overlap, the screen appears white.
Because they add together to make white light, **red, green and blue are called the primary colours of light**.

In your eye there are three types of colour-sensitive cells, called ***cones***. One type of cone detects red light, one type detects green light, and the third type responds to blue light.

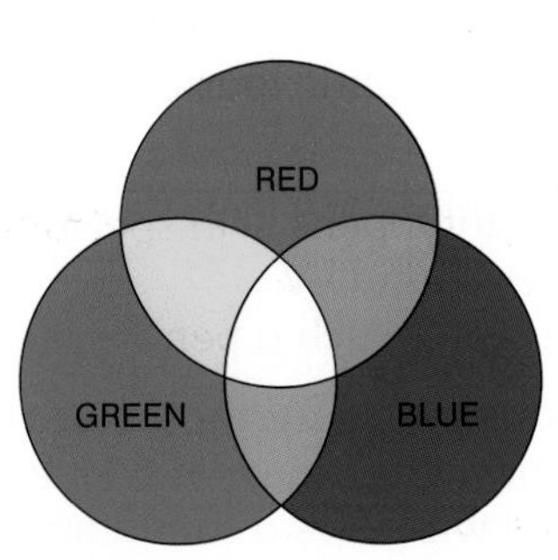

If you switch on the colours in pairs, you get 3 **secondary** colours: ***yellow***, ***magenta*** (purple), and ***cyan*** (greeny-blue).

Look closely at the screen of a colour TV set (see page 307).
Can you see the red, green and blue dots?
Why are these three colours used?
Which dots must light up to make the TV screen look yellow?

Summary

The spectrum of white light can be remembered by ROY G. BIV.
Red has the longest wavelength and is deviated least by a prism.

The full electromagnetic spectrum in order of increasing wavelength is: gamma-rays, X-rays, ultra-violet, visible light, infra-red, radio waves.

Subtraction of colours (absorption): filters, paints, pigments and inks subtract colours, e.g. a blue flower absorbs ROYGIV, and reflects blue light. In red light it looks black.

Addition of colours: the three primary colours (R,G,B) add to give white.
Secondary colours are yellow, magenta, cyan.

▷ Questions

1. Copy out and complete:
a) White light is composed of . . . . colours, in order: red, ... , ... , ... , ... , ... , ... , which form the visible . . . . . The colour with the longest wavelength is . . . . . The colour deviated through the largest angle by a prism is . . . . .
b) The full electromagnetic spectrum, in order, is: gamma-rays, ... , ... , ... , ... , ... The section with the longest wavelength is . . . . .
c) The most dangerous radiation is . . . . because it can . . . . atoms in your body.

2. The diagram shows white light being dispersed by a prism:

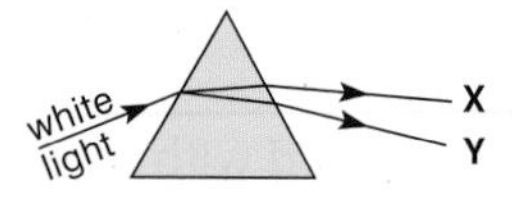

a) What colour would you see i) at X ii) at Y?
b) What may be detected i) above X ii) below Y?

3. Name a type of electromagnetic wave which:
a) can pass through metals,
b) can cause a suntan,
c) is used for radar,
d) is diffracted round hills,
e) is emitted by warm objects.

4. Name a type of electromagnetic wave which:
a) is used in a TV remote control,
b) is used to sterilise food,
c) is stopped by the ozone layer,
d) is used to locate rainclouds,
e) is radiated from mobile phone masts.

5. On the chart on pages 208–9, find the wavelength of your favourite radio station.

6. From the eye-sensitivity graph on page 210, which wavelength has the same sensitivity as red light of wavelength 650 nm?
On that graph, where is a) I-R b) U-V?

7. Use your Physics to explain the following:
a) Dark skins are partly protected against U-V radiation.
b) Babies need cream with a sun protection factor (SPF) of 50.
c) People skiing in the mountains are likely to get tanned.
d) Forged bank notes can be detected by using a U-V lamp.
e) It is important not to damage the Earth's ozone layer.

8. Use your Physics to explain the following:
a) Gamma rays are more dangerous than visible light rays.
b) Food stays fresher after it has been exposed to gamma radiation.
c) It is important that a microwave oven door has a safety switch.
d) Inside a microwave oven, food should not be wrapped in aluminium foil.
e) The inside of an electric toaster has a shiny surface.

9. You can buy 'security' pens with special ink which only shows up in U-V.
Suggest some uses for such a pen.

10. What are the dangers of U-V radiation? Use your scientific knowledge to write a letter of advice to a friend who uses a sun-lamp.

11. Describe two ways in which an X-radiologist in a hospital can work more safely.

12. Explain how a tumour in a patient can be given a high dose of γ-rays while minimising the damage to healthy tissue.

13. Suggest an experiment to show that infra-red travels at the same speed as visible light.

14. In the radar diagram on page 211, the wave takes 0.0001 s to travel to the plane and back. How far away is the plane?

15. Draw a scale diagram of the Earth (radius 6400 km) and a 'geostationary' satellite moving over a fixed point with the Earth, in an orbit of radius 42 000 km (see page 154). Show that three satellites spaced out in this orbit can cover communications over the whole Earth.
The microwave path from Britain to USA via such a satellite is about 90 000 km. Calculate the time taken for the waves to travel. Would you notice this in a telephone conversation?

16. What colour would a green book look:
a) in white light? b) in green light?
c) in red light? d) through a red filter?

17. Why does red lipstick look unpleasant under yellow street lighting?

18. Write essays or discuss the following topics:
a) The use of coloured spot lights for different stage effects.
b) The choice of colours for road signs, advertisements, clothes and rooms.
c) The use of colour in language, eg. feeling blue, red with rage, green beginner, etc.

19. *His face is red, he's feeling blue,*
Can't see his green mistake – can you?

20. Emma says, "The most dangerous thing about mobile phones is when drivers use them." (see page 216). Write an essay to discuss this. What is the law on this?

21. See the data on risks on page 216. The skull bones of young people are thinner than in adults. Discuss whether this is a problem. Does it mean that mobile phone risks are different for young people?

22. If the radiation at 1 m from a source is 1.0 mW/m^2, what is it at
a) 2 m, b) 3 m, c) 10 m?

23. Design a poster to persuade young people to use their phones less often or more safely.

24. Imagine a mobile phone mast (page 217) is proposed for near your school or home. Discuss how you would plan an action campaign about it.

25. 'Strobe' lights flashing at 15–20 Hz are known to cause brain disturbances.
Many mobile phone masts also vary their signals at this frequency.
Speculate on possible consequences arising from this. How could this be tested?

26. Discuss the need for communications in the modern world.
How would your life be affected if there were no radio, TV or telephones?

27. Explain, with diagrams,
a) the difference between analogue and digital signals,
b) the advantages of using digital signals in communications.

Further questions on page 238.

chapter 29

What is sound caused by?

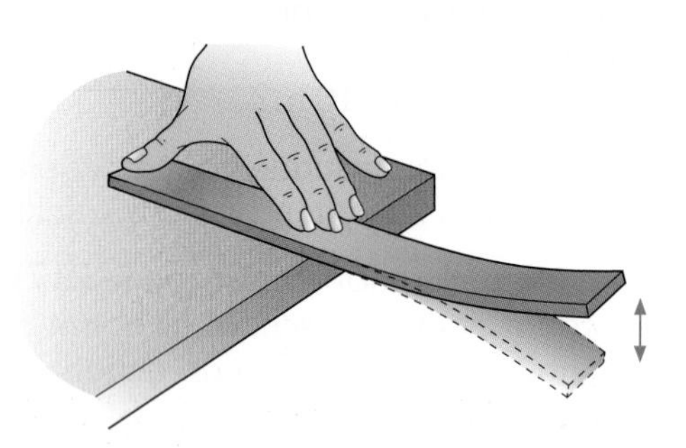

Experiment 29.1
Hold down one end of a ruler firmly to the bench. Flick the other end to make a sound.

Look at the end of the ruler. What is it doing? Does it make any sound when it has stopped vibrating?

Experiment 29.2
Place your fingertips against the front of your throat. What can you feel when you make a noise?

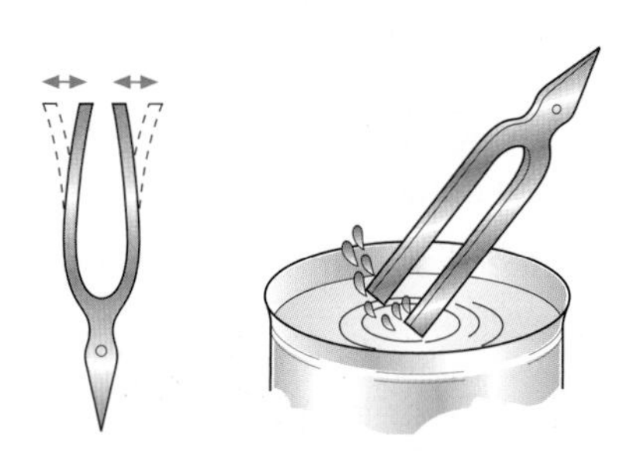

Experiment 29.3
Get a tuning fork and bang it on a cork to make it vibrate (it vibrates like two rulers fastened together). What can you hear?

Can you see that the ends are vibrating? If not, touch the ends to the surface of some water in a beaker. What do you see?

These experiments show that **sound is caused by vibrations** and is a form of kinetic energy (see page 98).
How are these vibrations caused in
a) a drum? b) a guitar?

How are the vibrations caused by a vinyl gramophone record?

Experiment 29.4
Look at an old vinyl record through a magnifying glass. Can you see the wobbly grooves?

Push a needle through a piece of card and then put the point of the needle on to the record as it revolves on a turntable.

What can you hear?
What are the wobbly grooves doing to the needle and card?

How do the vibrations travel to your ear?

Experiment 29.5
Stretch a long 'slinky' spring along a smooth bench and vibrate one end to and fro along the length of the spring, to send a longitudinal wave down the spring (see page 166 again).

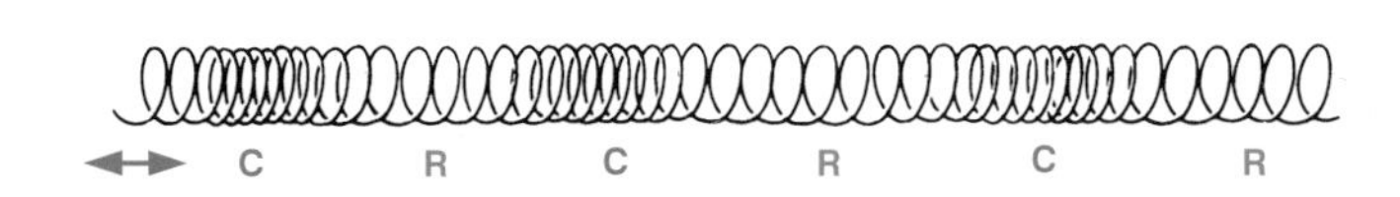

If you look closely at the spring you can see that, at any instant, some parts of the spring are pushed closer together (***compression***) and some parts are pulled farther apart (***decompression*** or ***rarefaction***).

It is the same with a sound wave in air. In some places the ***molecules*** of air are pushed together at a slightly higher pressure (compression) and in some places the molecules are farther apart at a slightly lower pressure (rarefaction).

These compressions and rarefactions shoot out across the room to your ear, travelling at the speed of sound. Molecules of air do not travel across the room – they just vibrate to and fro.

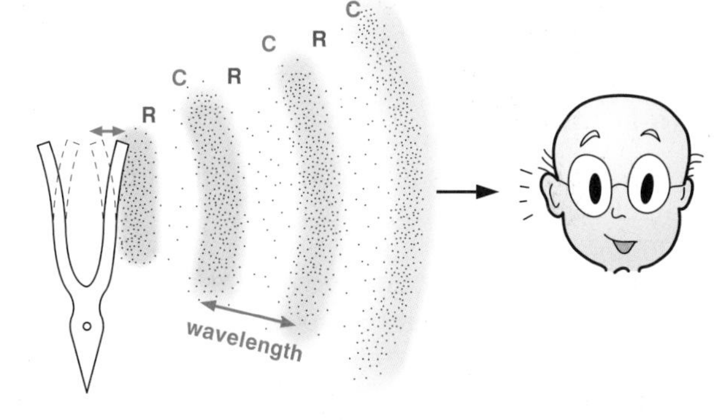

The ***wavelength*** of the sound is the distance between successive compressions (or rarefactions).

For sound waves, like all other waves (page 167):

speed (m/s)	=	**frequency** (Hz)	×	**wavelength** (m)

Sound waves can be reflected (page 168), and refracted (page 168) and diffracted (page 169).

What happens if there are no molecules?

Experiment 29.6 ⚠
Hang an electric bell inside a jar connected to a vacuum pump.
Switch on the bell. Can you hear it?

Start the pump to take the air molecules out of the jar. What happens to the sound of the bell?

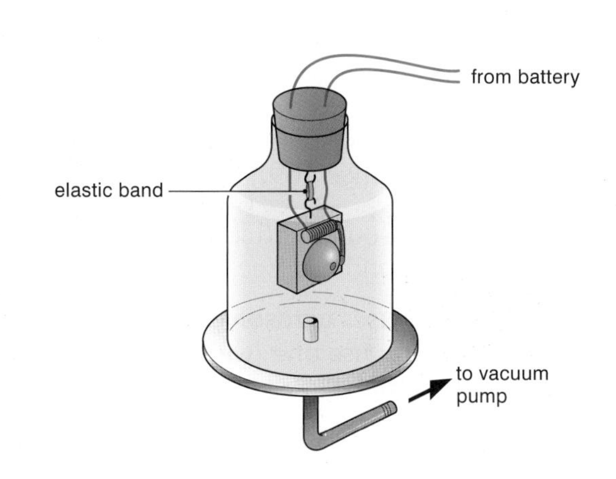

Sound cannot travel through a vacuum, because there are no molecules to pass on the vibrations. Why can't we hear the explosions on the Sun?

▷ Reflection of sound

Have you ever heard an ***echo***?

Can sound be reflected by walls?
Why do noises sound louder in a room than in the open air?

You may have seen that some of the compressions on the slinky spring (in experiment 29.5) were reflected back down the spring. In the same way, compressions and rarefactions are reflected back from the walls of a room.

If the distance is long enough you may hear a clear echo.

Echo-sounding

Ships can use echoes to find the depth of the sea.

Example
A ship sends out a sound wave and receives an echo after 1 second. If the speed of sound in water is 1500 m/s, how deep is the water?

Time for sound to reach the bottom $= \frac{1}{2}$ second (and $\frac{1}{2}$ second to return)

$\therefore$ Depth of water $= \frac{1}{2}\text{ s} \times 1500\text{ m/s}$

$= \underline{750\text{ metres}}$

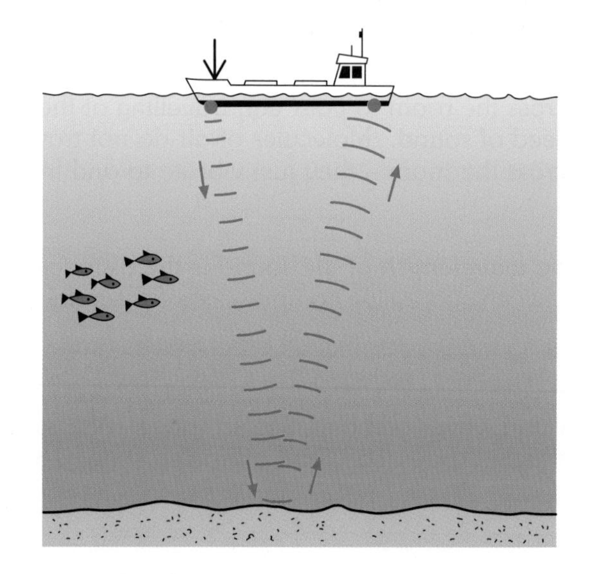

Fishing boats use this **sonar** to detect shoals of fish. If the shoal of fish in the diagram swim under the boat, how will the captain know?

This echo-sounder uses a very high frequency sound. This **ultrasound** is so high that we cannot hear it. Because ultrasound has a high frequency, it has a short wavelength. This means that it can be sent out as a narrow beam (if the waves are longer, they spread out more because of ***diffraction***, see page 169).

Bats also use ultrasound and listen to the echoes to 'see' their surroundings (see page 228).

A ***motion-sensor*** uses ultrasonic echoes in the same way. You can use one (with a computer) to draw motion graphs (pages 123–126).

There are more uses of **ultrasonics** on page 229.

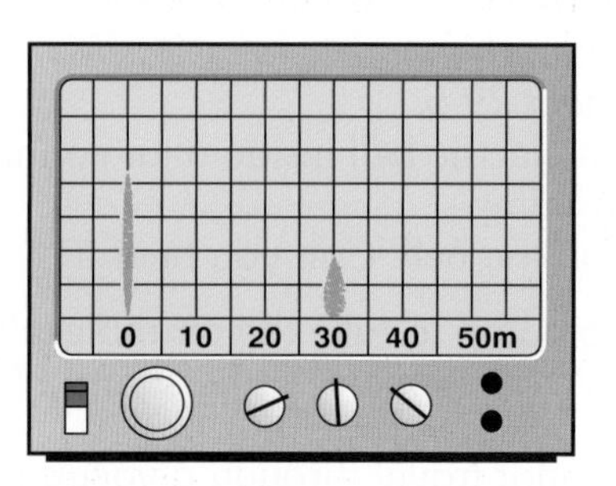

The sonar display on the captain's oscilloscope. It shows the transmitted pulse and the echo.

▷ Measuring the speed of sound

Experiment 29.7 Outdoors
Stand a measured 50 metres from a large wall. Clap, or bang sticks together, and listen to the echo. Then try to clap in an even rhythm of clap–echo–clap–echo–clap . . . while a friend times 100 of your claps with a stopwatch.

During the time from one clap to the next clap, the sound would have time to go to the wall and back, **twice** – that is a distance of 200 metres. In the time of 100 claps, the sound would travel 200 × 100 = 20 000 metres.

$$\therefore \text{Speed} = \frac{\text{distance travelled}}{\text{time taken}} = \frac{20\,000 \text{ metres}}{\text{time in seconds}}$$

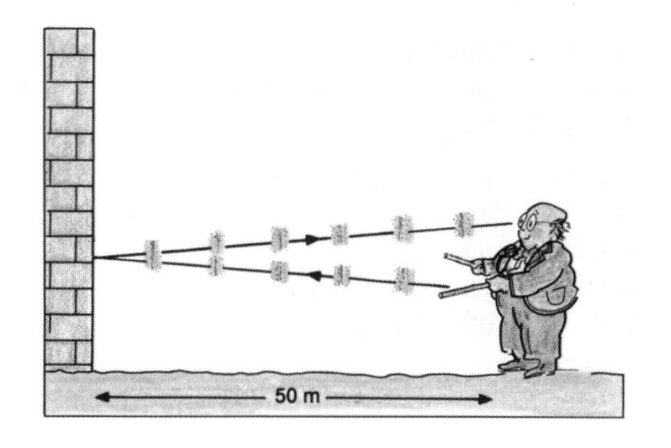

Experiment 29.8 Indoors
An alternative method, for indoors, is to use sound switches with an electronic timer:

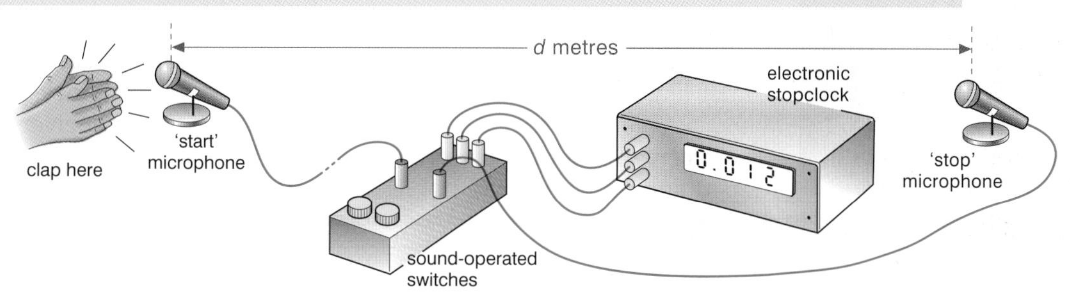

The sound arriving at the first microphone switches ***on*** the clock and the same sound arriving at the second microphone switches it ***off***. The clock shows the time for the sound to travel the distance ***d***.

$$\therefore \text{Speed of sound} = \frac{\text{distance } (\boldsymbol{d})}{\text{time taken}}$$

At 0 °C the speed of sound in air is 331 m/s (1200 km/h or 740 m.p.h.). At room temperature, its speed in air is faster, about 340 m/s.

If sound travels 1 kilometre in 3 seconds, how can you use lightning and thunder to find your distance from a storm?

Experiment 29.9
Measure the speed of sound in wood by clamping the microphones (of experiment 29.8) face down to a wooden bench. What do you find?

Sound travels faster through solids than through gases (or liquids) as you can see in the diagram. This is because the molecules are more tightly linked together in a solid.

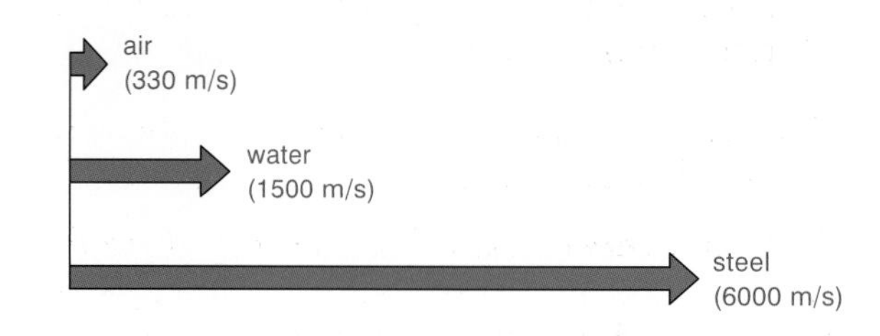

▷ Physics at work: Ultrasonic echoes

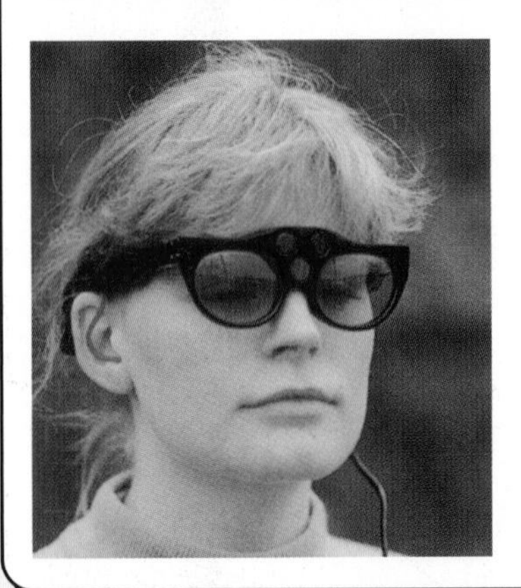

We cannot hear sounds which have frequencies above 20 000 Hz, but many animals can. Dogs will respond to an ultrasonic whistle even though it seems silent to us.
Bats and dolphins use ultrasound frequencies of about 150 000 Hz to 'see' by listening to the echoes. Echo-sounders on ships (see page 226) also use a very high frequency.

There is a reason for using a very high frequency: a high frequency means a ***short*** wavelength.
To see this we can use: **speed = frequency × wavelength** (see page 225) to calculate the wavelength.

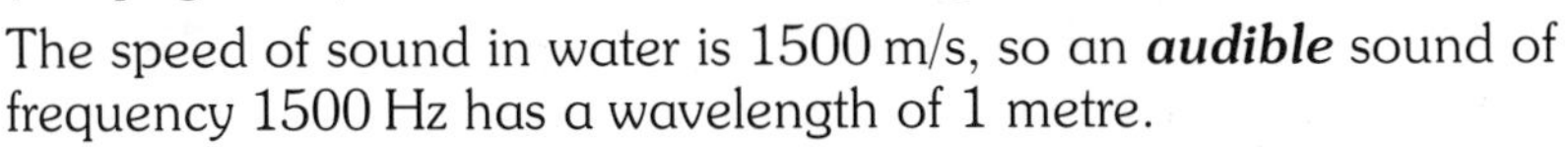

The speed of sound in water is 1500 m/s, so an ***audible*** sound of frequency 1500 Hz has a wavelength of 1 metre.
With this long wavelength it is difficult to detect small objects because the wave ***diffracts*** and bends round them, see page 169).
However if the sound has an ultrasonic frequency of 150 000 Hz, the wavelength is only 0.01 m (= 1 cm). With this shorter wavelength the sound is reflected back from smaller objects and so the echo can be timed (and the distance to the object calculated).

An ultrasound beam is also much more directional and can be aimed rather like a torch. This has been used in ultrasonic spectacles for blind people. The spectacles have a transmitter and a receiver. The receiver produces a high or low sound in the person's ear depending on whether the object causing the echo is near or far.

Echo sounding is also used to detect flaws inside pieces of metal. A transmitter sends out pulses of ultrasound and a receiver picks up the echoes from different parts of the metal and shows the results on a CRO (cathode ray oscilloscope, page 310):

Pulse A is the transmitted pulse; pulse B has been reflected by the flaw (b) in the metal; pulse C is the echo from the end (c) of the metal.

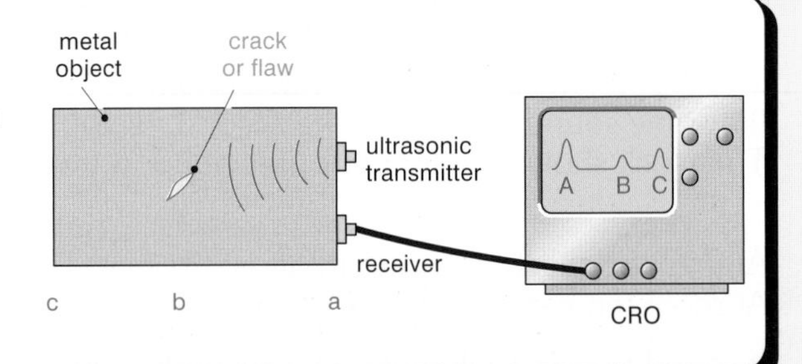

Geologists use echo-sounding to search for oil and gas:

In this diagram, which microphone will receive the sound first?

The speed of sound in rock is about 4000 m/s. If the sound arrives at **A** after $\frac{1}{2}$ s, estimate the depth to the hard rock.

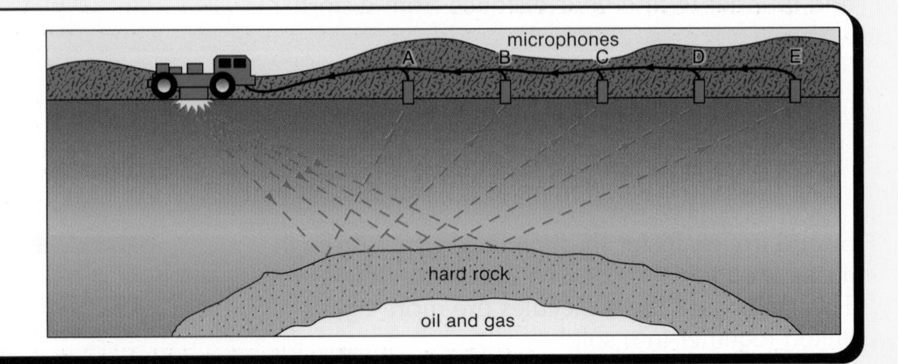

▷ Physics at work: Ultrasound

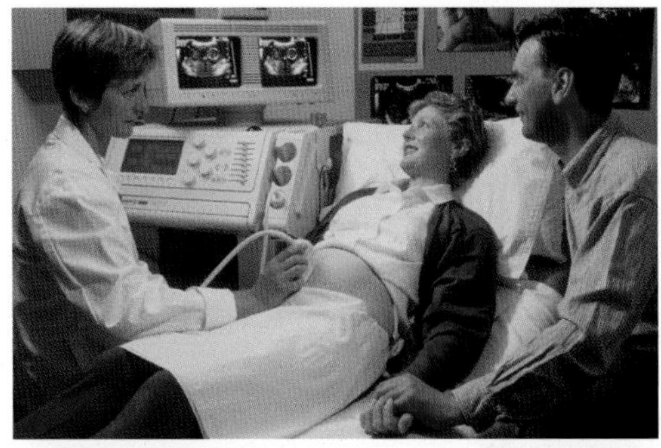

Ultrasonic echoes are also used in medicine, to 'see' inside the body. This is safer than X-rays, which can be very dangerous.
Unborn babies can be seen by moving an ultrasonic transmitter/receiver across the mother's stomach:

Whenever a sound wave travels from one substance to another (skin, muscle, bone), some of it is reflected back, to give echoes. The machine uses these echoes to build up a picture on a TV screen:

In order to see fine detail the wavelength of the sound waves must be short. By using a very high frequency of 1.5 MHz (1 500 000 Hz) the wavelength in water is only 1 mm (see the calculation opposite).

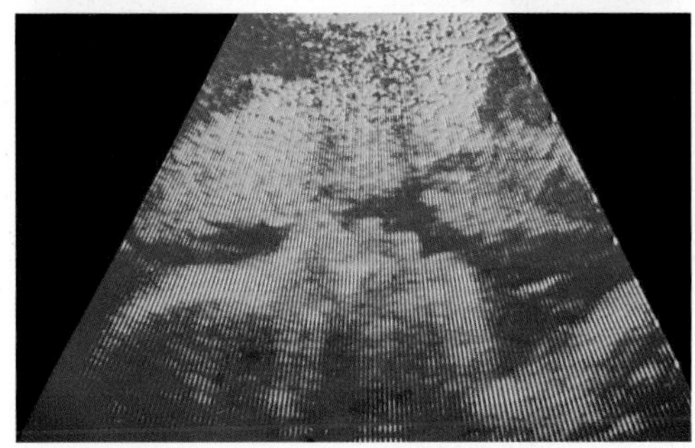

This picture shows a pre-natal ultrasound 'scan' of an unborn baby, 28 weeks after conception:
Can you see its face?

The same method can be used to investigate heart and liver problems, and to look for tumours.

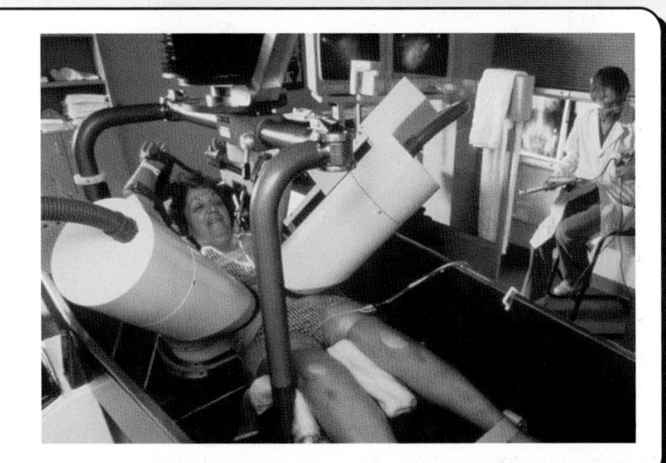

As you get older, stones can build up inside your kidneys. They are very painful. With ultrasound they can be removed without having to cut open your body!

The photo shows a high-energy shock-wave of ultrasound being used to smash the stones into tiny pieces.
Then they can come out with the urine.

An ultrasonic beam is used because:
- it is a narrow beam that can be focussed at the right place,
- it can transfer more energy than an audible sound wave.

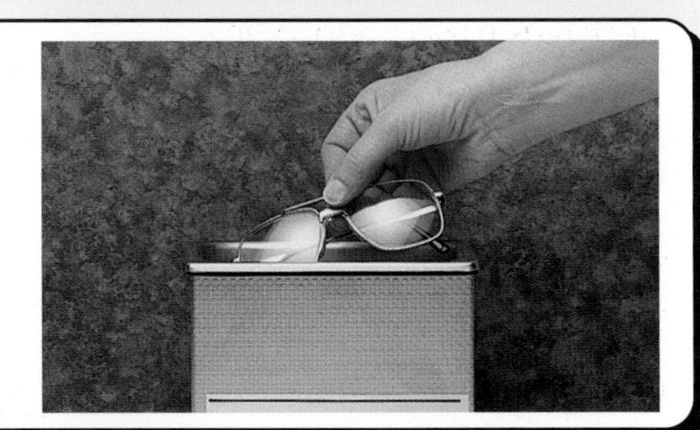

Opticians can use ultrasound to clean spectacles.
The water inside this tank is vibrated by ultrasound:
The vibrations shake the dirt loose.

The same idea is used to clean jewellery, clothes, and even street-lamp covers.

Ultrasonic beams are also used by dentists – the vibrations shake the dirt and plaque off your teeth.

The vibrations are also used to kill bacteria in milk.

▷ The ear

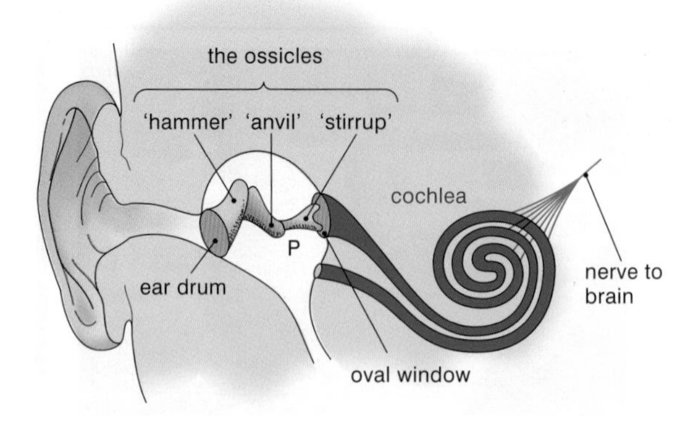

Sound waves are collected by your outer ear and passed in to the ***eardrum*** which is made to vibrate by the compressions and rarefactions.

These vibrations are passed to the ***oval window*** by three bones (called the ***hammer, anvil*** and ***stirrup***) which act as a lever (with the pivot at point P). This means that they magnify the force of the vibrations (see page 117).
Also, the oval window has a smaller area than the eardrum, so this increases the pressure on the oval window and on the liquid in the ***cochlea*** (see page 79).

The vibrations of the liquid in the cochlea affect thousands of ***nerves*** which send messages to your brain. This allows you to recognise the sound.

What range of *frequencies* can you hear?

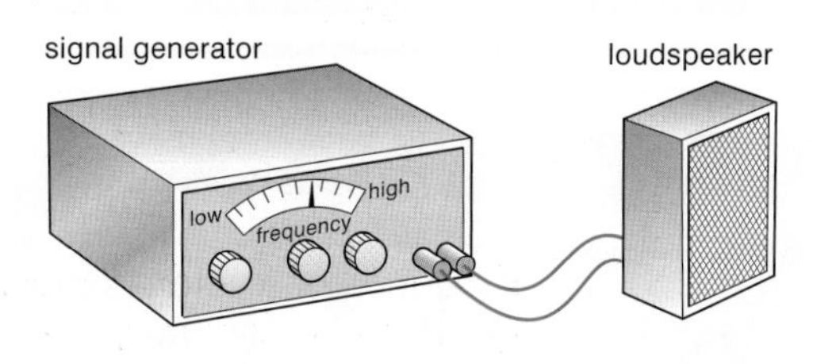

Experiment 29.10
Connect a loudspeaker to a signal generator (which produces different frequencies as the pointer is moved to different positions).

Turn the pointer to lower and lower frequencies. What happens to the ***pitch*** of the note from the loudspeaker?
What is the lowest frequency you can hear?
How can you make this a fair test?
What is the highest frequency you can hear?
Does this vary from one person to another?

Doctor: Have your ears been checked lately?
Professor Messer: No, they've always been a pinkish colour.

Young children can hear sounds as low as 20 hertz (20 cycles per second) and as high as 20 000 hertz, but as you get older, this range becomes less.

About 20% of the population has some kind of hearing defect. This may be due to old age, or an infection in the ear, or damage to the cochlea by very loud noises (for example, in a disco).

The graph shows a boy's ***audiogram***.

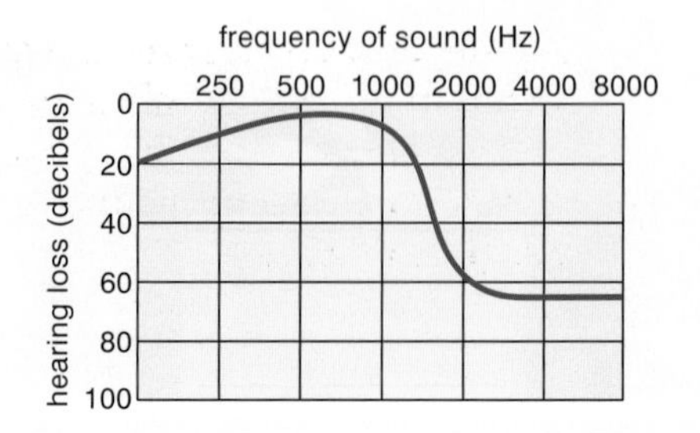

This shows that he has a hearing loss at high frequencies (above 2000 Hz). His hearing loss is about 60 decibels (see also page 234).
He should wear a hearing aid which is adjusted to boost the high frequency sounds.

▷ Resonance

A swing with someone sitting on it has a certain ***natural frequency*** of vibration. If you are asked to push the swing to make it go higher, you will obviously push it each time it comes near you. That is, the frequency of your pushes will be the ***same*** as the natural frequency of the swing. Then the swing vibrates with a large amplitude.

This is an example of ***resonance***. We say the swing is ***resonating***. If you push at a different frequency, it will not swing as high (and you might hurt your hand).

Resonance occurs when:

the applied frequency of the pushes	=	**the natural frequency of the object**

A short-sighted singer called Groat,
Could do wonderful things with his throat.
At his specs he aimed sound,
And with resonance found
That they cracked when he sang the right note.

Experiment 29.11
Hang a weight from a length of string (like a pendulum) and then use a straw to make it swing with a large amplitude,
a) by giving it one strong tap
b) by giving it lots of little taps at just the right frequency.

Which way gives the bigger swings?

In a similar way, if soldiers march in step over a bridge, they can make the bridge vibrate so much at its natural frequency that it may break.

A large bridge in America once fell down because it resonated with vibrations caused by the wind.

Have you heard windows or seats on a bus start to rattle (as they resonate) when the engine speeds up to a certain frequency?

If a singer sings near a wine glass with a frequency equal to the natural frequency of the glass, it may resonate so strongly that the glass breaks.

▷ Pitch, loudness and quality of musical notes

The notes from a musical instrument can vary in three ways:

1. in ***pitch***
2. in ***loudness***
3. in ***quality*** or ***tone***.

What do these three things depend on?

1 Pitch

We have already seen (in experiment 29.10) that ***the pitch of a note depends on the frequency***.

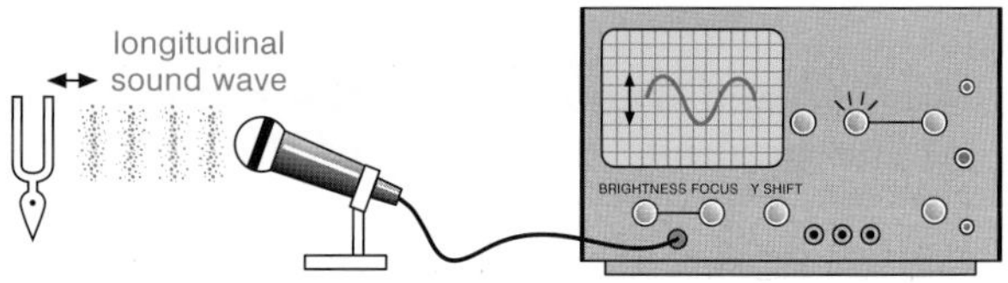

transverse wave on screen

longitudinal sound wave

microphone

oscilloscope

BRIGHTNESS FOCUS Y SHIFT

Experiment 29.12
An oscilloscope (a kind of television set, see page 310) is very useful for showing vibrations. Connect a microphone to an oscilloscope and hold a tuning fork nearby.

When the oscilloscope is correctly adjusted, it shows on the screen a ***transverse*** wave of the ***same frequency*** as the ***longitudinal*** sound wave from the tuning fork.

Experiment 29.13
Look at the waves on the screen caused by a low-pitched tuning fork and a high-pitched tuning fork. What do you notice?

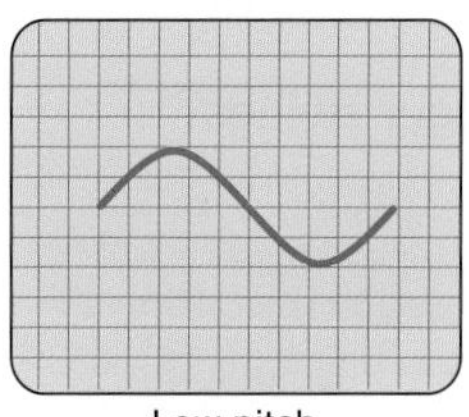

Low pitch
(low frequency)
long wavelength

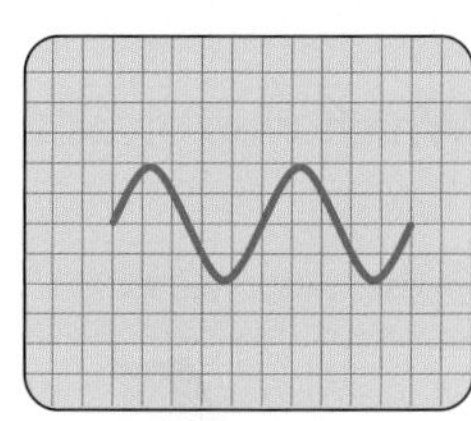

High pitch
(high frequency)
short wavelength

2 Loudness

Experiment 29.14
Whistle a soft note into the microphone and then the same note but louder. Look at the waves on the screen. What do you notice?

The loudness depends upon the amplitude of the wave.
A wave with a larger amplitude contains more energy.

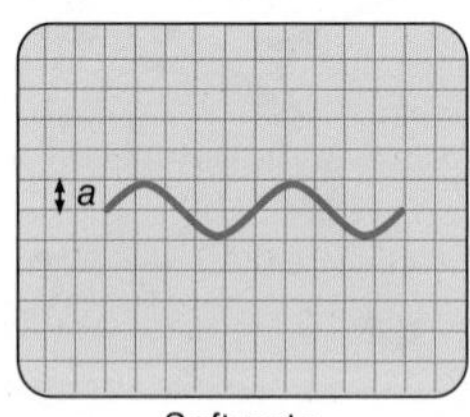

Soft note

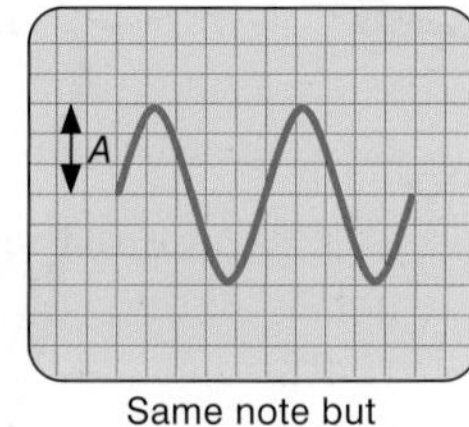

Same note but
LOUDER

The loudness of a sound also depends on how much air is made to vibrate.

Experiment 29.15
Repeat experiment 29.4 with the vinyl record, using a small piece of card and then a large piece of card. Which sounds louder?

Prof Messer: If you drop a piano down a mine-shaft, what key does it play in?
Mrs Messer: A flat minor?

3 Quality (or tone)

If a violin and a piano play the same note (at the same pitch and the same loudness), you can still tell them apart because the notes have a different ***tone*** or ***quality***.

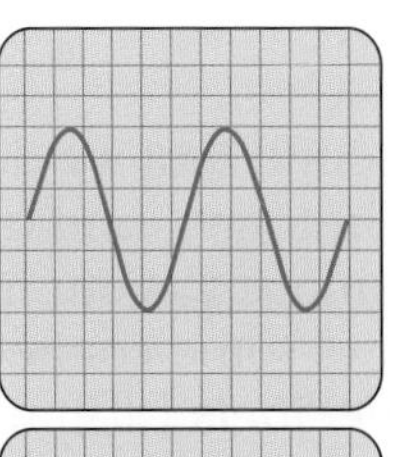
tuning fork

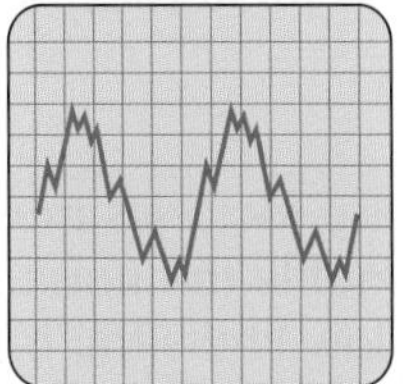
violin

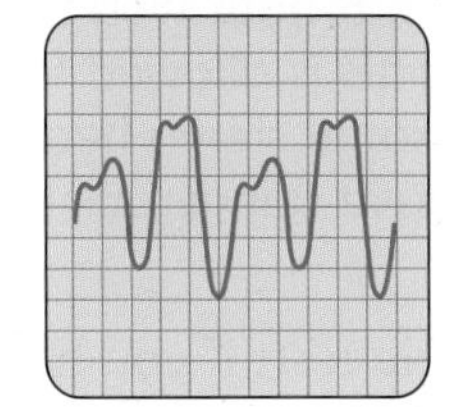
piano

> *Experiment 29.16*
> Play different musical instruments in front of a microphone connected to an oscilloscope.
> Play the same note on each instrument and sketch the ***waveform*** that you see.

What do you notice?

The quality of a sound depends upon the waveform.

What causes these different waveforms?

A tuning fork produces a pure note with only one frequency.
Other instruments usually produce many frequencies or ***harmonics*** at the same time.
All the harmonics add together to give a complicated waveform:

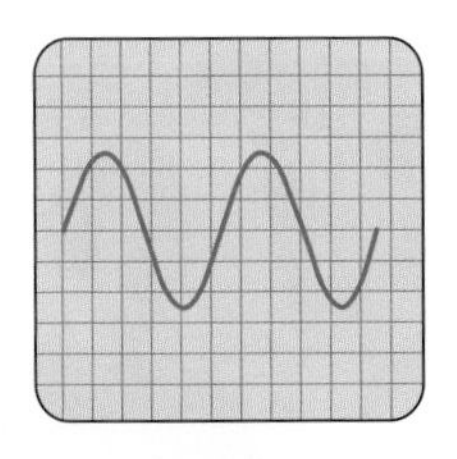
If this wave
(the first harmonic)

is added to

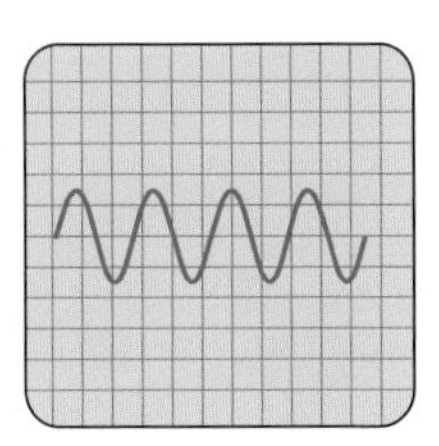
this wave of **twice** the frequency
(the second harmonic)

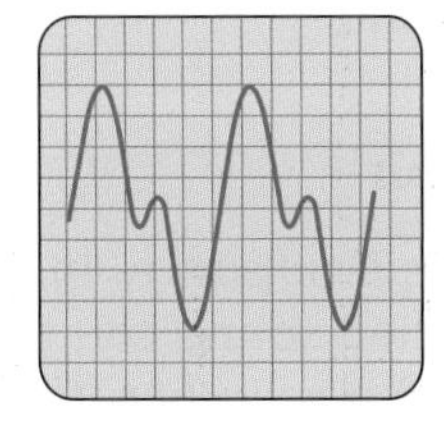
the result is this waveform

If more harmonics are added in, different waveforms are produced. Different instruments give different harmonics and different waveforms depending on the shape and size of the instrument, and so they give different sounds.

An electronic organ or ***synthesizer*** allows you to choose which frequencies you want to mix together.
You can use it to imitate other musical instruments, or you can create entirely new sounds:

> *Experiment 29.17*
> Look at the oscilloscope while you sing or hum into the microphone. Try to keep the same pitch and loudness but change the shape of your mouth with your lips and your tongue. What do you notice?

▷ Noise

Noise is any sound that we do not like.
Noise can be irritating, and a loud noise can do permanent damage to your hearing.
The chart shows some common noise levels, measured in ***decibels*** (dB), using a sound-level meter:

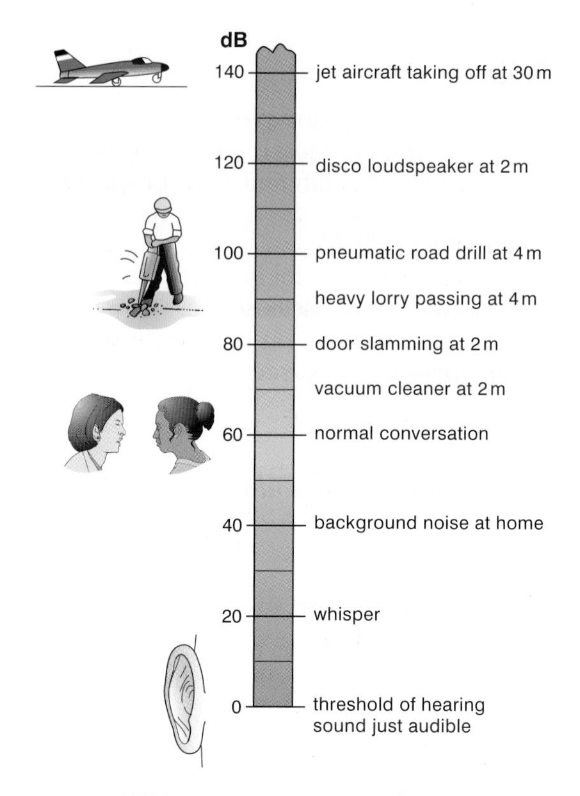

Workers in noisy factories must wear ear-protectors to muffle the noise.
In Britain the law limits the maximum noise dose to 90 dB for an 8-hour working day.
A level of 100 dB is allowed for only 48 minutes.
However, personal stereo headphones and discos often have dangerous noise levels of over 100 dB!

Here is a simple test: if you have to shout to be heard by someone at arm's length, then your environment is dangerously noisy and you should leave. Once damaged by noise, your ears cannot be mended – the nerves in your cochlea are dead.

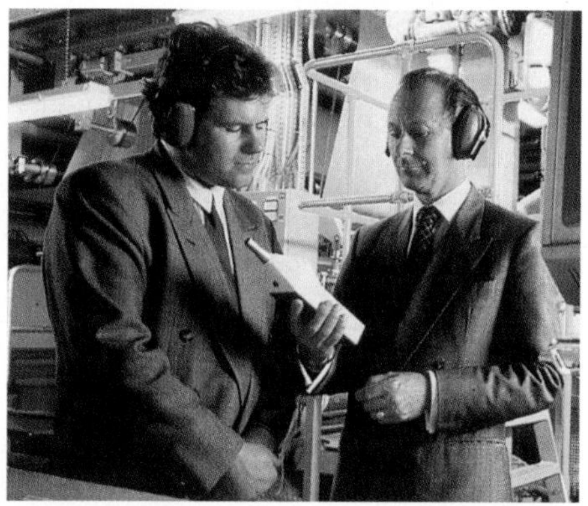

Measuring a high noise-level with a sound-meter

Reverberation

Echoes can be a nuisance. In an empty room or hall, the reflected sound can take a long time to die away. A long ***reverberation time*** makes it difficult for you to hear someone speaking.

The ***acoustics*** of a hall or a factory can be improved by using curtains, carpets and soft materials to absorb the sound. Concert halls are very carefully designed with absorbers to ensure the right amount of reverberation.

Summary

Sound is caused by vibrations and cannot travel through a vacuum.
A sound wave consists of compressions and rarefactions of the air.
Speed of sound = frequency × wavelength

Echoes are caused by the reflection of sound.
Resonance occurs when the applied frequency equals the natural frequency of the object.

Sound can be reflected, refracted, diffracted (see page 169).
The human range of hearing is 20–20 000 Hz.
Frequencies above this are ultrasonic and have many uses (see pages 228–229).

The pitch of a note depends on the frequency.
Loudness depends on the amplitude.
Quality depends on the waveform.

▷ Questions

1. a) Sound is caused by . . . . .
b) A sound wave consists of places at higher pressure (called . . . .) and places of . . . . pressure (called . . . .).
c) Wave speed (in metres per second) equals frequency (in . . . .) multiplied by . . . . (in . . . .).
d) Sound cannot travel through a . . . . .
e) Echoes are caused by the . . . . of sound.
f) The speed of sound in a solid is . . . . than the speed of sound in air.
g) Ultrasounds have a . . . . above . . Hz and so have a short . . . . . They have many uses, including pre-natal . . . . .
h) Pitch depends on . . . . .
Loudness depends on . . . . .
Quality depends on . . . . .

2. The speed of sound is 340 m/s. If thunder is heard 20 seconds after lightning, how far away is the storm? What must you assume?

3. What is the wavelength of a sound wave of frequency 100 Hz?
(Speed of sound = 340 m/s.)

4. a) What are the highest and lowest frequencies that the human ear can detect?
b) What are the shortest and longest wavelengths that the human ear can detect?

5. Draw a line graph of distance travelled against time (up to 5 seconds) for sound in air.
Use your graph to find the distance of the lifeboat crew from a sinking ship if they see a distress rocket and then hear it after 4.2 s.

6. Explain how echoes are used:
a) by a ship to find the depth of the water
b) by geologists looking for oil
c) by bats and dolphins
d) by blind people with special equipment.

7. A man fires a gun and hears the echo from a cliff after 4 seconds. How far away is the cliff? (Speed of sound = 340 m/s.)

8. A girl stands 90 metres from a wall and claps her hands to hear clap–echo–clap–echo at an even steady rate of 1 clap per second. What is her result for the speed of sound?

9. A sonar pulse sent out by a boat arrives back after 3 seconds. If the speed of sound in water is 1500 m/s, how deep is the water?

10. Explain, with diagrams, what is meant by:
a) amplitude b) wavelength
c) frequency d) quality of a note
e) resonance.

11. The diagram shows a graph of a sound wave given by a tuning fork.

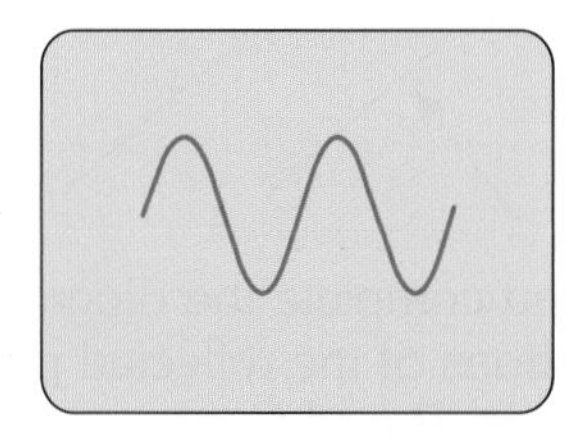

Copy this and then draw graphs to show:
a) a sound wave of higher pitch but the same loudness
b) a sound wave of the same pitch but louder
c) a sound wave of the same pitch but given by a different instrument.

12. Explain three methods by which the note from a guitar may be raised in pitch.

13. A sound-level meter is used at different distances from a disco loudspeaker:

Distance (m)	1	2	4	8	16
Noise level (dB)	130	120	110	100	90

Plot a graph and use it to find the noise level at 6 m. Is this distance safe for your ears?

14. Explain how ultrasound is used in pre-natal scanning.

15. Draw a poster to persuade your friends to reduce their personal stereo sound levels.

16. Discuss the risks of being in a disco for several hours, and how to reduce them.

17. A man is kidnapped, blindfolded and imprisoned in a room. How could he tell if he was in a) a town? b) the country? c) a bare room? d) a furnished room?

Further questions on page 239.

Further questions on waves: light and sound

▷ Waves

1. a) Here is an incomplete diagram which shows three successive straight waves A, B, and C on water, as they are being reflected at a straight barrier XY. Wavefront A is just about to be reflected while B and C have already been partly reflected:

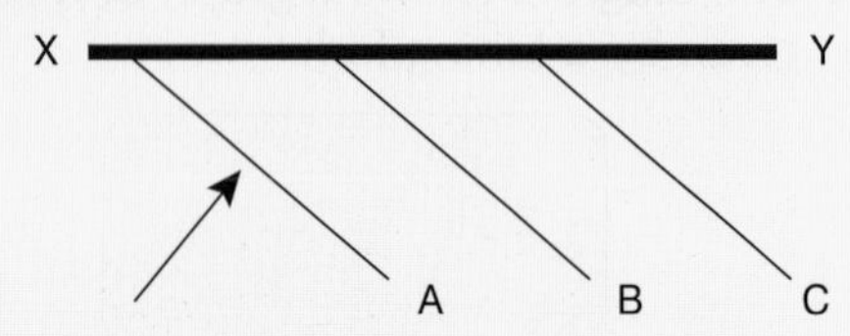

Copy and complete the diagram showing the positions of the reflected parts of the wave-fronts B and C. [4 marks]

b) If the wavelength of an incident wave is 1.5 cm and the frequency is 10 Hz, calculate
  i) the wavelength of a reflected wave, [1]
  ii) the speed of the waves. [3]

2. The diagram below shows wavefronts of light in air arriving at a glass block.

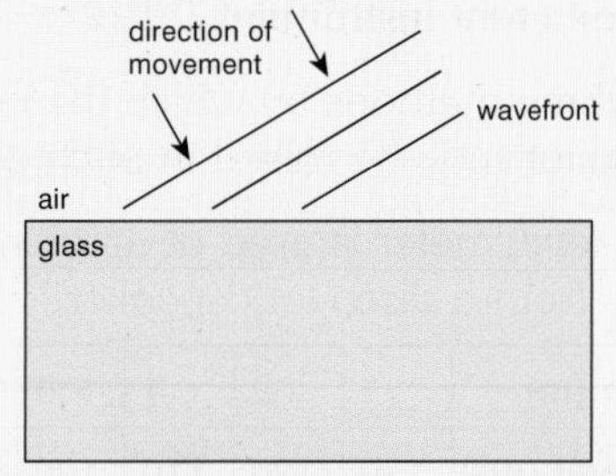

a) State what happens to the speed of the wavefront as it enters the glass block. [1]

b) Copy and complete the diagram to show the direction of the wavefronts once they have entered the glass. [1]

c) Explain what causes the wavefront to change direction as it enters the glass [1] (Edex)

3. Radio **X** is broadcast on wavelength 1500 m at a frequency of 200 kHz. Radio **Y** is broadcast on wavelength 250 m.
Calculate a) the velocity of radio waves, [3]
b) the frequency of Radio **Y**. [2]
Explain why these radio stations can be easily 'picked-up', wherever you live, even though there may be buildings or hills between you and the transmitter. [4] (WJEC)

▷ Mirrors

4. The diagram below shows the plan view of an object O in front of a plane mirror.

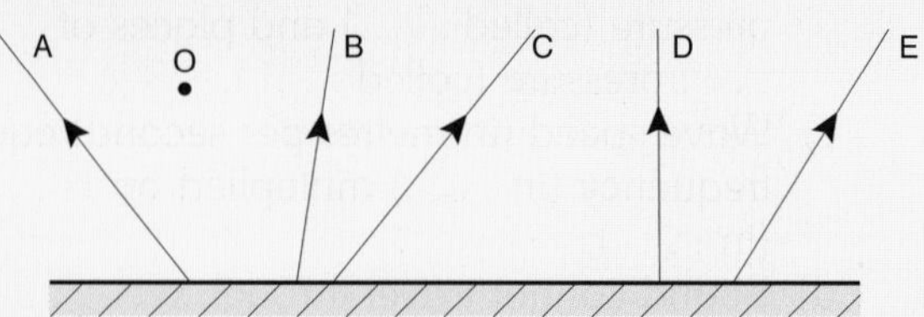

Which one of the reflected rays of light appears to come from the ***image*** of O? (NI)

5. The image of an object formed by a plane mirror is
A virtual **B** magnified **C** real
D diminished **E** upside-down (NI)

6. The diagram shows two rays of light shining on to a mirror from a torch labelled T.

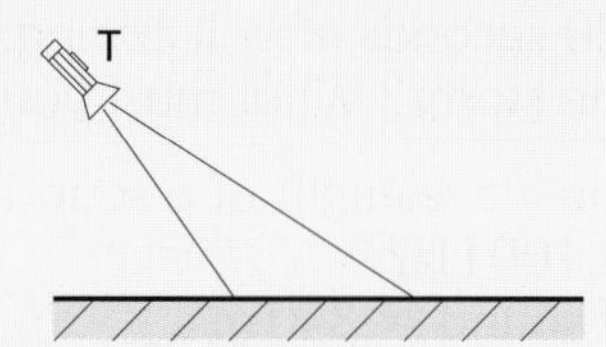

a) Copy the diagram and:
  i) draw the two rays of light after they have been reflected by the mirror.
  ii) draw the **image** of the torch in the correct position (label this **I**). [3]

b) Could you use a plane mirror on its own to project an image on to a screen? Explain. [2] (AQA)

7. a) The diagram shows a lamp bulb placed at the principal focus of a concave mirror. Copy it and draw **two** rays of light which leave the lamp and reflect from the mirror. [2]

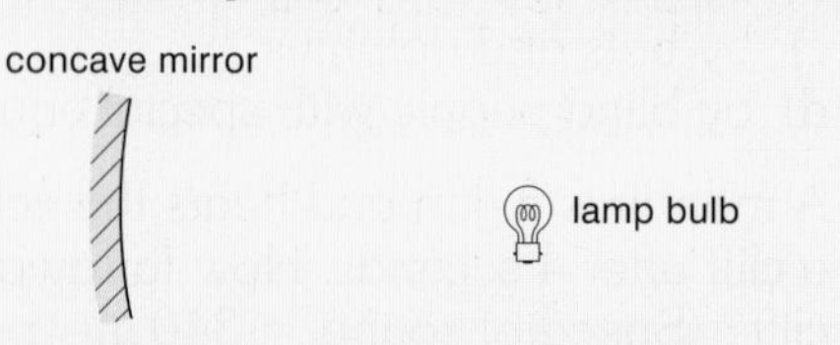

b) Give two examples of devices which use concave reflectors for electromagnetic waves other than light. In each case, state which part of the electromagnetic spectrum is involved. [4] (AQA)

 Note: although a question may be marked (AQA) etc, it will often be relevant to other examinations as well.

▷ Refraction

8. A ray of light is shone on to a glass block at point A.
Some of the light is reflected and the rest is refracted.

a) Copy and draw on the diagram the **paths** of the light as it leaves A. [3]

b) Light is a wave motion. What happens to the wavelength of the light as it is refracted by the glass? [1] (OCR)

9. a) The diagram shows a ray of red light incident on a glass prism. Copy and complete the diagram.

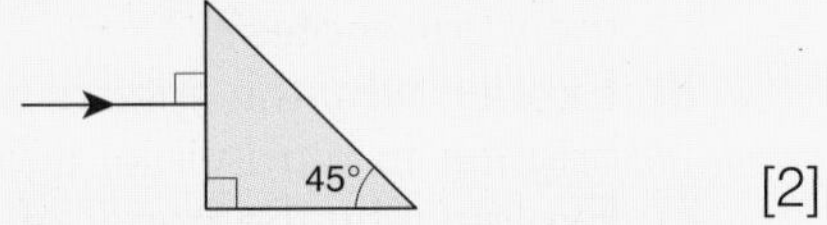

[2]

b) The diagram shows a ray of red light incident on a different prism. Copy and complete the diagram.

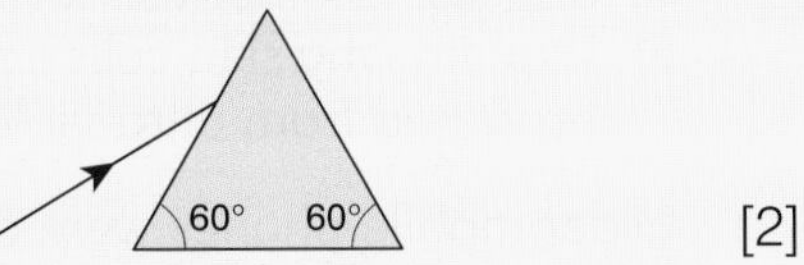

[2]

c) If white light were used instead of red light on this prism, what difference, if any, would you notice? [1] (OCR)

10. The critical angle for light going from glass into air is 42°.

a) Draw a diagram showing a ray of light travelling from glass to air at an angle of incidence of 42°.

b) Draw a diagram showing a ray travelling from glass to air with an angle of incidence less than 42°.

c) The diagram shows a short length of optical fibre.
Copy and complete the diagram to show the path of the narrow beam of light through the optical fibre. [3]

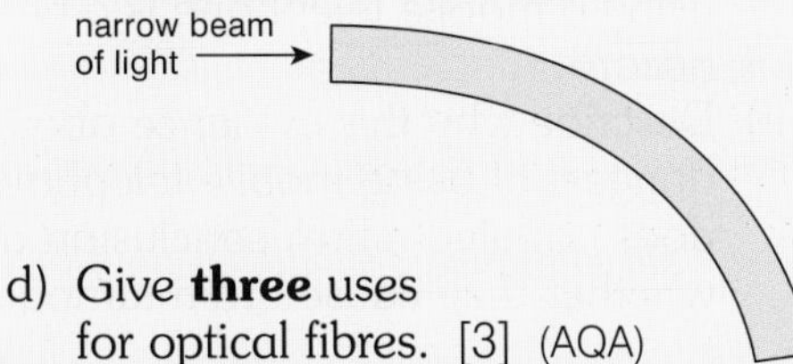

d) Give **three** uses for optical fibres. [3] (AQA)

11. Doctors use an endoscope to look at the inside of a patient's stomach. It has a lamp, some lenses and a bundle of optical fibres. Only part of the endoscope (the optical fibres and a tiny lens) needs to go inside the patient, via their mouth.

a) Explain why it is better to use an endoscope instead of using X-rays to look inside a patient. [1]

b) Draw a diagram of an optical fibre to show how the light from the lamp reaches the patient's stomach. [2]

c) Explain why the endoscope uses a bundle of fibres instead of a single one. [2]

12. Charlie watches cable television and uses the same cable for his computer internet connection. In the cable, an infra-red beam passes along an optical fibre.

a) Name a material used for optical fibres. [1]

b) Describe the structure of an optical fibre. [2]

c) Explain how the infra-red beam can pass along the optical fibre even though the cable is not straight. [2]

d) Explain how the beam carries the information for the TV and computer. [2]

13. Information can be carried by light travelling along an optical fibre.

a) Name the process by which the ray of light travels through the fibre. [1]

b) Explain, as fully as you can, why the light ray stays inside the optical fibre. [2]

c) Explain why information is sent by means of light in optical fibres rather than by electrical signals in metal cables. [2]

d) Give one other use of optical fibres. [1]

14. A light beam in an optical fibre can be used to carry information as a digital signal.

a) Draw a diagram to show what is meant by a digital signal. [1]

b) The signal from a microphone is an analogue signal. How does an analogue signal differ from a digital signal? [1]

c) When signals are sent through optical fibres they lose energy. State what happens to the brightness of the light beam as it loses energy. [1]

d) State one disadvantage of losing energy as the light travels through the fibre. [1]

▷ The electromagnetic spectrum

15. The diagram shows the electromagnetic spectrum:

Radio		Visible	Ultra-violet		γ-rays

a) Copy it and fill in the missing names.
b) Which region:
  i) has the longest wavelength?
  ii) has the highest frequency?
  iii) causes a suntan?
  iv) is used in burglar alarms? (AQA)

16. a) Give **three** properties which are the same for all electromagnetic waves. [3]
b) Which parts of the electromagnetic spectrum:
  i) will cause a suntan?
  ii) can be used both to cook and in communications?
  iii) can be used to sterilise fruit?
  iv) can be detected from warm objects? [4] (AQA)

17. Ultra-violet, gamma rays, radio waves are parts of the electromagnetic spectrum. Write down
a) which has the longest wavelength,
b) which has the highest frequency,
c) which is emitted by the nucleus of an atom,
d) how you could detect ultra-violet,
e) **two** properties they have in common. (WJEC)

18.

visible

			blue	green			microwaves	

a) i) Use the list below to copy and complete the electromagnetic spectrum chart above. [2]
gamma rays, infra-red, red light
ultra-violet, X-rays, radio waves
  ii) State **two** properties common to all members of the spectrum. [2]
b) Name the part of the spectrum that:
  i) has the longest wavelength; [1]
  ii) is emitted by all hot objects. [1]
c) i) Calculate the wavelength of microwaves of frequency 10^{10} Hz (10 000 000 000 Hz) given that the velocity of microwaves is 3×10^8 m/s (300 000 000 m/s). [3]
  ii) State **two** practical uses of microwaves. [2] (WJEC)

19. This is a list of types of wave.
gamma infra-red light microwaves
radio ultra-violet X-rays
Choose from the list the type of wave which best fits each of these descriptions.
a) Stimulates sensitive cells on the retina. [1]
b) Used inside fluorescent lamps. [1]
c) Used for rapid cooking in an oven. [1]
d) Used to take a photograph of the bones in a broken leg. [1]
e) Emitted by a remote control unit. [1] (OCR)

20. When banknotes are printed, special ink is used for some parts. The banknotes may be inspected using ultra-violet light.
a) Why do banks use this special ink on their banknotes? [1]
b) Explain why the special ink appears brighter in ultra-violet light. [2]
c) Explain why the paper of the banknote does not appear bright, even though some of the ultra-violet rays are reflected. [1]
d) Suggest another use for U-V rays. [1]
e) Explain why skin cells need to be protected from ultra-violet radiation. [2]

21. Older mobile phones were described as ***analogue***, while newer ones are ***digital.***
a) Using diagrams, explain the difference between analogue and digital signals. [2]
b) Explain the advantage of using digital signals. [2]

22. a) Mobile telephones transmit information using microwaves, with a frequency of 1200 MHz (1200 000 000 Hz).
They travel at a speed of 3.00×10^8 m/s (300 000 000 m/s). Calculate the wavelength of the microwaves used. [3]
b) Microwaves that are used to cook food have a frequency of 2500 MHz. Some users of mobile telephones have developed brain tumours.
  i) Suggest why some people think that using a mobile telephone is a health hazard. [1]
  ii) Describe why this evidence about the dangers of using mobile telephones does not give a firm conclusion about whether they cause brain tumours. [3] (Edex)

▷ Sound

23. In an experiment Colin studies sound waves. He sets up a loudspeaker to produce sound as shown below:

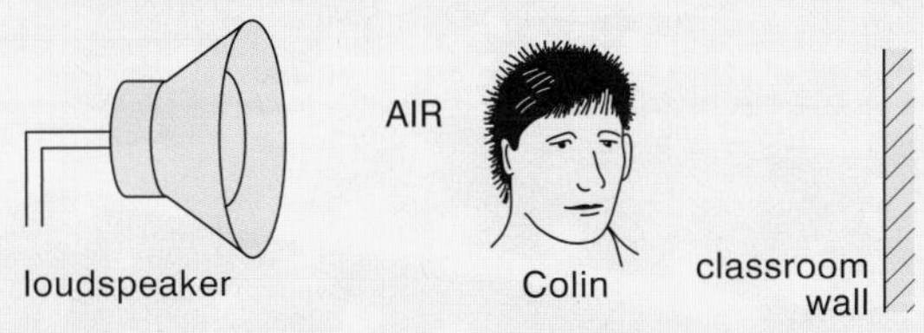

Colin adjusts the signal to the loudspeaker to give a sound of frequency 200 Hz.

a) What happens to the air in between Colin and the loudspeaker? [2]

b) Explain how Colin receives sound in **both** ears. [2]

24. A girl stands some distance from a high wall and claps her hands.

a) What two measurements would need to be made in order to calculate the speed of sound? [2]

b) Describe how you would make these measurements. [4]

c) The speed of sound in air is 330 m/s. How far from the wall would you stand? Choose an answer from these distances:
10 m ***200 m*** ***500 m*** [1]
Give reasons why you did ***not*** choose each of the other two distances. [2]

(AQA)

25. The diagram shows a fishing boat using sonar to detect a shoal of fish.

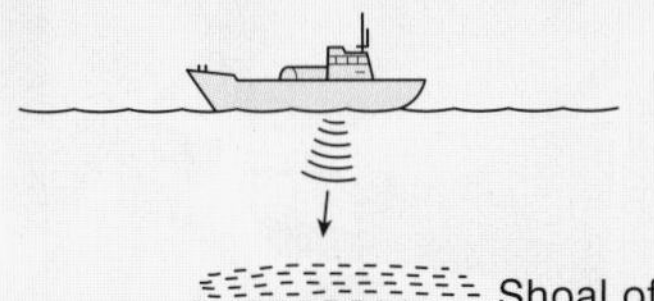

A short pulse of sound waves is emitted from the boat, and the echo from the shoal is detected $\frac{1}{10}$ s later. The sound waves travel through sea-water at 1500 m/s.

a) How far has the pulse travelled in $\frac{1}{10}$ s? [1]

b) How far below the boat is the shoal of fish? [1]

c) The reflected pulse lasts longer than the emitted pulse. Suggest a reason for this. [2] (OCR)

26. David and Melissa stand 600 m apart, both 200 m from the foot of a long, high cliff. David fires a pistol.

David
Melissa

Copy the diagram to scale (2 cm ≡ 100 m) and use it to

a) explain why Melissa hears the shot and an echo of the shot, [2]

b) find the distance travelled by the sound that formed the echo. [1]

c) Given that the sound travels at 340 m/s, calculate the time difference between Melissa hearing the shot and the echo. [2]

(WJEC)

27. a) What are ultrasonic waves? [1]

b) Explain how ultrasonic waves can produce an image of the fetus in a pregnant woman. (Draw a diagram to help your answer.) [4] (AQA)

28. Humans can hear sound with frequencies between 20 Hz and 20 000 Hz.

a) Suggest a frequency of an ultrasound wave. [1]

b) Ultrasound passes through body tissue as a longitudinal wave. Describe how the particles move as ultrasound passes through tissue. [2]

c) Ultrasound can be used to shatter kidney stones. Use your ideas about the properties of ultrasound to explain why it can be used to do this. [3]

d) Describe one non-medical use of ultrasound. [1]

29. Ultrasonic waves are used for pre-natal scans.

a) The frequency of some ultrasonic waves is 3 MHz (3 000 000 Hz). The waves travel at a speed of 1500 m/s in the human body. Calculate the wavelength of the ultrasonic waves. [4]

b) Explain how ultrasonic waves produce an image of the fetus. [3]

c) Explain, as fully as you can, why X-rays are not used for pre-natal scans. [3] (AQA)

ELECTRICITY
and magnetism

The photographs show some of the many uses of electricity and magnetism.

What difference would it make to your life if electricity and magnetism had not been discovered?

chapter 30

STATIC ELECTRICITY

Experiment 30.1
Rub a plastic pen or comb on your sleeve and then hold it near some tiny pieces of paper.

What happens?

Experiment 30.2
Rub a plastic object on your sleeve and then hold it near (but not touching) a thin stream of water from a tap. What happens?

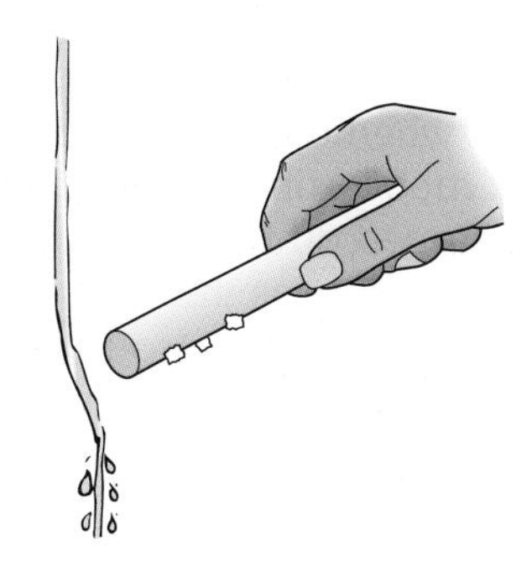

Because of friction with your sleeve, the plastic object has been ***charged*** with stationary or ***static electricity***.

Have you ever heard a crackling noise as nylon clothes rub together when you undress?
In the dark you can see the tiny sparks of electricity (like tiny flashes of lightning).

Experiment 30.3
Rub a strip of polythene (a grey plastic) on a dry woollen cloth and hang it in a paper stirrup as shown:

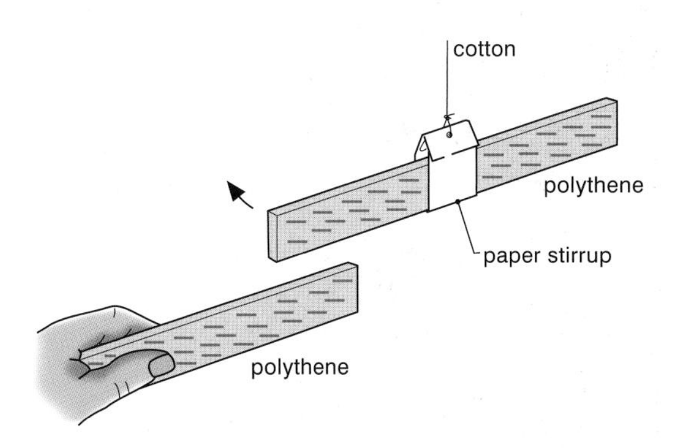

We say the electricity on polythene is a ***negative*** charge.

Rub another strip of polythene (so it is also negatively charged) and bring it close.
What happens?

Repeat the experiment with two strips of cellulose acetate (a clear plastic). What happens?

Now repeat the experiment with one strip of charged polythene and one strip of charged cellulose acetate. What happens?

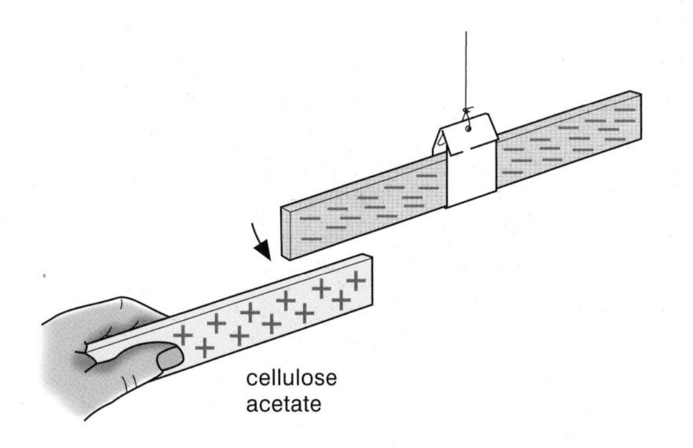

Because it behaves differently, we say that cellulose acetate is charged ***positively***.

We find that: **Like electric charges repel.**
Unlike electric charges attract.

The closer the charges, the greater the force.

▶ The electron theory

Objects become charged because the atoms of all substances contain both negative and positive charges (see also page 342).

The positive charges (called **protons**) are in the central core or ***nucleus*** of the atom.
The negative charges (called **electrons**) are spread in orbits round the outer part of the atom.

In an uncharged or ***neutral*** atom, the number of protons (+) is equal to the number of electrons (–).

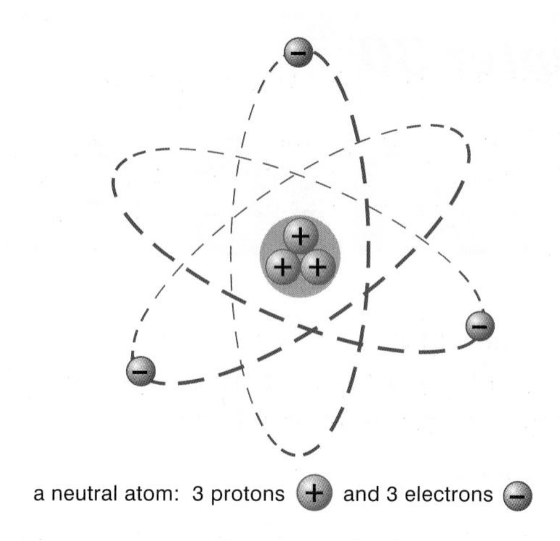

a neutral atom: 3 protons + and 3 electrons –

We believe that when a polythene strip (or an ebonite rod) is rubbed on wool, some of the outer electrons are scraped off the wool and move on to the polythene. This means that the polythene has an extra number of electrons (so it is negatively charged). And the wool then has fewer electrons than protons (so it is positively charged).

Notice that only the negative electrons can move – the positive protons remain fixed.

The rubbing does not make the charges – it simply separates them.

When cellulose acetate (or glass) is rubbed, it becomes positively charged. Draw a diagram to show the movement of electrons in this case.

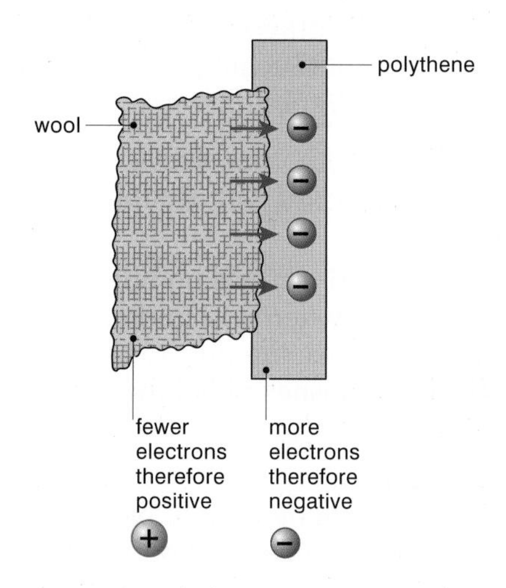

Electrostatic induction

> *Experiment 30.4*
> Rub a balloon on your clothes and then put it on the wall or ceiling so that it stays there!

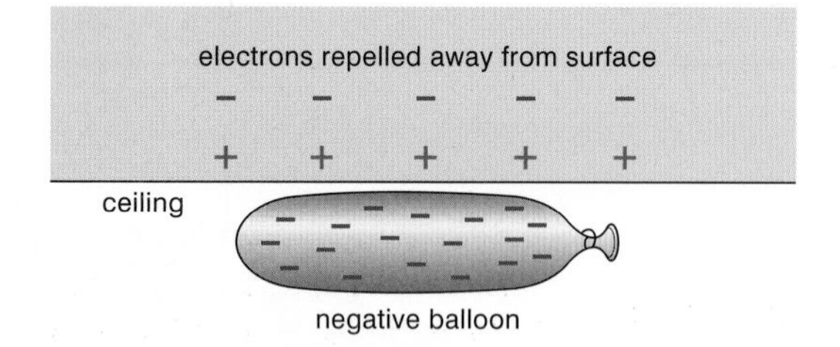

This happens because the negative charge on the rubbed balloon repels some of the electrons in the ceiling away from the surface. This leaves the surface positively charged and so the negative balloon is attracted to the ceiling.

The separated charges in the ceiling are called **induced** charges.

This also explains why the paper and the water were attracted in experiments 30.1 and 30.2.

Why do your CDs tend to gather a lot of dust?
Why is it difficult to clean nylon carpets?

The gold-leaf electroscope

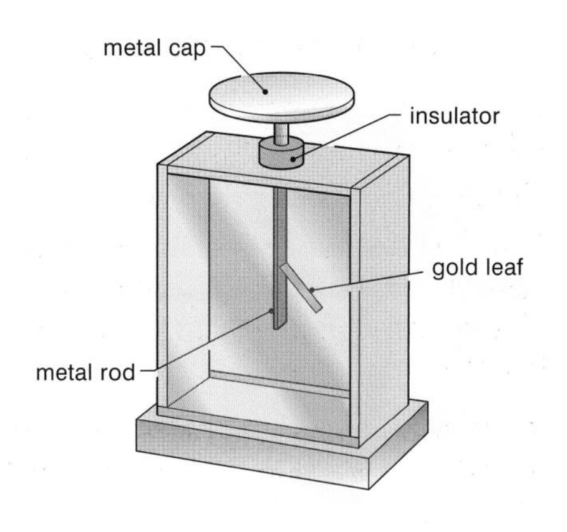

This instrument can be used to tell whether an object is charged. The metal cap is connected to a metal rod and a strip of very thin metal (the gold leaf). This gold leaf is hinged at the top so that it can move.

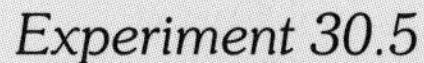

Experiment 30.5
Rub a polythene strip and bring it close to, but not touching, the cap of an electroscope.

What happens to the leaf?

The negative polythene induces charges in the electroscope by repelling negative electrons down to the metal rod and the leaf. Because like charges repel each other, the light leaf is repelled and lifted.

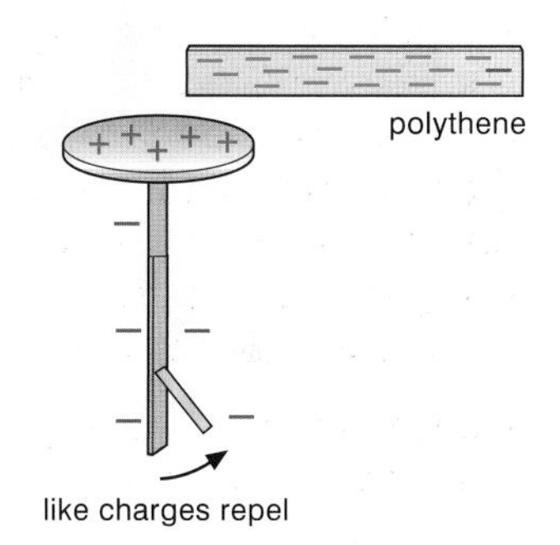

What happens when you remove the polythene? Why?

What happens if you bring up a positively-charged strip? Draw a diagram to show how the electrons move in this case.

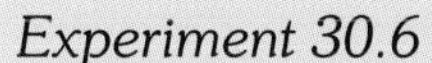

Experiment 30.6
Charge a polythene strip and then slide the strip firmly across the edge of the metal cap of the electroscope.

What happens? Does the leaf stay up even if you remove the polythene?

In this case, some of the negative electrons moved from the polythene to the electroscope, where they spread out over all the metal parts and made the leaf rise by repulsion.

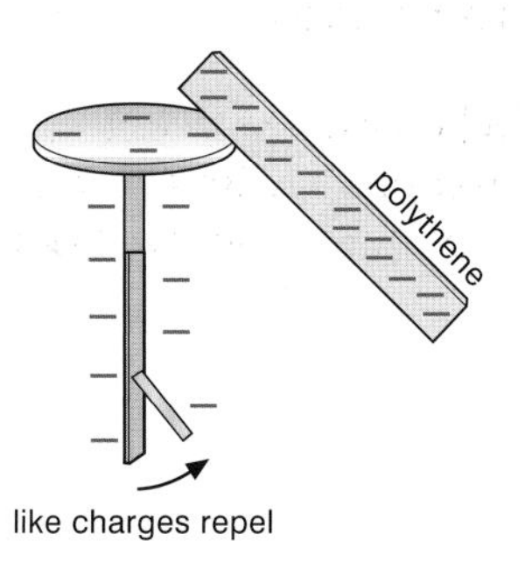

What happens if you now bring up a positively-charged strip?

Experiment 30.7 Conductors and insulators
Charge an electroscope as in experiment 30.6, then touch the cap with your finger.

What happens to the leaf? Why does it fall?
What happens to the electrons on it?

Your skin is a **conductor** of electricity because it allows electrons to move through it to Earth. You '***earthed***' the electroscope.
The moving electrons are an **electric current**.

Charge the electroscope again and then touch it with a plastic pen. Does anything happen this time?
Plastic is an **insulator** because it does not allow electrons to pass through it. What happens with other materials?

Conductors	Insulators
Skin	Plastic
Metals	Air
Water	Rubber

▶ Experiments with a Van de Graaff generator

This is a machine used to charge objects easily.

These experiments should be done only under the supervision of a teacher – be careful you do not get an unexpected shock from the machine (it is safer to approach the dome while holding a pin in front of you – see experiment 30.12). ⚠

Experiment 30.8
Place a piece of fur on the dome and start the machine running.

What happens to the hairs on the fur?
Why is this? (Remember all the hairs have the ***same*** kind of charge.)

Experiment 30.9 A hair-raising experience
If someone in the class has soft dry hair, ask them to stand on a thick sheet of polythene (to insulate them from the Earth) while they touch the dome.

Experiment 30.10
Without touching the machine, blow soap bubbles near it.

What happens? Why are the bubbles attracted to the dome? How is this used in electrostatic paint spraying? (Page 246).

There is a force on the bubbles even though they are not touching the dome. We say there is an **electric field** near the dome. We imagine that there are electric ***field lines***, as shown in the diagram:

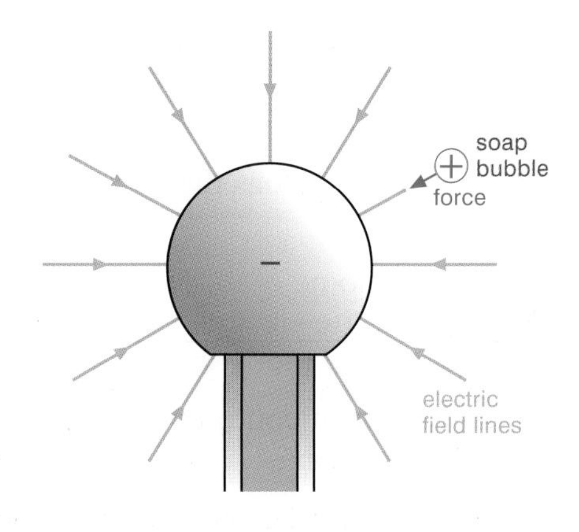

The electric field is strongest near the dome: the force on the bubble is strongest there.

Experiment 30.11 An electric wind
While standing on an insulator and touching the dome as in experiment 30.9, aim the point of a pin at a candle flame.

Which way does the flame move?

This happens because, if the point is positively charged, it attracts to it some of the electrons from atoms in the air. This leaves the atoms positively-charged.
We say the air has been ***ionised***.
The positively-charged atoms (called **ions**) are repelled away from the point, by the electric field.

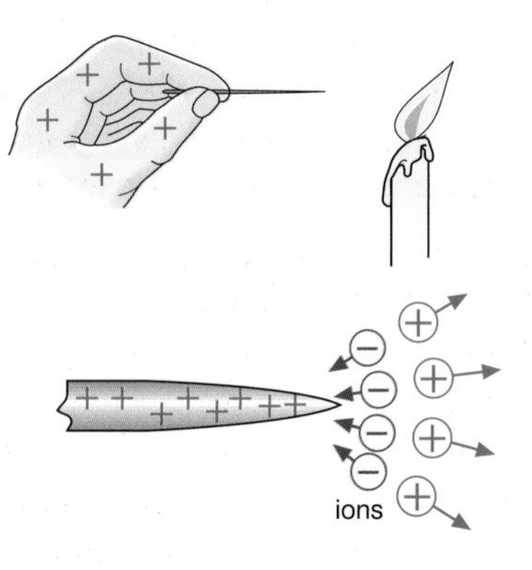

Experiment 30.12 A lightning conductor
Place a metal ball near a Van de Graaff generator so that a spark jumps between them.
This is a tiny lightning flash (some flashes of lightning may be up to 6 km long).

The spark makes a sharp cracking sound because it is heating and expanding the air very rapidly. What is thunder?

Now approach the dome with a pointed metal pin in your hand. Why does the machine stop sparking?

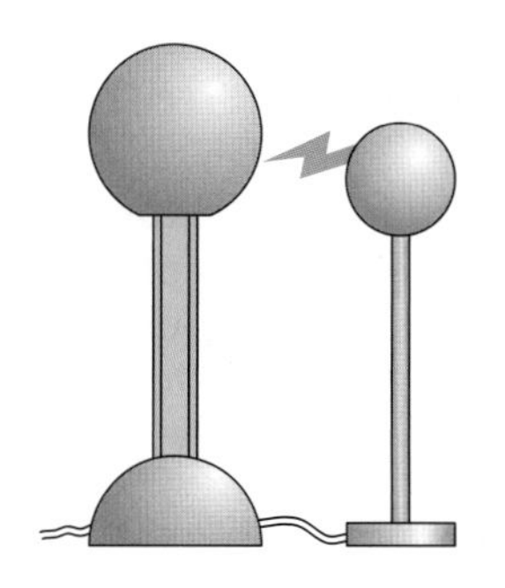

This is how a **lightning conductor** works – if a negatively-charged cloud passes overhead, it induces a positive charge on the point at the top of the lightning conductor (as in experiment 30.4). The point then repels positive ions to the cloud (as in experiment 30.11) to neutralise it, so that it is less likely to produce a lightning flash.

The electrons that are attracted to the conductor travel down it to the Earth. An electric current flows through the wire.

When is it dangerous to use an umbrella?

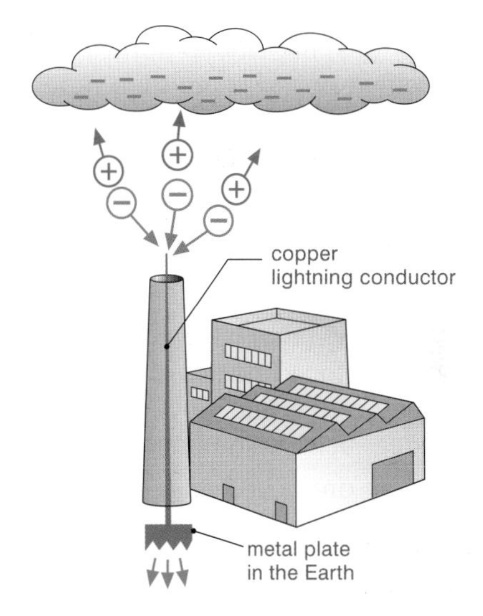

▶ Capacitors

Even after you switch off the machine you can get a shock from it. It has ***stored*** some electric charge. Objects that store charge are called **capacitors**.

Capacitors are used in radio and TV sets, and many electronic circuits (see page 325). The size of their capacitance is usually measured in micro-farads (μF).

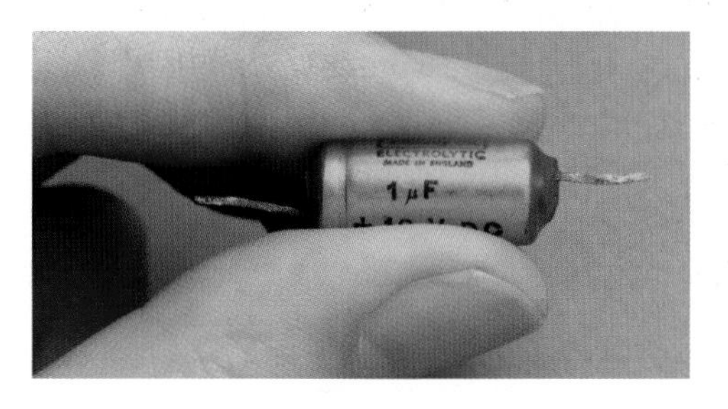

A common capacitor is made from two long strips of metal foil separated by a strip of insulator (waxed paper or plastic), and then rolled up like a swiss roll:

If one metal strip is charged + and the other is –, the capacitor stores electrical charge. It is storing energy, rather like a small rechargeable battery.

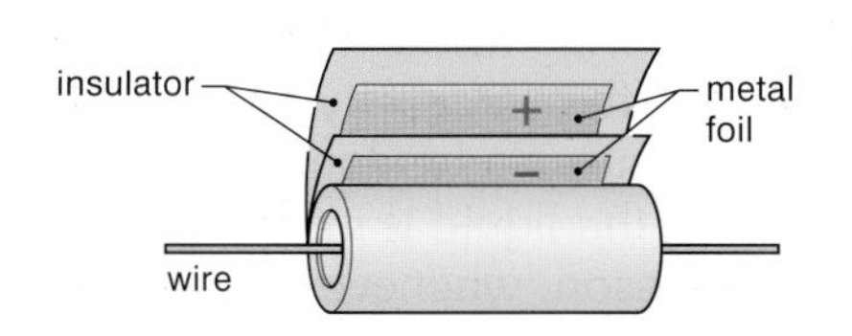

The amount of charge that is stored is measured in units called **coulombs**, usually shortened to C.

A coulomb is a very large amount of charge (about 6 million million million electrons), so we often use micro-coulombs (1 μC = 10^{-6} C, one millionth C).

The dome of the Van de Graaff generator stores about 1 μC, a lightning flash might contain 10 C.

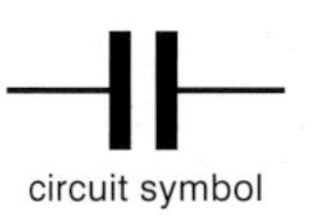
circuit symbol

▷ Physics at work: Static electricity

An electrostatic precipitator

Coal-burning power stations produce huge amounts of smoke pollution. This smoke is a cloud of small dust particles or ash. It can be removed by using static electricity.

Some thin wires are stretched across the centre of the chimney: These wires are charged positively to about 50 000 V and they cause the gas around them to be charged or ***ionised*** (see also experiment 30.11 on page 244).
Because of this, the smoke particles become positively charged. These positive particles are then repelled by the wires, towards the earthed metal plates, where the dust sticks.

A mechanical hammer hits the plates every few minutes and the ash falls down into a bin. Later, it is used to make house-bricks.

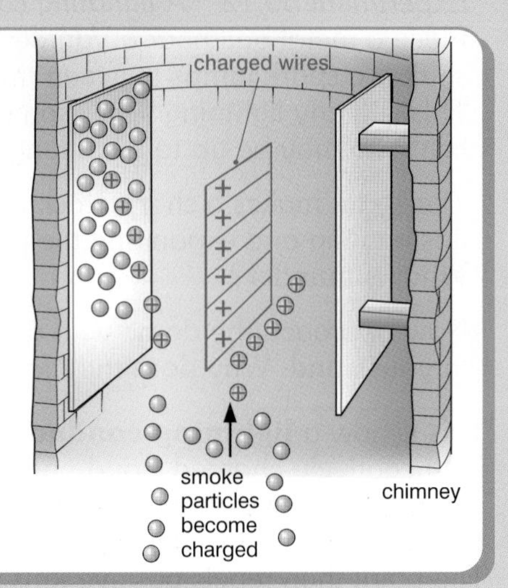

Fingerprinting

A similar method can be used to show up fingerprints on paper: The paper is placed near a charged wire like one of the plates in the chimney and a fine black powder is used instead of smoke. The powder sticks to the fingerprint but not to the clean paper.

Photocopiers are based on a similar method (see page 315).

Paint spraying

Bicycles and cars are painted using an electrostatic paint spray. The nozzle is given a charge and this makes a better spray – the droplets all have the same charge and repel each other so that the paint spreads out to form a large even cloud.

Less paint is needed because the charged droplets are all attracted to the object (even the back of it) because it has an opposite charge (see page 241).

The same idea is used to make crop-spraying more efficient for farmers.

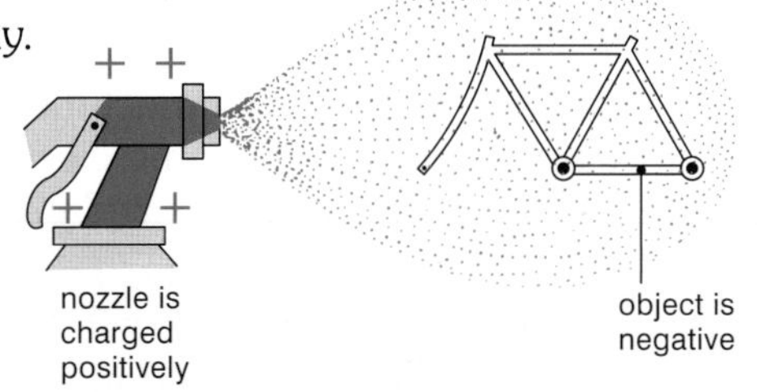

The droplets of paint repel each other but are attracted to the object

Preventing fires

A liquid flowing through a pipe can become charged in the same way as two objects rubbed together. This can be dangerous if it causes a spark and the liquid is inflammable.
For this reason, whenever an aeroplane is being re-fuelled by a tanker lorry, they are always connected together by a copper wire.

For the same reason, spare petrol for cars should always be carried in metal cans, never plastic.

Large ships have been known to explode because their tanks were being cleaned out by a high speed water-jet which became charged and caused a spark. Nowadays the tanks are filled with an inert gas before cleaning starts.

Walking on nylon carpets can give you a shock when you touch the door handle or a radiator. Modern carpets can be made conducting to prevent this.

Summary

When rubbed, polythene (or ebonite) becomes charged negatively (it has an excess number of electrons).
Cellulose acetate (or glass) becomes charged positively (it has a shortage of electrons).

Like charges repel; unlike charges attract.

A gold-leaf electroscope can be used to detect charges. Charges are measured in coulombs.

Charged objects can separate (induce) charges in other objects (by attracting or repelling electrons).

A conductor allows electrons to move through it; an insulator does not.

Sharp points lose their charge easily. This is used in a lightning conductor.
Capacitors are used to store charges.

▶ Questions

1. a) Like charges . . . . , unlike charges . . . . .
b) When polythene is rubbed with wool, some . . . . leave the . . . . and move on to the . . . . so that the polythene becomes . . . . charged and the wool is left . . . . charged.
c) When rubbed, cellulose acetate becomes . . . . charged because it . . . . some electrons.
d) A negatively-charged object attracts a piece of paper because it . . . . electrons away from the surface of the paper. This leaves the surface of the paper . . . . -charged so that it is . . . . to the object. These separated charges are called . . . . charges.
e) In a gold-leaf electroscope, the leaf rises because like charges . . . . .
f) Conductors (unlike . . . .) allow . . . . to travel through them.
g) Sharp points . . . . their charge easily by . . . . the air. The . . . . at the top of a lightning conductor loses . . . . to the clouds to neutralise them so that they are less likely to produce a flash of . . . . .
h) Charges are measured in . . . . and can be stored in . . . . .

2. Use your Physics to explain the following:
a) Nylon clothing crackles as you undress.
b) In dry weather, people walking on nylon carpets may get a shock if they touch a radiator or a metal door knob.
c) Passengers sliding off a car seat sometimes get a shock from the door.
d) Petrol road-tankers usually have a length of metal chain hanging down to touch the ground.
e) Because some anaesthetics are explosive, the floor tiles in an operating theatre are made of a conducting material.

3. Use your Physics to explain the following:
a) A rubbed balloon will stick to the wall for some time.
b) A mirror or window polished by a dry cloth on a dry day soon becomes dusty.
c) Cassette-cases, CDs and other plastic objects soon become dusty.
d) As sellotape or 'cling film' is pulled off a roll it is attracted to your hand.
e) When spraying an object with paint, less paint is wasted if the object is charged.
f) TV screens soon become very dusty.
g) The amount of smoke leaving a factory can be reduced by placing a charged object inside the chimney.

4. Explain the following:
a) It might be dangerous to raise an umbrella in a storm.
b) The dome of a Van de Graaff machine is spherical with no sharp edges on the outside.
c) It is safer to approach it while holding a pin in front of you.
d) On high voltage electrical equipment, the blobs of solder should be rounded and smooth.

Further questions on page 279.

Have you started revising for your examinations? See page 382.

chapter 31

CIRCUITS

Experiment 31.1
Connect a cell or a battery to a lamp. This can be done by loose wires or on a ***circuit board***:

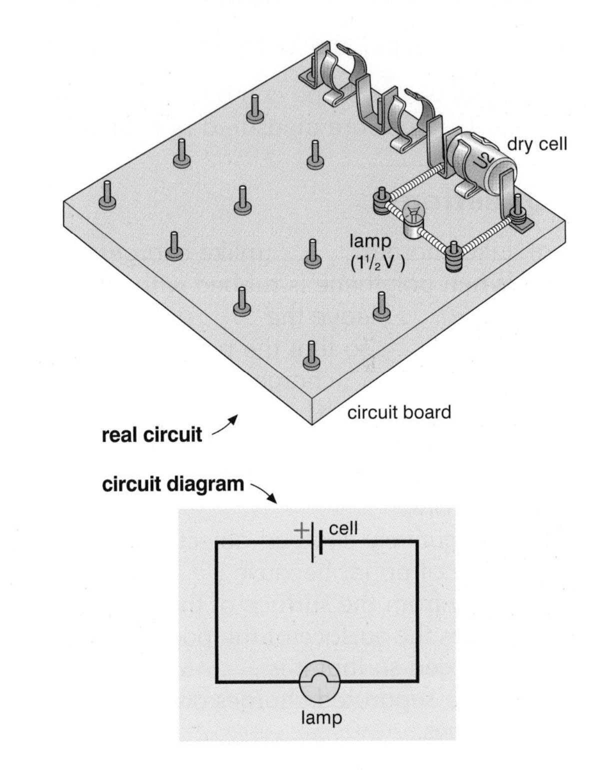

When the lamp lights, we say an electric ***current*** is flowing round the ***circuit***.

In fact, negative electrons are being pushed out of the negative pole of the cell and are drifting slowly round the circuit, from atom to atom in the wire, to the positive pole of the cell.
A current flowing in one direction like this is called a **direct current** (d.c.).

When drawing diagrams of circuits we use symbols for each component as shown here:

Now make a gap in your circuit.
What happens?
How can you switch the lamp on and off?
Does it matter where you make the gap?

For an electric current to flow, there must be a complete circuit, with no gaps.

The diagram shows the circuit with a switch added. Look at some switches of different kinds and include one of them in your circuit:

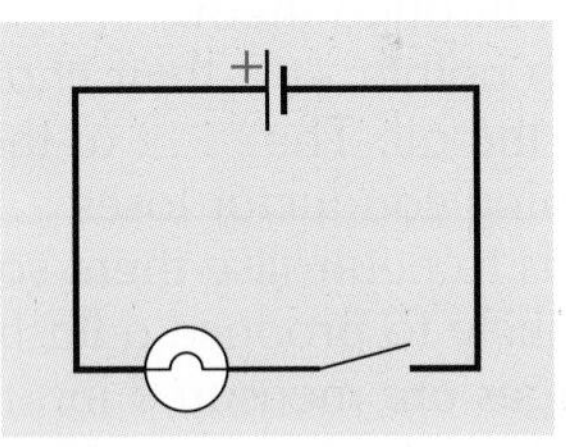

How can you use this circuit to send messages?
How was the Morse Code used in the old days?

Here are some other symbols used in electric circuit diagrams:

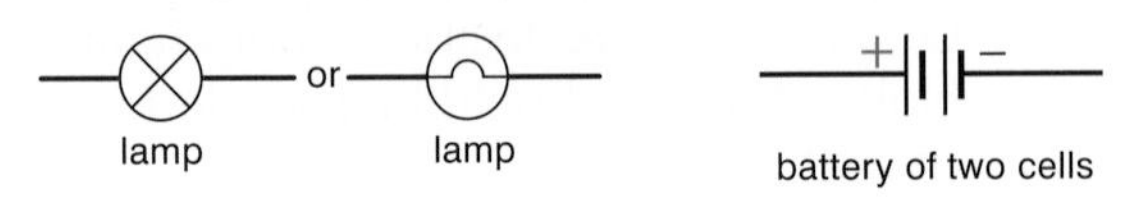

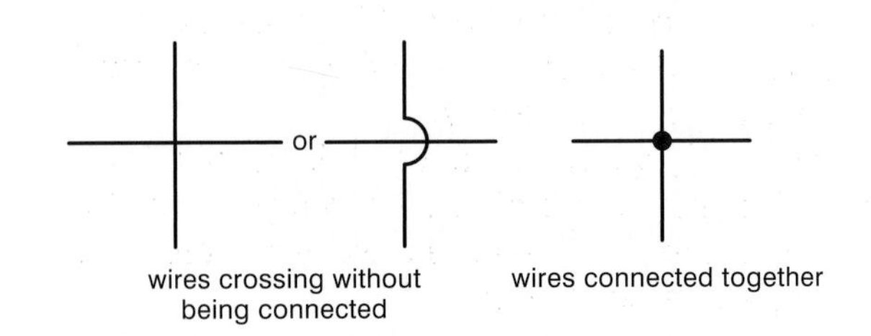

Experiment 31.2 Conductors and insulators
Connect a circuit as shown (very like the last experiment but with a gap in it).

Place a piece of copper across the gap.
What happens?
Place a piece of plastic across the gap.
What happens?

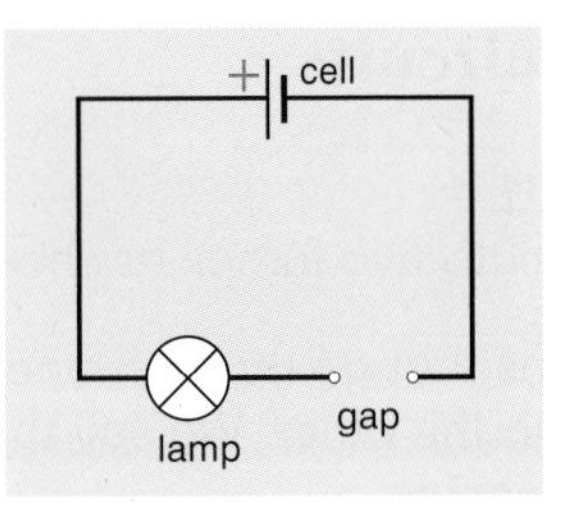

Some materials resist the flow of electrons more than others. You could draw up a table:

Conductors	Insulators
Copper Aluminium	Plastic Rubber

All metals are good conductors.
This is because inside them they have a large number of free electrons that can move easily from atom to atom, so that a current flows.
Insulators do not have free electrons inside them.

Is air an insulator or a conductor?
Why are the handles of electricians' screwdrivers made of plastic?
Why are copper wires usually covered with plastic?

Experiment 31.3 ⚠
Connect a dry cell to a lamp as shown.
Use a length of thick copper wire to connect one side of the lamp to the other side for just a second.
(!Do not do this with anything but a dry cell!)

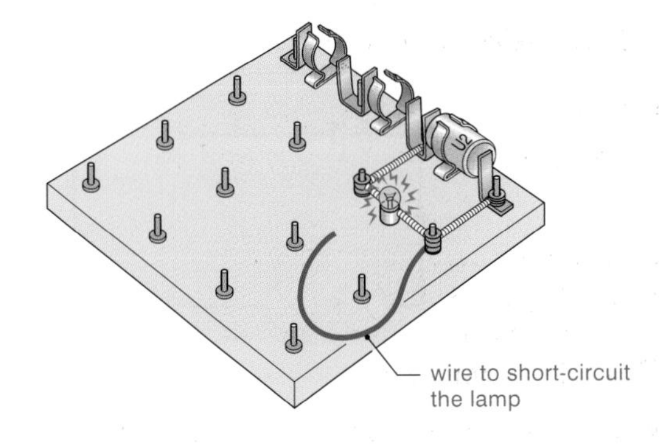

What happens? If the current is not flowing through the lamp, where is it going?

We say you have ***shorted*** the lamp or caused a 'short circuit'. Your wire has ***less resistance*** to the flow of electrons than the thin wire inside the lamp. **Electricity tends to travel by the easiest path**, not necessarily the shortest path.

Conventional current and electron flow

We know now that an electric current is really a flow of electrons from negative to positive.
However when cells were first invented scientists guessed, ***wrongly***, that something was moving from positive to negative and they marked arrows this way on their diagrams.
Unfortunately we have kept to this convention when drawing circuit diagrams, but you should remember that in a metal the electrons are really moving the opposite way.

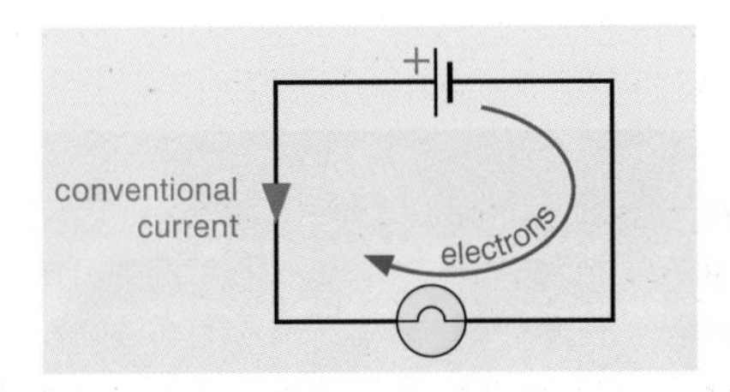

▶ Series circuits

Experiment 31.4
Connect a cell to two lamps as shown:

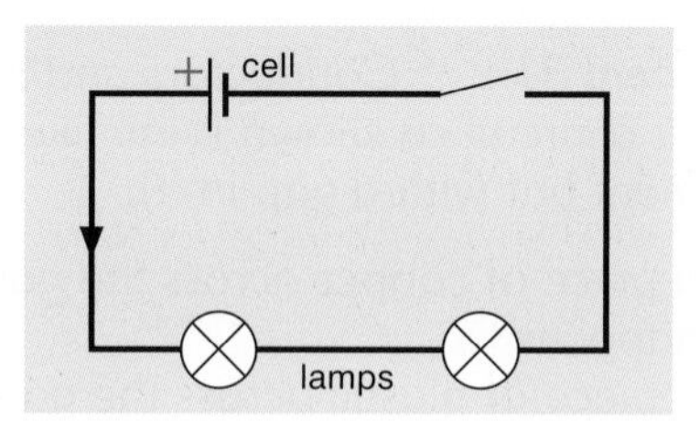

All the electrons that go through one lamp must also go through the other. We say the lamps are connected **in series**.

The electric current flowing in a circuit is measured in ***amperes*** or ***amps*** (usually shortened to A). 1 ampere is a flow of about 6 million million million electrons per second past each point!

The size of a current can be measured using an ***ammeter***.

In experiments it is important to connect an analogue (pointer) ammeter so that its red (+) terminal is always nearer to the positive (+) pole of the cell than to the negative pole. Otherwise the pointer moves the wrong way.

analogue ammeter

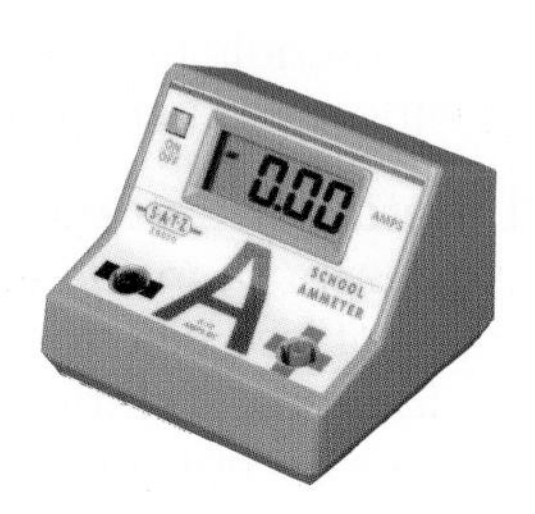

digital ammeter

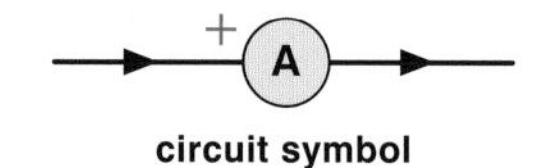

circuit symbol

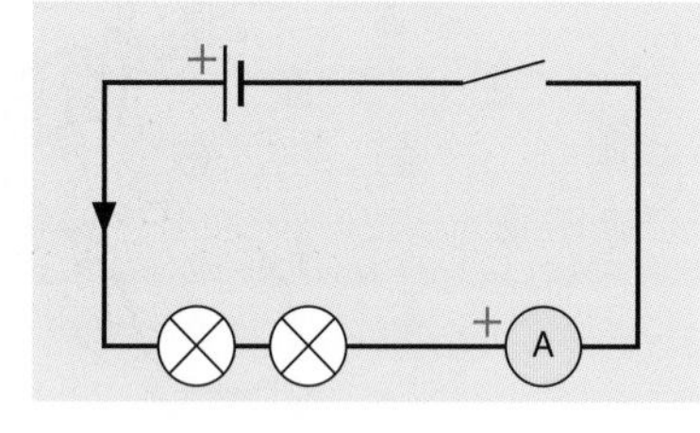

Experiment 31.5 Using an ammeter
Add a suitable ammeter to the circuit of the last experiment.
Read the scale. (On some ammeters, the numbers may have to be scaled up or down. Your teacher will explain if this is necessary.)

How much current is flowing in your circuit?

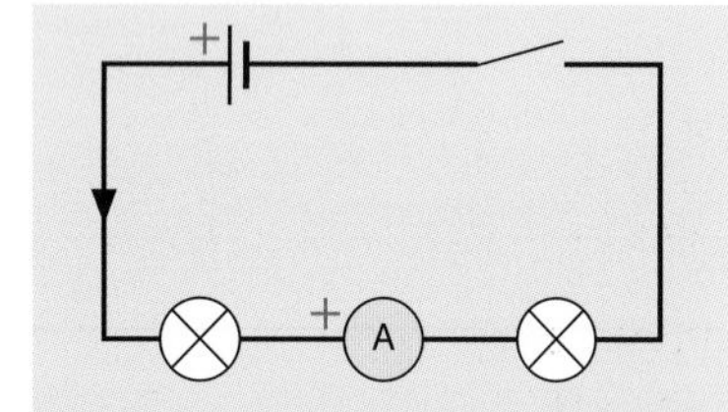

Experiment 31.6
Move the ammeter to different positions in the series circuit. What do you find?

The same current flows through each part of a series circuit.

Experiment 31.7
If you unscrew one lamp in your series circuit, what happens to the other lamp?

Christmas tree lamps are usually wired in series. What happens if one lamp breaks?

Are the lights in your house wired in series?

▶ Parallel circuits

Experiment 31.8
Connect a cell (or a power-pack) to two lamps in the way shown here.

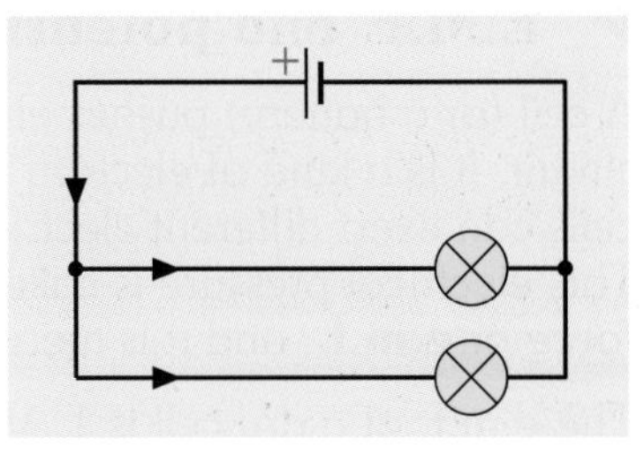

We say the two lamps are connected **in parallel**.

Experiment 31.9
Add switches to your circuit as shown (or loosen the wires instead).

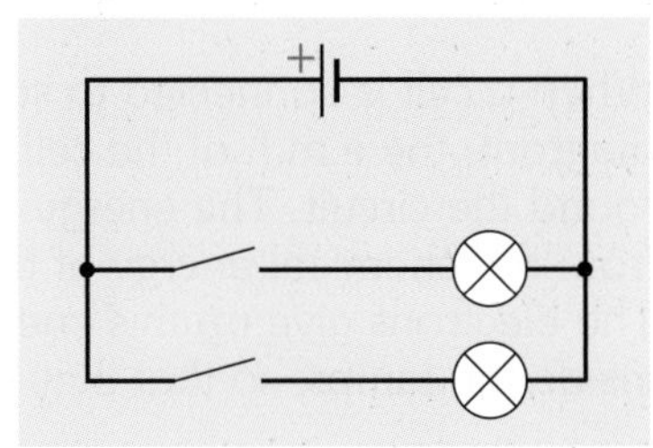

Can you switch the lamps separately?

Are the lamps in your home wired in series or in parallel?

Where in this circuit would you put a switch to control ***both*** lamps together? Try it.

Experiment 31.10
If you have 3 ammeters, connect the circuit shown here:
(With just one ammeter, place it in each position in turn.)

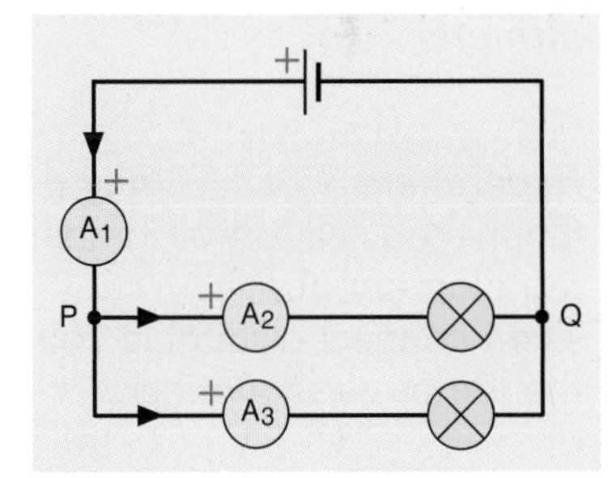

Read the ammeters. What do you notice about the reading on A_1 compared with the total of the readings on A_2 and A_3?

The current in the main circuit is the sum of the currents in the separate branches (sometimes called Kirchhoff's First Law).

As the electrons (from the negative pole of the cell) reach point Q, they divide between the two branches of the circuit until they reach P where they join together again.

Experiment 31.11 Different parallel circuits
There are many different ways of drawing the same parallel circuit!

Because bending the connecting wires does not change the movement of the electrons, all the circuits shown below are ***electrically the same*** as each other! (In each case the current divides to go through the lamps and then rejoins.)
Try connecting some of these circuits yourself.

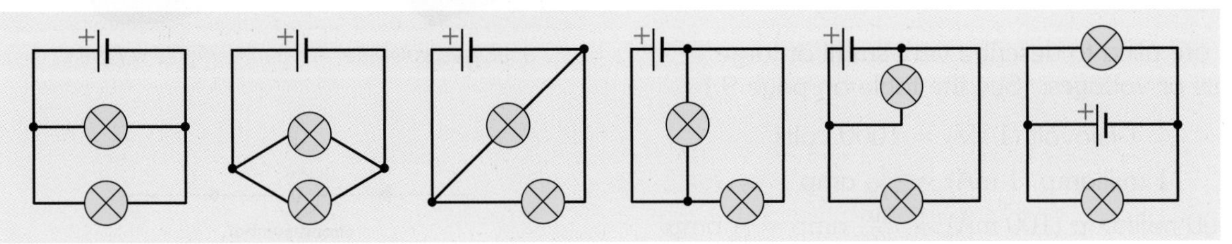

▶ E.M.F. and potential difference

A cell (or a battery) pushes electrons round a circuit. It is a kind of electron pump. Different cells can exert different electrical pressures.
This electrical pressure is called ***electromotive force*** or **e.m.f.**, and it is measured in **volts** (V).

The e.m.f. of a dry cell is $1\frac{1}{2}$ V; the e.m.f. of a Van de Graaff generator may be 100 000 V.

When a cell is connected to some lamps, as in the diagram, the e.m.f. of the cell pushes electrons round the circuit. The energy to do this comes from the chemical energy of the cell.
The electrons give up this energy to the thin wires inside the lamps, so that they get hot and glow.

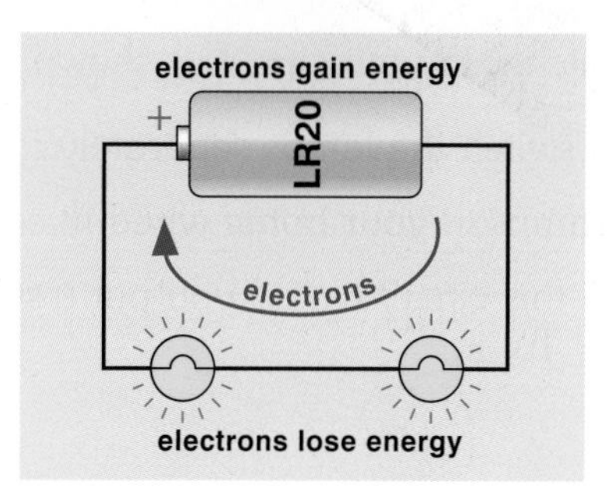

Across each lamp there is an electrical energy difference, called a ***potential difference (p.d.)*** or 'voltage'. It is measured in ***volts*** (V) using a ***voltmeter***.

Experiment 31.12 Using a voltmeter
Connect a cell to two lamps in series as shown in the diagram above.
Then connect a suitable voltmeter ***across*** one of the lamps as shown here:

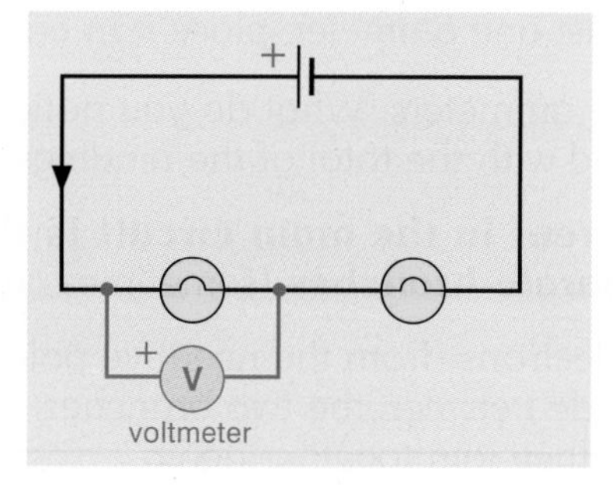

As with an ammeter, a pointer voltmeter must be connected the correct way round (+ to +).
What is the reading on your voltmeter?

Reconnect the voltmeter (always in ***parallel, across*** a component) to find the potential difference across the other lamp. Then use your voltmeter to find the potential difference across the cell.
What do you notice about your three readings?

The ***total*** voltage across the lamps ***equals*** the voltage across the cell.

analogue voltmeter

digital voltmeter

Prefixes

These are used to describe very small or large currents or voltages. (See the table on page 9.)

eg. 1 kilovolt (1 kV) = 1000 volts

1 milliamp (1 mA) = $\frac{1}{1000}$ amp

100 milliamp (100 mA) = $\frac{100}{1000}$ amp = $\frac{1}{10}$ amp

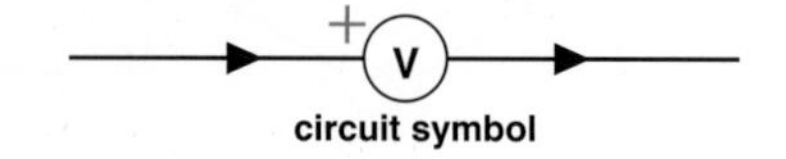

circuit symbol

▶ Ohm's Law and resistance

Experiment 31.13
Connect a single cell (or power pack) in series with a 6 V lamp and a suitable ammeter.
Note the ammeter reading.
Connect another cell in series to assist the first cell (or you can double the p.d. of the power pack).
What happens to the ammeter reading?
Use a third cell to increase the p.d. again.

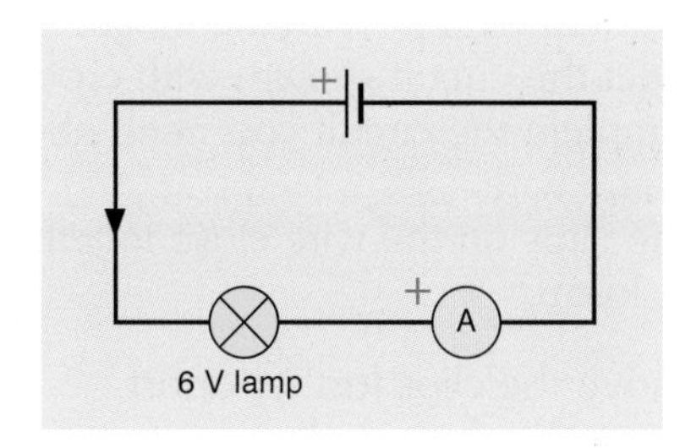

What do you notice about the current as you increase the potential difference?

In 1826, Georg Ohm discovered that:

the current flowing through a **metal** wire is ***proportional*** to the **potential difference** across it (providing the **temperature** remains **constant**).

'Proportional' means that if you double the p.d., the current is doubled (see page 390).

Ohm's Law doesn't apply precisely to a lamp because its temperature changes (see p. 259).

Resistance

The thin wire in a lamp tends to resist the movement of electrons in it. We say that the wire has a certain ***resistance*** to the current. The greater the resistance, the more voltage is needed to push a current through the wire. The resistance of a wire is calculated by:

$$\textbf{Resistance, } R = \frac{\textbf{p.d. across the wire } (V)}{\textbf{current through the wire } (I)}$$

or, in symbols: $R = \frac{V}{I}$

or, rearranging: $I = \frac{V}{R}$

or, easiest to remember: $V = I \times R$

where V = p.d. in volts (V)
I = current in amps (A)
R = resistance in a unit called an ***ohm*** (Ω).

This fact you should know evermore,
(the formula links to Ohm's Law)
Overcome all resistance,
Learn by dogged persistence
That $V = I \times R$

Example
A 2 V accumulator is connected to a wire of resistance 20 ohms.
What current flows in the circuit?

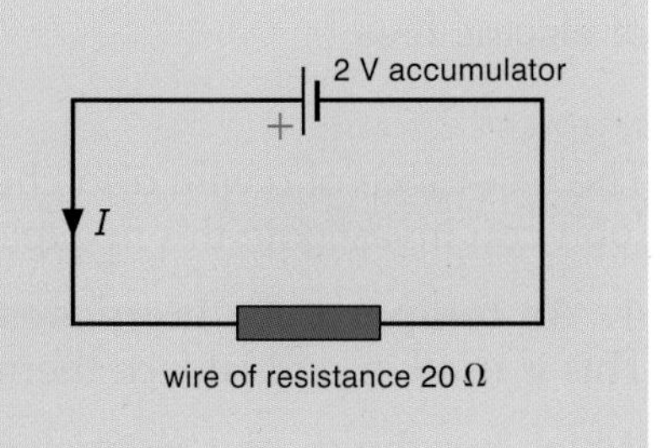

Formula first: $V = I \times R$

Then put in the numbers: $2 = I \times 20$

∴ The current, $I = \frac{2}{20} = \frac{1}{10}$ amp = $\underline{0.1 \text{ A}}$

What resistance would give you twice the current with the same battery?

▶ Four factors affecting resistance

Experiment 31.14 Varying length
Connect the circuit shown, with crocodile clips at A and B.
To complete the circuit you need about 30 cm of 30-gauge nichrome wire.
Put the clips on the wire close together and note the brightness of the lamp.

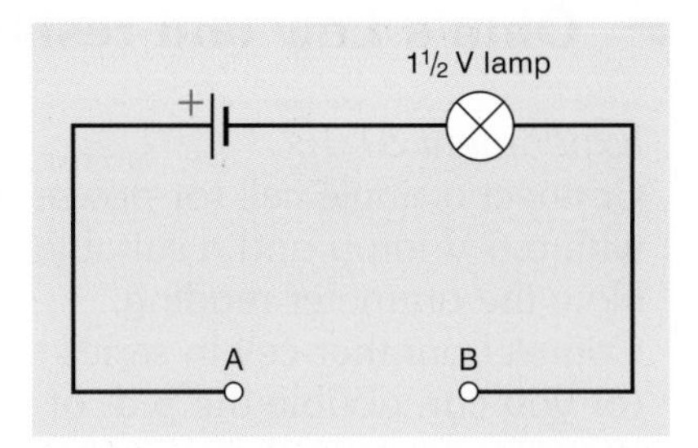

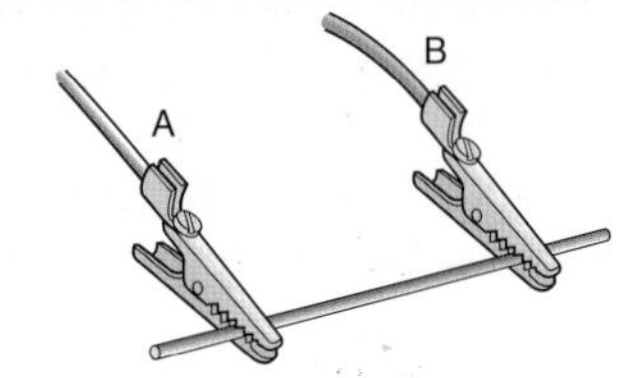

Now move the clips farther apart.
Is the lamp brighter or dimmer?
Is the current greater or less?
Has the resistance decreased or increased?

1. As the length increases, the resistance increases.
This fact is used in a rheostat (see next page).

Experiment 31.15 Varying cross-sectional area
Using the same circuit, place your clips about 20 cm apart on the nichrome wire. Note the brightness of the lamp.

Now add another (identical) length of nichrome wire between the clips so that the cross-sectional area is doubled.

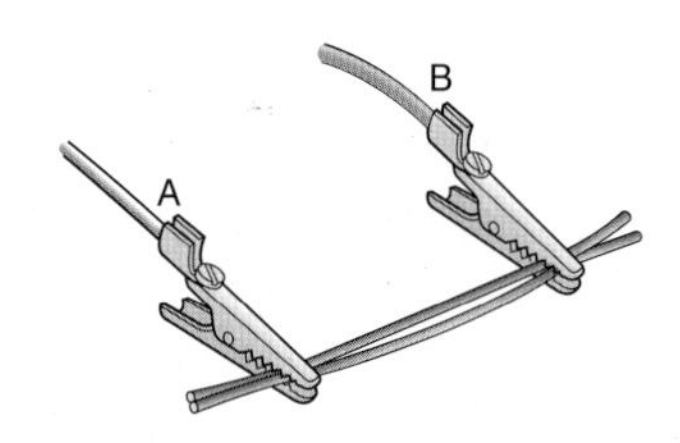

Is the lamp brighter or dimmer?
Is the current greater or less?
Has the resistance decreased or increased?

2. As cross-sectional area increases, the resistance decreases.

Experiment 31.16 Changing the substance
Using the same circuit, place your clips 20 cm apart on a nichrome wire. Note the brightness of the lamp.

Now replace the nichrome wire by a ***copper*** wire of the same length and the same area (30-gauge).

Which substance has the greater resistance?

3. Copper is a good conductor and is used for connecting wires. Nichrome has more resistance and is used in the heating elements of electric fires.

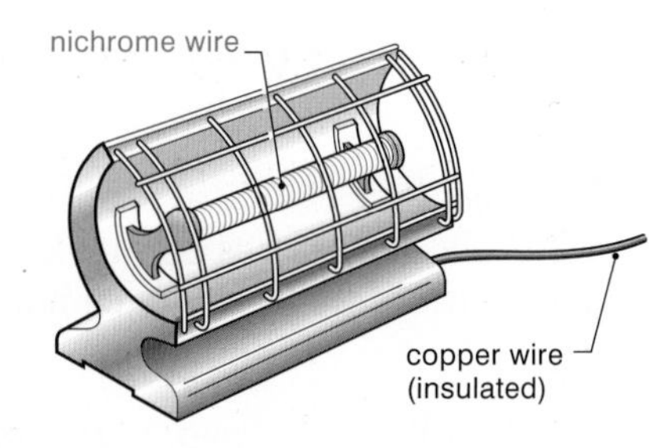

The resistance of a wire also changes as the temperature changes (you can investigate this in experiment 31.23).

4. As temperature increases, the resistance of a wire increases.
This is used in a resistance thermometer.

Which has the greater resistance – a long, thin, hot nichrome wire or a short, thick, cool copper wire?

▷ Resistors

Resistors (sometimes made of a length of nichrome wire) can be used to reduce the current in a circuit.

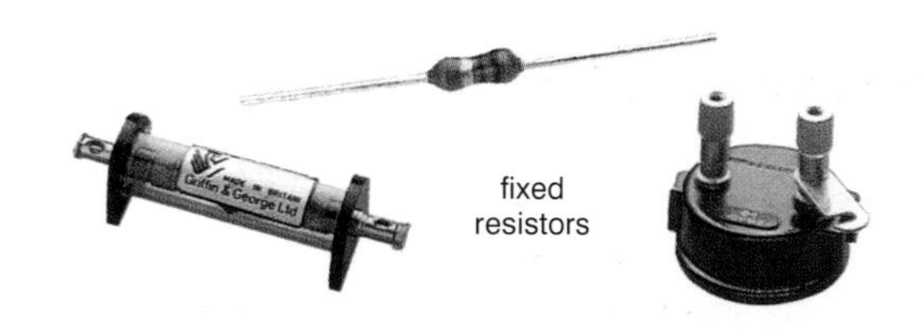
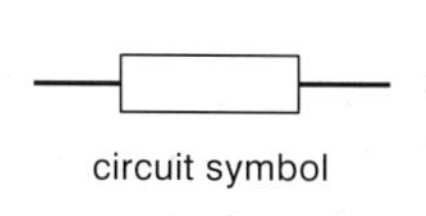

A variable resistor or ***rheostat*** is used to vary the current in a circuit. As the sliding contact moves, it varies the length of wire in the circuit.

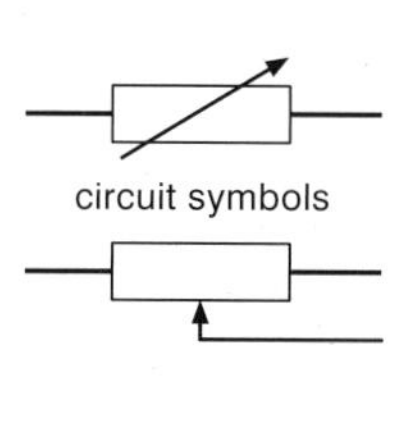

> *Experiment 31.17 Using a rheostat*
> Connect a rheostat in a circuit as shown and use it to control the brightness of the lamp.

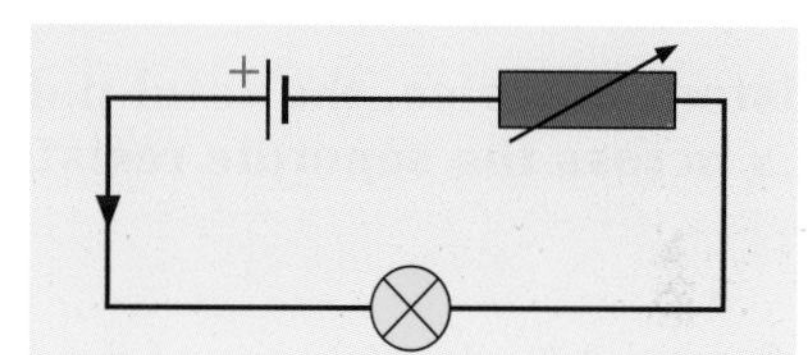

How are spotlights dimmed in a theatre?
How is a rheostat used in the speed control of a model racing car?

▷ Measuring resistance

> *Experiment 31.18 Measuring resistance*
> To find the resistance of a resistor, use the circuit shown here. Connect all the series components first (the black parts of the diagram) and then add the voltmeter (in parallel) last of all.
>
> Use the rheostat to adjust the current to a convenient value and note the readings of the ammeter and voltmeter.
>
> Move the rheostat and take more readings.

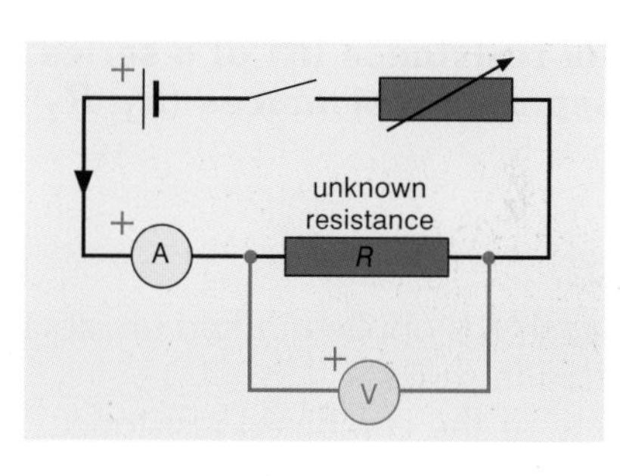

Calculate the resistance from the formula (page 253).

$$\textbf{Resistance} \text{ (in } \Omega\text{)} = \frac{\textbf{potential difference} \text{ (in V)}}{\textbf{current} \text{ (in A)}}$$

The average of the last column of the table should give you an accurate value of the resistance.

Evaluate your experiment.
Comment on whether your evidence is
a) reliable and b) valid.

Some sample results:

p.d. (in V)	Current (in A)	Resistance (in Ω)
2.0	1.0	2.0
3.0	1.5	2.0
4.0	2.0	2.0
6.0	3.0	2.0

▶ Resistors in series

Experiment 31.19
Get two resistors whose values are marked (in Ω) and connect them in series as shown here (in black).

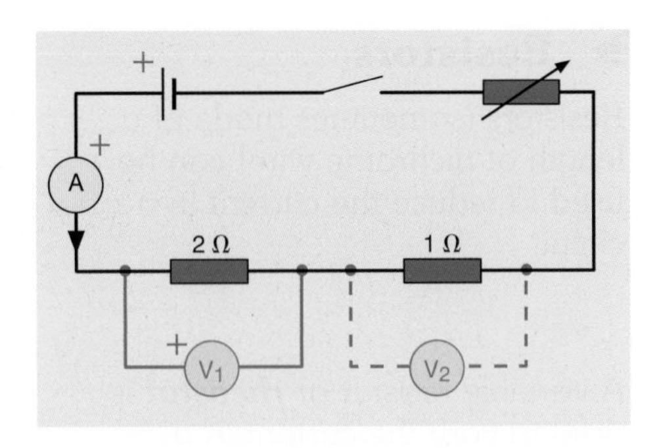

We know already (from page 250) that **the same current flows through each part of a series circuit.**

Now investigate the potential differences in a series circuit by connecting a suitable voltmeter:
a) in position V_1 across one of the resistors
b) in position V_2 across the other resistor
c) in position V_3 across both together.
What do you notice about the voltmeter readings?

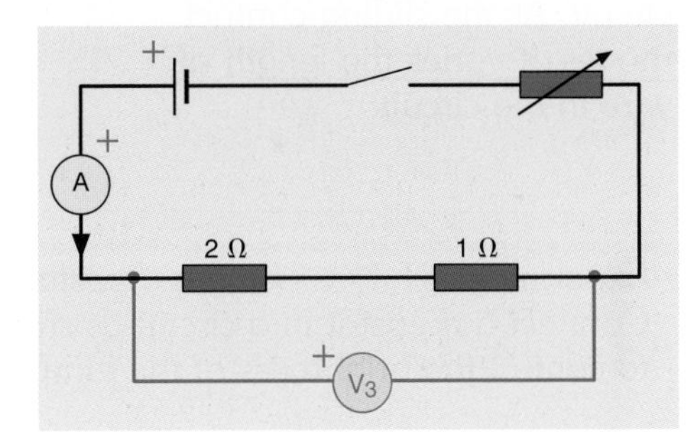

For resistors in series, the total p.d. is the sum of the p.d.s across the separate resistors ($V_3 = V_1 + V_2$).

What is the combined resistance of the two resistors in series? To find out, use the reading of V_3 and the ammeter reading to calculate the combined resistance, using $R = \frac{V}{I}$

Do you find that a 2 Ω resistor and a 1 Ω resistor in series combine to act like a 3 Ω resistor? Try this with other resistors.

The total resistance (R) of a series circuit is equal to the sum of the separate resistances (R_1, R_2):

$$R = R_1 + R_2$$

Example
A p.d. of 4 V is applied to two resistors (of 6 Ω and 2 Ω) connected in series.
Calculate a) the combined resistance,
b) the current flowing,
c) the p.d. across the 6 Ω resistor.

a) Formula first: Combined resistance $= R_1 + R_2$
Then numbers: $= 6\,\Omega + 2\,\Omega$
$= \underline{8\,\Omega}$

b) Formula first: $V = I \times R$ for the combined resistance
Then numbers: $4 = I \times 8$
$\therefore I = \frac{4}{8} = \frac{1}{2}$ amp $= \underline{0.5\text{ A}}$

c) Formula first: $V = I \times R$ for the 6 Ω only
Then numbers: $V = 0.5 \times 6$
$\therefore V = \underline{3\text{ V}}$ across the 6 Ω resistor.

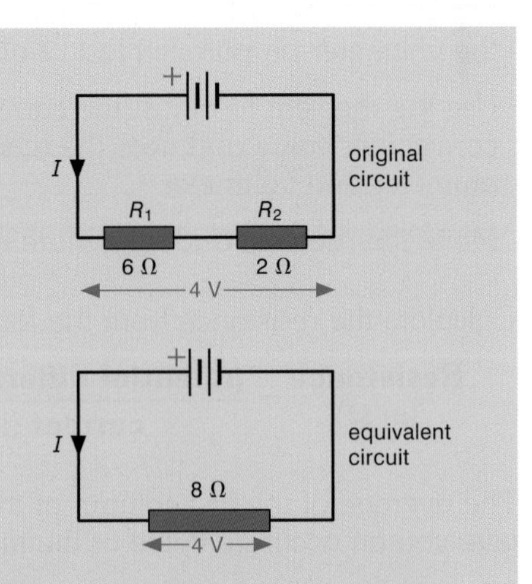

What is the p.d. across the 2 Ω resistor?

▶ Resistors in parallel

Experiment 31.20
Get two resistors whose values are marked (in Ω) and connect them in parallel as shown (in black):

Connect a voltmeter in position V_1 across one resistor, and then in position V_2 across the other.

What do you notice about the readings?

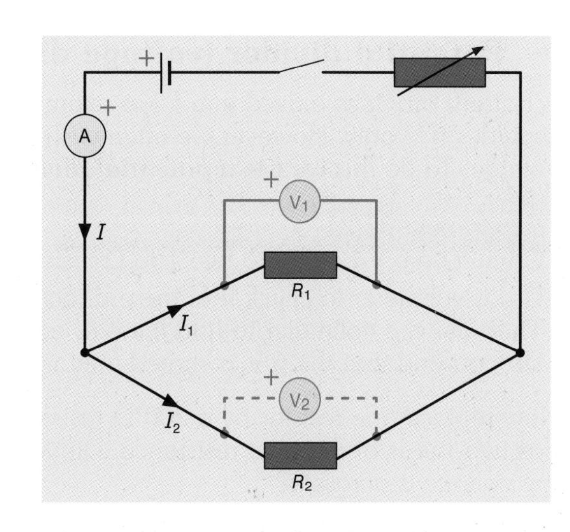

For resistors in parallel, the potential difference across each is the same.

We already know (from page 251) that
the current in the main circuit is the sum of the currents in the parallel branches ($I = I_1 + I_2$).

What is the combined resistance of the two resistors in parallel? Use your voltmeter reading with the ammeter reading to find the value of the combined resistance (using $R = \frac{V}{I}$).

Notice that the **combined resistance (*R*) is *less* than either of the separate parallel resistances (R_1, R_2)** because the electrons find it easier to travel when they have more than one path they can take.

The combined resistance ***R*** can be found from: $$\frac{1}{R} = \frac{1}{R_1} + \frac{1}{R_2}$$ or $$R = \frac{R_1 \times R_2}{R_1 + R_2}$$

When the two resistors are ***equal***, the combined resistance is ***half***.

Example
A p.d. of 6 V is applied to two resistors (of 3 Ω and 6 Ω) connected in parallel. Calculate a) the combined resistance,
b) the current flowing in the main circuit,
c) the current in the 3 Ω resistor.

a) Formula first: $\frac{1}{R} = \frac{1}{R_1} + \frac{1}{R_2}$

Then numbers: $\frac{1}{R} = \frac{1}{3} + \frac{1}{6} = \frac{2+1}{6} = \frac{3}{6} = \frac{1}{2}$

∴ Combined resistance R = 2 Ω

b) Formula first: $V = I \times R$ for the combined resistance
Then numbers: $6 = I \times 2$
∴ $I = \frac{6}{2} = 3$ A in the main circuit.

c) Formula first: $V = I_1 \times R$ for the 3 Ω resistor only
Then numbers: $6 = I_1 \times 3$
∴ $I_1 = \frac{6}{3} = 2$ A in the 3 Ω resistor.

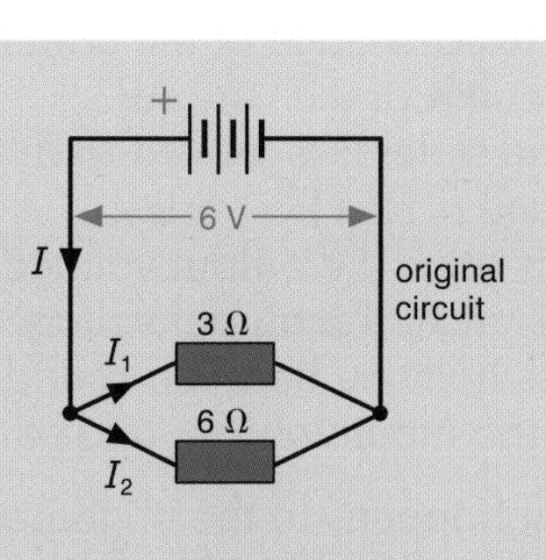

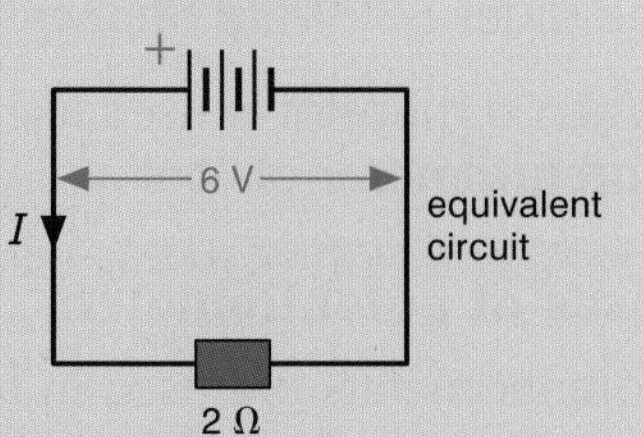

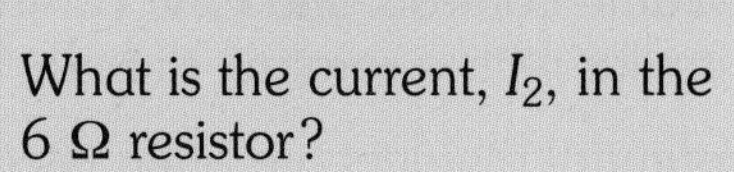

What is the current, I_2, in the 6 Ω resistor?

▶ Potential divider (voltage divider)

A battery provides a fixed e.m.f. For example, a 6 V battery provides 6 V only. However we often wish to provide a different voltage. To do this we use a ***potential divider*** circuit.

Experiment 31.21
Connect a 6 V battery to two 100 Ω resistors in series, as shown:

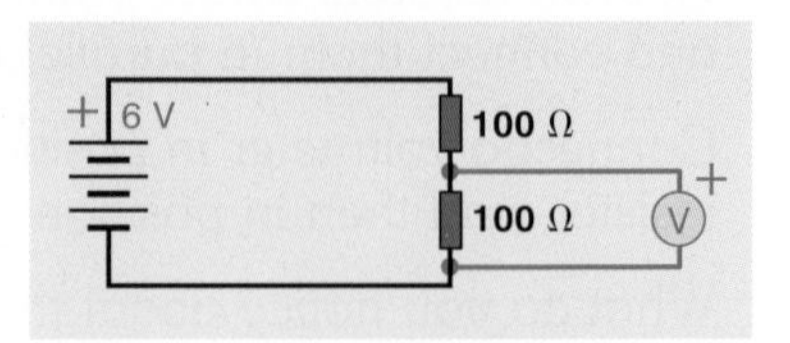

Use a voltmeter to check that the p.d. across the 2 resistors is 6 V. Then use the voltmeter to find the p.d. across each resistor in turn. Do you find that the 6 V is shared equally between the 2 resistors?

Now replace one resistor by a 200 Ω resistor. The 200 Ω resistor has two-thirds of the total resistance. Do you find two-thirds of the voltage is across it?

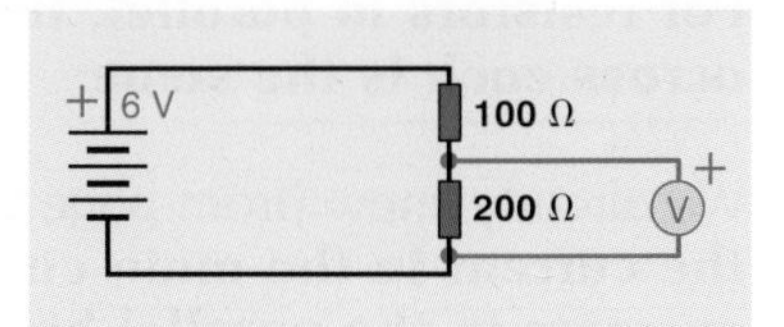

If we apply Ohm's Law to this circuit:

$V_2 = IR_2$ for resistor R_2 ... (1)

$V = I(R_1 + R_2)$ for the resistors in series ... (2)

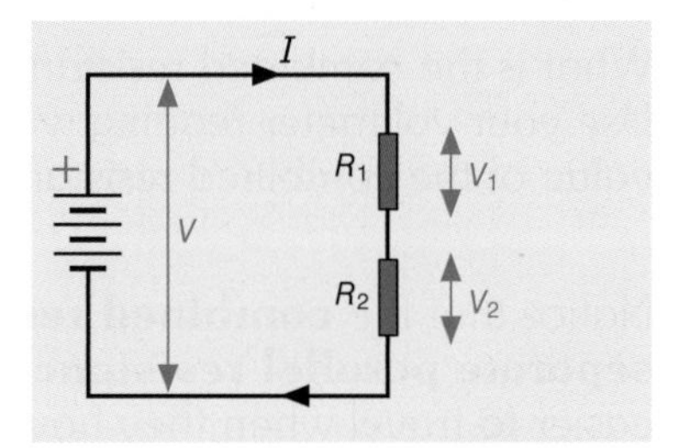

Dividing equation (1) by equation (2):

$$\frac{V_2}{V} = \frac{R_2}{R_1 + R_2} \quad \text{or} \quad V_{out} = \frac{R_2}{R_1 + R_2} \times V_{in}$$

This is the potential divider equation.

Experiment 31.22
Connect a battery to a rheostat (see page 255) so that all three of its terminals are used:

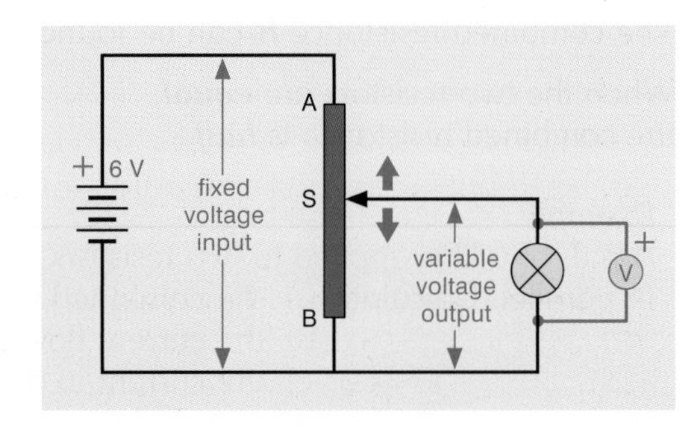

The fixed voltage of the battery is applied across the full length of the wire AB.
As the slider S is moved to different positions, the voltage across BS varies.
If S is at B, the output voltage is zero.
If S is at A, the output voltage is the full input voltage.
If S is halfway between B and A, the output voltage is half the input voltage; and so on.

What happens to the brightness of the lamp as you move the slider S? What happens to the voltmeter reading?

When a rheostat is used like this it is called a ***potentiometer*** or 'pot'.

Uses of potentiometers

Potentiometers are used in radios and CD-players as volume controls and tone controls. They are often circular with a rotating slide or 'wiper':

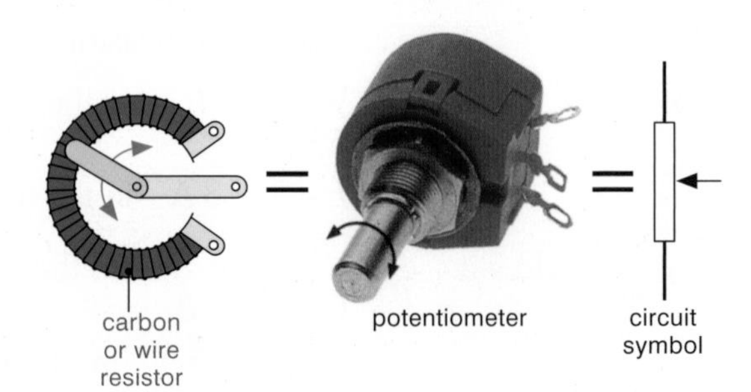

Potential dividers are often used in electronics (p. 329).

▶ Current : Voltage graphs

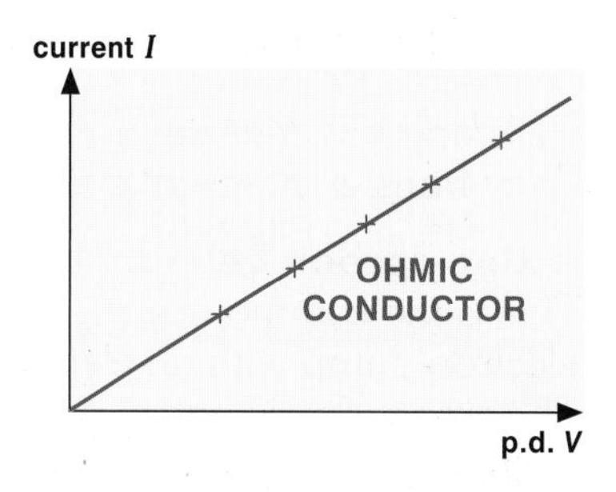

In experiment 31.18 (on page 255), you measured several values of the **current *I*** through a resistor, and the **voltage *V*** across the resistor. If you plot a graph of these values, you get a ***straight line through the origin:***

This means that the current ***I*** is ***proportional*** to the voltage ***V***. This is Ohm's Law (see page 253).
The steeper the graph, the lower the resistance.
The flatter the graph, the higher the resistance.

A substance that gives a straight graph like this is called an ***ohmic conductor***. Copper wire and all other metals give this shape of graph, unless they change temperature.

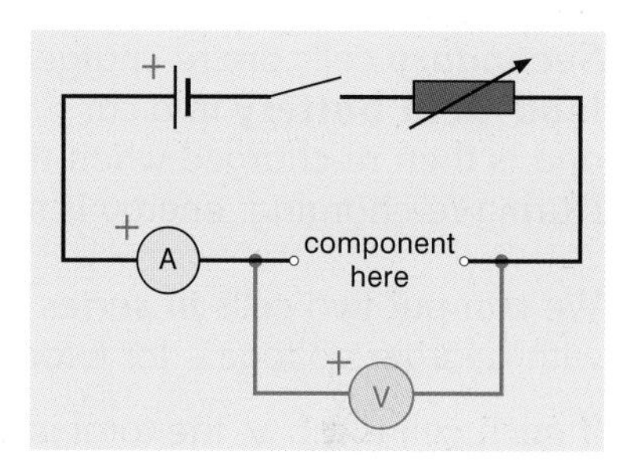

Not all objects and substances give a straight graph like this. You can use the same circuit as before (experiment 31.18) to investigate what happens with other objects and materials. Alternatively, you can use a ***current sensor*** and a ***voltage sensor***, connected to an interface and a computer. The computer will plot the graph for you.

Experiment 31.23 ***A filament lamp***
Use your circuit to investigate how the current ***I*** varies with the voltage ***V*** in a filament lamp (e.g. a torch bulb).

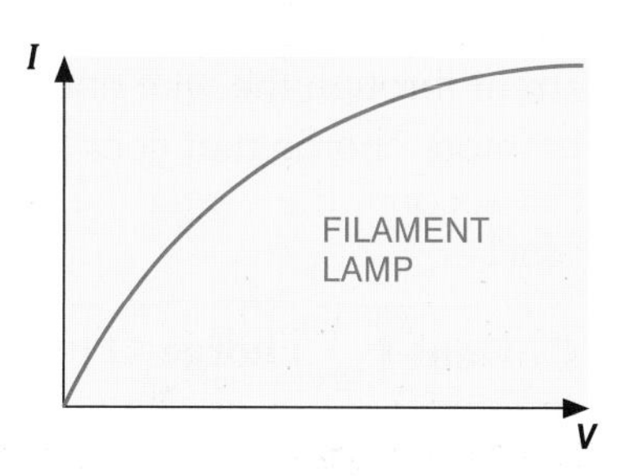

You can see that the lamp is a ***non-ohmic*** conductor because the graph is not a straight line. It does not obey Ohm's Law.

As more current flows, the metal filament gets hotter and so its resistance ***in***creases. This means the graph gets flatter, as shown:

Experiment 31.24 ***A thermistor***
A thermistor is used in electronics and is made of a ***semi-conductor*** substance (see page 319).

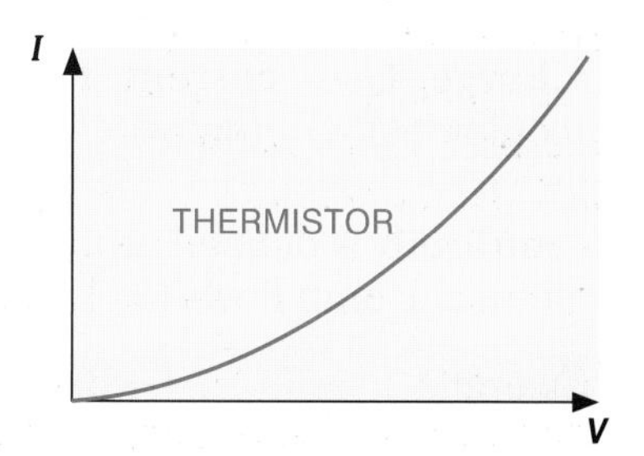

Look at the graph. Is a thermistor an ohmic conductor? Why does the graph bend the opposite way to the lamp?

As more current flows, the thermistor gets hotter and so its resistance ***de***creases (see page 319). The graph gets steeper.

Experiment 31.25 ***A semi-conductor diode***
A diode is used in electronics (see page 316 for details).

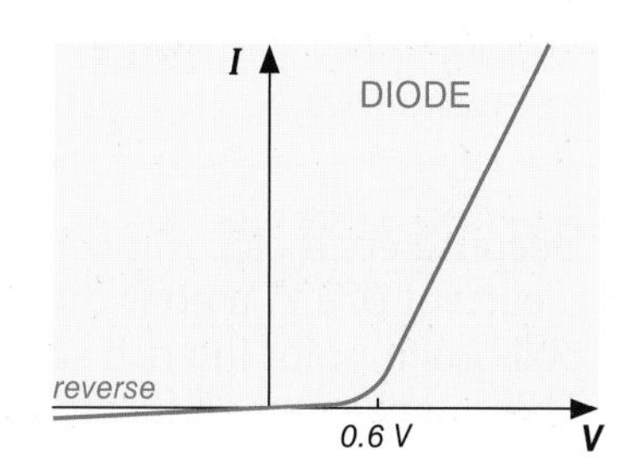

Look at the graph. Almost no current flows if the voltage is applied in the reverse direction.

A current flows in the forward direction whenever the voltage is more than about 0.6 volts.

▶ Energy transfers in a circuit

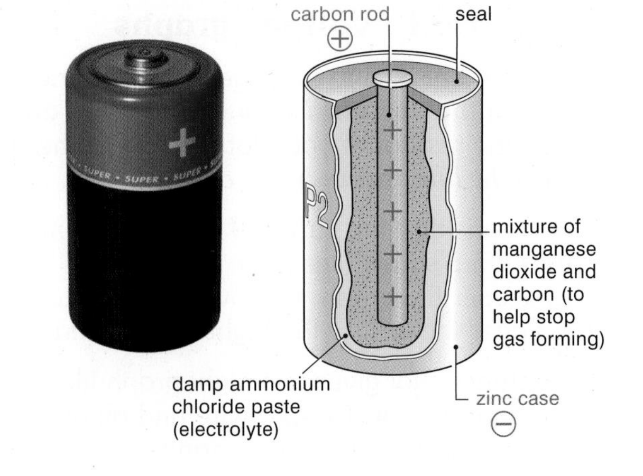

Cells and batteries are useful sources of electricity. They transfer chemical energy to electrical energy.

A **zinc-carbon cell** (dry Leclanché cell):
This is the common cell used in torches. It is an electron pump with an electrical pressure of 1.5 volts. The zinc is slowly eaten away as the chemical energy of the zinc is transferred to electrical energy.
This is a ***primary*** cell – once the chemicals are used up you throw it away.

Secondary cells are re-chargeable. For example, a **lead–acid battery** in a car turns the starter motor and is then re-charged when the engine is running. During re-charging, energy is stored in the battery.

We can put two cells ***in series*** to make a ***battery*** with a larger voltage – for example, in a torch:

If each cell is 1.5 V, the total is 3 volts.
Electrons (negative charges) are repelled from the negative terminal of the battery, and flow as a current through the wire and the lamp.
The more charge that goes through the wire in each second, the bigger the current.
In fact:

$$\text{Current } I \text{ (in amps)} = \frac{\text{charge } Q \text{ (in coulombs)}}{\text{time } t \text{ (in seconds)}} \quad \text{or} \quad I = \frac{Q}{t}$$

Re-arranging this, we get:

$$\text{Charge, } Q \text{ (coulombs)} = \text{current, } I \text{ (amps)} \times \text{time, } t \text{ (seconds)} \quad \text{or} \quad Q = I \times t$$

1 coulomb is defined as the charge that passes a point if 1 amp flows for 1 second.

Example
A current of 1 A flows through a lamp for 1 minute.
How much charge passes through the lamp?

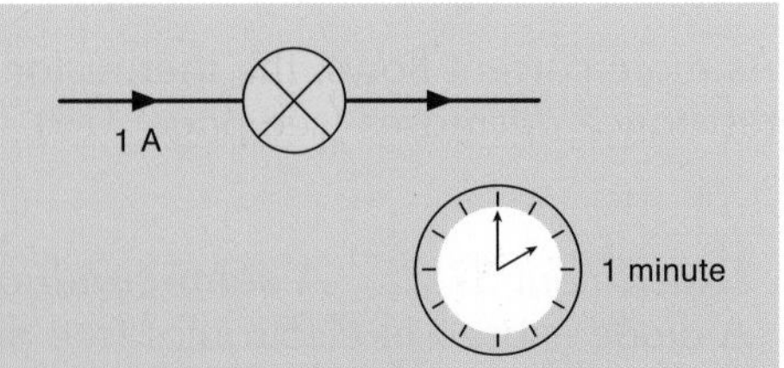

Formula first: **Charge, Q = current, I × time, t**
Then numbers: = 1 A × 60 s
= 60 C (60 coulombs)

Because each electron carries only $\frac{1}{6\,000\,000\,000\,000\,000\,000}$ of a coulomb, a current of 1 A means that more than 6 000 000 000 000 000 000 electrons pass each point in each second!
This is only a small fraction of the electrons in the wire!

As the electrons flow round the circuit, they gain potential energy in the cell and then lose this energy in the lamp. The amount of energy depends on the potential difference (p.d.) or 'voltage'.

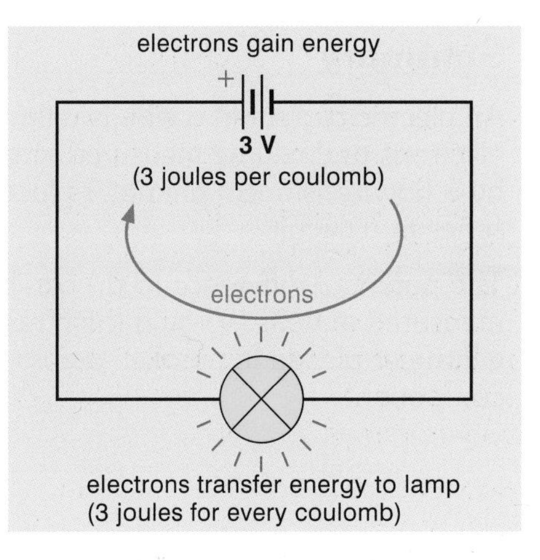

1 volt is defined as the p.d. between two points if 1 joule of energy is transferred (from electrical energy to other forms) when 1 coulomb of charge flows between the two points.

That is: **1 volt** = 1 joule of energy per coulomb of charge.

In this circuit, the voltage across the battery is 3 V.
This means that each coulomb of charge going through the battery receives 3 joules of energy.
As this charge goes through the lamp, it transfers the 3 joules of energy to the filament (to heat it).

So, if Q coulombs pass through a p.d. of V volts, then:

Energy transferred =	**p.d., V** ×	**charge, Q**
(joules)	(volts)	(coulombs)

or $$E = V \times Q$$ or $$V = \frac{E}{Q}$$

From the opposite page, $Q = I \times t$, and so:

Energy transferred =	**p.d., V** ×	**current, I** ×	**time, t**
(joules)	(volts)	(amps)	(seconds)

or $$E = V \times I \times t$$

Example
A 6 V battery passes a current of 1 A through a lamp for 1 minute.
How much energy is transferred from the battery to the lamp?

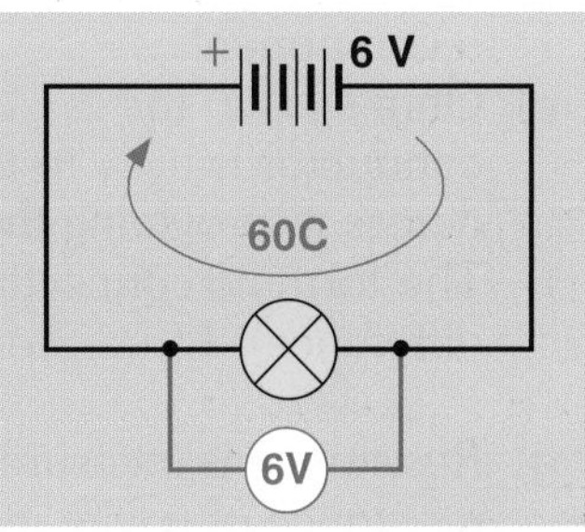

From the example opposite: charge passed by 1 A for 60 s = 60 C.

Formula: **Energy transferred = p.d., V × charge, Q**
= 6 V × 60 C = 360 joules

This is the energy transferred from the battery to the lamp.

Tips for solving circuit problems

1. Simplify the circuit as much as you can. Sometimes it helps to re-draw it without any ammeters or voltmeters.

2. Use the equation:
Use a finger to find the formula for the one you want.

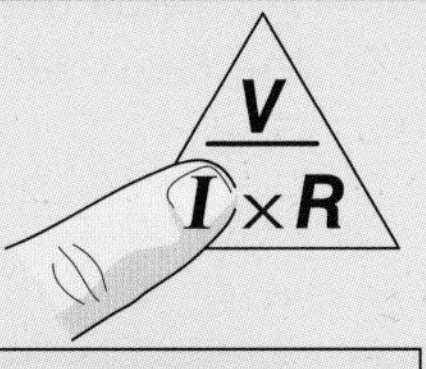

3. When the resistors are in series:
a) Total resistance = $R_1 + R_2$.
b) The current is same through each one.
c) The larger resistor has the larger p.d. across it.
d) The voltage across each resistor can be found by: $V = I \times R$.
e) The voltages across the series resistors add up to the voltage across the cell.

4. When the resistors are in parallel:
a) Total resistance of two resistors = $\frac{R_1 \times R_2}{R_1 + R_2}$.
b) Parallel resistors have same p.d. across them.
c) The smaller resistor carries the larger current.
d) The current through each resistor can be found by: $I = V/R$.
e) The currents in the parallel branches add up to the current in the main circuit.

Summary

An electric current in a wire is a flow of electrons and can be measured in amperes (A) by a (low resistance) ammeter placed in series in a circuit.

The potential difference (p.d.) can be measured in volts (V) by a (high resistance) voltmeter placed in parallel, across a component.

The resistance of a wire increases with length and temperature; it decreases as the cross-sectional area is increased. It also depends on the substance.

Ohm's Law: the current flowing in a wire is proportional to the p.d. across it, providing the temperature is constant.

$$V = I \times R \quad \text{or} \quad I = \frac{V}{R} \quad \text{or} \quad R = \frac{V}{I}$$

▷ Questions

1. Copy out and complete:

a) A current is a flow of . . . . . For this to happen there must be a . . . . circuit.

b) Copper is a good . . . . .
Plastic is an . . . .

c) Current is measured in . . . . using an . . . . placed in . . . . in a circuit.

d) Potential difference (p.d.) is measured in . . . . using a . . . . placed in . . . . across a component.

e) Ohm's Law: the . . . . flowing through a conductor is . . . . to the . . . . difference across it, providing the . . . . is constant.

f) The formula connecting ***V*** (. . . . difference), ***I*** (. . . .) and ***R*** (. . . .) is . . . . .

g) Resistance is measured in . . . . . The resistance of a wire increases as the length . . . . ; as the temperature . . . . ; and as the cross-sectional area . . . . .

2. *The Prof. is lost, can't see the light,*
So help him get his circuits right:

3. Copy and complete the diagrams below:

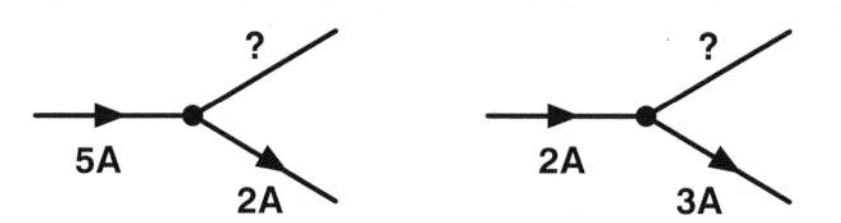

4. Draw a circuit diagram to show how two 4 V lamps can be lit brightly from two 2 V cells.

5. Calculate the combined resistance of each case:

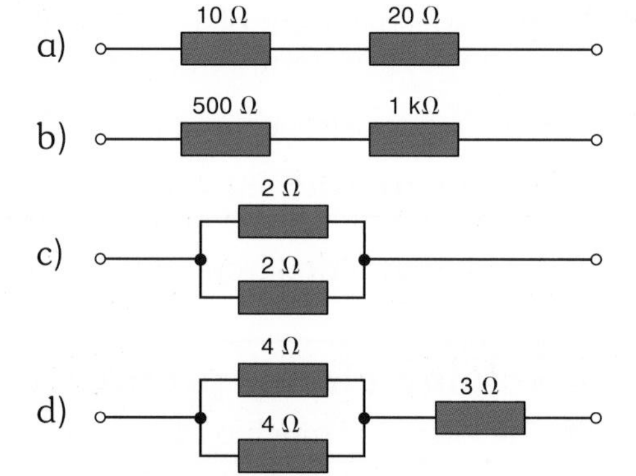

6. Explain what happens to the identical lamps X and Y in the circuit shown, when a) only switch S_1 is closed b) only S_2 is closed c) both S_1 and S_2 are closed.

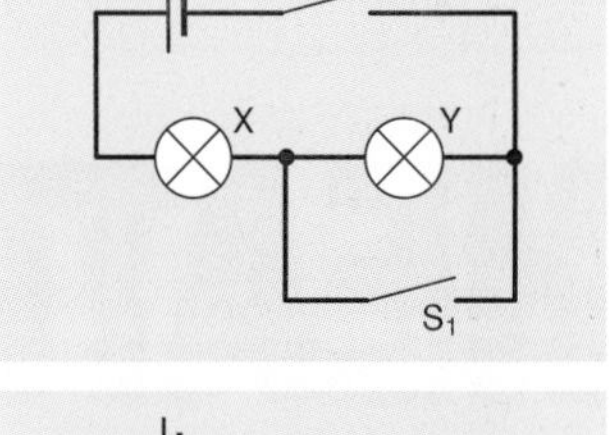

7. Draw a circuit diagram to show how 3 lamps can be lit from a battery so that 2 lamps are controlled by the same switch while the third lamp has its own switch.

8. In the diagram, the lamps are identical and ammeter A_2 reads 0.5 A.
   a) What are the readings on A_1, A_3, A_4?
   b) Redraw the diagram to include a switch to control both lamps together.
   c) Redraw the diagram to include two switches to control the lamps separately.
   d) Redraw it to include a rheostat to control lamp P only.

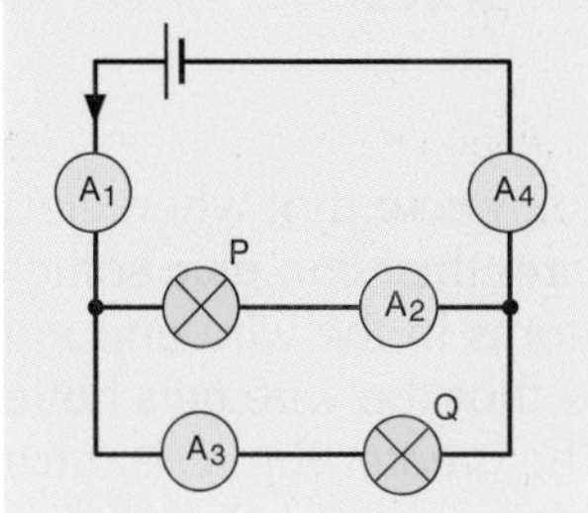

9. a) What p.d. is needed to send 2 A through a 5 Ω resistor?
   b) A p.d. of 6 V is applied across a 2 Ω resistor. What current flows?
   c) What is the value of a resistance if 20 V drives 2 A through it?

10. In the circuit shown, the p.d. across the 4 Ω resistor is 8 V.
   a) What is the current through the 4 Ω resistor?
   b) What is the current through the 3 Ω resistor?
   c) What is the p.d. across the 3 Ω resistor?
   d) What is the p.d. applied by the battery?

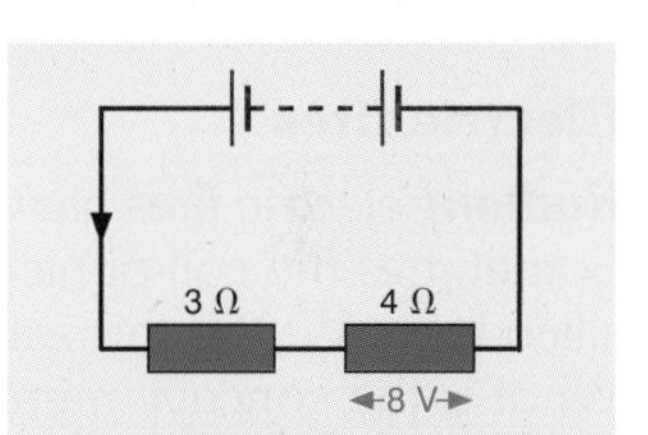

11. In the circuit shown, the voltmeter reads 10 V.
   a) What is the combined resistance? b) What current flows?
   What is the p.d. across c) the 2 Ω resistor? d) the 3 Ω resistor?

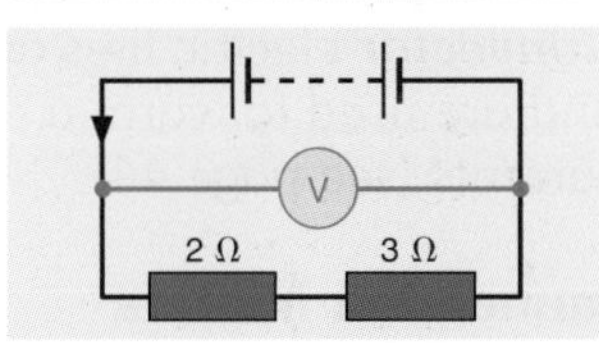

12. If a p.d. of 10 V causes a current of 2 A to flow, for 1 minute, how much energy is transferred?

13. In the circuit shown: a) What is the combined resistance?
   b) What is the p.d. across the combined resistance?
   c) What is the p.d. across the 3 Ω resistor?
   What is the current in d) the 3 Ω resistor? e) the 6 Ω resistor?

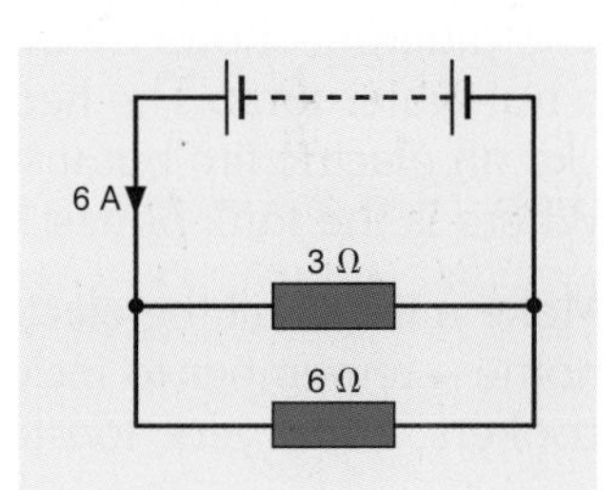

14. A flash of lightning carries 10 C of charge which flows for 0.01 s. What is the current? If the voltage is 10 MV, what is the energy?

15. *It seems the Prof's gone out of his mind. How many errors can you find?*

chapter 32

HEATING effect of a current

You know that when electrons pass through a wire, they can give some of their energy to the atoms in the wire and make them vibrate more, so that the wire gets hotter.
The greater the resistance, the hotter it becomes – this is why an electric fire glows red-hot while the connecting wires stay cool (see also page 254).

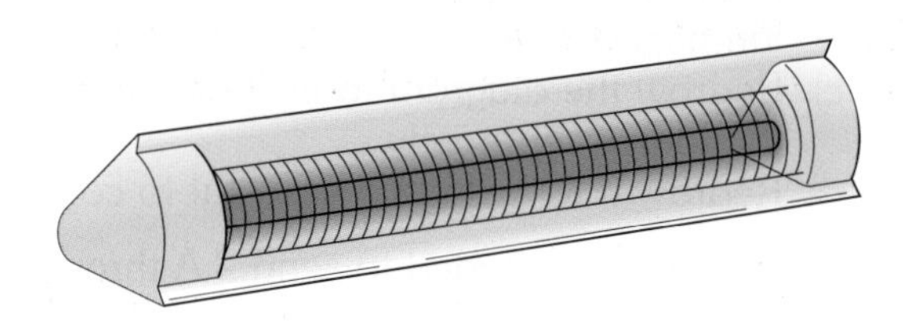

Electric fires

Radiant electric fires glow red-hot at about 900 °C. Sometimes the coil of nichrome wire is inside a silica tube. Why is this safer?
What is the concave mirror for?

Convector electric fires are cooler at about 450 °C, and are used to warm a room by convection currents (see page 44).

Alec Trishan once plugged in his fire,
To raise the room temperature higher,
But he soon got too hot,
And he thought he should not,
So he tied a large knot in the wire.
(Find two things wrong with this.)

Immersion heaters

The diagram shows an ***immersion heater*** fitted into a hot-water tank. The heater contains resistance wire, like an electric fire but insulated from the water.
Where is the tank hottest? Why?

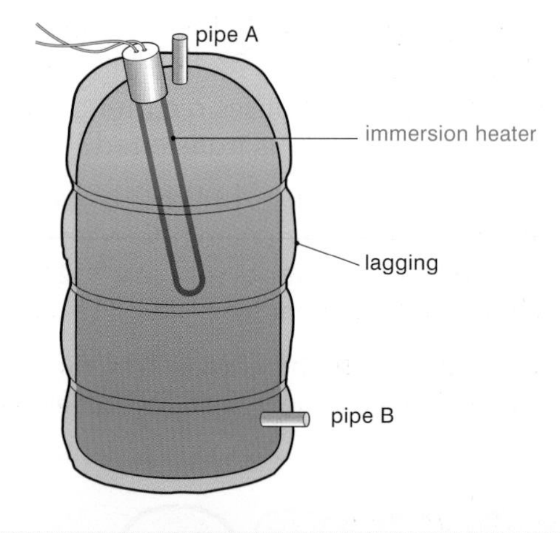

Make a list of all the electrical heaters in your home – remember to include electric kettles and cookers, hairdryers, toasters, electric blankets, etc.

The ***efficiency*** of a heater is usually high.

Example
An electric kettle, rated at 2000 W, transfers 2000 joules per second electrically to heat some water. It is switched on for 100 seconds and the water receives 180 000 joules. What is the efficiency of the kettle?

Formula first: (page 102) $\text{Efficiency} = \dfrac{\text{useful energy output}}{\text{total energy input}} \times 100\%$

Then numbers: $= \dfrac{180\,000}{2000 \times 100} \times 100\% = \underline{90\%}$

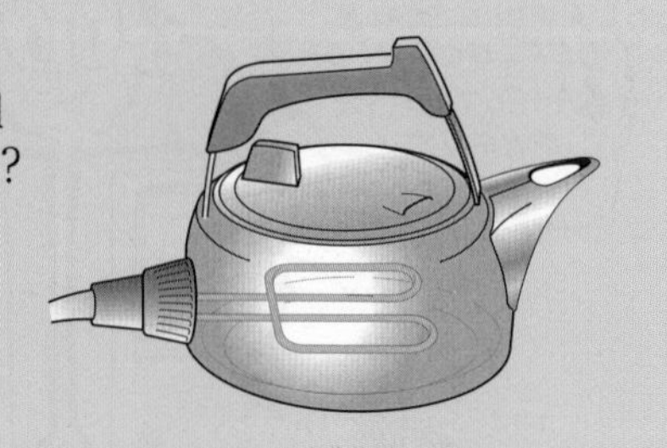

▶ Lamps

Electric filament lamp

Look closely at a light bulb – can you see that the filament is a coil of wire which has been coiled again to give a hotter 'coiled coil'?
The filament (heated to 2500 °C) is made of ***tungsten*** which has a high melting point.

These lamps are not very efficient. Less than 10% of the electrical energy is converted into light energy – the rest is heat (see page 102).

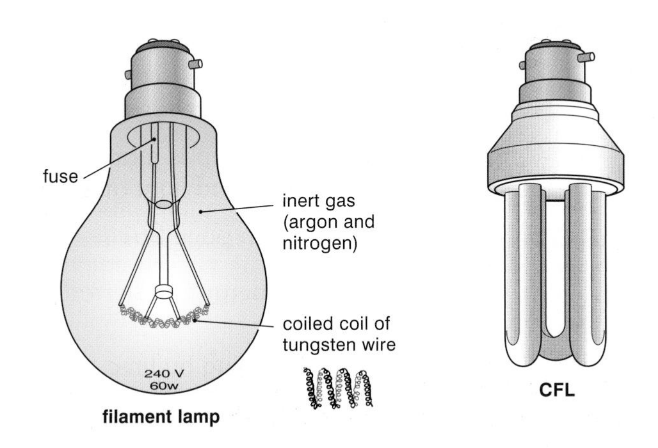

Energy-saving lamp

Modern energy-saver lamps are called CFLs (**c**ompact **f**luorescent **l**amps, see pages 103, 212). They are 5 times as efficient and last much longer (about 8000 hours).

Look at the graph:
How much money is saved with a CFL after 8000 hours of use? Why is this good for our planet?

LED lighting (see page 318) can also be used.

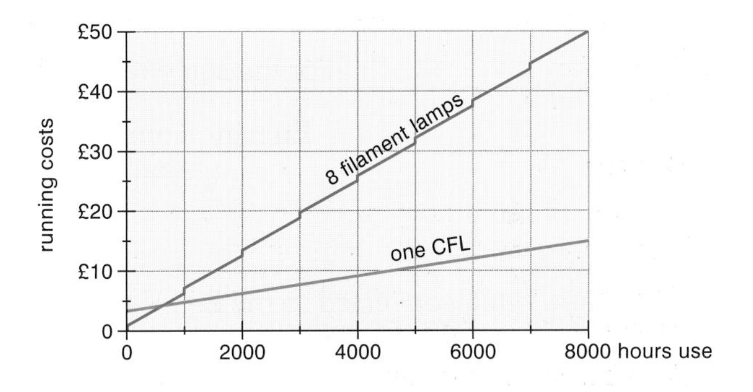

The circuit diagram shows how a lamp is wired to a ceiling rose and a switch (L and N refer to the 'live' and 'neutral' wires from the a.c. mains supply – see also page 268).

The switch ***must*** be in the live wire for safety when you are changing lamps.
NB *Never* inspect any mains wiring without first switching it off at the mains meter switch.

Why should wall switches not be fitted in bathrooms? Why are string pull switches safer?

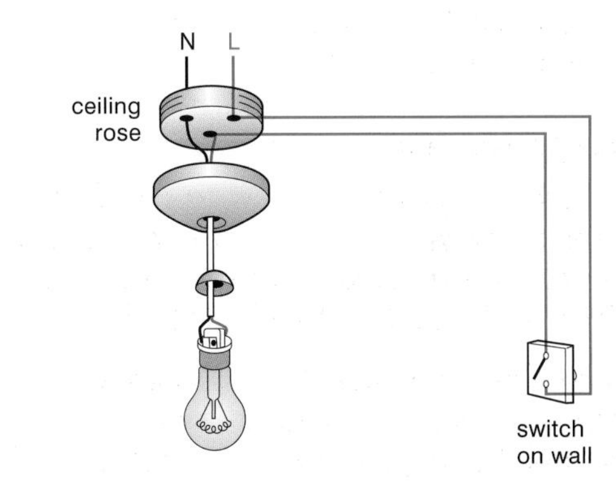

Two-way switches

With a two-way switch as shown in the diagram wire C can be connected to either A or B.

These switches can be used to switch a lamp from two positions – for example, the top and bottom of the stairs:

Can you control the lamp with either switch?
In the diagram, would the lamp be on or off?

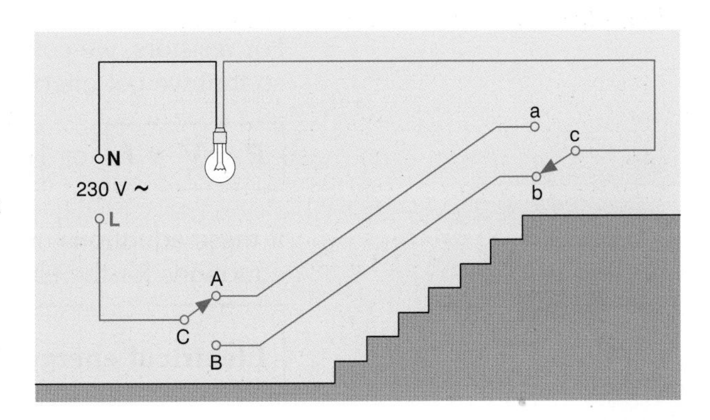

▶ Electrical power

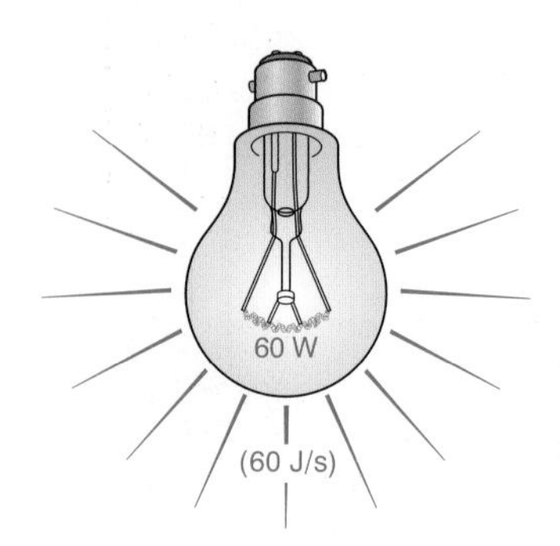

In Physics, we always measure energy in joules (see page 97). Therefore the ***rate*** of transferring energy, called **power**, is measured in joules per second, called ***watts*** (W). See also page 110.

If a lamp is marked 60 W, it means it is transferring electrical energy to heat and light at the rate of 60 joules per second.

From page 261,

Energy transferred (in joules) = p.d., $\boldsymbol{V}$ × current, $\boldsymbol{I}$ × time, $\boldsymbol{t}$

Dividing by time, to find the power:

Power, P = potential difference, V × current, I
(in watts) (in volts) (in amps)

or $\boldsymbol{P = V \times I}$

For the same reason (see also page 110) it follows that:

Energy transferred, E = power, P × time, t
(in joules) (in W) (in s)

or $\boldsymbol{E = P \times t}$

Example
An electric kettle connected to the 230 V mains supply draws a current of 10 A. Calculate:
a) the power of the kettle,
b) the energy transferred in 1 minute.

a) Formula first: **Power, P = p.d., V × current, I**
Then numbers: = 230 V × 10 A
= 2300 W (=2.3 kW) (where 1 kW = 1 kilowatt = 1000 W)

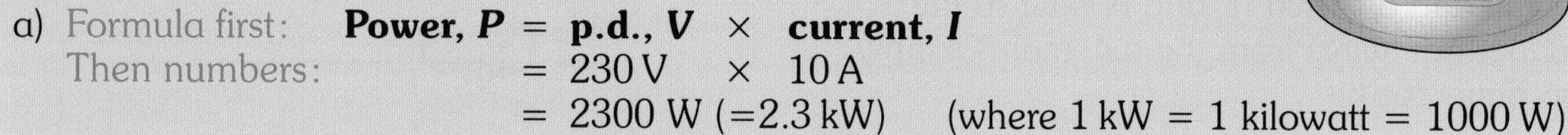

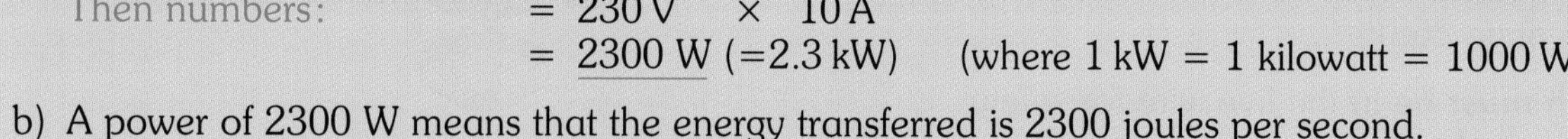

b) A power of 2300 W means that the energy transferred is 2300 joules per second.

Formula first: **Energy transferred, E = power, P × time, t**
Then numbers: = 2300 W × 60 s
= 138 000 joules (= 138 kJ)

Suppose all this energy is given to 2 kg of water in the kettle.
You could now work out the rise in temperature, by using the example on page 37.

For resistors, we can combine $\boldsymbol{P = V \times I}$ with $\boldsymbol{V = I \times R}$ (Ohm's Law), so that we get alternative equations:

$\boldsymbol{P = V \times I}$ or $\boldsymbol{P = I^2 \times R}$ or $\boldsymbol{P = \frac{V^2}{R}}$ **watts**

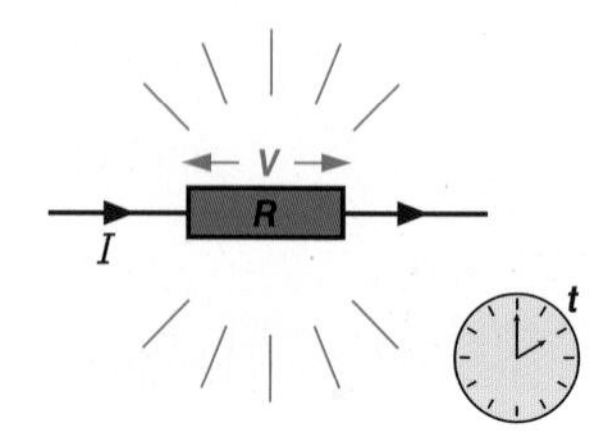

If these equations are multiplied by the time t, then they become equations for the electrical ***energy*** transferred to heat.

Electrical energy transferred $= \boldsymbol{VIt = I^2Rt = \frac{V^2t}{R}}$ **joules**

Power of appliances in the home

This can vary widely, from 3 watts for an electric clock to 5 kilowatts for an electric cooker. The power is usually marked on a label on the back of the appliance. Look carefully at some in your home and at the values shown in the table.

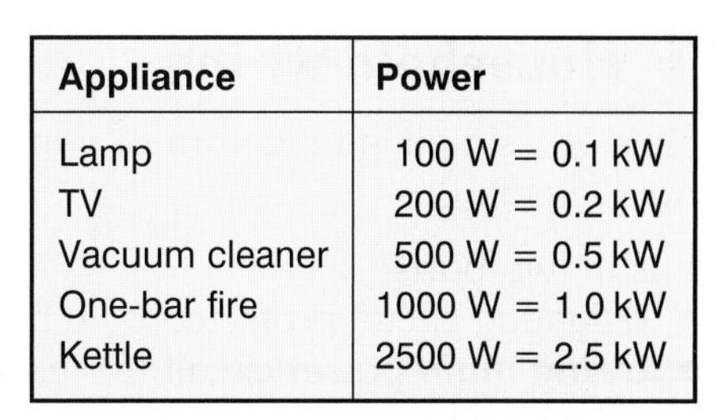

Appliance	Power
Lamp	100 W = 0.1 kW
TV	200 W = 0.2 kW
Vacuum cleaner	500 W = 0.5 kW
One-bar fire	1000 W = 1.0 kW
Kettle	2500 W = 2.5 kW

The cost of electricity

If you look at your electricity meter at home you will see it is marked in **kilowatt-hours** (kW h).

The ***kilowatt-hour*** is a unit of ***energy*** (***not*** power) and is calculated by:

Energy transferred (in kW h) = **power** (in kW) × **time** (in hours)

Since 1 kW = 1000 W and 1 hour = 3600 seconds, it follows that 1 kW h = 3 600 000 joules.

We used a joulemeter on page 36.

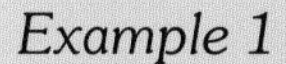

Example 1

A 2000 W electric fire is used for 10 hours.
What is the cost, at 8p per kW h?

Formula first: **Energy transferred** (in kW h) = **power** (in kW) × **time** (in hours)

Then numbers: = 20 kW × 10 hours
= 20 kW h

∴ Cost at 8p per kW h = 20 × 8p
= 160p = £1.60

The Prof. got a bill for his heater,
The shock turned him into a cheater.
He tried to reverse
The effect on his purse
By reversing the wires to the meter.

Can you explain why this would not work?

Example 2

Here's part of a bill from an electricity company: Unfortunately it has been chewed by Professor Messer's dog. What is the total due to be paid?

∴ Number of units used = 2586 – 1486
= 1100 kW h

∴ Cost at 8p per unit = 1100 × 8p
= 8800p = £88.00

∴ Total = £88 + £11 ('standing charge' – this is a fixed rental charge)
= £99

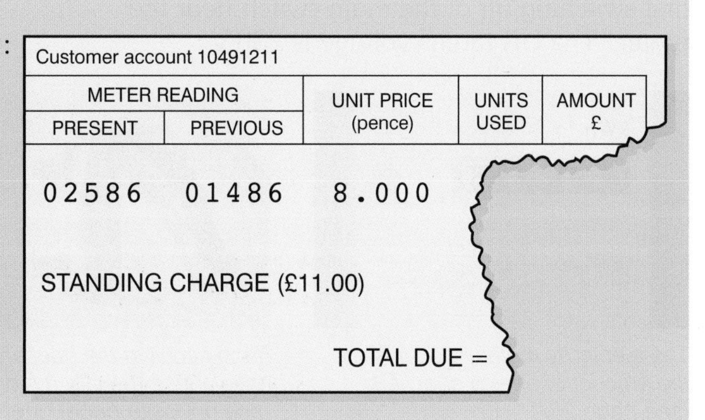

Customer account 10491211

METER READING		UNIT PRICE (pence)	UNITS USED	AMOUNT £
PRESENT	PREVIOUS			
02586	01486	8.000		

STANDING CHARGE (£11.00)

TOTAL DUE =

▶ Household wiring

Here is a simplified diagram of a modern household circuit.

It is in two parts:

- a lighting circuit protected by a 5-amp fuse,
- a ***ring main*** power circuit (for electric fires, etc.) protected by a 30-amp fuse.

Each power point is also protected by a fuse inside the plug (see page 270).

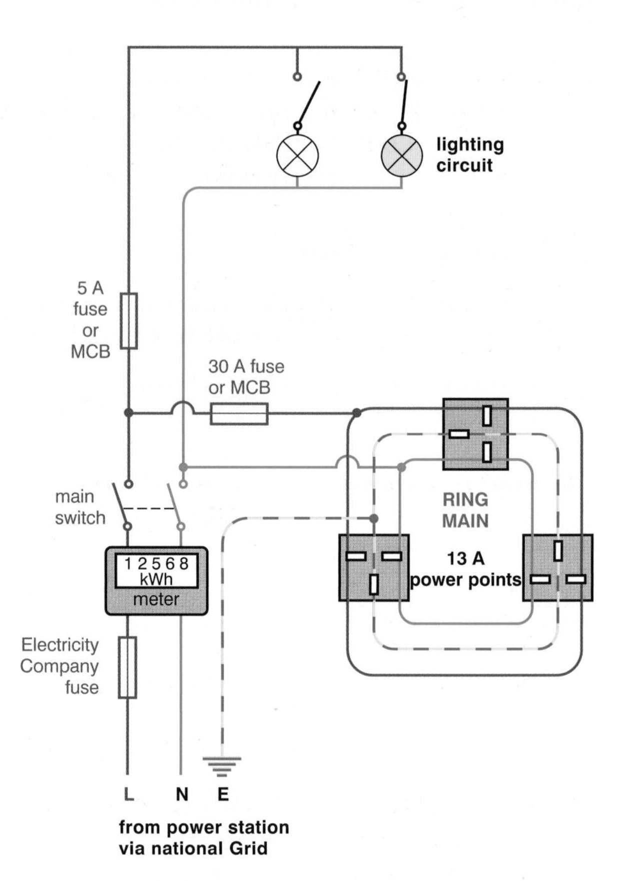

The electricity is generated in a power station (see pages 104, 349) and then supplied to your home via the national Grid (page 303).

L and **N** in the diagram refer to the Live and Netural wires coming from the power station. The third wire (**E**) is connected to Earth.

The ***Live*** wire is the most dangerous – the mains voltage may easily kill you.
The ***Neutral*** wire completes the circuit, but because it is earthed back at the power station, its voltage is usually not as high as the live wire.
The ***Earth*** wire is for safety and only carries a current when there is a fault, so that a fuse melts (see opposite page).

The current supplied to your home is **a.c.**, ***alternating current*** (see page 299).
This means that the current flows one way and then the opposite way, again and again.
It goes each way 50 times in each second.

The Live voltage alternates between positive and negative compared to the neutral wire.

!***Never*** touch any part of a mains circuit without first switching off at the main switch near the meter! The UK mains voltage is 230 V.

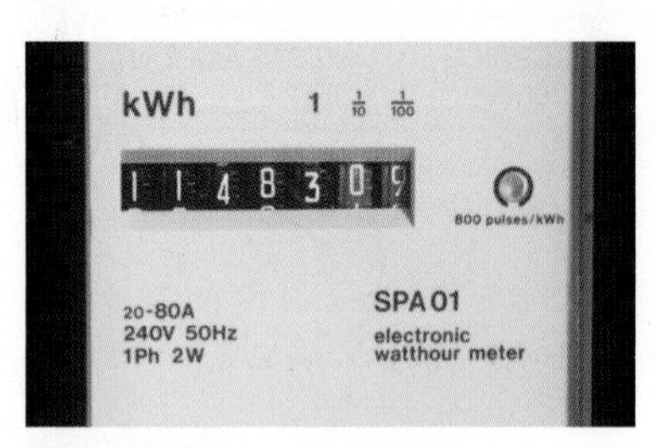

The electricity meter measures the amount of energy supplied (see page 267)

Each circuit in the house will have a fuse (usually 5 A or 30 A) (see opposite page) or else…

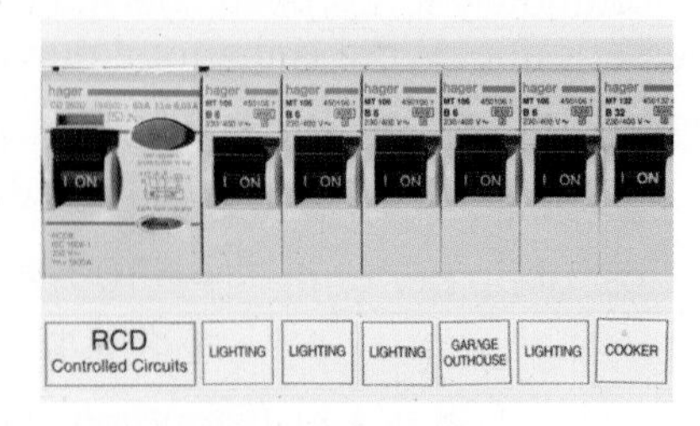

*…or else each circuit will have an MCB (**m**iniature **c**ircuit **b**reaker) (see page 272)*

▷ Fuses

> *Experiment 32.1*
> Connect the circuit shown in black in the diagram, with crocodile clips at A and B.
> Complete the circuit with a short length (about 3 cm) of 36-gauge tin wire, placed between A and B.
> Check that the lamp is on.
>
> Now imitate a fault in the circuit by using a piece of copper wire to 'short' out the lamp (as shown in red).
> What happens to the thin wire?

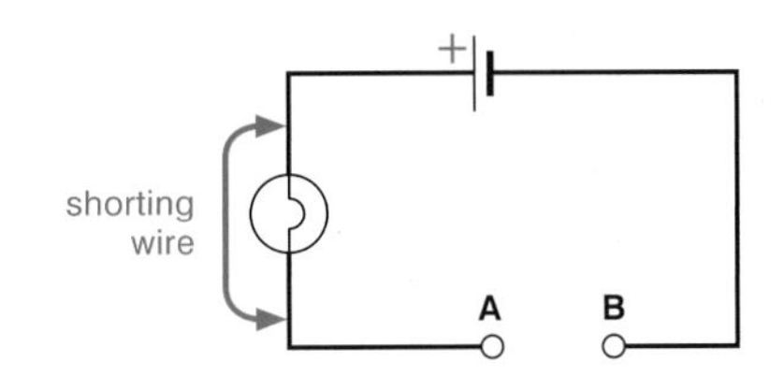

The thin tin wire melted or ***fused*** and so stopped the current flowing.

Fuses are included in circuits as deliberate weak links for safety, so that if a fault occurs and too much current flows, the fuse wire melts before anything else is damaged or starts a fire.

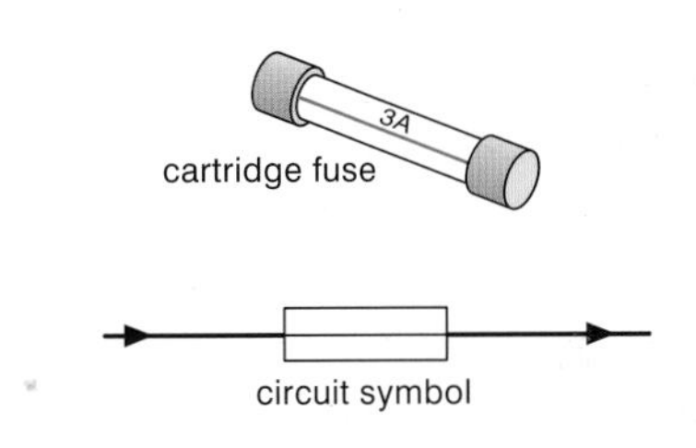

If a fuse melts, you must:
- switch off that circuit, (if necessary use the main switch near the meter, see opposite),
- find the reason for the fault,
- then replace the fuse with one of the correct value.

Often a **circuit-breaker** (see page 272) is better than a fuse, because:
- It acts more quickly.
- It is more reliable.
- It can be re-set.
- One type (RCCB) is much more sensitive and can prevent you being electrocuted.

A young electrician called Hyde,
Learned nothing because of his pride.
He got his wires wrong,
And before very long,
He cried – then he sighed – then he died!

The Earth wire

If a fault develops in an appliance and a live wire touches the metal case, a large current flows from the live wire to the earth wire and the fuse breaks. This makes the appliance safe until the fault is mended and then the fuse replaced.

Notice that the fuse (like a switch) ***must*** be placed in the ***live*** wire so that it can isolate the appliance from the live wire. This system only works if the appliance is properly earthed.

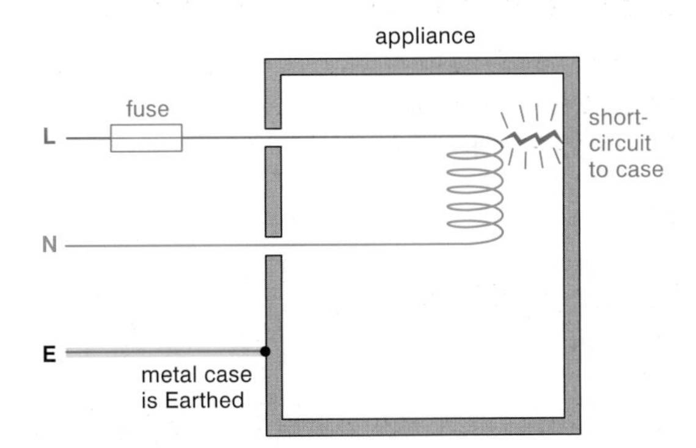

Your hairdryer probably does not have an earth wire connected to it. This is because it has an insulating plastic case. It is ***double-insulated*** and marked with the symbol: ⧈

▶ Fused plugs

It is absolutely essential that a plug is wired with the coloured wires going to the correct places:
Live – brown wire,
Neutral – blue wire,
Earth – green and yellow wire.

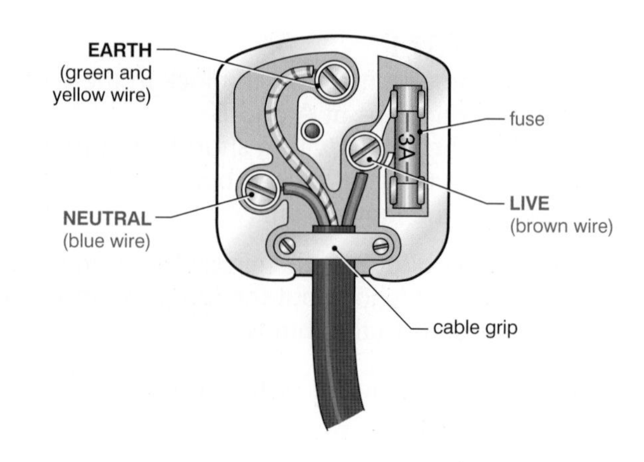

Which parts of the plug are:
a) conductors, b) insulators?

Practise wiring a plug, taking care not to damage the insulation or the wires by cutting too deeply. Do not make the bare ends longer than necessary, and make sure that the cord-grip screws are firm, and are clamping the outer cable.

Make sure that you fit a fuse of the correct value.

Fuse rating

The fuse rating is the maximum current that the fuse can carry without melting.
Only certain values are available: eg. 3 A, 5 A, 13 A.

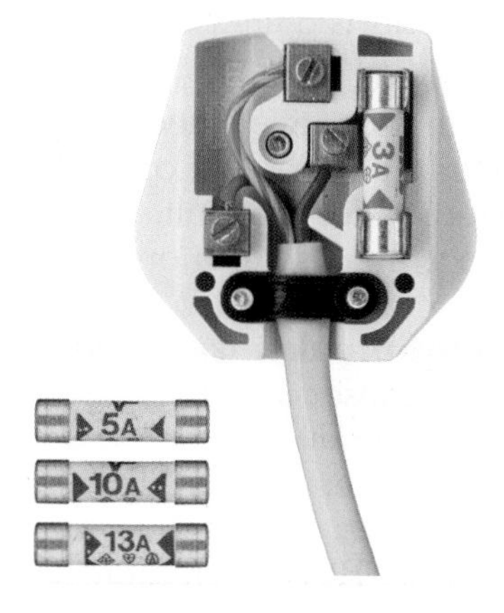

Example
A vacuum cleaner has a rating of 460 W on the 230 V mains.
What fuse should be fitted in the plug?

Formula first: **Power** = **p.d.** × **current** (p. 266)

$$460\text{ W} = 230\text{ V} \times \text{current}$$

$$\therefore \text{Current} = \frac{460}{230} = \underline{2\text{ A}}$$

A 3 A fuse should be fitted. A 13 A fuse could allow a damaging current to flow before melting.

What fuse rating would you choose for:
a) a table lamp of 100 W, b) a kettle of 2400 W?

A careless young student called Hughes,
Once fitted the wrong size of fuse.
This fault caused a wire
In his house to catch fire.
– His death made the 6 o'clock news!

Summary

Power = IV watts (or I^2R or $\frac{V^2}{R}$)

where I is in amperes, V in volts, R in ohms.

Energy (in joules) = **power** (in W) × **time** (in seconds)

Energy (in kW h) = **power** (in kW) × **time** (in hours)

A fuse (or a circuit-breaker) is used to protect a circuit and is always placed in the live wire.
Mains wiring colours: brown (live), blue (neutral), green and yellow (earth).

▶ Questions

1. Copy out and complete:

a) The power (in watts) in an electric circuit is equal to the current (in . . . .) multiplied by the . . . . (in . . . .).

b) To calculate the energy (in joules), the power (in . . . .) is multiplied by the . . . . (in . . . .).

c) To calculate the energy (in kilowatt-hours), the . . . . (in . . . .) is multiplied by the . . . . (in . . . .).

d) A fuse (or a . . . . -breaker, or a light switch) should always be placed in the . . . . wire of a mains circuit.

e) When wiring a fused plug, the brown wire should go to the . . . . terminal, the blue wire to the . . . . terminal and the green and yellow wire to the . . . . terminal.

f) The earth wire should be connected to the . . . . of an appliance.

2. A 20 W CFL lamp is switched on for 1000 seconds and transfers 15 kJ as heat.

a) How much total energy is transferred electrically?

b) How much energy is transferred as useful light energy?

c) What is its efficiency?

3. A vacuum cleaner is labelled 250 V 500 W. When connected to a 250 V supply, how much current does it take?

4. A 60 W lamp is connected to a 240 V supply.

a) What current does it take?

b) What is the resistance of the filament?

5. *The Prof. has got a mental block*
– Now tell me why he gets a shock.

6. A 2 kW fire is switched on for 6 hours. What is the cost if one unit (kW h) costs 10p?

7. Here are the readings on an electricity meter:
1st January 28016
1st April 29984

a) How many Units were used?

b) If one Unit costs 8p, what is the total cost?

8. A 2 kW fire, a 200 W TV and three 100 W lamps are all switched on from 6 p.m. to 10 p.m. What is the total cost at 8p per kW h?

9. The diagram shows the inside of a 3-pin plug:

a) What is the colour of each of the wires attached to A, B, C?

b) Which of these is i) the earth wire ii) the live wire?

c) What is placed in the gap CD?

d) Why is this component necessary?

e) How is its value chosen?

f) Why is the direction of the wire round screw A not as satisfactory as those round B and C?

10. Copy out and complete the table:

Appliance	Power (W)	p.d (V)	Current (A)	Correct fuse (from 3 A, 5 A, 10 A, 13 A)
Car headlamp	48	12		
TV	230	230		
Hairdryer		230	2	
Iron	920	230		
Kettle		230	10	

11. Discuss the most effective ways of reducing electricity bills, a) at home, and b) at school.

12. Discuss the advantages of using energy-saver CFL lamps, including economic and environmental aspects.

MY SPECIAL PLUG WILL BE EXTRA SAFE!
FUSE
FUSE
FUSE
WHAT DID I DO WRONG ?

Further questions on page 278.

▷ Physics at work: Circuit-breakers

There are 2 main types of circuit-breaker:

- **MCB** (standard or **m**iniature **c**ircuit **b**reaker)
  If the current exceeds a certain value (usually 3 A, 13 A or 30 A) the MCB breaks the circuit, like a fuse, to avoid a fire.
  Compared with a fuse, an MCB is faster, more reliable, and can be re-set easily.
  However these currents are too big for your safety – they will easily electrocute you.

- **RCCB** (**r**esidual **c**urrent **c**ircuit **b**reaker) or **RCD** (**r**esidual **c**urrent **d**evice)
  This is a more sensitive device and can switch off the current very quickly (in $\frac{1}{30}$ s) even if the fault current is as low as $\frac{1}{30}$ amp.
  You should always use an RCCB in any situation where you could be electrocuted (eg. using an electric lawnmower).

MCB

There are many different designs.

Example 1

This one uses the heating effect of a current:

What happens to the bi-metal strip if too much current flows?
Which way does it bend? (See page 23.)
How does this break the circuit?

Example 2

Many circuit-breakers use an electromagnet:
(See pages 287–289.)
What happens to this electromagnet when the current is flowing?
If the current is too big, what happens to the iron bar?
What does this do to the current?

How would you re-set this circuit-breaker?
How could you design it to be more sensitive?

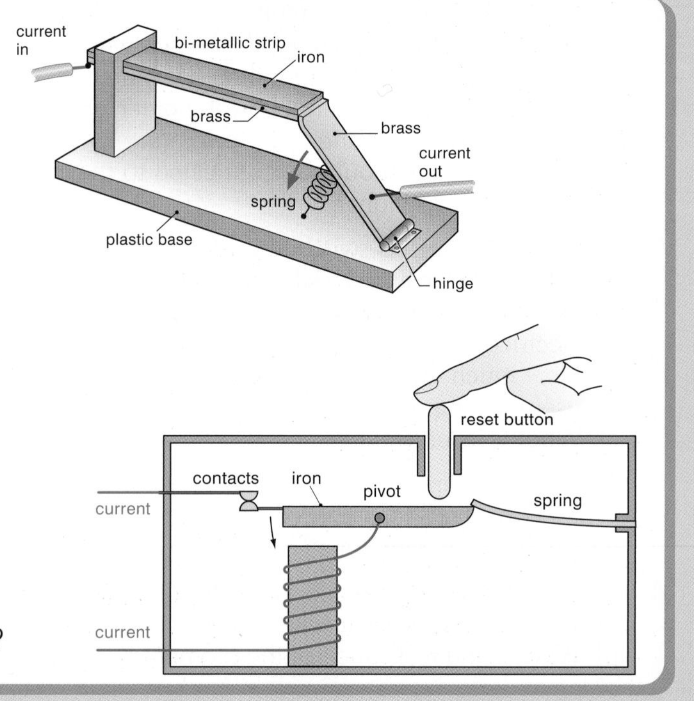

RCCB (RCD)

The live wire and the neutral wire are wound on an iron core in opposite directions:
So normally the 2 magnetic fields cancel out.

However, if there is a fault, and some of the live wire current passes through your body, then the live and neutral currents will not be equal.

So then the 2 fields will not cancel out, and there will be a current induced in the third coil.

This trips a relay (page 321) and breaks the circuit very quickly (in $\frac{1}{30}$ s), to keep you safe.

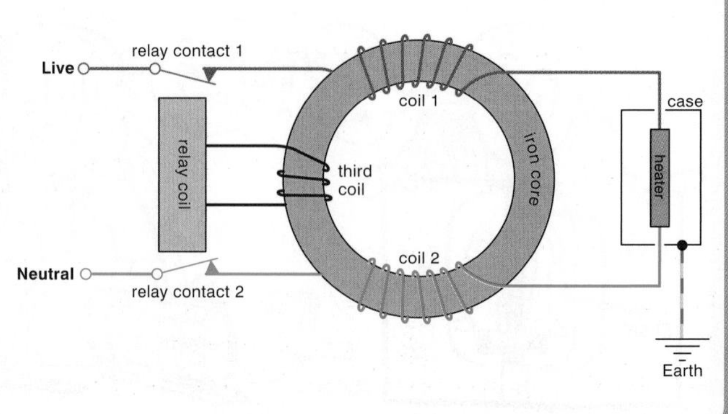

chapter 33

CHEMICAL effect of a current

What happens if you try to pass electricity through a liquid?

> *Experiment 33.1* ⚠
> Connect the circuit, as shown, to two electrodes in a beaker.
> Pour a liquid into the beaker and use the lamp to see if the liquid can conduct electricity.
> Try this with distilled water, salt water, vinegar, paraffin, acids and copper sulphate solution.
> (Remember to wash out the apparatus each time you change the liquid.)

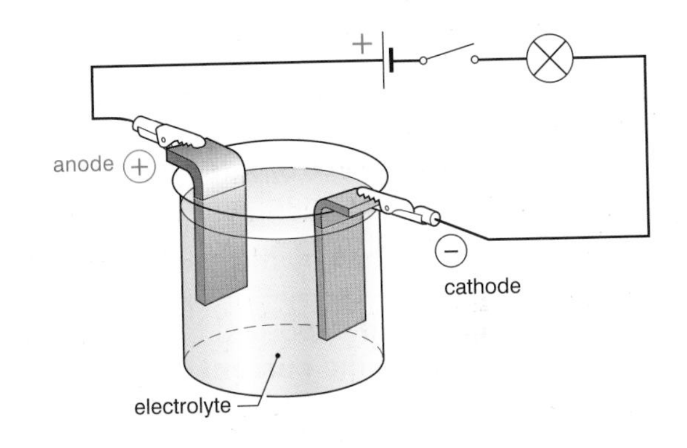

Liquids which conduct electricity are called ***electrolytes***. Show your results in a table of electrolytes and non-electrolytes.

When electricity is passed through a liquid, it may change the liquid – this is called **electrolysis**. The apparatus used for electrolysis is called a ***voltameter*** (not a voltmeter).

> *Experiment 33.2 Electrolysis of water* ⚠
> Pass electricity through water, using a water voltameter with two platinum electrodes connected to a battery as shown. Water is a poor conductor of electricity, so some sulphuric acid must be added to make an electrolyte.

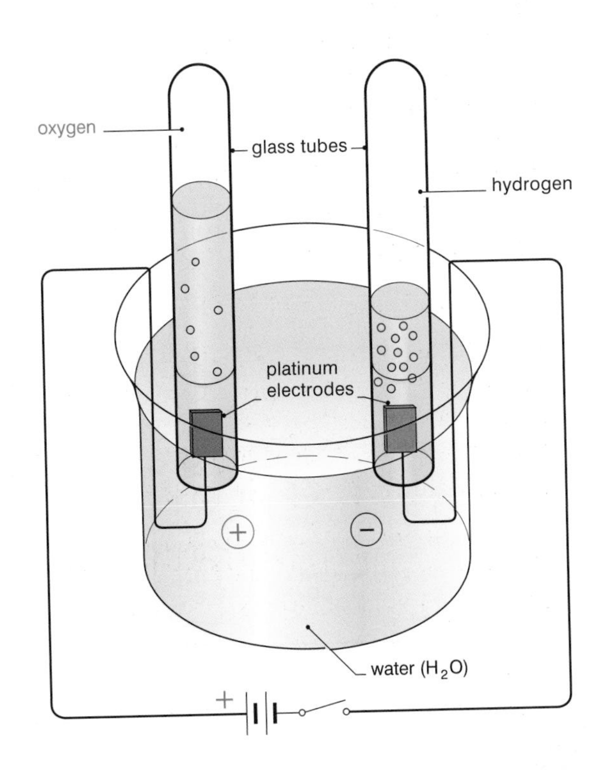

What happens?

When a current is flowing, bubbles appear at both the anode (+) and the cathode (−).
If you can collect and test these gases, you will find that they are oxygen at the anode and hydrogen at the cathode.

What do you think has happened?

The electricity has a chemical effect on the water and has split it into oxygen and hydrogen.
(The volume of hydrogen is twice that of oxygen and so the formula for water is H_2O.)

▶ Electroplating

Experiment 33.3 Electrolysis of copper sulphate
Connect the circuit shown here to a copper voltameter consisting of a beaker of copper sulphate solution and two clean copper electrodes. ⚠
Label and find the mass of each electrode (to 0.1 gram) before putting them in the electrolyte.

Pass a current of about $\frac{1}{2}$ A for about $\frac{1}{2}$ hour.

Then carefully dry the electrodes and find the new mass of each one. What do you find?

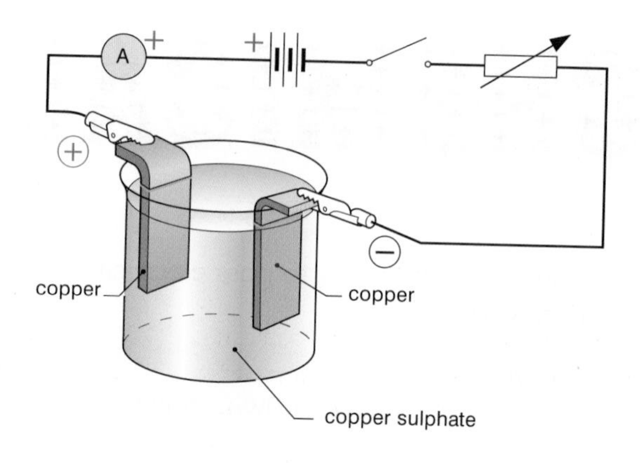

The cathode (−) has gained mass and is covered with a bright new coating of copper.
We say it has been ***electroplated*** with copper. The anode (+) has lost an equal mass of copper.

Experiments (first done by Michael Faraday) show:

the **mass of metal** deposited in electrolysis is
- ***proportional*** to the **current**, and
- ***proportional*** to the **time** for which it flows.

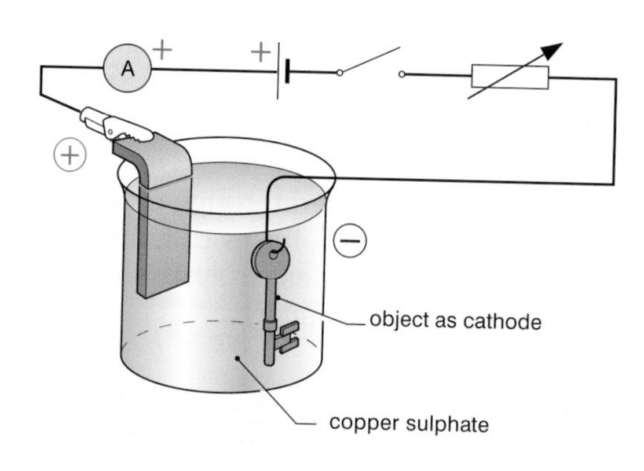

Experiments also show that:
the metal (or hydrogen if it is present) is always deposited at the cathode.

Experiment 33.4 Copper-plating an object ⚠
Replace the copper ***cathode*** of experiment 33.3 by a clean metal object (eg. a key or a badge).

Pass a very small current for several minutes until your object becomes plated with copper.

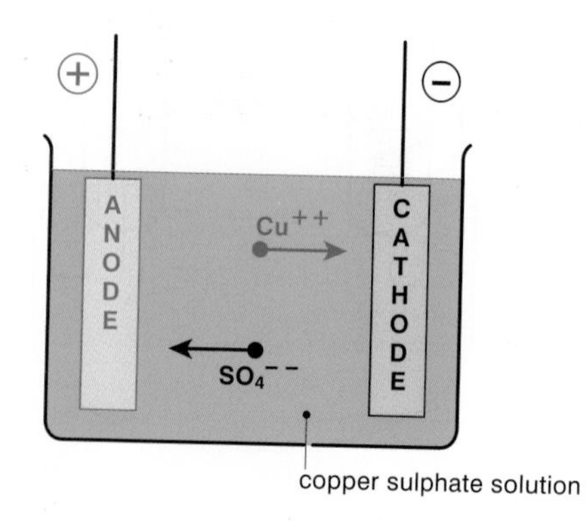

The ionic theory

A molecule of copper sulphate has the chemical formula $CuSO_4$ (= 1 atom of copper Cu, 1 atom of sulphur S, 4 atoms of oxygen O_4).
When copper sulphate is dissolved in water, it splits into two charged parts (Cu^{++} and so SO_4^{--}).
These are called ***ions*** (see page 244).

The copper atom Cu loses 2 electrons to become a positive ion Cu^{++}.
As unlike charges attract, this Cu^{++} ion drifts toward the negative cathode where it gains 2 electrons to become an atom of copper-plating on the cathode.
The SO_4^{--} ion drifts to the positive anode.

▶ Uses of electrolysis

1. Electroplating
In a similar way to your copper-plating in experiment 33.4, spoons can be silver-plated if they are used as the cathode of a voltameter containing silver. Chromium plating for cars and bicycles is achieved by a similar method.

2. Refining copper
In experiment 33.3, the copper transferred to the cathode was very pure even though the anode may have been made of impure copper. This process is used to produce pure copper for electric cables.

3. Extraction of sodium and aluminium
These metals are obtained by electrolysis (of common salt for sodium, and of aluminium oxide for aluminium).

▶ Questions

1. Copy out and complete:
   a) A liquid which conducts electricity is called an . . . . .
   b) The positive electrode of a voltameter is called the . . . . and the negative electrode is called the . . . . .
   c) In the electrolysis of water, hydrogen forms at the . . . . , and . . . . forms at the . . . . .
   d) In the electrolysis of copper sulphate, ... is deposited on the . . . . .
   e) A metal or hydrogen is always deposited at the . . . . .
   f) Michael Faraday discovered that the . . . . of metal deposited is proportional to the . . . . and proportional to the . . . . .
   g) The current in the electrolyte is carried by charged atoms called . . . . .
   The positive . . . . move towards the . . . . .
   In copper sulphate solution, the atoms of copper have lost . . . . and so have a . . . . charge which causes them to move towards the . . . . .

2. Explain why Professor Messer had difficulty in electroplating a plastic toy.

3. The diagram shows some apparatus used to copper-plate an object at B.
   a) What is the substance at A?
   b) What is the liquid C?
   c) Which would be connected to the positive of the battery?
   d) What factors determine the mass of copper deposited?

4. In a particular voltameter, 1 gram of copper is deposited. The experiment is repeated at three times the current for twice the time. What mass of copper is deposited?

5. The mass of copper deposited on a cathode when 1 amp flows for 1 second is 0.000 33 g. How much copper is deposited
   a) if 2 A flow for 500 s
   b) if 2000 coulombs are passed?

How much revision have you done?
See page 382.

Further questions on electricity (1)

▷ Circuits

1. A pupil made an electric circuit like the one below in order to measure the current through two lamps.

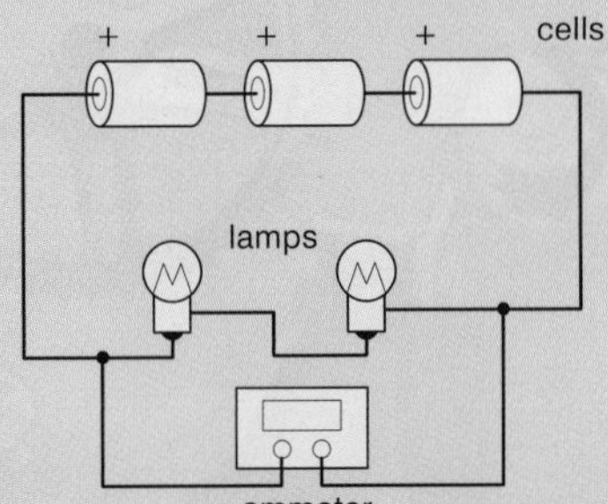

a) Are the lamps in ***series*** or ***parallel?*** [1 mark]
b) The pupil has made a mistake in this circuit. What is the mistake? [2]
c) Draw a diagram showing the ***correct*** way to connect the circuit. Use the proper circuit symbols in your diagram. [4] (AQA)

2. Jim is a do-it-yourself enthusiast. He fitted a fog-lamp to his car and wired it into the supply to the headlamp. He wired it incorrectly. The circuit he used for the fog-lamp is shown:

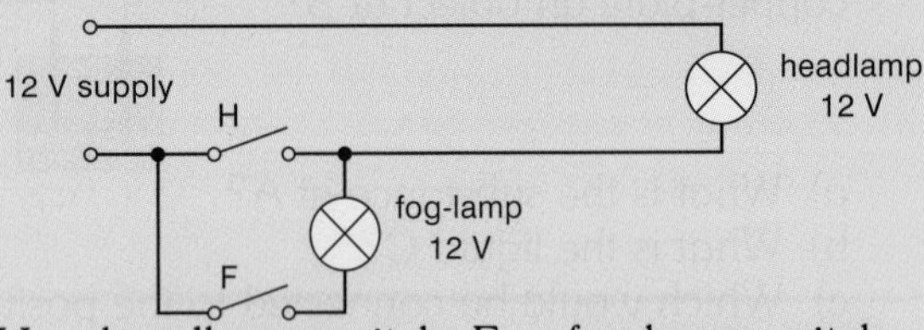

H = headlamp switch, F = fog-lamp switch.

a) Jim saw that the fog-lamp was lit when he pressed its switch, F, but it was not at full brightness. Use the circuit to explain why. What would Jim notice about the brightness of the headlamp? [3]
b) Jim saw that the fog-lamp did not work when he turned on the headlamp switch as well. Why? [2]
What would Jim notice about the brightness of the headlamp now? [1]
c) Jim tried to find out what was wrong. He removed the headlamp bulb.
What happened when he then turned on the switches? [1]
d) Sketch a diagram of the circuit Jim should have wired to make both the headlamp and the fog-lamp work properly. [2] (OCR)

3. a) Explain why a current flows when there is a complete circuit, by explaining what happens to the free electrons. [3]
b) Draw a circuit diagram to show how you could find the ***resistance*** of a lamp. [3]
(Edex)

4. The diagram shows two resistors connected in series. The p.d. between A and C is 9 V.

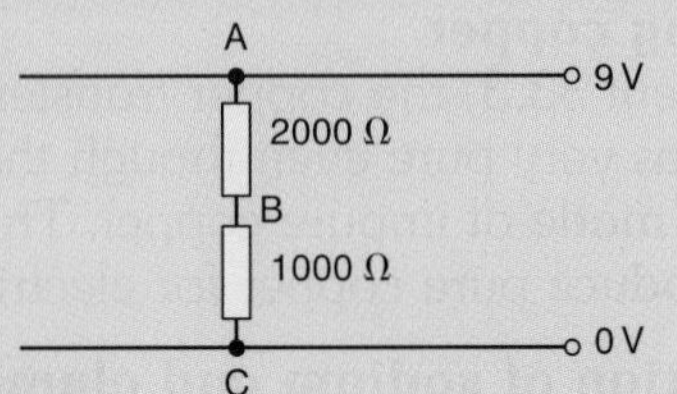

Calculate:
a) the total resistance between A and C, [1]
b) the current flowing in the resistors, [3]
c) the p.d. across the 1000 Ω resistor. [3]
(AQA)

5.

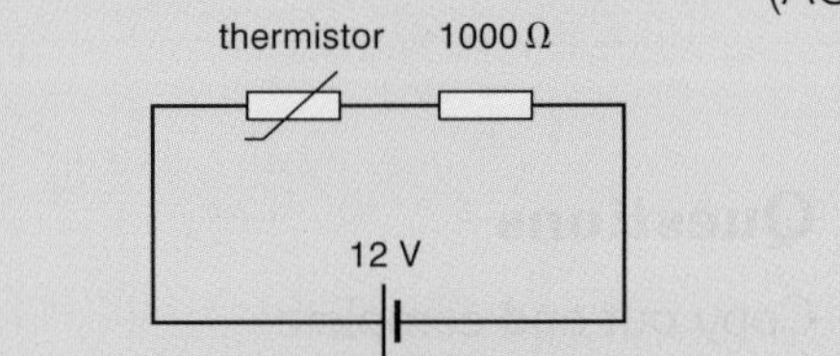

a) In the circuit shown, the thermistor has a resistance of 200 Ω.
i) Calculate the current through the resistor and thermistor. [3]
ii) Calculate the potential difference across the 1000 Ω resistor. [2]
b) The temperature of the thermistor decreases so that the potential difference across the 1000 Ω resistor becomes 4 V. Calculate the new value of the resistance of the thermistor. [2] (OCR)

6. a) State **three** safety features present in modern mains plugs. [3]

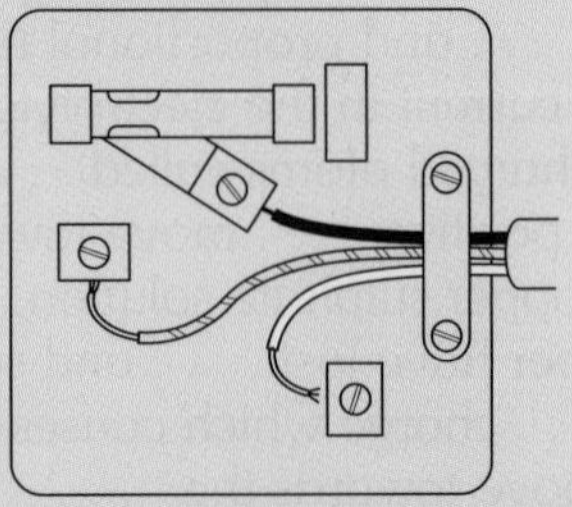

b) Look at this plug. What **four** changes should be made to make the plug safe? [4]
c) Explain the purpose of the Earth wire. [2]

 Note: although a question may be marked (OCR) etc, it will often be relevant to other examinations as well.

7. The table shows corresponding values of potential difference across a torch bulb and the current passing through it.

Potential difference (V)	0	0.02	0.1	0.5	1.0	1.65	2.3	3.1	4.0
Current (A)	0	0.04	0.08	0.12	0.16	0.20	0.24	0.28	0.32

a) Draw a diagram of a circuit which could have been used to obtain these data. [5]

b) On graph paper plot a graph of current on the y-axis against p.d. on the x-axis. [3]

c) i) Use the graph to find the potential difference across the bulb when the current through it was 0.25 A.

ii) Calculate the resistance of the bulb filament when the current through it was 0.25 A. [4] (AQA)

8.

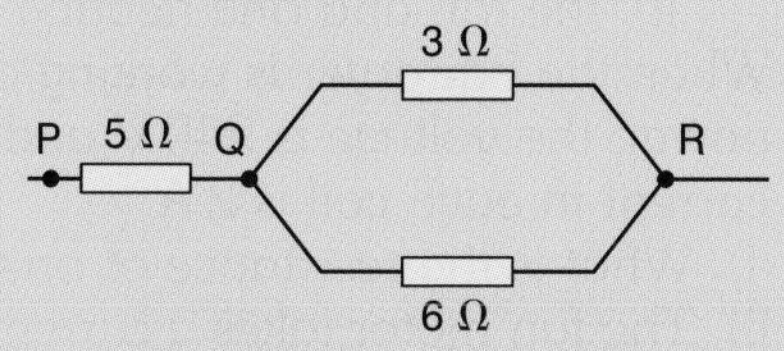

The p.d. across PR is 14 V. The equivalent resistance of network QR is 2 Ω. Calculate:

a) the resistance of network PR,

b) the current flowing through PQ,

c) the potential difference across PQ,

d) the current through the 3 ohm resistor.

9. An experiment was set up to investigate the heating effect of an electric current.
A constant current I was passed through a coil of constant resistance immersed in water.
T, the temperature rise per minute, was measured for various values of current I.
The values obtained are given in the table.

I (in A)	0	1.00	2.00	2.50	3.00	3.50	4.00
T (in K/min)	0	0.06	0.26	0.40	0.57	0.74	0.93

a) Draw up a table of values I^2 and T. [2]

b) Draw a graph of T (y-axis) against I^2 (x-axis). [3]

c) Determine the slope of the straight line portion of the graph. [2]

d) Determine the slope of the graph at a rate of temperature rise of 0.8 K/min. [2]

e) Suggest a reason for the change in the slope of the graph. [2]

10. a) For a particular specimen of wire, a series of readings of the current through the wire for different potential differences across it is taken and plotted as shown. Describe how the resistance is changing. Suggest the most likely explanation. [4]

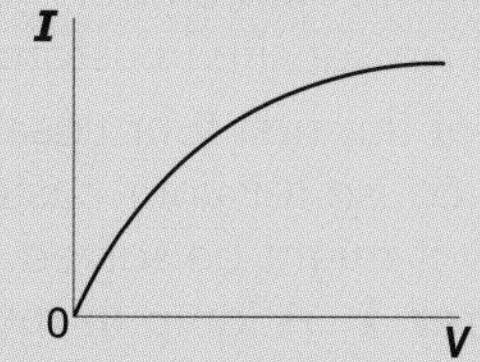

b) How would the resistance of a piece of wire change if

i) the length were doubled? [2]

ii) the diameter were doubled? [2] (AQA)

11.

In this circuit, ammeter A_1 reads 2 A and the voltmeter reads 12 V. Switch S is open.

a) What is the reading of ammeter A_2? [1]

b) Calculate the resistance of the lamp. [3]

c) Calculate the power of the lamp. (Use $P = VI$) [2]

Switch S is now closed.

d) What happens to the readings of:

i) Ammeter A_1? [1] ii) ammeter A_2? [1]

e) Explain your answers to d). [2] (OCR)

12. The diagram shows a circuit containing a lamp L, a voltmeter and an ammeter.

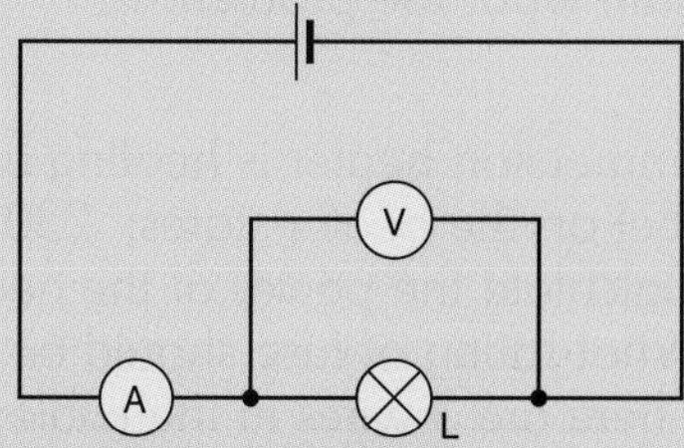

The voltmeter reading is 3 V and the ammeter reading is 0.5 A.

a) What is the resistance of the lamp? [3]

b) What is the power of the lamp? [3]

(Edex)

▷ Electric power

13. Imagine that you are an electrician and that the following problems and questions are brought to you by some of your customers. In each case describe how you would answer them. Remember, your business depends on satisfied customers.

In each of the first four cases, explain a possible cause for your customer's trouble and how it might be solved.

Customer 1 'A lamp in my house has gone out.' [2]

Customer 2 'The lights downstairs have all gone out.' [2]

Customer 3 'All the lights in the house have gone out.' [2]

Customer 4 'All the lights in my street have gone out.' [2]

The next customer would like a clear wiring diagram.

Customer 5 'I would like two lamps in the dining room both worked from the same switch.' Draw a suitable wiring diagram. [4]

The next customer is quite bewildered by the problem and you must give this customer a clear, full explanation of how you make the calculation.

Customer 6 'How much a week will it cost me to run my 1 kW electric fire and 2 kW electric radiator if they are both used for 4 hours each evening? (1 kW h costs 5p) [4]

The last customer does not even know there is a problem!

Customer 7 'I should like you to come and fit a mains socket in the bathroom so that I can have the electric fire on when I have a bath.' Explain what the problem is. [3]

(AQA)

14. An immersion heater is heating a water tank. A label on the heater states: 230 V 10A

a) Calculate the power of the heater. [2]

b) What rating of fuse should be used? [1]

c) There are 3 wires in the cable; two go to the heater, the third is connected to the copper tank. Explain the function of this third wire and the fuse in the circuit. [3]

d) Calculate the resistance of the heater. [2]

e) Why is the hot-water outlet at the top? [1]

15. The diagram shows a 240 V electric hairdryer with a plastic case.

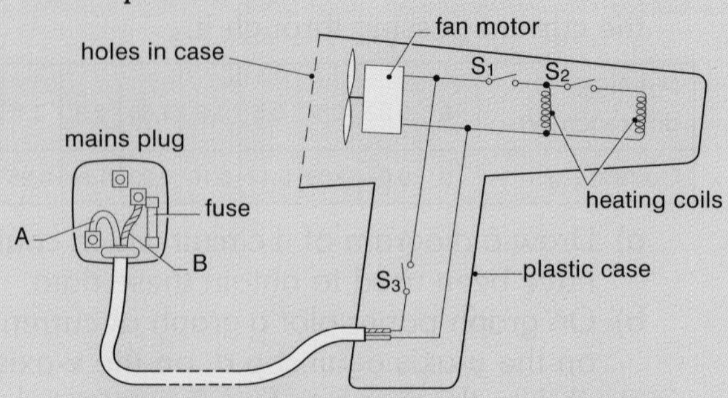

a) i) Write down the colours the cables in the plug should have.

ii) One pin in the plug is not being used. Does this make the dryer dangerous?

b) Which switches need to be closed to put on: i) the fan alone?

ii) the fan and one heating coil?

c) When the hairdryer is working at full power, the voltage is 240 V and the current in each coil is 2 A.

i) What is the resistance of one coil?

ii) The fan motor takes a current of 0.5 A. What is the total current from the supply when both coils are in use?

iii) What fuse is needed in the plug? Choose from: 1 A 3 A 7 A 10 A

iv) Suppose you find a 13 A fuse in the hairdryer! Explain why this might be dangerous.

(AQA)

16. Some information about two electric lawn-mowers – 'Supatrim' and 'Powermo' – is shown:

	Supatrim	Powermo
Cost	£60	£80
Power rating	500 W	1000 W
Time to mow lawn	30 min	20 min
Recommended fuse	3 A	13 A
Width of cut	30 cm	50 cm

The cost of electrical energy is 6p/kW h.

a) Explain the meaning of 'kW h'. [1]

b) Calculate the cost of the electrical energy used when mowing the lawn with i) the Supatrim, ii) the Powermo. [2]

c) Explain why a different fuse is recommended for each of the lawnmowers. [1]

d) Give ***two*** reasons why a person might choose to buy the Powermo rather than the Supatrim. [2] (Edex)

17. When a current of 4.00 A passes through a certain resistor for 10 minutes, 28 800 J of heat are produced. Calculate:
a) the power of the heater, [2]
b) the voltage across the resistor. [3] (WJEC)

18. An electric heater is rated at 230 V, 2300 W.
a) Calculate the working current. [3]
b) What is the correct fuse rating? [1]
c) Explain the purpose of a fuse. [3]
d) The heater runs for 20 hours. How much does it cost, if 1 kW h costs 8p? [3]

19. A car battery is used to light a 12 V lamp. A constant current of 3 A passes round the circuit.
a) Explain what happens to the energy of electrons as they flow in the lamp wire. [4]
b) How much energy is transferred by the lamp in 20 seconds? [3] (AQA)

20. A lamp is marked '3.0 V, 0.2 A'.
Writing down the formula that you use in each part and, showing your working, calculate
a) its resistance when it is working at its normal brightness, [3]
b) its power, [3]
c) the charge passing through it in 600 s, [3]
d) the energy it transfers in 10 minutes. [3]
(WJEC)

21. Shaun is cutting a large area of grass with an electric mower. He is worried that he might cut through the mains cable of the mower. He decides to use a residual current circuit breaker (RCCB) as well as a fuse.
a) Describe how the fuse gives protection. [2]
b) Explain how an RCCB provides additional protection. [2]
c) The table shows how different currents are likely to affect people when they receive an electric shock.

Current in mA	Effect on people
1	none
5	tingling effect
10	could become harmful
100	probably fatal

Shaun's RCCB switches off in just 0.02 s when it detects residual current of 30 mA. Use the data in both columns of the table to suggest how effectively this RCCB would protect Shaun. [2] (Edex)

▷ Electrostatics

22. When Darren takes his jumper off, it becomes negatively charged as it rubs against his shirt. This is because the jumper has
A lost electrons **B** gained electrons
C lost protons **D** gained protons (Edex)

23. An aircraft flies just below a negatively-charged thunder cloud as shown. Electrostatic charges are induced on the aircraft.

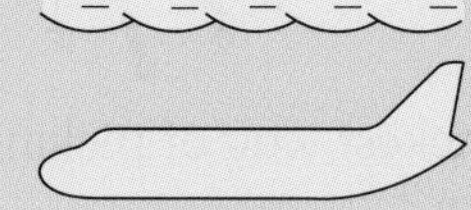

a) Copy it and show the positions and signs of the induced charges on the aircraft. [2]
b) Explain, in terms of the movement of electrons, the distribution of the charges you have shown. [3]
c) What would happen to the induced charges when the aircraft flies away from the cloud? [1] (Edex)

24. a)

A balloon, after rubbing with a woollen cloth, is found to be negatively charged and is attracted to the cloth. Explain:
i) why the balloon becomes negatively charged when it is rubbed, [2]
ii) why the balloon is attracted to the cloth. [1]

b) Name ***one*** practical use of static electricity. [1]
c) i) State ***one*** situation where static electricity is dangerous. [1]
ii) Explain what precautions can be taken to reduce the danger. [1] (WJEC)

25. Electrostatic paint spraying is the most efficient way of spraying car bodies.
a) Explain why the paint drops spread out after becoming negatively-charged. [1]
b) Explain why the car body panel is given an opposite (positive) charge. [1]
c) Suggest two reasons why this method of spraying is very efficient. [2]

chapter 34

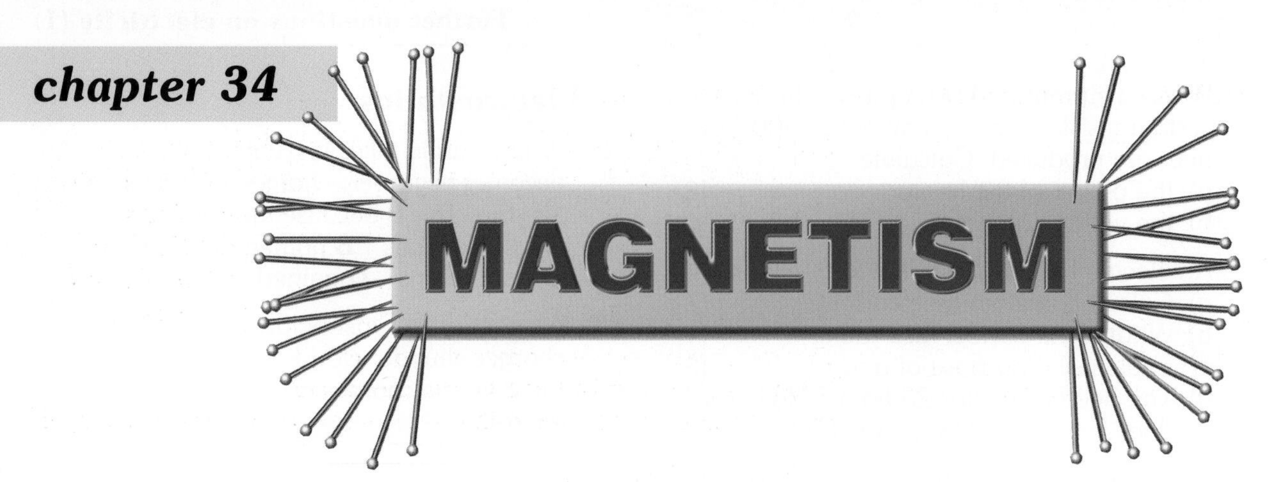

MAGNETISM

What is the quickest way of picking up some pins that have been spilt on the floor?

If a magnet is used, pins stick to the ends or ***poles*** of the magnet. Can you think of any other uses for magnets?

Experiment 34.1
Place a bar magnet in a paper stirrup as shown and hang it from a wooden table or a wooden retort stand.

Leave it hanging until it stops moving. From a map or from the position of the Sun, find out which direction is North.

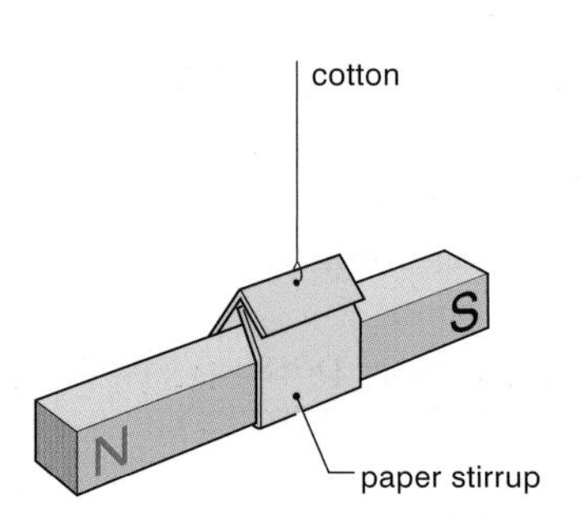

Which direction does the magnet point to?

You have made a **compass**. A freely moving magnet comes to rest pointing roughly North–South.
The end pointing North is called the ***North-seeking pole*** or ***North pole (N-pole)*** of the magnet.
The other end is the ***South-seeking pole (S-pole)***.

Experiment 34.2
Mark the N-pole of the magnet in experiment 34.1. Then take that magnet well away and repeat the experiment with another magnet.

Now bring the N-pole of the first magnet close to the N-pole of the hanging magnet. What happens?

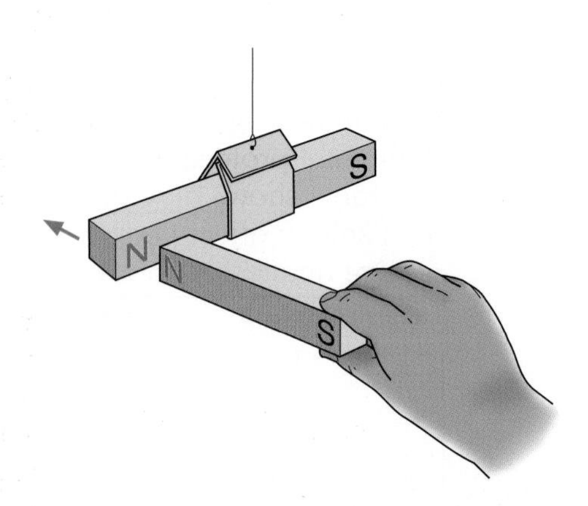

What happens if you bring the S-poles close together?
What happens if you bring a N-pole near a S-pole?

We find that: **Like poles repel each other.**
Unlike poles attract each other.

All physics students should learn well
This scientific fact:
Like magnetic poles repel
And unlike poles attract.

▶ Making a magnet

A. Stroking method

Experiment 34.3
Mark one end of a piece of unmagnetised steel. Check that it is not magnetised by dipping it into a dish of small iron nails.

Stroke the piece of steel with the N-pole of a magnet. Start at the marked end of the steel and lift the magnet clear at the end of each stroke. Do this about 10 times in the same direction.

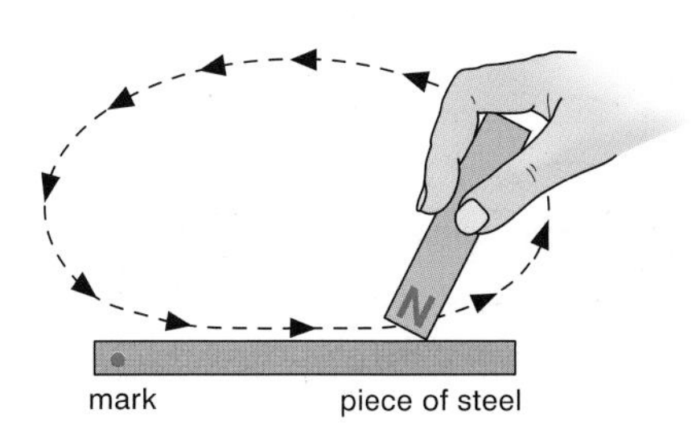

How many nails can the steel pick up now?
Do you find the marked end is now a N-pole?
(Use a small compass to test this, remembering that like poles repel.)

B. Electrical method

Experiment 34.4 An electromagnet
Wind about 1 metre of insulated copper wire round and round a large nail. Then connect the ends of the wire to a battery, as shown.

How many nails can it pick up?
This is an **electromagnet** (see also page 288).

Disconnect the battery and test the large nail again.
Is it magnetised less strongly than before?

Why are cranes in steelworks often fitted with electromagnets?

Destroying magnetism

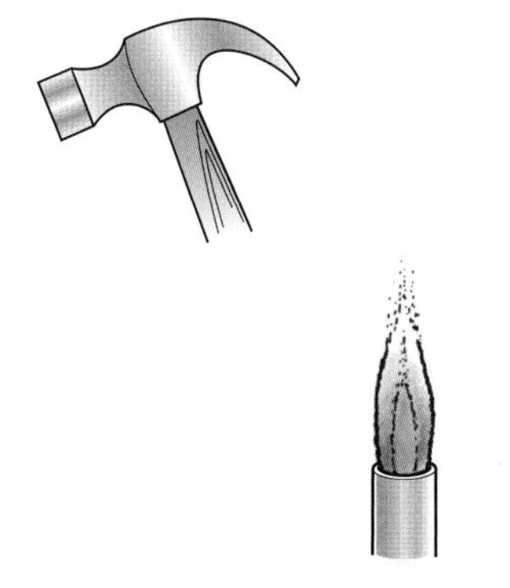

A. Hammering

Experiment 34.5 ⚠
Magnetise the piece of steel strongly (as in experiment 34.3) and then hammer it. Test it with nails.
Is it magnetised as strongly as before?

B. Heating

Experiment 34.6
Magnetise the piece of steel as before and then heat it in a Bunsen flame until it is red hot.
When it cools down, test it. Is it demagnetised?

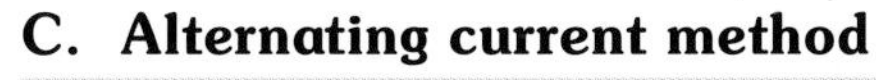

C. Alternating current method

Experiment 34.7
This can be done only if your teacher can provide you with a safe supply of alternating current.
DO NOT use mains voltage – it is dangerous! ⚠
Pass alternating current (see page 299) through the coil you used in experiment 34.4 and then slowly reduce the amount of current in the coil (or gradually pull the nail out of the coil).
Then test the nail. Is it completely demagnetised?

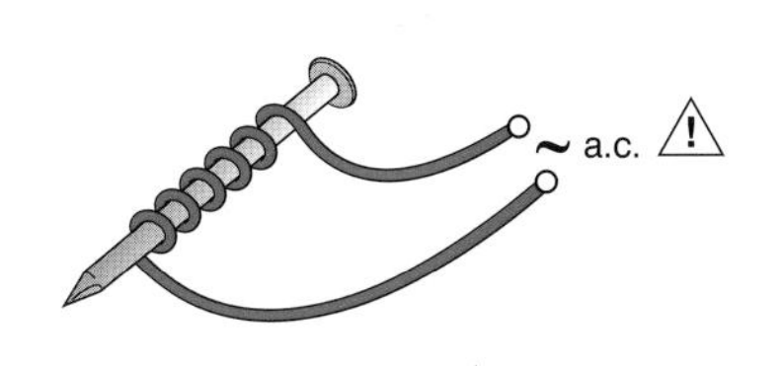

Which of these three methods would you use to demagnetise a watch?

▷ Magnetic fields

The region round a magnet where it has a magnetic effect is called its ***magnetic field***.

Experiment 34.8
Place some small plotting compasses near a weak bar magnet (preferably on an overhead projector):

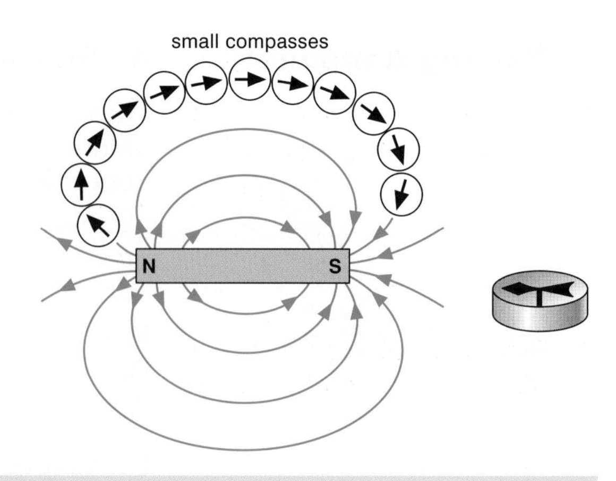

The compasses point along curved lines as shown in the diagram. These lines are called ***lines of flux*** (or ***lines of force***) and they show the direction of the magnetic field at each point.

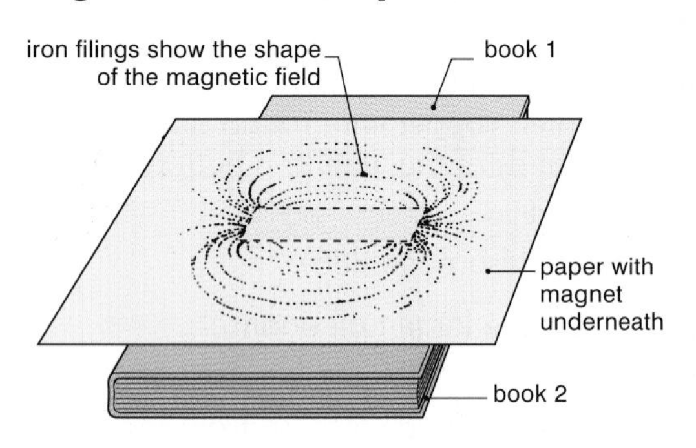

Experiment 34.9 ⚠
Place a sheet of paper over a bar magnet.

Sprinkle a thin layer of iron filings over the paper and then tap the paper gently. The iron filings act like thousands of tiny compasses and point themselves along the lines of flux.

Look carefully at the pattern that appears. Can you see that it is the same shape as in the diagram at the top of the page?

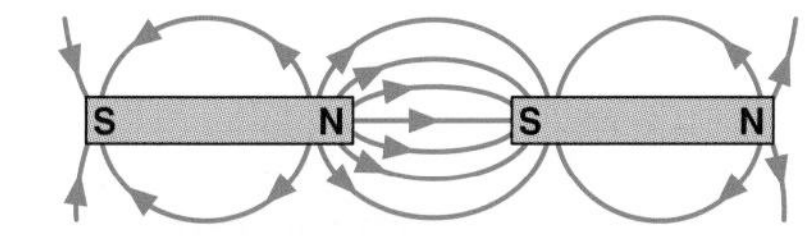

Experiment 34.10
Use the iron filings method to find the shape of the magnetic field round two magnets placed so that the N-pole of one is near the S-pole of the other.

Where do the iron filings gather most thickly? Where is the field strongest? Where is it most uniform?

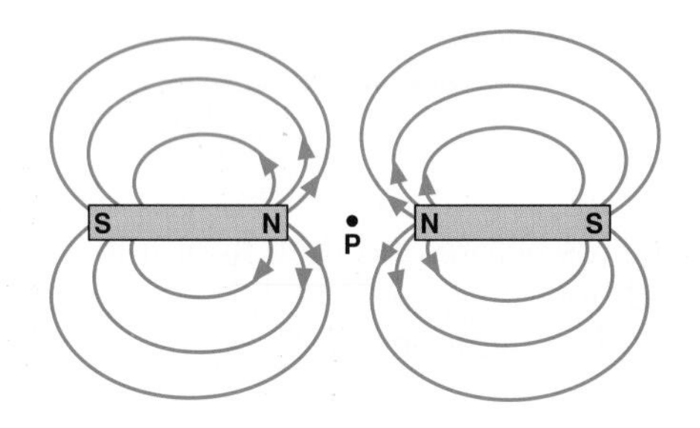

Experiment 34.11
Use the same method to find the shape of the field when a N-pole is placed near another N-pole.

At point P in the diagram, there is no magnetic field. P is called a ***neutral point***.

The magnetic field of the Earth

How does a compass show you that the Earth has a magnetic field?

In fact, the Earth acts as if there is a bar magnet inside it (although such a magnet cannot really exist because the centre of the Earth is too hot).

Notice in the diagram that the ***S-pole*** of the imaginary magnet is in the northern hemisphere so as to attract the N-pole of a compass.

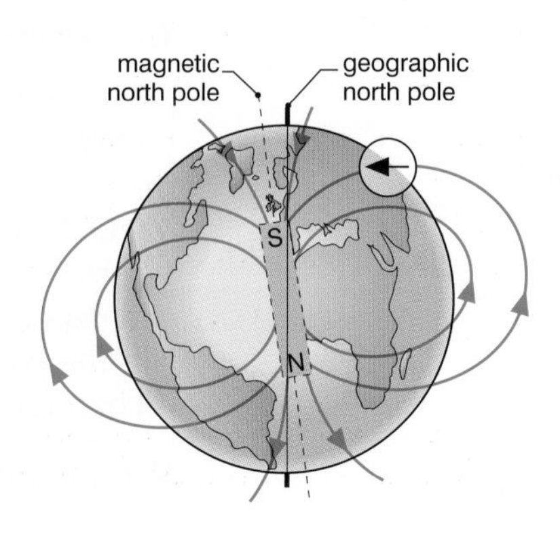

▷ Theory of magnetism

In iron and steel, there are atomic magnets which line up with each other in groups, called ***domains***.

In an unmagnetised piece of iron, the magnetic domains are pointing in all directions and so cancel out each other:

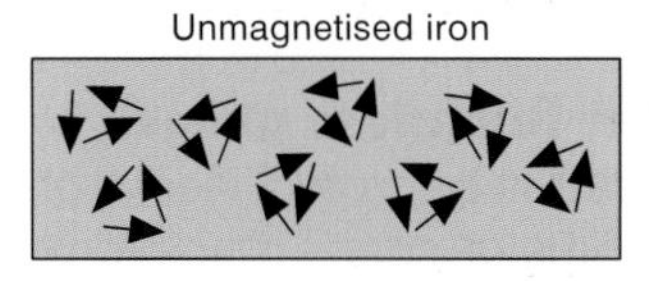

If the magnetic domains can be turned round to point the same way, then the piece of iron becomes magnetised. This is because all the tiny N-poles add up at one end and all the S-poles add up at the other end:

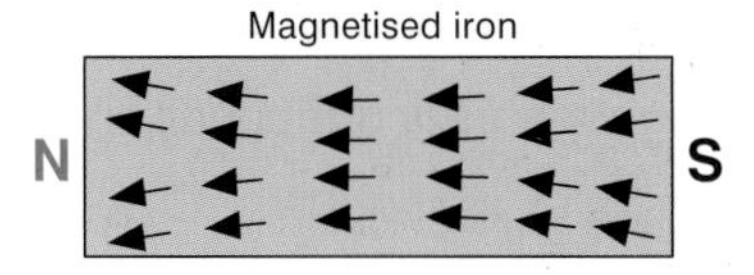

▷ Magnetic induction

Experiment 34.12
Hold an unmagnetised piece of iron near some nails. If it is unmagnetised, it will not attract the nails.

Now bring near a strong magnet as shown:

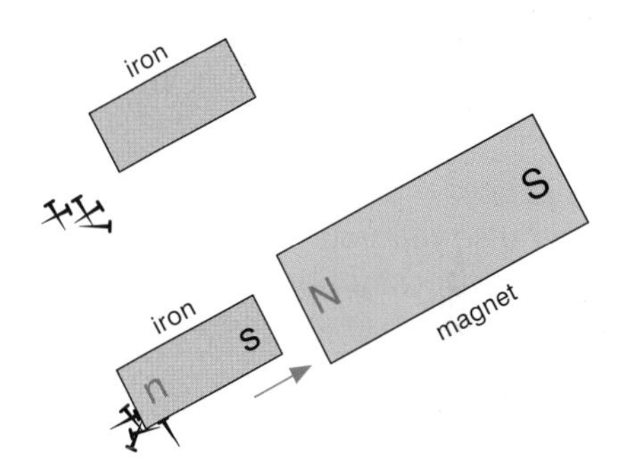

What happens to the nails?
This means that without being touched or stroked, the piece of iron has magnetism ***induced*** into it.

What do you think has happened to the magnetic domains in the piece of iron?

▷ Magnetic materials

Iron is called a 'soft' magnetic material because it is easy to magnetise and also loses its magnetism easily. Iron is used in electromagnets (page 288) and transformers (page 302). Iron is also used as the magnetic material in a reed switch (see page 320).

The diagram shows another kind of magnetic switch:

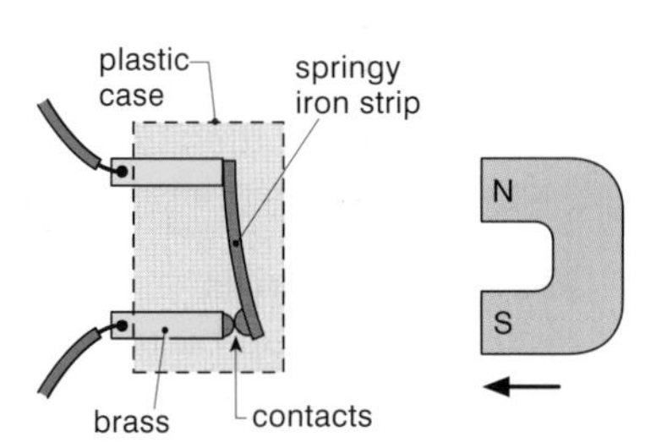

What happens if the magnet moves closer?
How could you use this switch, with a battery and an electric bell, to make a burglar alarm for your door?

Steel is called a 'hard' magnetic material because it is harder to magnetise and also does not lose its magnetism so easily. It is used to make permanent magnets.

Permanent magnets are used in compasses and as magnetic catches to keep cupboard doors shut.
Fridges and freezers have magnetic strips all round the door to keep the door completely closed and so keep the cold air inside.

A permanent magnet is placed in the oil sump tank of a car engine so that it collects any bits of steel that are worn off the engine:

A magnetic drain plug removes bits of steel worn off a car engine

Iron oxide is another 'hard' magnetic material. It is used to make computer discs, videotapes, and tape for tape-recorders (page 307).

▷ Physics at work: Magnets

Paint thickness gauge

If paint is to protect steel from rusting – for example, on ships or cars – it has to cover all the surface. The paint is usually about $\frac{1}{5}$ mm thick and this can be checked with a paint thickness gauge:

The gauge uses a magnet and a spring-balance.

Experiment

Hang a bar magnet from a spring-balance (using sellotape). Touch the magnet to a piece of steel and measure the force (in newtons) needed to pull it away. Then measure the force with one or more pieces of paper ('the paint') between the magnet and the steel. What do you find?

This is how a paint thickness gauge works. (It can also tell you if the steel has rusted away or been replaced by fibreglass.)

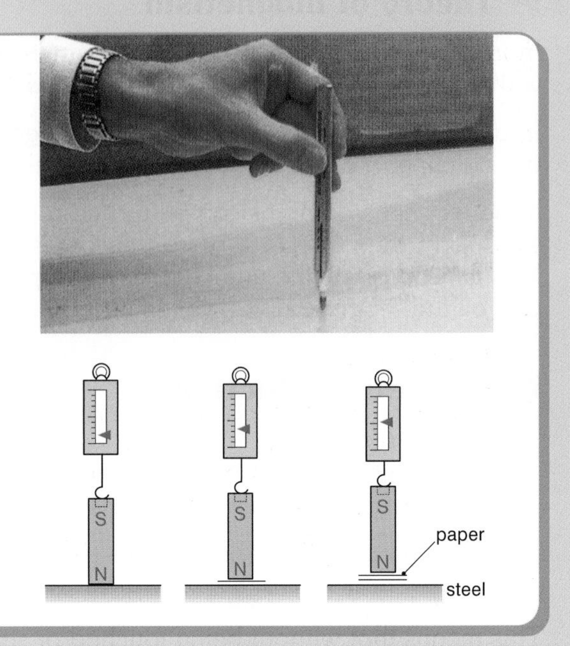

Magnetic inks and paints

It is possible to make magnetic inks by mixing very small particles of a magnetic substance with a liquid. The mixture (a dark brown colour) can be used as a paint to make magnetic tape for tape-recorders (see page 307), and hard-discs for computers.

Banks use magnetic ink on cheques so that the cheques can be sorted automatically by a machine which detects the magnetic field round each number:

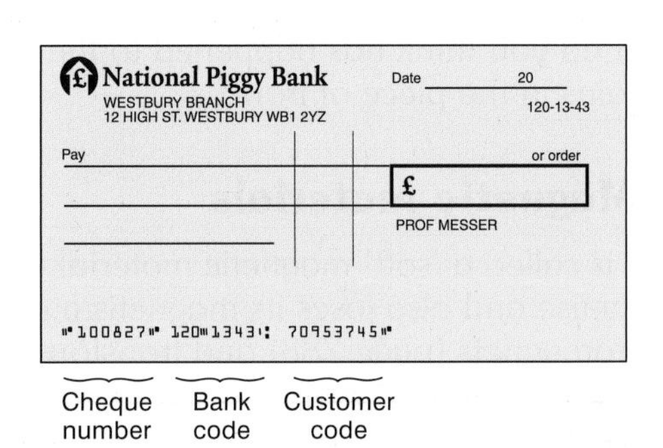

Automatic cash-cards also use magnetic ink to store information on the card.

Magnetic liquids

Magnetic ink can be made with oil so that it does not dry out. This magnetic liquid can be used to detect very small cracks in the surface of steel pipelines. These cracks can be very dangerous if left untreated.

The pipe is magnetised by a simple coil and painted with magnetic liquid:

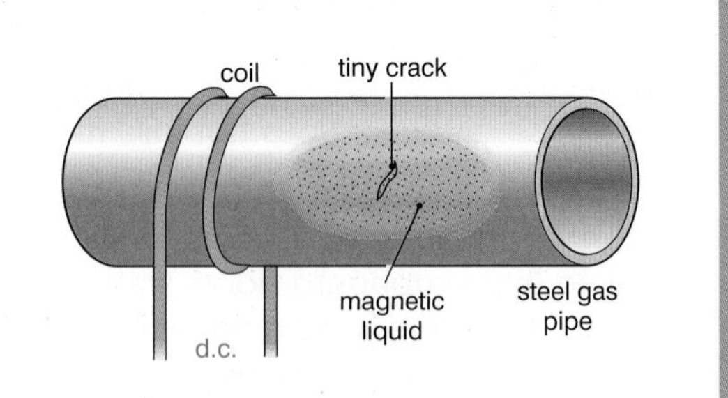

If there is a tiny crack, some of the magnetic lines of force leak out of the pipe at the crack. Then the magnetic liquid shows up the shape of the crack.

Magnetic oil is also used to lubricate the shafts of motors and the coils of loudspeakers. A magnet is used to keep the magnetic oil in the right place.

Summary

Like magnetic poles repel. Unlike poles attract.

Strong magnets can be demagnetised by
a) hammering b) heating
c) using a coil with alternating current.
Iron (unlike steel) is easily magnetised and easily demagnetised.

Magnets can be made by
a) the stroking method, or
b) the electrical method
(more details on page 287).

Lines of flux show the direction of the field round a magnet.

▶ Questions

1. Copy out and complete:
a) Like poles . . . . ; unlike poles . . . . .
b) A magnet can be made by . . . . it or by using a current in an . . . . .
c) Strong magnets can be demagnetised by
i) a coil carrying . . . . current,
ii) . . . . or
iii) . . . . .
d) The Earth's magnetic field is rather like that of a . . . . magnet with its . . . . pole in the northern hemisphere.
e) Iron is easily . . . . and easily . . . . .
Steel is used to make . . . . magnets.
f) If the N-pole of a magnet is brought near an unmagnetised nail, then magnetism is . . . . into the nail. The nail is then . . . . to the magnet.

2. Explain the following:
a) Magnets are often fitted to the doors of refrigerators and cupboards.
b) A crane in a steelworks is fitted with a large electromagnet (see page 288).
c) Flour is usually passed near a magnet before being packed.
d) A steel ship becomes magnetised as it is constructed.
e) The sump plug on a car is usually magnetised.
f) A magnetised screwdriver can help to place screws in holes that are hard to reach.

3. You are given a small plotting compass and three grey metal bars. One is aluminium, one is unmagnetised iron, one is a magnet. How can you identify each one?

4. Use the idea of magnetic domains to explain:
a) If a magnet is broken in half, each half is a magnet.
b) Heating or hammering destroys magnetism.
c) Magnetism is felt strongly at the poles but not at the centre of a magnet.
d) There is a limit to how strongly an iron bar can be magnetised (***saturation***).
e) Magnetism is induced into an iron bar placed near a magnet.

5. Choose a material, giving a reason, for each of the following: a) the core of an electromagnet, b) a compass needle, c) the case of a compass.

6. Draw diagrams showing the magnetic fields round a) a single bar magnet, b) two magnets with their unlike poles close together, c) two magnets with their N-poles close together, d) the Earth.

7. Professor Messer says that 'like poles must attract' because the North pole of the Earth attracts the N-pole of a compass. Explain why he is wrong.

8. *Look, here's a new compass the Prof. has designed. But how many silly mistakes can you find? (8?)*

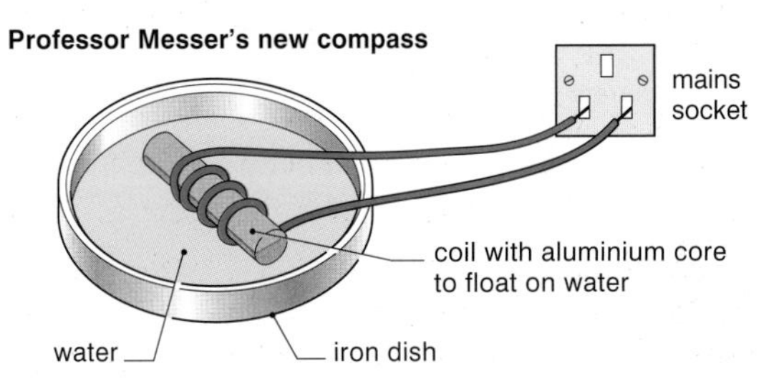

chapter 35

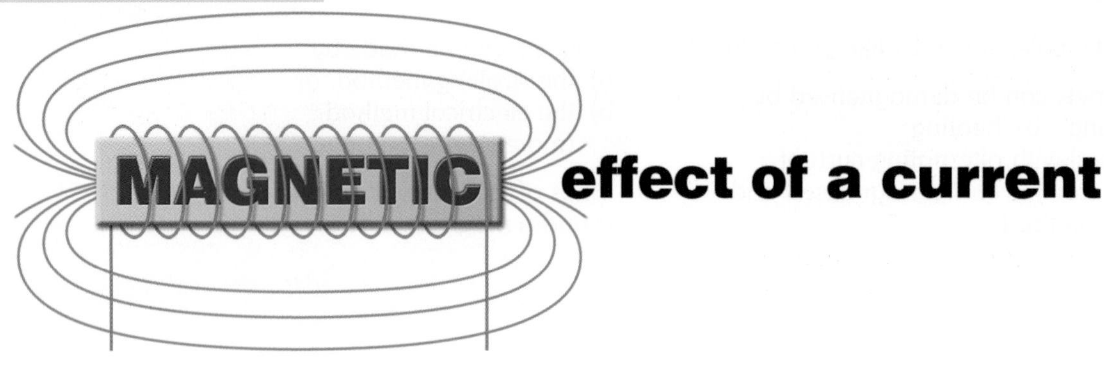

effect of a current

In 1819, Hans Christian Oersted discovered that electricity has a ***magnetic*** effect.

Experiment 35.1 Oersted's experiment
Let a compass settle in a North–South direction. Then, with the circuit shown, hold the wire over the compass and press the switch (for a few seconds only).

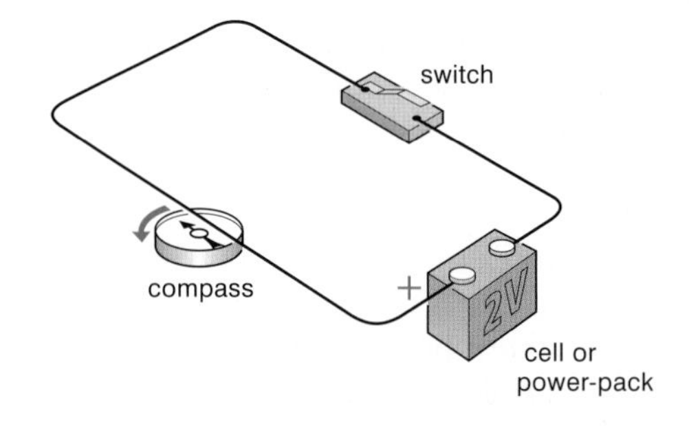

What happens to the compass needle when the current flows? Can you see it move?
What happens to the deflection of the compass if you reverse the direction of the current?

We find that **a wire carrying a current has a magnetic field round it.**

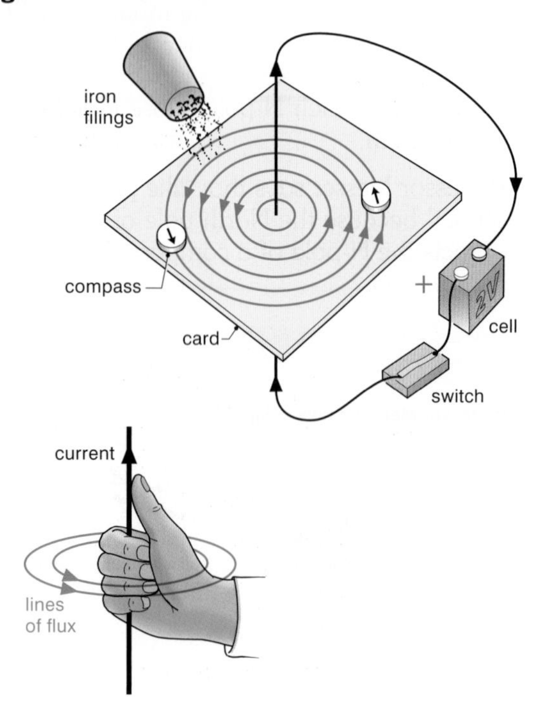

Experiment 35.2 Finding the shape of the field
Using the apparatus shown, sprinkle some iron filings on to the horizontal card.
Pass a large current through the vertical wire (briefly), while tapping the card so that the iron filings show the shape of the magnetic field.

Can you see that **the magnetic field round the wire is in the shape of circles?**

Place a small compass on the card to find the direction of the magnetic field lines.

Does your result agree with the diagram?
What happens if you ***reverse*** the current?

The diagram shows how to remember this result using the ***Right-hand Grip Rule***: imagine gripping the wire, with your right thumb pointing the same way as the current … then your fingers are curling the same way as the magnetic field lines.

▷ The magnetic field near a coil

Experiment 35.3 A short coil
Sprinkle iron filings round a coil as in the diagram. Pass a large current and tap the card.

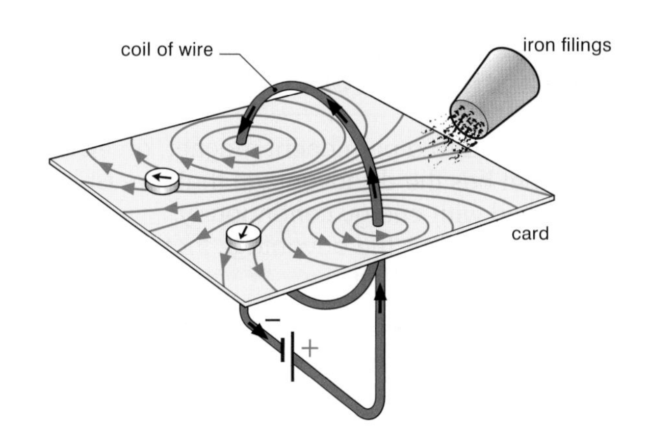

Study the shape of the magnetic field:
Use a small compass to find the direction of the field.
Does the direction of the field round each wire agree with the result of experiment 35.2?

Experiment 35.4 A long coil (or 'solenoid')
Repeat the experiment but with a long coil or **solenoid**.

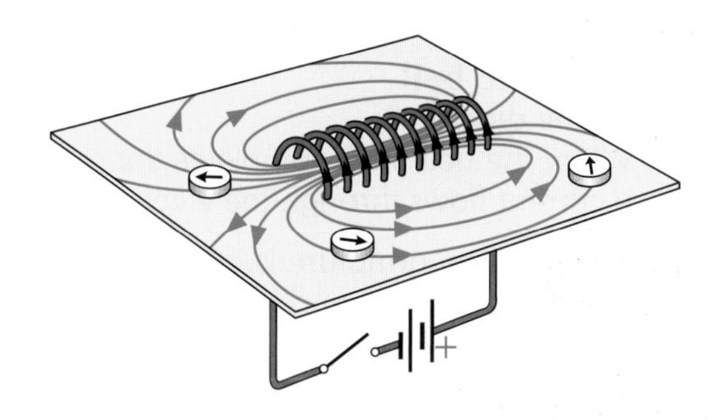

Study the shape of the field. Does it look familiar? (Look at page 282 if you are not sure.)

The magnetic field round a solenoid has the same shape as the field round a bar magnet.

Notice that the field ***inside*** the solenoid is very strong and uniform. It can be used to magnetise objects (see also page 281).

Use a small compass to find which end of the solenoid behaves like a N-pole.
Now look at this N-pole end of the solenoid. Which way is the current flowing – clockwise or anticlockwise?

Now do the same for the S-pole end of the solenoid. What do you find?

A rule for the polarity of a coil
The diagram shows an easy way to remember this result. Use this rule to check the diagrams above.

Experiment 35.5 Adding an iron core
Using the same apparatus, move a small compass away from the coil until its direction shows that it is only just affected by the magnetic field. Now slide an iron bar into the solenoid. What is the effect on the compass?

Why does an iron core make a stronger field?

The strength of the magnetic field can be ***in***creased by:
- using a larger current,
- using more turns of wire on the coil,
- using a 'soft' iron core,
- bringing the poles together, as shown:

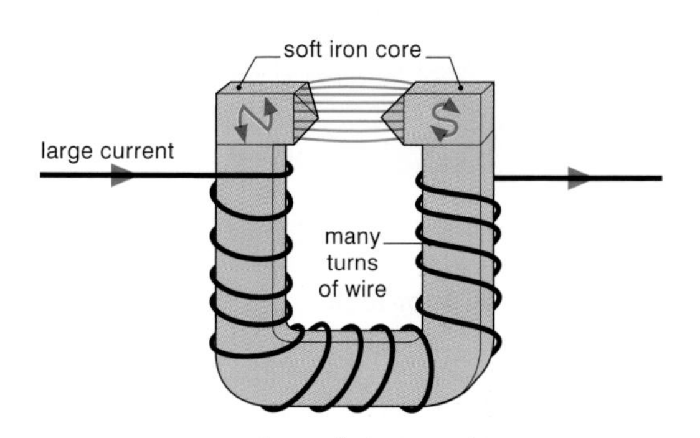

a strong electromagnet

▷ Physics at work: Uses of electromagnets

Compared to a permanent magnet, an electromagnet has the disadvantage that it must be supplied with energy continuously.
However, it has the advantage that it can be switched on and off.

1. Electromagnets are used on cranes in steelworks and scrapyards (see also page 240).

2. In hospitals dealing with eye injuries, electromagnets are used to remove steel splinters.

3. **The electric bell**
   Study the diagram:
   a) When the bell-push completes the circuit, a current flows through the electromagnet.
   b) The soft-iron armature is attracted toward the electromagnet and the hammer hits the gong.
   c) This movement breaks the circuit at point X, so that the current stops flowing and switches off the electromagnet.
   d) The spring pulls the armature back so that contact is made and the sequence begins again.

   The 'make-and-break' contact screw at X is adjustable to give the best sound. If the gong is removed, the bell becomes a buzzer.

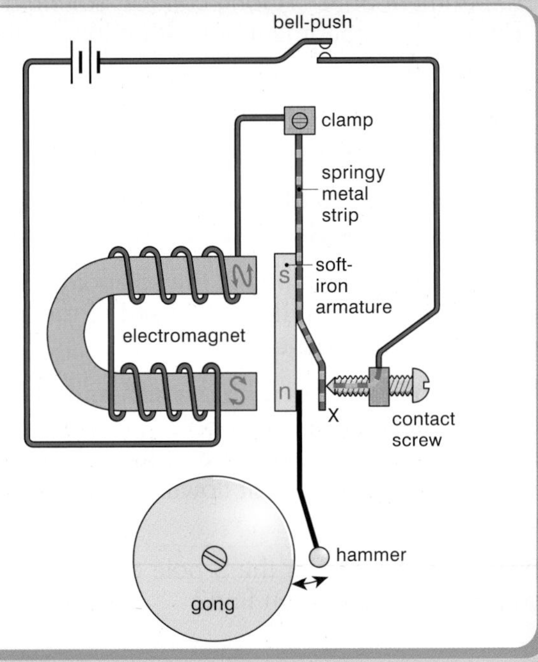

wall
solenoid
top view of the door
iron
to the operator

A door lock

4. **An electromagnetic door lock**
   The diagram shows a top view of a security door, perhaps in a bank. There is an iron bar keeping the door locked.
   If the security officer agrees to let you in, he presses a switch, so that a current flows through the solenoid, which becomes magnetised.

   What happens to the iron bar?
   What are the springs for?

   This idea is often used in factories to operate valves and other machinery by remote control.

5. **A circuit-breaker** (p. 272) can use a similar idea.

▷ Physics at work: Uses of electromagnets

6. The electromagnetic relay

A relay is a magnetic switch, used so that a ***small*** current can switch on a large current.

For example, in a car, the starter motor needs a very large current (100 A or more), which should be carried in thick wires kept as short as possible. A relay is used so that this large current is switched on by a smaller current in the driver's key switch inside the car.

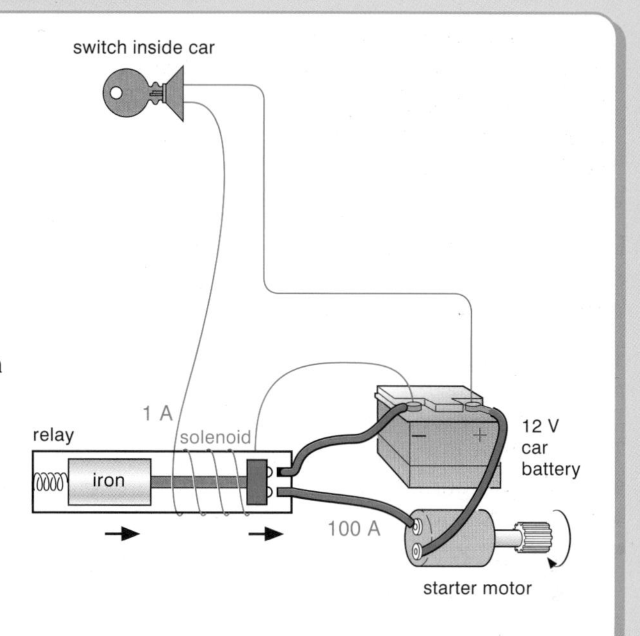

In the diagram, a small current in the long thin wires to the key switch on the dashboard is used to energise the solenoid.
The solenoid attracts a soft-iron armature, which moves into the coil and so completes the motor circuit. What is the spring for?

Relays are often used in electronic circuits, to control a large current (see pages 320–321).

A computer can use a small current in a relay to switch on and control a large machine.

7. The telephone earpiece (or radio earphone)

This converts varying electric currents into sound waves. The varying electric currents pass round the coils of an electromagnet, which attracts an iron disc:

As the currents vary, the movement of the disc varies. This makes a sound wave in the air (see also page 224).

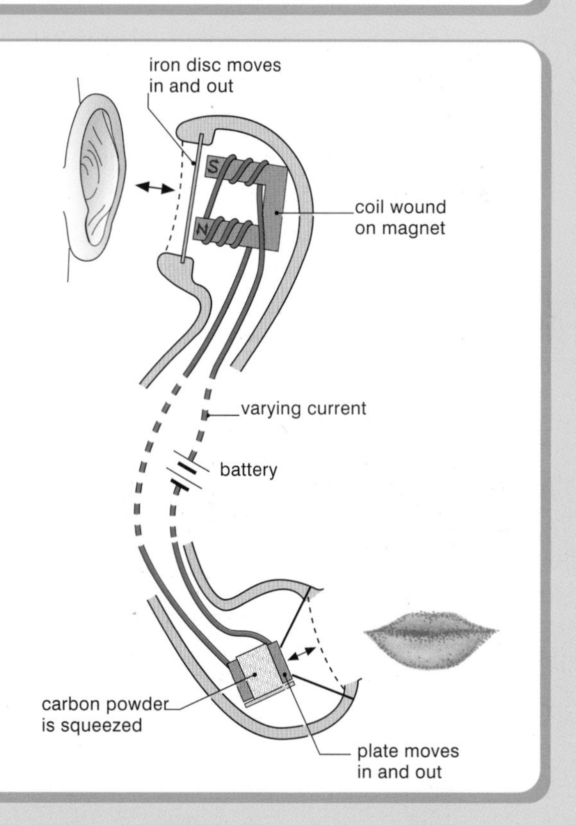

The varying electric currents can be produced by a **carbon microphone** in the mouthpiece of another telephone.
The sound waves force a plate in and out, so that it squeezes some carbon powder. When the carbon powder is squeezed, its resistance decreases and so the current from the battery varies in the same way that the sound wave varies.

▶ The motor effect

Experiment 35.6
Hang a flexible wire between the poles of a powerful magnet.
Connect the wire to a cell and a switch as shown:

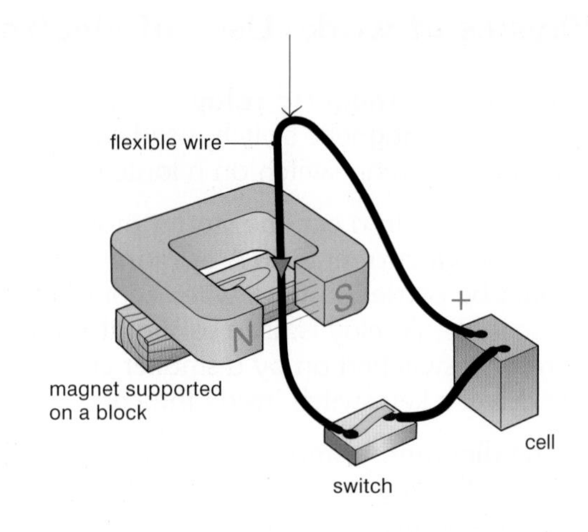

What happens when you press the switch briefly?

Does the movement reverse if you reverse the direction of the current?
What happens if you turn over the magnet?

This movement is put to practical use in electric motors (page 298).

The force on the wire can be increased by:
- using a larger current,
- using a stronger magnetic field,
- using a greater length of wire in the field.

The wire moves because the magnetic field of the permanent magnet reacts with the magnetic field of the current in the wire.

The diagram shows the combined field of the magnet and wire. The wire tends to be 'catapulted' outwards.

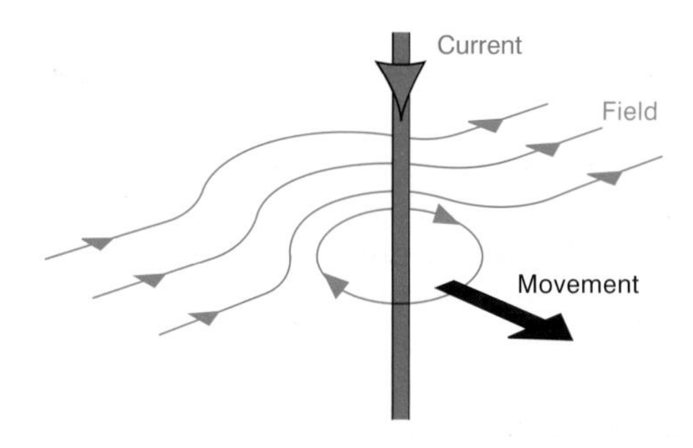

To remember the direction of movement we use **Fleming's Left-hand Rule**:
Hold your ***left*** hand in a fist and then spread out the thumb, first finger and second finger so that they are at right-angles to each other.

– Point your **F**irst finger in the direction of the magnetic **F**ield (from N to S).

– Rotate your hand about that finger until your se**C**ond finger points in the direction of the **C**urrent (conventional current, from + to –).

– Then your thu**M**b points in the direction of the **M**ovement of the wire.

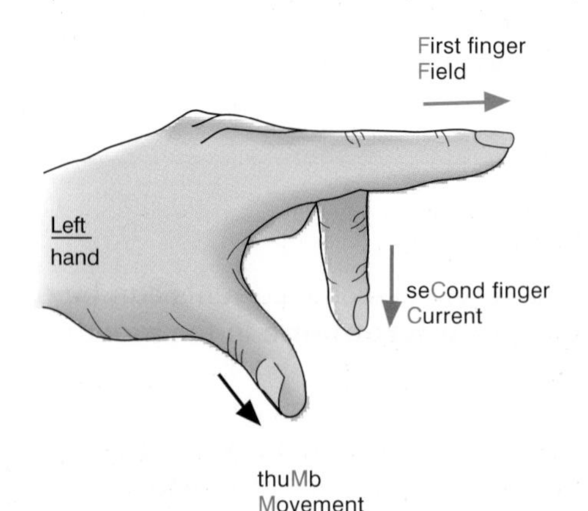

Check this rule with the movement in your experiment.

The moving-coil loudspeaker

A loudspeaker is used in your radio, TV, and in hi-fi systems. It converts electrical energy to sound energy. It changes electrical vibrations into vibrations of the air molecules.

Floor loudspeakers help them to monitor the sound

It contains a movable coil attached to a large cone. The coil fits loosely over the centre of a cylindrical permanent magnet so that the coil is in a strong magnetic field.

If a current flows in the direction shown, the coil will move to the right (use Fleming's Left-hand Rule to check this). If the current reverses, the coil moves the opposite way.

As the current (from an amplifier) varies rapidly, the coil and the cone vibrate rapidly.
As the cone vibrates out and in, it produces the compressions and rarefactions of a sound wave (see page 225).
If the current varies at a frequency of 1000 hertz, then you will hear a note of frequency 1000 hertz.

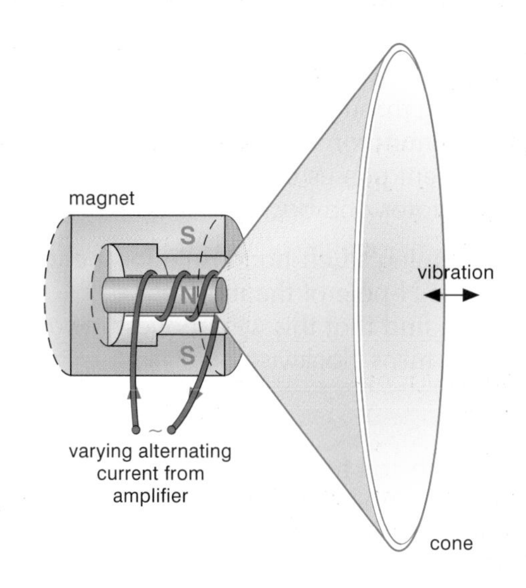

The same apparatus can be used in reverse as a moving-coil microphone (see page 301).

A coil pivoted in a magnetic field

> *Experiment 35.7*
> Wind several turns of wire round a cork and hang it, from a clamp, between the poles of a magnet as shown. Pass a current briefly through the coil.

What happens? Does the coil turn the opposite way if you reverse the current?

This twisting movement is used in electric **motors** and in moving-coil ammeters and voltmeters.

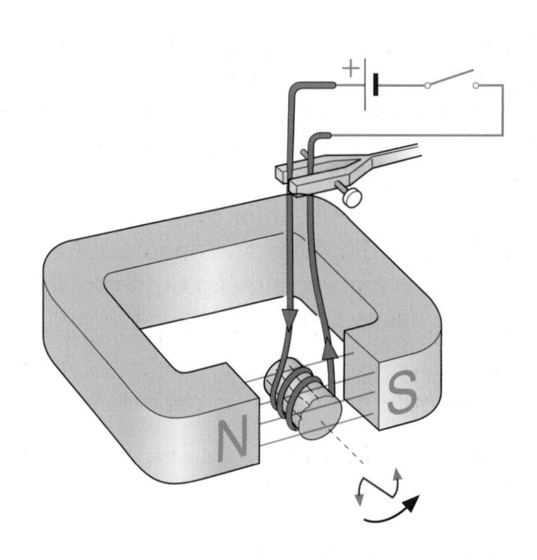

There are two ways to find the direction of rotation in the diagram:
a) Use Fleming's Left-hand Rule for each side of the coil.
b) Find which end of the coil becomes a N-pole. This end will be attracted to the S-pole of the magnet.

Apply both these methods to the diagram and check that they give the same result.

▶ The simple electric motor

The coil in the last experiment turns so that its N-pole moves to the S-pole of the magnet but then stops.

To keep the coil turning round toward the other pole as in a motor, we have to reverse the current in the coil at just the right moment.
In a direct current motor, the current is reversed every half turn by a ***commutator*** which looks like a copper ring cut into two halves:

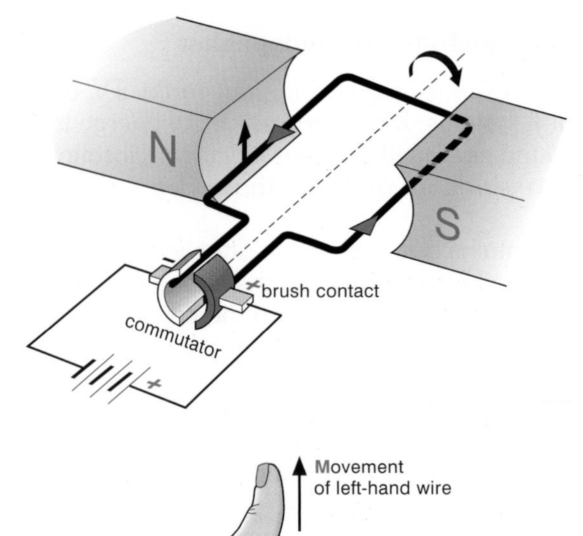

The diagram shows the current from the battery travelling through a brush contact to one half of the commutator. After travelling round the coil, the current passes through the other half of the commutator and back to the battery.

Use Fleming's Left-hand Rule for the wire that is near the N-pole of the magnet.
Do you find that this wire moves upwards, so that the coil turns clockwise?

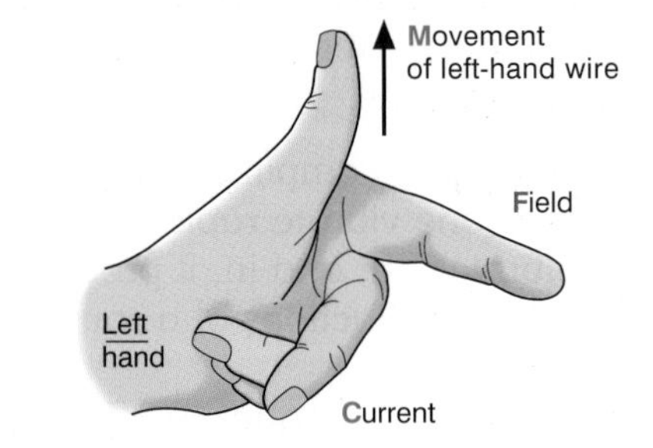

When the coil turns through 90° (so that the coil is vertical), the current stops flowing because the gaps in the commutator break the circuit.
However, the coil keeps turning because of its own momentum.

When the brushes make contact again, both the commutator and the coil have turned over so that the wire which is ***now*** nearest the N-pole has current coming out towards us.
This means that the force on it is upwards and so the coil still turns clockwise.

That is, the commutator makes sure that whichever wire in the diagram is nearest the N-pole, it always has current moving out toward us and so the wire keeps turning clockwise.

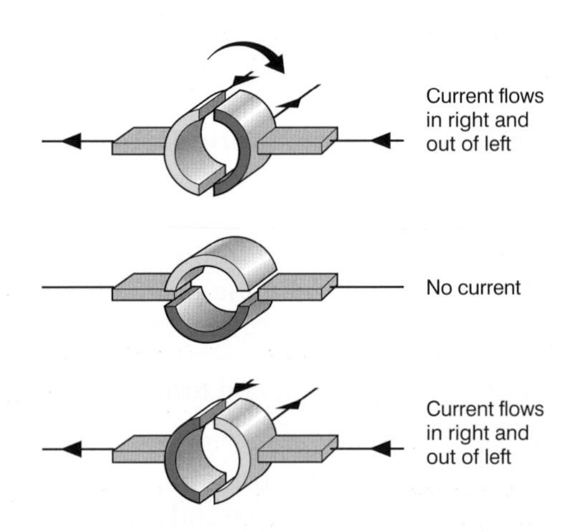

What would happen if you reversed the battery?

Of course, in practice the coil would have more than one turn of wire.

The brushes are usually made of carbon and held against the commutator by springs.

A designer of lifts, L. E. Vator,
Built a motor with no commutator,
It turned, with a wheeze,
Through just 90 degrees,
And trapped, between floors, its creator.

Experiment 35.8 Building an electric motor

a) Push a short knitting needle through a large cork.

b) Push 2 pins in one end of the cork as shown, to make a simple commutator.

c) Now wind round the cork about 30 turns of thin insulated copper wire, starting at one pin and finishing at the other.
It helps if you can cut channels in the cork to take the wire.

d) Scrape the insulation off the ends of the wire (using emery paper or a knife).
Wrap each end of the wire round a pin, making good electrical contact.

e) Push 2 pairs of long pins into a baseboard to support the axle at each end.

f) Strip the insulation from the ends of 2 wires and use drawing pins to hold them in position so that they just touch the commutator.

g) Use plasticine to support magnets on each side of the coil (with opposite poles facing).

h) Connect the wires to a 2 V or 4 V supply and give the coil a flick to start it.

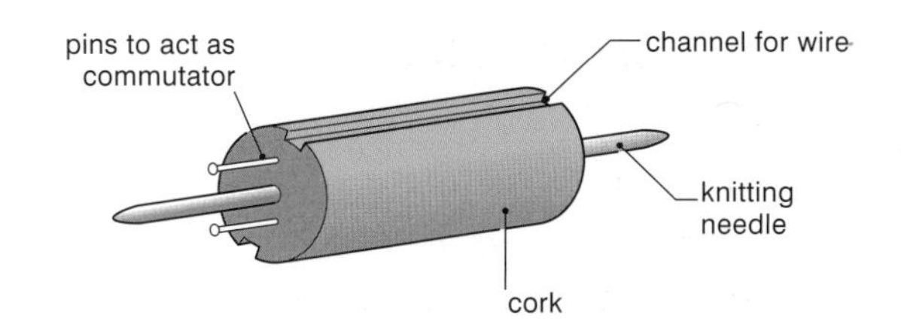

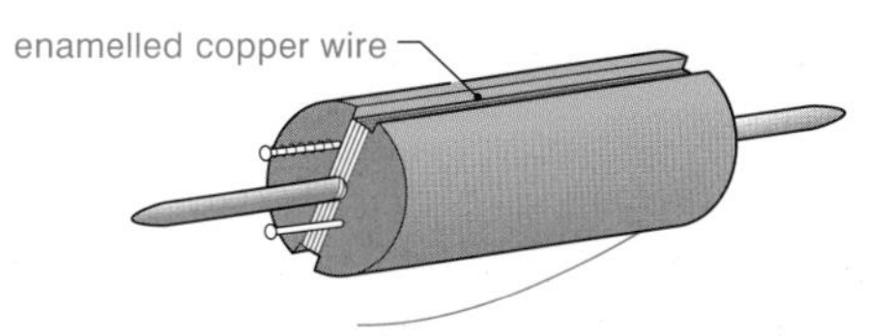

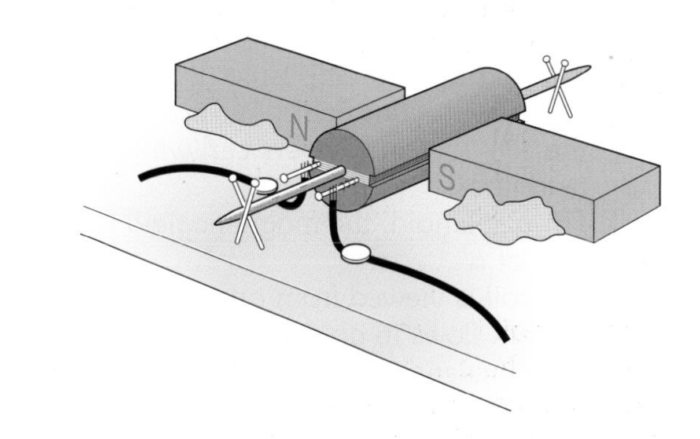

How could you improve your motor?

Practical electric motors

These can be made more powerful in 4 ways:

1. A large number of turns are wound on the coil.
2. A soft-iron core is used (instead of cork) so that the magnetic field is stronger.
3. The permanent magnets can be replaced by electromagnets which give a stronger field. (This also allows the motor to work from an a.c. supply as well as a d.c. supply, because when the current reverses, ***both*** the magnet's field and the coil current reverse at the same time.)
4. The motor will be more powerful (and run more smoothly) if extra coils are added round the core. This means that the commutator must be split into more than two parts: 4 coils would need a commutator with 8 segments.

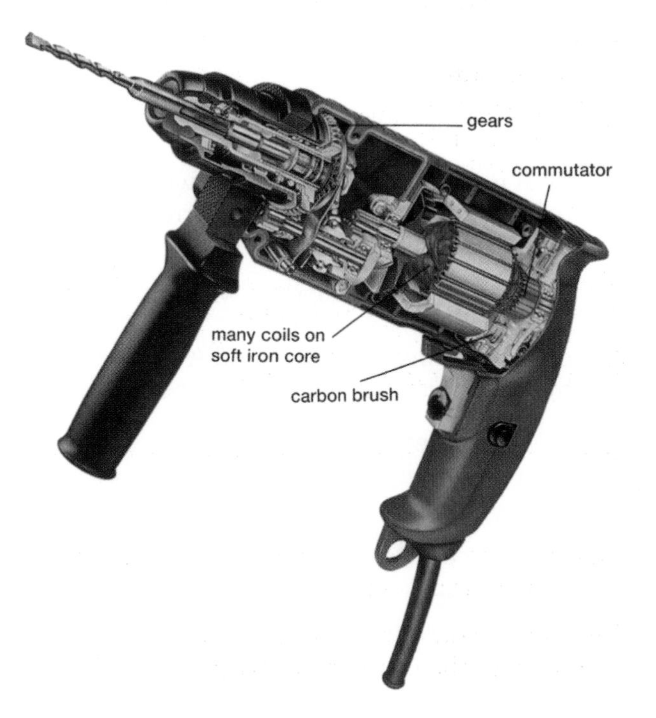

Summary

An electric current produces a magnetic field.

For a straight wire, the lines of force are circular.
Right-hand Grip Rule:
Right thumb pointing with the current, your fingers curl with the magnetic field lines.

For a coil, the lines of force are like those of a bar magnet. The poles can be found by the N–S rule.

Electromagnets are used in electric bells, relays.

If a wire carrying a current is placed across a magnetic field, there is a force on the wire.
Fleming's Left-hand Rule:
First finger for **F**ield (N to S)
se**C**ond finger for **C**urrent (+ to –)
thu**M**b then gives **M**ovement.

A coil carrying a current in a magnetic field tends to twist. This is used in an electric motor.

▶ Questions

1. a) The lines of . . . . round a straight current-carrying conductor are in the shape of . . . . .
b) The Right-hand Grip Rule: if you grip the . . . . . with your right . . . . pointing with the . . . . . , your fingers curl with the . . . . field lines.
c) For a solenoid, the magnetic field is like that of a . . . . .
d) If a coil is viewed from one end and the current flows in an anticlockwise direction, then this end is a . . . . pole.
e) The magnetic effect of a coil can be increased by increasing the number of . . . . , increasing the . . . . , or inserting an . . . . core.

2. a) Fleming's Rule for the motor effect uses the . . . . hand.
b) A motor contains a kind of switch called a . . . . which reverses the current every half . . . . .

3. For the coil in the diagram, when the switch is pressed:
a) What is the polarity of end A?
b) Which way will the compass point then?

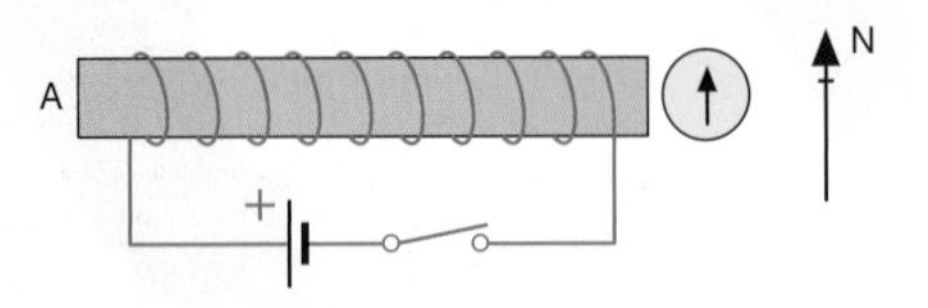

4. Draw a circuit diagram to show how a small current through a relay can be used to switch a large current. How is this used in cars?

5. a) Draw a large circuit diagram of an electric bell and explain, step by step, how it works.
b) Draw a wiring diagram to show how one electric bell can be rung by switches at both the front and back doors, using only one battery.

6. The diagram shows a relay circuit for a burglar alarm. Explain, step by step, how it works. In what way is it 'fail-safe'?

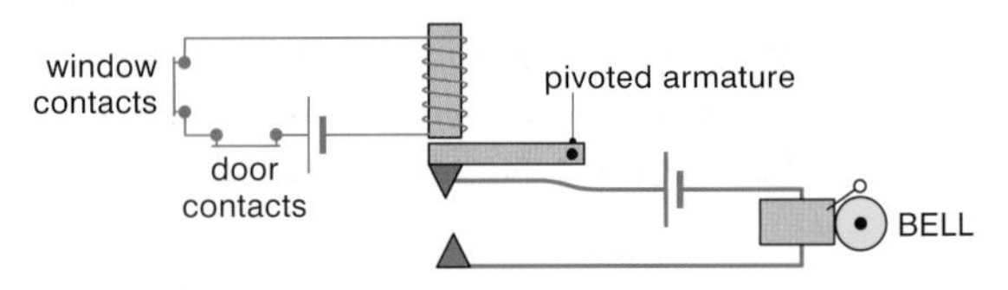

7. The diagram shows a design for an electrically operated model railway signal.
a) Explain, step by step, how it works.
b) Explain how, by adding a scale and a spring, you could use it as an ammeter.

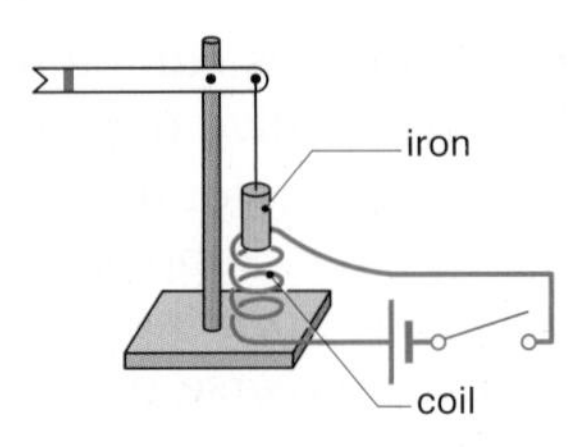

8. Look at the **ticker-timer** on page 123. The electromagnet is supplied with alternating current, of frequency 50 Hz. Describe carefully how this machine works.

The best way to revise? See page 382.

9. Which way does the wire in the diagram tend to move?

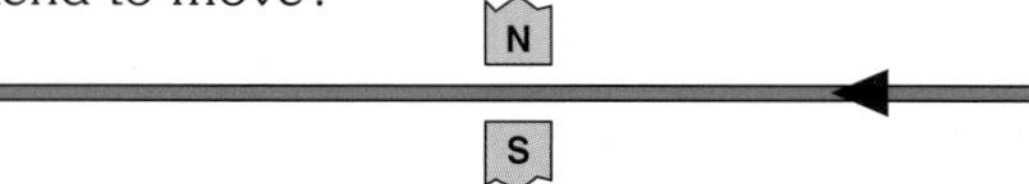

10. a) Explain, with the aid of a diagram, how an electric motor works.
b) Show on your diagram which way your motor would rotate.
c) In what ways may a motor be made more powerful?

11. *A motor has two parts, you'll note,*
A rotor and a stator.
So now please state (and learn by rote)
Which has the commutator.

12. The diagram shows an electromagnetic **circuit-breaker**. Study the diagram:

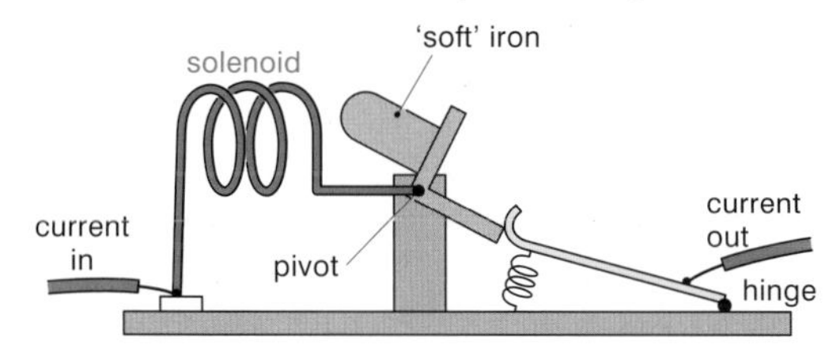

a) Describe, step by step, what happens if too much current flows. (See also p. 272.)
b) Why does this make the circuit safer?
c) How could you reset the circuit-breaker?
d) How could this be combined with a 'thermal' circuit-breaker (see page 272)?

13. *Professor Messer's not too bright,*
So help him get his circuits right:

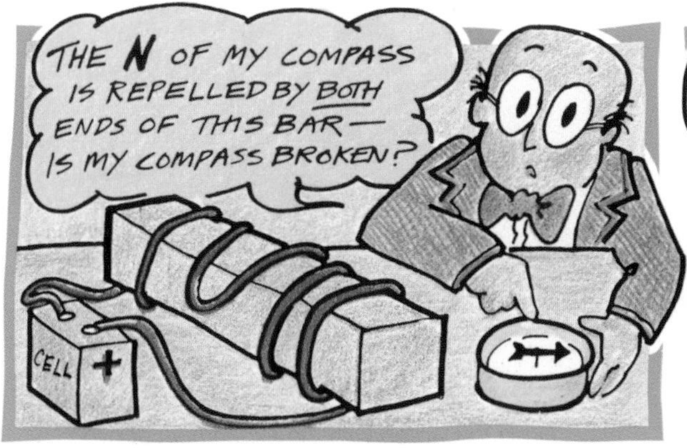

14. Two students investigated how the strength of an **electromagnet** depended on the current:

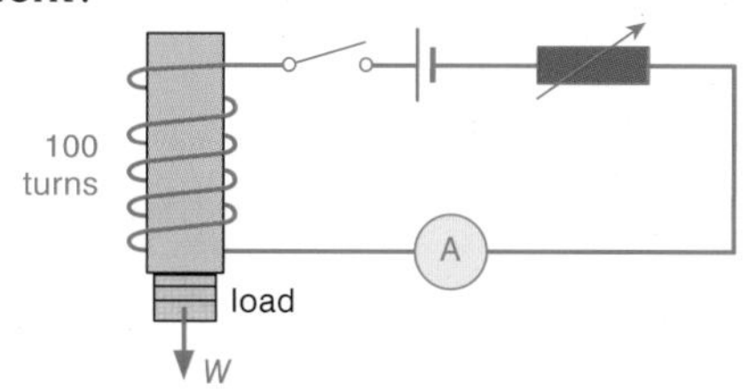

Their results are shown in the table:

Current (A)	0	0.5	1.0	1.5	2.0	2.5	3.0	3.5	4.0	4.5	5.0
Load (N	0	0.2	0.8	1.6	3.0	5.4	10.0	13.6	14.6	14.9	15.0

a) Use the data to plot a graph of load (y-axis) against current (x-axis).
b) What load do you think could be supported if the current was i) 2.75 A ii) 6.0 A?
c) Sketch the graph you would expect to get if the coil had only 50 turns.
d) What is happening to the domains in the iron (see page 283), and why does the graph level off at the top?

15. A motor draws a current of 2 A at 10 V and lifts a load of 10 N through a distance of 3 m in 2 seconds.
a) What is the work done on the load? (See page 97.)
b) What is the rate of working on the load (power output)?
c) What is the electrical power input?
d) What is the efficiency of the motor?

Further questions on page 332.

chapter 36

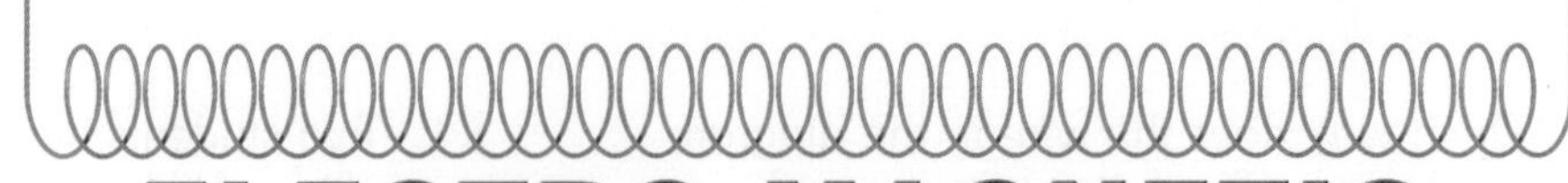

ELECTRO-MAGNETIC INDUCTION

As soon as Oersted discovered that electricity produced magnetism, many people began to look for the reverse effect.
Eventually, in 1831, Michael Faraday discovered how to make electricity using magnetism.

Experiment 36.1
Connect a straight piece of wire to a sensitive ammeter. Then ***move*** the wire across a strong magnetic field.

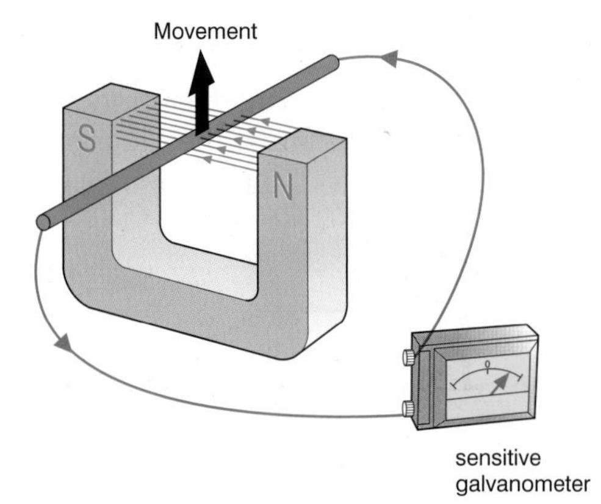

What does the ammeter show you?

We call this current an ***induced*** current.

Faraday found that **if a wire is moved to cut across lines of flux, then a current is induced in the wire** (if there is a complete circuit).
This is called **electromagnetic induction**.

The electrons in the wire have been given a push as the wire moves across the magnetic field.
If the wire moves along the magnetic field lines, there is no current. It has to cut ***across*** the field.

The current can be increased by:
- using a stronger magnetic field,
- moving the wire faster.

If the wire is not moving there is no current.

To remember the direction of the induced current, we use **Fleming's Right-hand Rule:**
Hold your ***right*** hand in a fist and then spread out the thumb, first finger and second finger so that they are at right-angles to each other.
– Point your **F**irst finger in the direction of the magnetic **F**ield (from N to S).
– Rotate your hand about that finger until your thu**M**b points in the direction of the **M**ovement of the wire.
– Then your se**C**ond finger points in the direction of the **C**urrent.

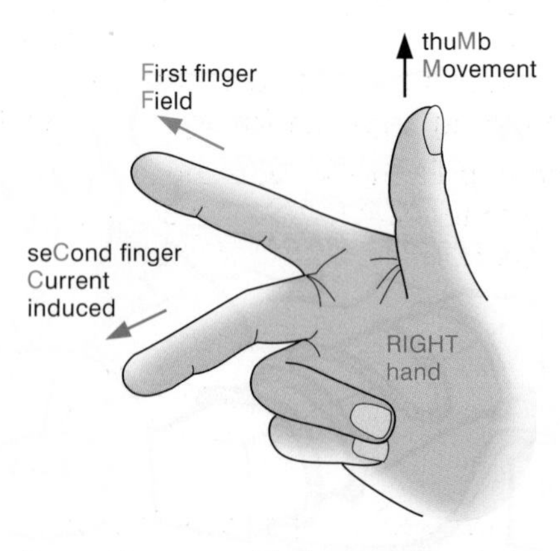

Does this agree with the experiment above?

Experiment 36.2
Connect a coil with a large number of turns to a centre-zero ammeter.

a) Then move a bar magnet quickly into the coil. What happens?
This is shown on the diagram:

b) Hold the magnet still, inside the coil. Does any current flow?

c) Then pull the bar magnet out of the coil. Which way does the current flow?

d) Move the magnet in and out repeatedly. What do we call this kind of current?

e) Move the magnet very slowly and then quickly. Which way gives more current?

f) Move in a weak magnet and then a strong magnet at the same speed. Which is better?

g) Try the experiment with another coil having only a few turns. Do you get less current?

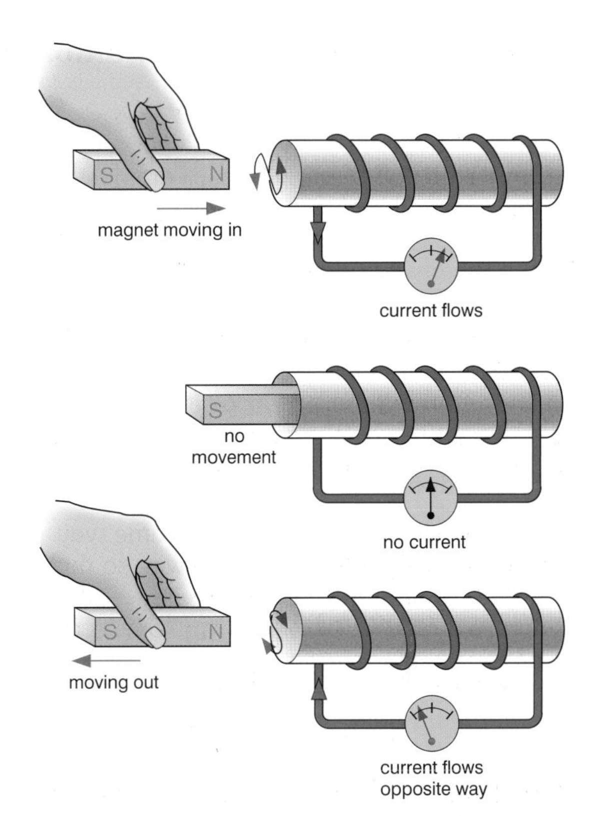

Faraday's Law: He found that the induced voltage (and the current) can be increased by:
- using a stronger magnet,
- moving the magnet faster,
- increasing the number of turns on the coil.

The ***direction*** of the current in the coil can be found by **Lenz's Law**:

the direction of the induced current is such that it opposes the change producing it.

This is shown in the diagrams above. When the N-pole of the bar magnet is moving ***in***, the current flows so as to produce a N-pole at that end of the coil. This means that the current is opposing the movement of the magnet (by trying to repel it).

When the N-pole moves ***out***, the current flows the other way to make a S-pole which again ***opposes*** the movement (by trying to attract the magnet).

These induced currents are used in a moving-coil microphone (see page 301):

Singers rely on induced currents in a microphone

They are also used in tape-recorders (page 307), video-recorders, and hard-discs in computers.

▷ Current generators (dynamos)

Experiment 36.3
Place a magnet on a record-player turntable. Connect a coil to a centre-zero ammeter and hang the coil between the poles of the magnet:

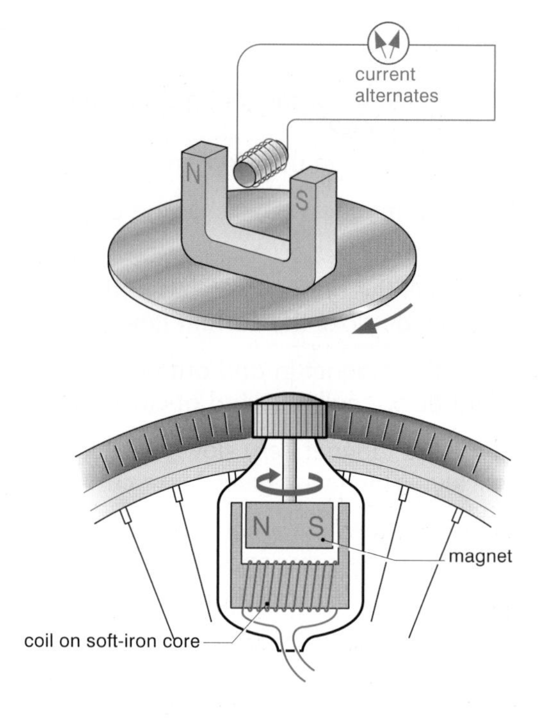

What happens to the pointer of the ammeter when the magnet rotates?
What happens if you speed up the turntable?

A current that reverses to and fro like this is called ***alternating current (a.c.)***.

Bicycle dynamo

A ***bicycle dynamo*** works in the same way.
A magnet rotates near a coil of wire so that the lines of flux are cut by the wire:
The coil is wound on a soft-iron core so that the magnetic field is stronger.

Experiment 36.4
Connect a bicycle dynamo to a lamp and turn it slowly and then quickly. What happens?

The simple a.c. generator (an alternator)

A generator can be built rather like a motor, with a rotating coil and a fixed magnet, as shown:

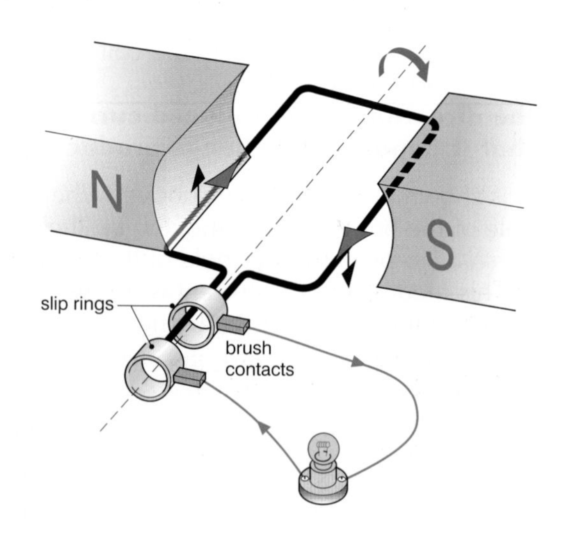

As the coil rotates, it cuts lines of force and so a current is induced.

Use Fleming's ***Right***-hand Rule to find the direction of the current at the instant shown in the diagram (if the coil is rotating clockwise). Do you agree with the arrows on the diagram?

As the coil rotates, the current is conducted in and out by way of ***slip-rings*** and carbon brushes.

Sketch a diagram of this alternator when the coil has turned through half a revolution, and mark in the direction of the current.
Which way does the current flow through the lamp now? What does 'alternating' mean?

Can you find three ways in which this generator could be improved to increase the current?

▶ Alternating current (a.c.)

Experiment 36.5
Connect an a.c. generator with slip-rings (or a bicycle dynamo) to an ***oscilloscope***.
An oscilloscope is a kind of television set which will display the voltage output of the generator.

Rotate the generator steadily so that the oscilloscope draws a graph of the voltage against time. Sketch the shape of the graph.

Now connect the oscilloscope to a safe low voltage a.c. supply, so that it is really connected by way of the mains cable to the Electricity Company generator in the power station.

oscilloscope
brilliance
focus
Y-gain
volts/cm
Y-shift
time-base
ms/cm
input
X-shift
dynamo

Look at the shape of the graph on the oscilloscope (this shape is called a ***sine curve***):

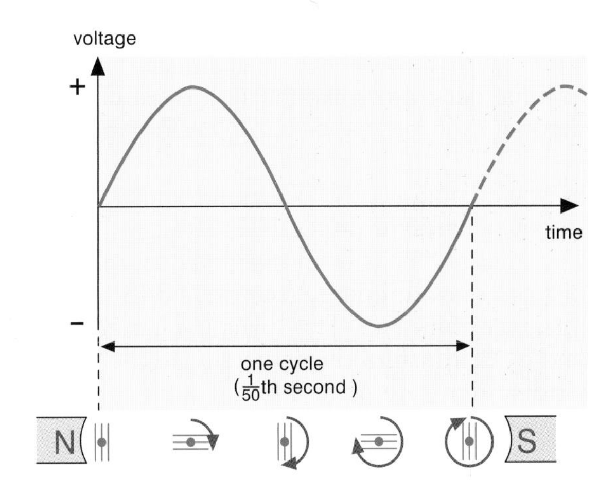

Each complete ***cycle*** of the graph corresponds to one complete revolution of the generator.
For the mains supply (in Britain), each cycle takes $\frac{1}{50}$th second. That is, the ***frequency*** of the mains supply is 50 cycles per second (50 Hz).

The diagram shows how the varying voltage corresponds to the different positions of the coil:

The voltage is largest when the coil is horizontal.
The voltage is zero when the coil is vertical.

Peak value and effective value

The ***peak*** voltage occurs only for a moment.
The ***effective value*** is the value which is equivalent to a steady direct current and gives the same heating effect in an electric fire.
It is found that:

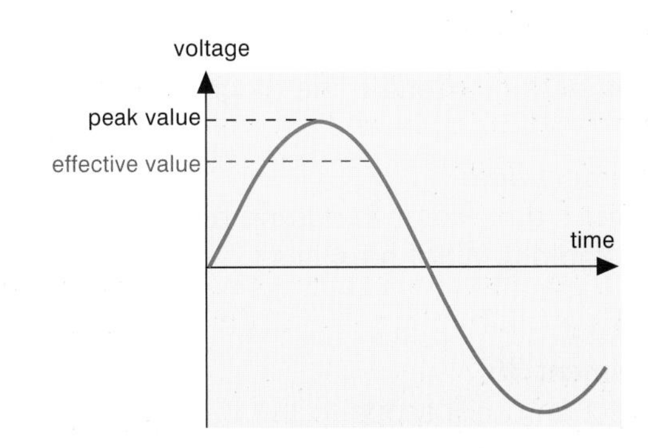

Effective value = 0.7 × peak value
or
Peak value = 1.4 × effective value

Example
The effective mains value in Britain is 230 V.
∴ Peak value = 1.4 × 230
= <u>320 V</u>

This means that mains wiring must be insulated to at least 320 V. The **L**ive wire in a mains circuit varies between +320 V and −320 V during each a.c. cycle.

▶ Practical generators

The voltage (and the current) of a generator can be ***increased*** by:

- using a coil with more turns,
- using a stronger magnet, or using a powerful electromagnet to make the field stronger, or winding the coil round a soft-iron core so that the field is stronger,
- rotating the coil faster,
- using a coil with a larger area.

Inside a power station

The photograph shows a very large alternator in a power station. It makes electricity for your home.

In large generators like these, it is usually the magnet that rotates, as in a bicycle dynamo.

In power stations, the alternator is usually driven by steam power (see pages 101, 104).
The energy comes from burning gas, oil or coal, or from nuclear energy (see page 348). The steam turns a turbine fan. This turns the generator and the energy is transferred electrically via the national Grid system (see page 303).

▶ The simple d.c. dynamo

This is built in just the same way as a d.c. motor. It has a ***commutator*** so that the current in the brush contacts ***always flows the same way***:

Use Fleming's Right-hand Rule to find the direction of the current in the diagram.

> *Experiment 36.6*
> Connect a d.c. motor to a lamp and spin the motor rapidly. What happens?

> *Experiment 36.7*
> Connect an oscilloscope to this d.c. dynamo. Sketch the graph of the output voltage:

There is a smoother output if the dynamo has many coils and the commutator is split into many segments:

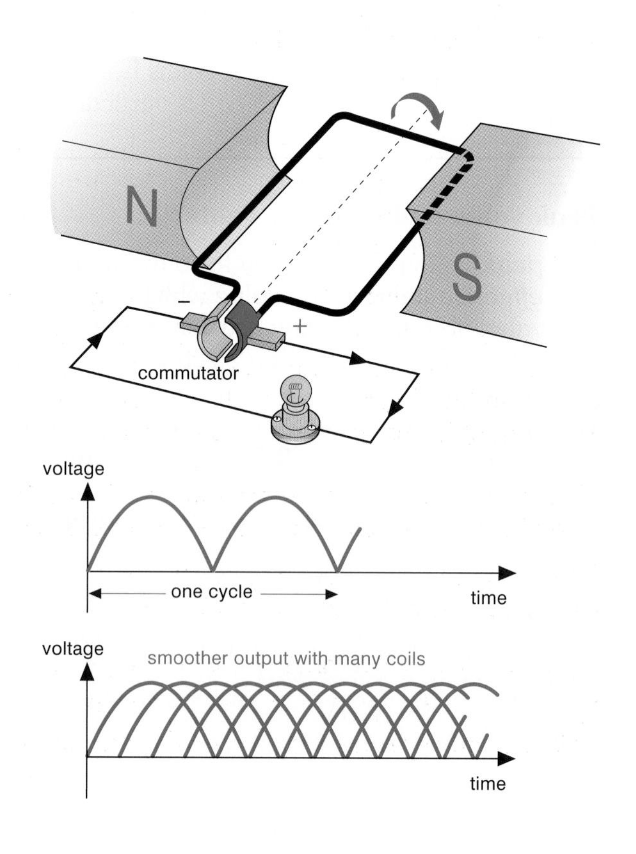

Moving-coil microphone

A moving-coil microphone is built like a loudspeaker (page 291) in reverse:

The sound waves from your throat make the thin diaphragm vibrate. This makes the coil vibrate.

Because the coil is in a magnetic field, the movement induces a voltage across the coil (by electromagnetic induction).
It is like experiment 36.2 on page 297, except the coil is moving, not the magnet.

The vibrations produce a varying voltage.
You can view the waveform on a CRO (see the experiments on page 232).

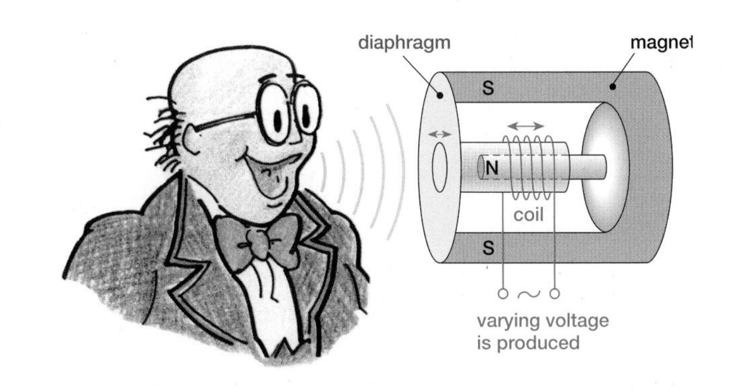

▶ Two coils linked by magnetism : a transformer

> *Experiment 36.8 Mutual induction*
> Place two coils side by side as shown:
> Connect coil **A** to a battery and a switch.
> Connect coil **B** to a centre-zero galvanometer.

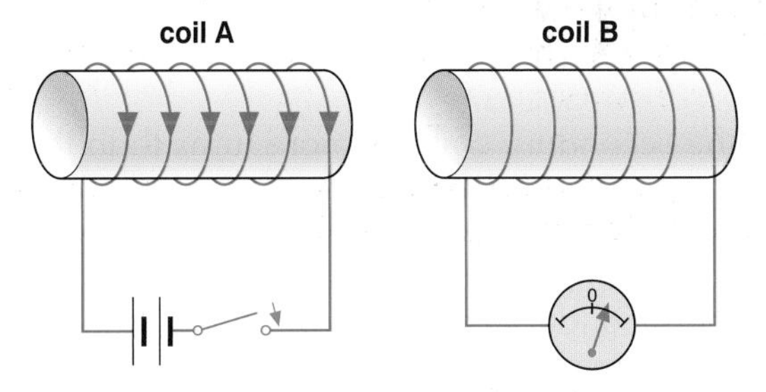

What happens as you switch on the current?
Does anything happen if you keep it on?
What happens if you switch off coil **A**?

A current is induced here, even though the coils are not moving.
When you switch ***on*** coil **A** it becomes an electromagnet and it puts a field round coil **B**. The effect is just the same as pushing a magnet quickly into coil **B**.
An induced current flows for a moment.

When the magnetic field is steady, the current stops.

When you switch ***off*** coil **A**, the field stops quickly. This is the same as pulling a magnet out of coil **B**, and an induced current flows the opposite way.
The current is only induced by a ***changing*** field.

Now try it with a soft-iron core through **A** and **B**:
Is the induced current larger now? Why is this?

If the current is switched on and off repeatedly, what happens in the second coil?

If the first coil carried alternating current, what would you expect to find in the second coil?

This is a simple **transformer** (see the next page).

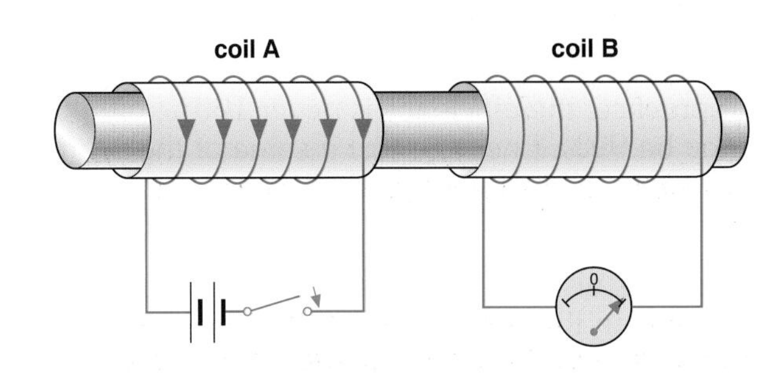

▶ The transformer

Demonstration experiment 36.9 ⚠
(!Danger – high voltages – demonstration only!)
Two coils are placed on a soft-iron core as shown:
One coil (the ***primary*** coil) is connected to an a.c. voltmeter and a source of alternating current. The ***secondary*** coil is connected to an a.c. voltmeter.

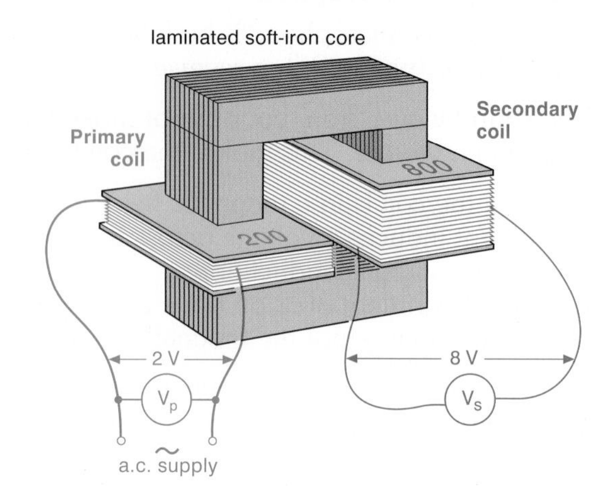

The two coils are connected to each other by a magnetic field (as in the last experiment, page 301). This means that there is an induced alternating current in the secondary coil.

If you measure the voltages you will find that there is a connection between these voltages and the number of turns on each coil. In fact:

$$\frac{\textbf{Secondary voltage}}{\textbf{Primary voltage}} = \frac{\textbf{Number of turns on secondary coil}}{\textbf{Number of turns on primary coil}} \quad \text{or} \quad \frac{V_s}{V_p} = \frac{N_s}{N_p}$$

If the secondary coil has more turns than the primary coil, then it is a **step-up** transformer, because the secondary voltage is bigger than the primary voltage:

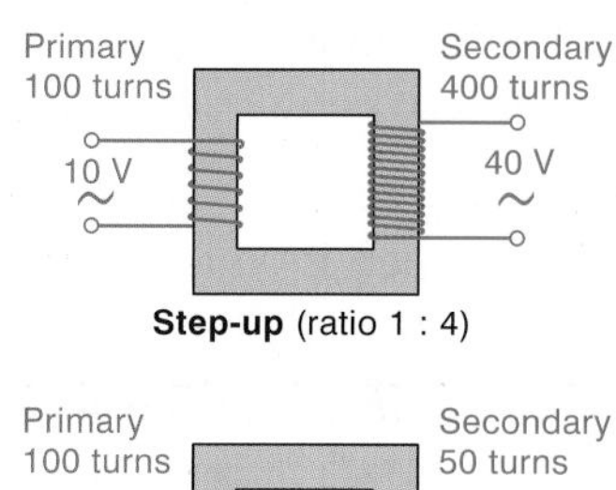

Step-up (ratio 1 : 4)

If the secondary coil has fewer turns than the primary, it is a **step-down** transformer:

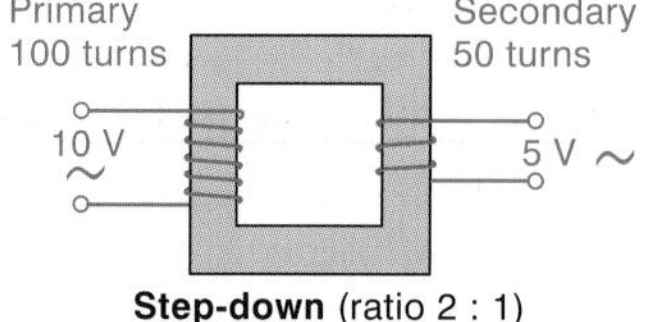

Step-down (ratio 2 : 1)

Transformers are very efficient machines.
If a transformer was 100% efficient, then:

Power supplied to primary coil = **Power delivered in secondary coil**

or $\quad V_p \times I_p = V_s \times I_s \quad$ (from page 266)

This means that if the voltage is stepped ***up*** by a certain ratio, then the current in the secondary is stepped ***down*** by the same ratio.

In practice, the efficiency is never 100%, although it may be 99%. Energy is lost if some of the primary magnetic field does not pass through the secondary. In practice, the coils are wound on top of each other:

Energy is also lost because some ***eddy currents*** are induced in the soft-iron core as well as in the coil. These eddy currents can be reduced by making the core laminated from thin metal strips, or 'lamina'.

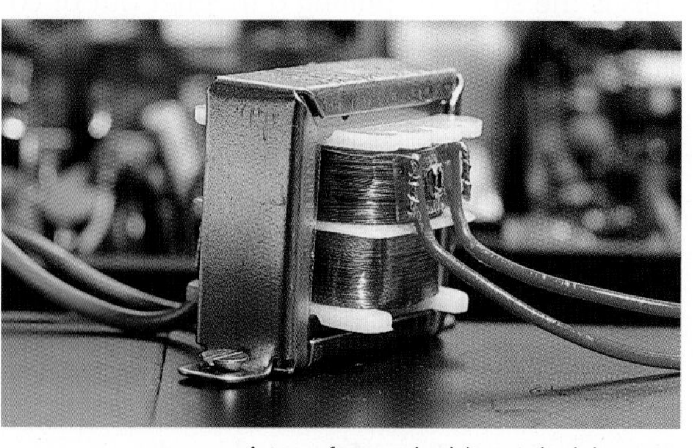

A transformer inside a television set

Example
A step-down transformer is required to transform 240 V a.c. to 12 V a.c. for a model railway. If the primary coil has 1000 turns, how many turns should the secondary have?

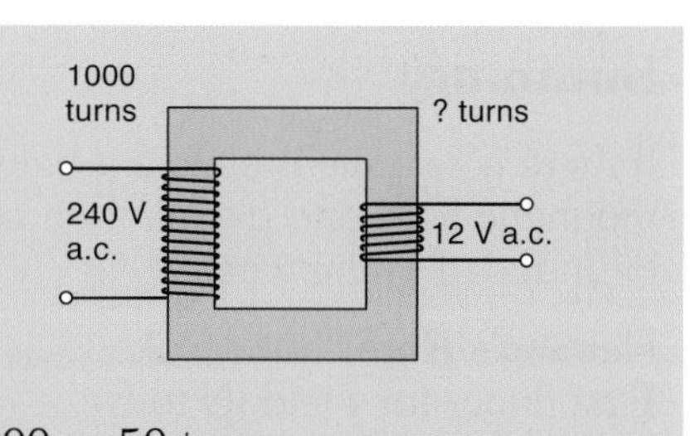

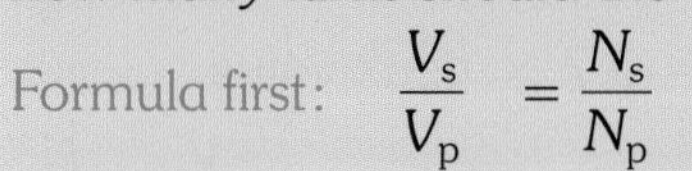

Formula first: $\frac{V_s}{V_p} = \frac{N_s}{N_p}$

Then put in the numbers: $\frac{12}{240} = \frac{N_s}{1000}$ $\quad \therefore N_s = \frac{12}{240} \times 1000 = \frac{1}{20} \times 1000 = \underline{50 \text{ turns}}$

Because the secondary voltage is $\frac{1}{20}$th of the primary voltage, the secondary current will be 20 times the primary current.

Uses of transformers

circuit symbol

1. Transformers are used to get low voltages from the 230 V a.c. mains, for door bells, radios, computers, hi-fi equipment, etc. In your TV set, a step-up transformer produces a very high voltage.

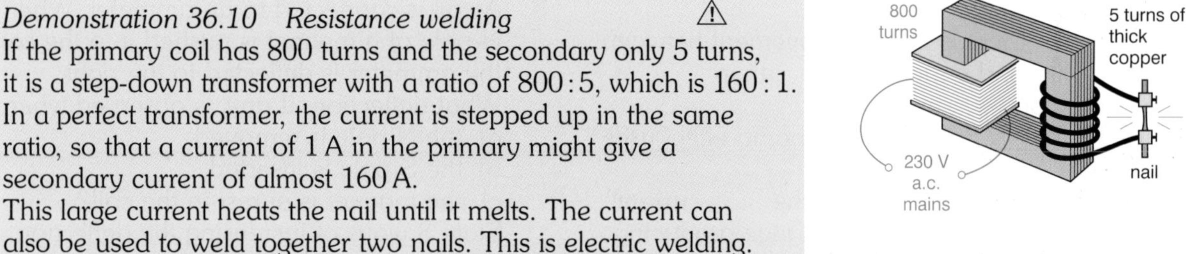

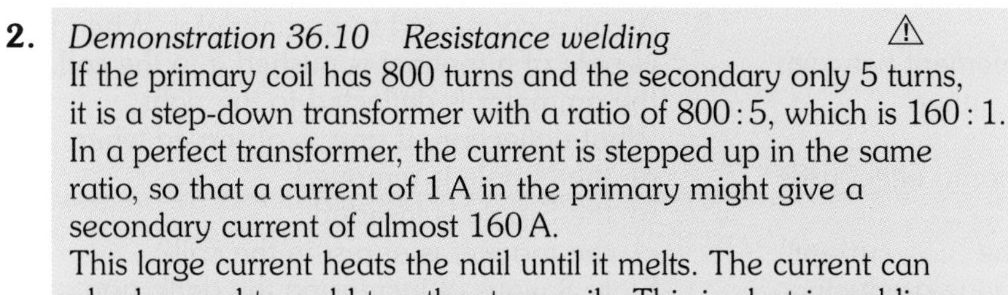

2. *Demonstration 36.10 Resistance welding* ⚠
If the primary coil has 800 turns and the secondary only 5 turns, it is a step-down transformer with a ratio of 800 : 5, which is 160 : 1.
In a perfect transformer, the current is stepped up in the same ratio, so that a current of 1 A in the primary might give a secondary current of almost 160 A.
This large current heats the nail until it melts. The current can also be used to weld together two nails. This is electric welding.

3. *Demonstration 36.11 An induction furnace* ⚠
If the secondary coil has just one turn, then the secondary current is even greater. This large current will quickly boil water or melt solder. 'Induction heating' is sometimes used in industry.

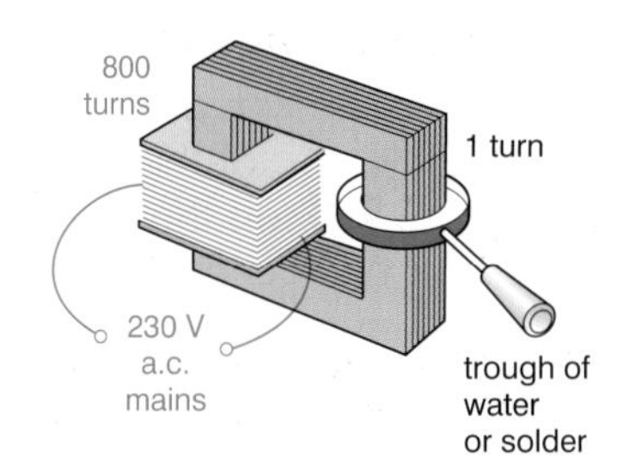

4. **The national Grid system**
To transfer energy from the power station to your home we could use ***either*** a) a low voltage and a high current
or b) a high voltage and a low current.
Method a) is wasteful as the high current heats the power lines.
Method b) is used, with transformers at each end of the Grid system to change the voltage. This is the main reason for using a.c. in the Grid (because transformers do not work with d.c.).

A Grid engineer called Sid,
Thought he'd change what he usually did,
He very soon found,
With transformers changed round,
That he melted the national Grid.

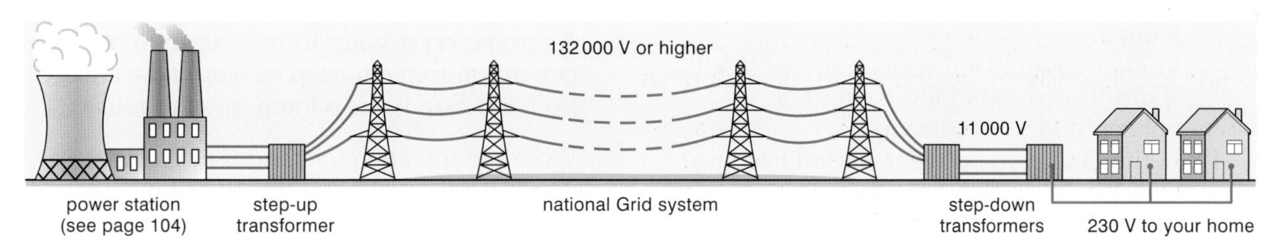

Summary

If there is relative movement between a magnetic field and a wire, then a p.d. may be induced in the wire.

Fleming's ***Right***-hand Rule:
First finger for **F**ield (N to S),
thu**M**b for **M**ovement,
se**C**ond finger then gives direction of **C**urrent.

Lenz's Law: the direction of the induced current is such that it ***opposes*** the change producing it.

A simple a.c. generator has a rotating coil with slip-rings. A d.c. generator has a commutator.

Effective value of a.c. = 0.7 × peak value.

Transformers change alternating voltages: $\frac{V_s}{V_p} = \frac{N_s}{N_p}$

If the voltage is stepped-down, the current is stepped-up in almost the same ratio.
Energy losses are reduced by a laminated core.

The Grid system: electricity is distributed at high voltage and therefore low current.

▶ Questions

1. Copy out and complete:
a) When there is relative movement between a wire and a magnetic . . . . then . . . . may be . . . . in the wire.
b) Fleming's Rule for the dynamo effect uses the . . . . hand.
c) Lenz's Law: the . . . . of the . . . . current is such that it . . . . the change producing it.
d) The strength of the induced current depends on the . . . . of the movement, the . . . . of the magnet and the number of . . . . on the coil.
e) An a.c. generator is fitted with . . . . whereas a d.c. generator is fitted with a . . . . .

2. Copy out and complete:
a) A transformer will work only with . . . . current.
b) The formula for a transformer is . . . . .
c) When the voltage is stepped-down, the current is stepped- . . . . in the . . . . ratio (if the transformer is . . % efficient).
d) The core of a transformer is . . . . to reduce . . . . currents.
e) In the national Grid system, the electricity is distributed round the country at . . . . voltage and . . . . current.
f) In the Grid, ... current is used because transformers will not work on . . . . current.

3. A coil is connected to an ammeter. When the N-pole of a magnet is pushed into the coil, the ammeter is deflected to the right.
What deflection, if any, is observed when
a) the N-pole is removed,
b) the S-pole is inserted,
c) the magnet is at rest in the coil?
State 3 ways of increasing the deflection on the ammeter.

4. a) Explain, with the aid of a diagram, the action of a simple a.c. generator.
b) Sketch a graph to show how the voltage varies during one revolution of the coil.
c) Mark on the graph where the coil is
i) horizontal ii) vertical.
d) On the same graph, sketch and label the graphs you would get if the coil rotated
i) at twice the speed
ii) at the same speed but in the opposite direction.

5. Prof. Messer says that he is careful with the 230 V a.c. mains because he does not want to be killed by a 320 V shock. Explain this.

6. A turbine on a 'wind-farm' is used to turn a generator. Explain, step by step, how energy from the Sun lights a lamp in your house.

7. What are the energy changes in the process which begins with coal arriving at a power station and ends in a lamp in your home?

8. An electric guitar has steel strings vibrating next to a small magnet and a 'pick-up' coil connected to an amplifier:

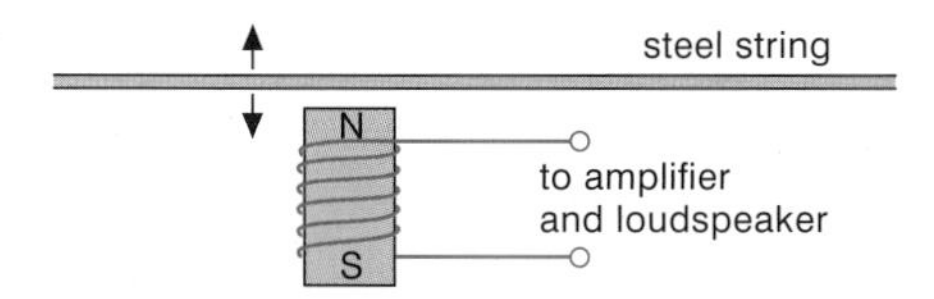

If you pluck the string so that it vibrates at a frequency of 500 Hz,

a) what happens to the magnetic field near the coil?

b) what happens in the coil?

c) what do you hear?

Electric guitars can't use nylon strings. Why?

9. Here are some data (for a transformer) that a student measured in an experiment:

primary voltage (V)	1.1	1.5	2.2	2.7	2.9	6.0
secondary voltage (V)	2.2	3.0	4.4	5.2	5.8	12.0

a) Draw a graph of these values.

b) Comment on the data values.

c) Is this a step-up or step-down transformer? How do you know?

d) If the primary p.d. is 5 V, what is the secondary voltage?

e) If the secondary current is 2 A, what is the primary current?

10. A 23 V lamp, needing 2 A, is supplied from a transformer connected to the 230 V mains.

a) What is the turns ratio?

b) How much power is supplied to the lamp?

c) How much power is supplied by the mains? What have you assumed here?

d) How much current is taken from the mains?

11. Complete the following table of transformers:

Primary p.d.	Secondary p.d.	Primary turns	Secondary turns	Step-up or step-down
100 V a.c.		10	100	
100 V a.c.		100	10	
240 V a.c.	12 V a.c.	200		
11 000 V a.c.	132 000 V a.c.		12 000	

12. Explain what is wrong in each diagram:

13. Explain why the core of a transformer is a) iron and b) laminated.

14. a) 230 000 W of power might be delivered through the National Grid at i) 230 V or ii) 230 000 V. Calculate the size of the current in each case.

b) Why would 230 000 V be used, despite the dangers? c) Why is a.c. used?

15. The overhead power lines could be made thicker. Explain one advantage and one disadvantage of doing this.

16. Research and discuss the health issues associated with overhead power cables.

17. National Grid power lines are usually overhead, hanging from pylons. Some people would prefer underground cables; others say this would be worse. Discuss reasons for and against each method, referring to economic and environmental issues.

Further questions on page 333.

The best way to revise? See page 382.

▷ Physics at work: In your home

Vinyl discs (records)

Music was recorded on old LP record discs by making a narrow groove in the plastic (see also page 224). The vibrations of the sound were copied into the wobbles of the groove. The groove was an **analogue** of the sound (see page 218). A louder sound gave a larger wobble (with a larger amplitude). A higher note gave a more rapid wobble (with a higher frequency).

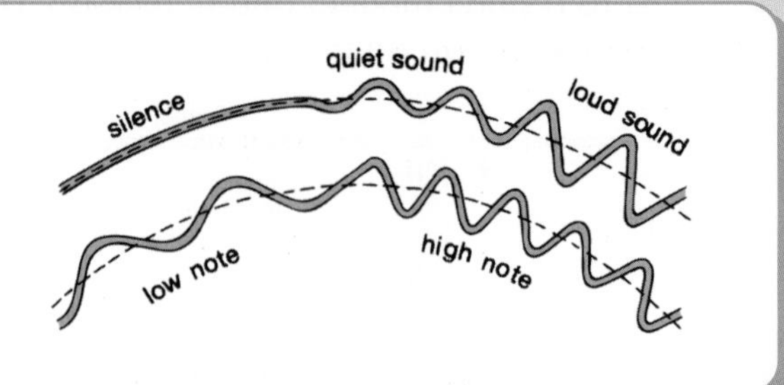

Compact discs, CD-ROMs, DVDs

Compact discs are different: they are **digital** discs. Analogue and digital are explained on page 218.

Instead of the wobbling grooves there is a pattern of '***pits***' marked on the shiny disc. There are billions of pits, with a varying length and spacing, depending on the sound. Between each pit there is a shiny area called a 'flat' or a 'land':

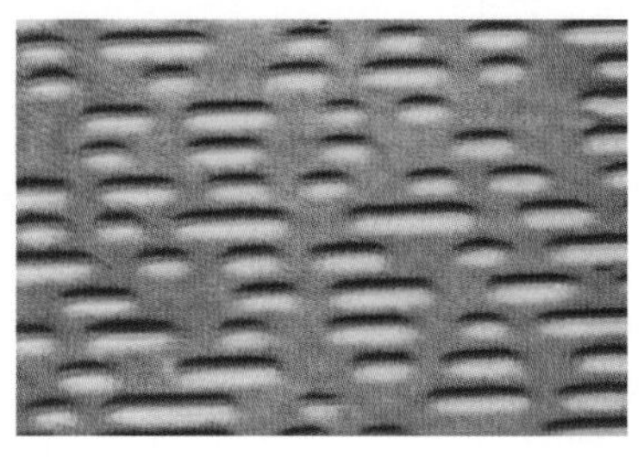

The pits and flats, magnified 2000X

The pits are laid down in a thin spiral track, with a width about $\frac{1}{30}$th of the thickness of a human hair. On an ordinary CD the track is about 5 km long!

The pattern of flats and pits is read by using a weak **laser** beam (see page 193). The laser beam is focussed on the spiral track and moves slowly across the disc as the CD turns:

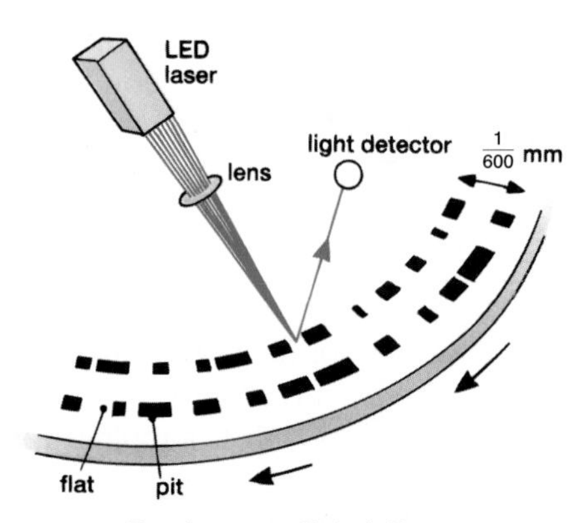

Tracks on a digital disc

The flats and pits reflect the light differently, and this produces a reflected beam which is ***on*** or ***off***. The reflected beam shines on to a light detector. This sends a digital signal to a DAC de-coder (page 218), which converts it to a varying current (an analogue signal). This is passed to an amplifier (see page 325) and then to a loudspeaker (page 291).

Most CDs and DVDs are **ROM** (**R**ead-**O**nly **M**emory) because they store data that you can 'read' but you cannot 'write' to them. Some CDs are called CD-RW because (with a laser) your computer can write to them as well as read them.

CDs can store about 700 mega-bytes (MB) of data. DVDs can store about 7× as much, because the pits are even closer together and the spiral is even longer.

The data on the disc is compressed and encoded in 'MPEG' format. Your DVD-player has a computer to de-code it at high speed before displaying it on screen.

▷ Physics at work: In your home

Cassette tape-recorder

A tape-recorder uses a 'tape head' to magnetise the iron oxide on the tape. The tape head is an electro-magnet with a very small gap (about $\frac{1}{1000}$ mm).

The strong magnetic field near the gap magnetises the tape (with the same frequency as the sound). The electromagnet changes the direction of the tiny magnetic domains (see page 283).

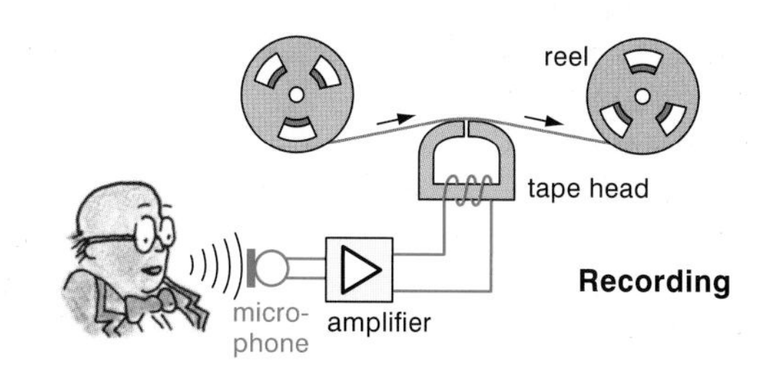

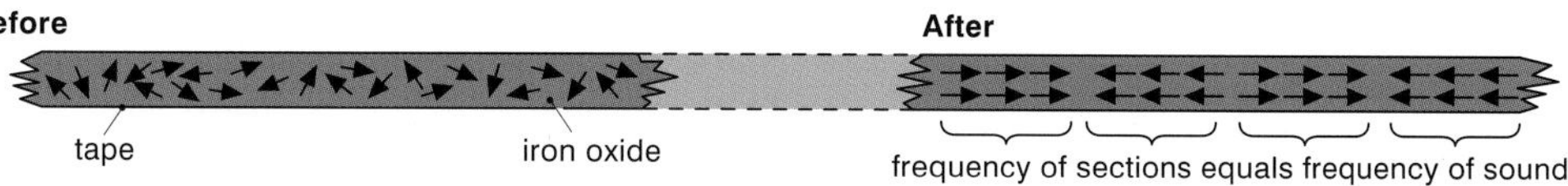

On playback, the magnetised tape goes past a tape head again. The magnetic field of the tape induces a small current in the coil (by ***electromagnetic induction***, see page 296). This current is amplified and passed to the loudspeaker.

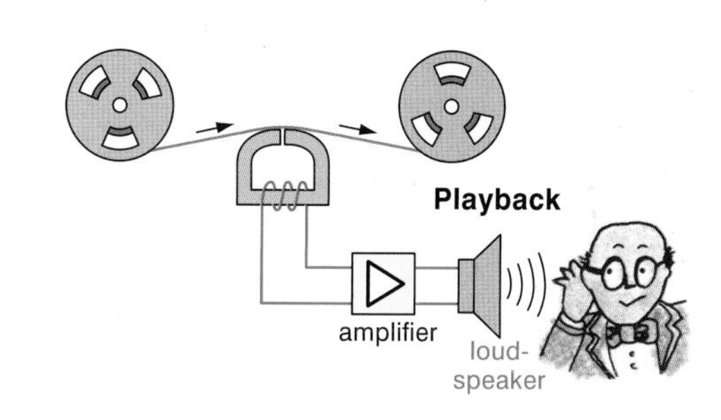

A recording can be erased by a similar 'erase head'. This carries an alternating current at a very high frequency (60 kHz). It demagnetises the tape (so the tiny magnets point in many directions).

Television

A television tube is built like a CRO (page 310). However, in a TV, the electron beam is deflected by currents flowing in the ***deflection coils***:

The magnetic field of a coil bends the electron beam to different parts of the screen (see page 309). The beam moves to build up a new picture every $\frac{1}{25}$ second (see page 311).

In a colour TV set there are ***three*** electron guns. One gun varies in brightness according to the ***red*** signal from the TV station. The other two guns vary with the ***green*** and ***blue*** signals. Because these are the three primary colours of light (see page 221) a complete range of colours can be produced.

If you look closely at a TV screen you will see it is covered with groups of red, green and blue dots. These dots glow when the electrons hit them.

A ***shadow mask*** is fitted just behind the screen: This has a hole in it for each group of coloured dots. It ensures that the beam from the 'red' gun only hits the red dots; and similarly for the green and blue.

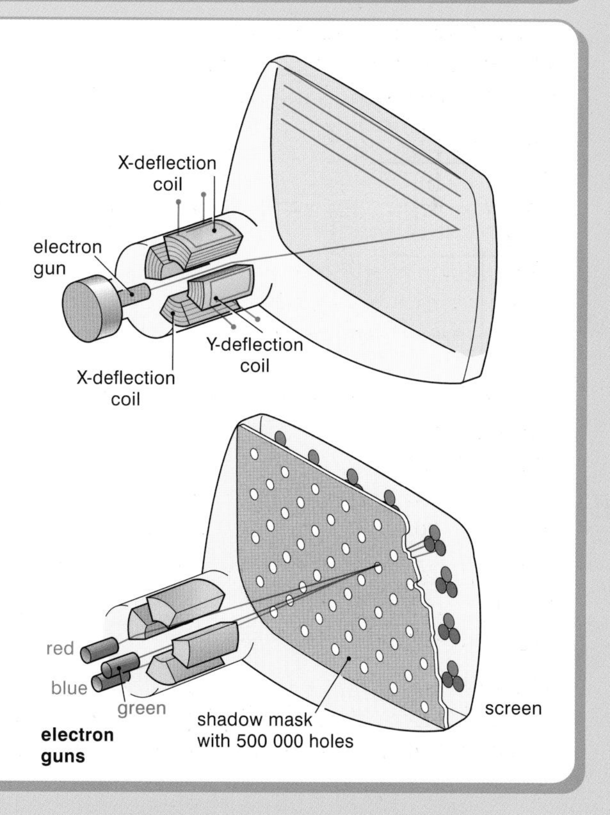

chapter 37

ELECTRON BEAMS

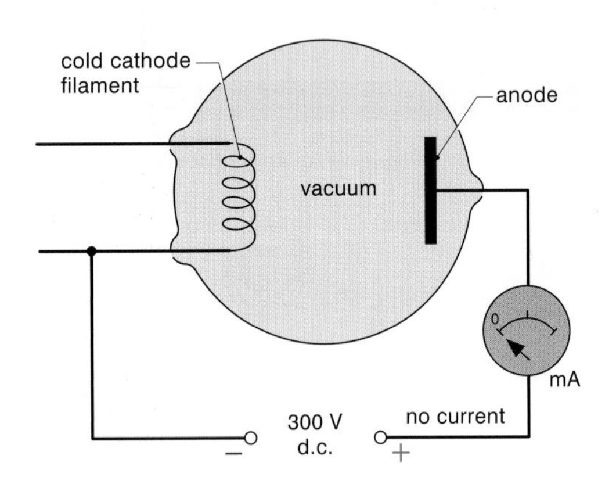

> *Experiment 37.1 Diode value* ⚠
> (!Danger – high voltage – demonstration only!)
> The apparatus, called a ***diode***, consists of a glass bulb containing two metal electrodes in a vacuum. One electrode, a metal plate, is called the anode. The other electrode (the cathode) is a filament which can be heated as in an electric lamp.
>
> Connect a sensitive ammeter and a high-voltage battery (or power-pack) across the diode as shown in the diagram.

Does any current flow? Why not?
Reverse the battery connections. Does any current flow through a vacuum?

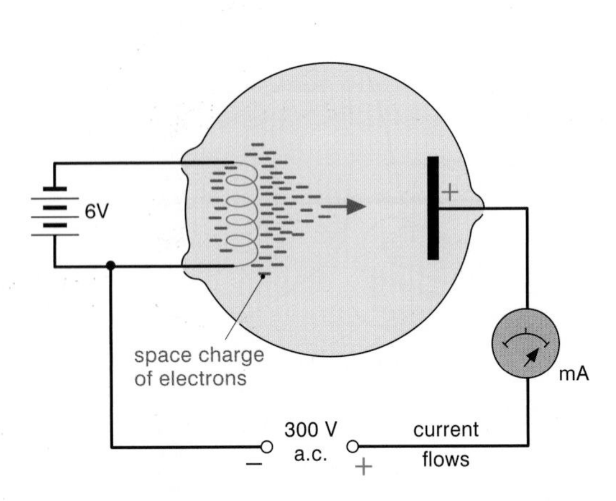

> Now connect a 6 V battery across the cathode filament so that it is heated and glows.

Does a current flow through the ammeter now?
Reverse the connections to the high-voltage battery, as before. Does a current flow through the ammeter now?

A current can flow only when the cathode is hot. We believe that **electrons** are 'boiled' off the hot cathode because at this high temperature they have a lot of energy.
This process is called **thermionic emission**.

The negative electrons form a ***space charge*** near the cathode.
If the anode is ***positive***, these negative electrons are attracted across the gap and a current flows.
If the anode is ***negative***, the negative electrons are repelled and no current flows.
This shows that ***electrons are negative***.

Because the current can flow only one way through the apparatus, it is called a diode ***valve***.

▶ Cathode-ray tubes

This glass bulb has a hot cathode with an anode close to it. The anode has a hole in it so that when it is positive and attracts electrons, some electrons pass through the hole and shoot across the vacuum. This arrangement is called an ***electron gun***.

In the middle of the bulb is a second anode in the shape of a Maltese cross. The end of the tube is a ***fluorescent screen*** as in a television set.

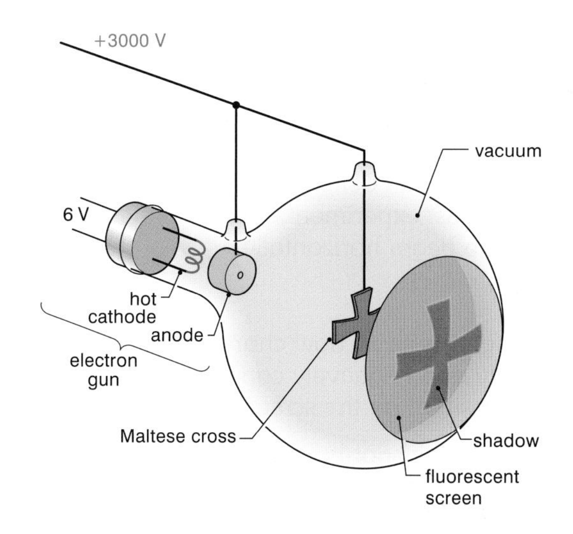

> *Experiment 37.2 Maltese cross tube* ⚠
> Switch on the cathode so that it glows white hot. Now switch on the 3000 V supply to shoot electrons across the vacuum.

What do you see on the screen?

The streams of electrons leaving the cathode and shooting across the vacuum are called ***cathode-rays***. The edges of the cathode-ray shadow on the screen are sharp. This is because the electrons are travelling in straight lines.

Mrs Messer: How do you make a Maltese cross?
Professor Messer: Pour water down his trousers.

Bring a strong magnet near the tube.
What happens to the shadow? Why is it distorted?
What happens if you reverse the magnet?

> *Experiment 37.3 Cathode-ray deflection tube*
> In this tube an electron gun sends a beam of electrons (cathode-rays) through the vacuum so that the beam cuts across a fluorescent screen. The screen is placed between two horizontal metal plates (labelled Y_1 and Y_2). ⚠

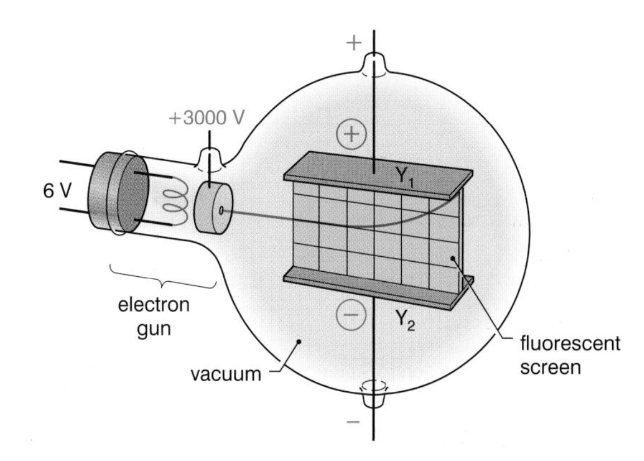

The electron beam can be deflected in two ways:

a) By **magnetic deflection**, using bar magnets as in the last experiment.
Alternatively, electromagnets (coils) can be used (as in a TV set, see page 307).

b) By connecting plate Y_1 to the positive (+) terminal of the power supply and plate Y_2 to the negative (−) terminal, as shown.
Why do the electrons deflect away from the negative plate and move nearer the positive plate?
What happens if you reverse the connections?
This **electrostatic deflection** is used in a cathode-ray oscilloscope.

Some (old) TVs and computers use cathode-ray tubes

▶ The cathode-ray oscilloscope (CRO)

This uses a cathode-ray tube very like the one in the last experiment: an electron gun sends electrons through the vacuum to a screen:

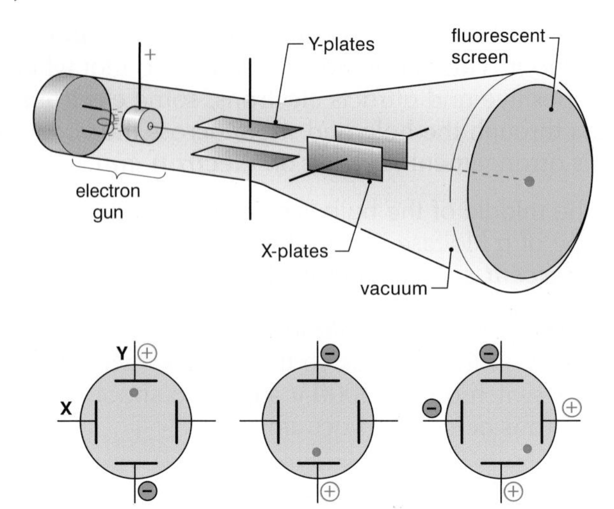

Two Y-plates can deflect the beam vertically (as in the last experiment). Two X-plates can deflect the beam horizontally.

The diagrams show how charging the plates positively and negatively can deflect the beam to any position on the screen:

How could the spot be deflected to the bottom left-hand corner?

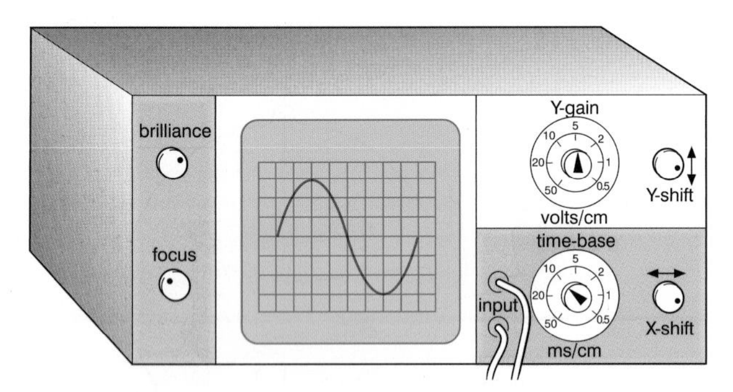

From the control settings shown, can you see that this waveform has:
peak voltage = 15 volts (3 cm on the screen)
time-period = 80 milliseconds (8 cm on the screen)

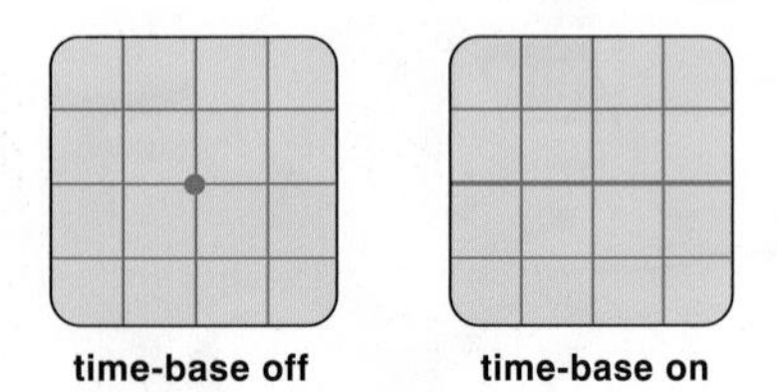

time-base off time-base on

The controls

The complete oscilloscope has several controls to adjust, usually in the following order:

1. The **brightness** and **focus** controls.

2. A voltage ***signal*** (perhaps from a microphone) is applied to the input terminals. This signal is amplified before it is applied to the Y-plates. The **Y-gain control** varies the amount of amplification. A small signal needs more gain.

   In the diagram the Y-gain is set to 5 V/cm. This means that the spot moves 1 cm up (or down) for every 5 V across the input terminals.

3. When we use an oscilloscope (CRO) to draw graphs, we must use the **time-base control**.

   When this control is switched on, the spot moves at a constant speed across the screen and then jumps back very quickly to start again. The time-base control is used to vary the speed of the spot. At high speeds it appears as a line, due to your persistence of vision (page 201).

   In the diagram the time-base is set to 10 ms/cm. This means that it takes the spot 10 milliseconds to travel 1 cm across the screen.

▷ Uses of an oscilloscope

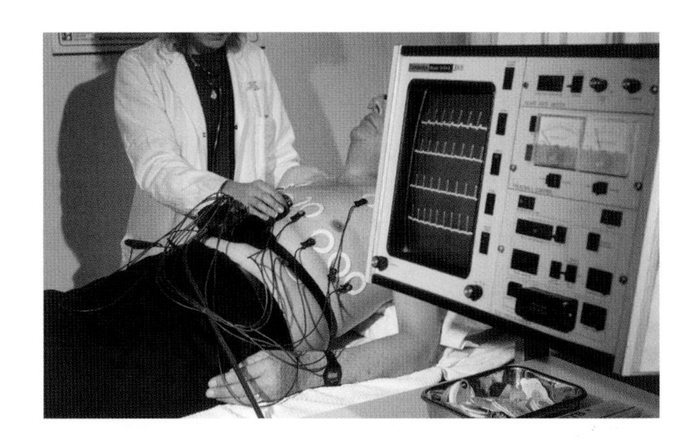

1. Measuring voltages

The deflection of the spot depends on the ***voltage*** applied to the deflection plates. A CRO is a good voltmeter, taking almost no current.

Experiment 37.4
With the time-base switched ***off***, connect one 1.5 V dry cell to the input terminals.
Measure the size of the deflection.

Now connect 2 dry cells in series to the input terminals. What happens to the deflection?

What happens with three cells?
What happens if you reverse the battery?

What happens if you connect a low-voltage ***alternating*** current supply?

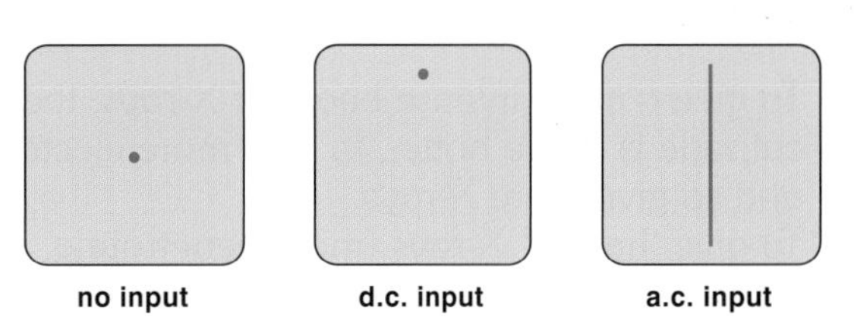

2. Studying waveforms

With an alternating input, the time-base should be used to move the spot horizontally at the same time, so that we can see the shape of the voltage waveform:
A CRO with a microphone can be used to show the waveforms of sounds (see page 232).
Hospitals use CROs to monitor heart-beats.

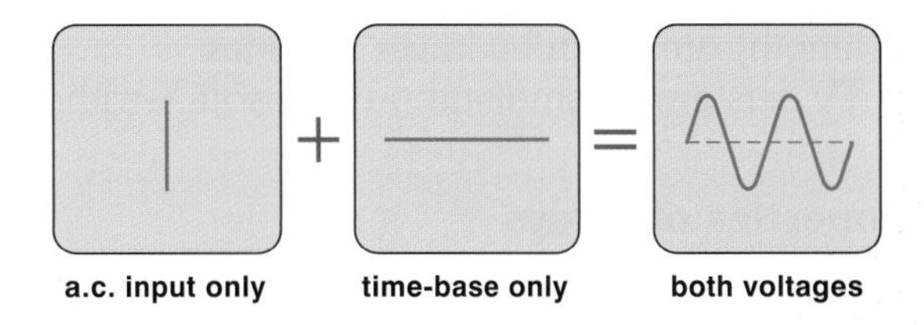

3. Television

A TV set is a CRO with two time-bases – one moves the spot across the screen, the other moves it vertically (like the way your eyes move when you are reading).
The spot marks out 625 lines on the screen and it does this 25 times in each second.
The signal from the aerial varies the brightness of the spot so that a picture is built up of bright and dark spots (as in a newspaper photograph).

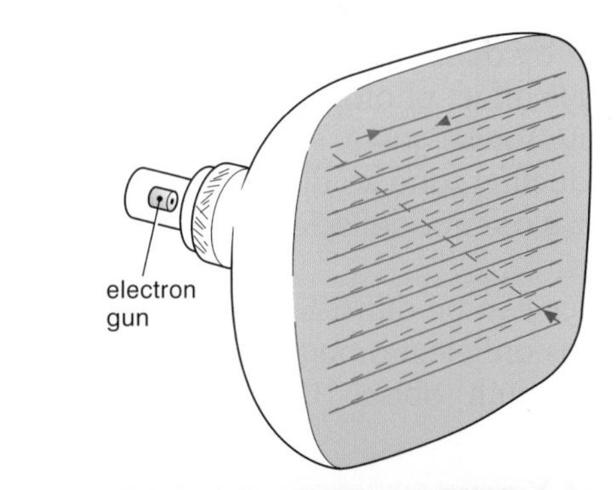

A VDU (Visual Display Unit) for a computer is the same.

In a colour TV set there are ***three*** electron guns (see the diagram, page 307). These build up a full-colour picture using just the 3 primary colours: red, green and blue.

In a radar set, a CRO displays a picture of the radio echoes received by a rotating aerial (see page 211).

LONDON'S RIVER
The River Thames from Gallions Reach to Southend Pier
A composite picture from the five Radars of the Thames Navigation Service

▶ X-rays

X-rays were discovered by W. Röntgen in 1895.

An X-ray tube is really a high-voltage diode valve. A hot cathode emits electrons which are accelerated to high speed by a high voltage (perhaps 100 000 V).

When the fast-moving electrons hit the tungsten target, some of their kinetic energy is converted to X-radiation (most of their energy is converted to heat which is conducted away to the cooling fins).

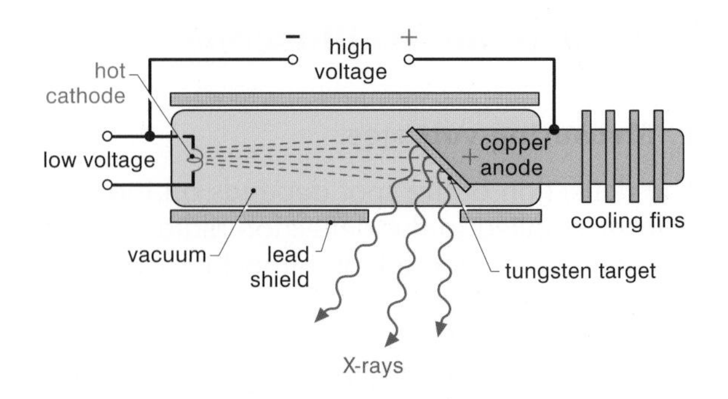

Controls for the X-ray tube

a) To give a more intense beam of X-rays, the cathode is made hotter, to give more electrons and so give more X-rays.

b) To give 'harder' X-rays (more penetrating, with shorter wavelength), the voltage across the tube is increased to give the electrons more kinetic energy.
'Softer' X-rays (less penetrating, longer wavelength) are given by lower voltages.
TV sets emit a small amount of soft X-radiation.

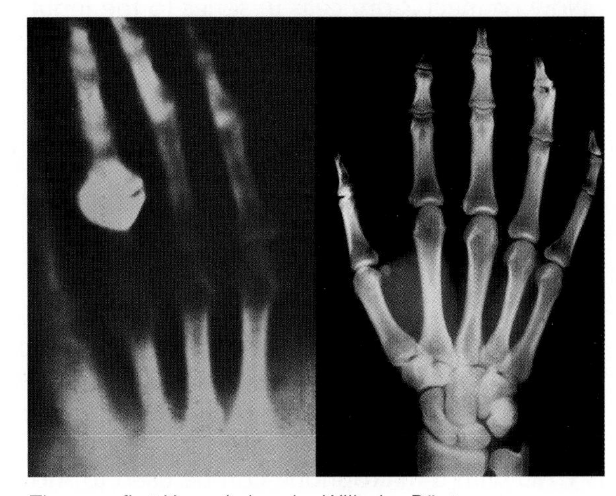

The very first X-ray (taken by Wilhelm Röntgen of his wife's hand and ring) and a modern X-ray

Properties of X-rays

1. ! X-rays are dangerous! They can cause cancer.
2. X-rays pass through substances but are absorbed more by dense solids (e.g. bones or lead).
3. X-rays affect photographic film.
4. X-rays ionise gases, so that the gases become conducting (and so discharge electroscopes).
5. X-rays are not deflected by magnetic or electric fields.
6. X-rays can be diffracted (so they are ***waves***).
7. X-rays are electromagnetic waves of very short wavelength (about 10^{-10} metre). See page 208.

The snake that swallowed two light bulbs:

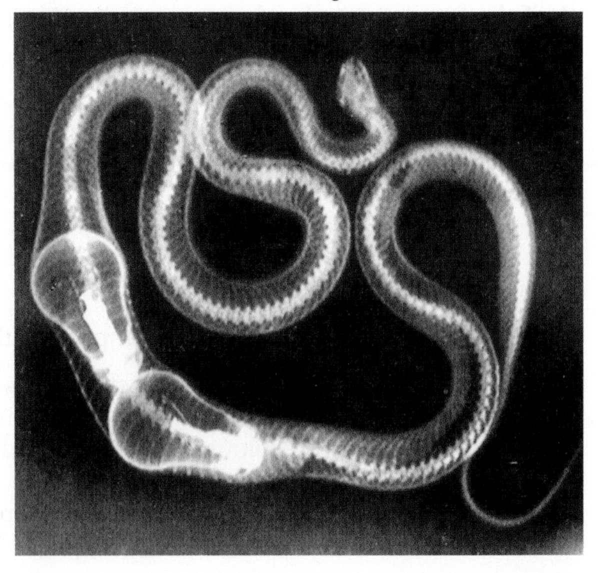

Uses of X-rays

1. In ***medicine***: to inspect teeth, broken bones, etc. With care, X-rays can be used to kill cancer cells.

2. In ***industry***: to inspect metal castings and welded joints for hidden faults. See also page 210.

3. In ***science***: photographs of X-rays diffracted by crystals give information about the arrangements of the atoms in different substances.

Summary

A hot cathode produces a space charge of negative electrons by thermionic emission. The electrons are attracted by a positive anode.

This is used in an oscilloscope (CRO).

An electron beam ('cathode rays') can be deflected by a magnet or by charged plates (as in a CRO).

When fast electrons are stopped by a target, X-rays may be produced.

▷ Questions

1. Copy out and complete:

a) When a wire is heated in a vacuum, it emits . . . . . This process is called . . . . emission.

b) A current flows in a diode valve only when i) the cathode is . . . . and ii) the anode is . . . . to attract the . . . . electrons.

c) A beam of electrons ('. . . . rays') can be deflected by i) a . . . . or ii) . . . . . The . . . . electrons are deflected away from a . . . . -charged plate.

d) In a cathode-ray oscilloscope (CRO) the vertical deflection of the spot is proportional to the . . . . applied across the Y-plates. When studying waveforms, the spot is moved horizontally by a . . . . circuit.

e) When fast electrons are stopped, ... of very . . . . wavelength may be produced.

2. Draw a simple labelled diagram of a cathode-ray tube for a CRO. Name and explain the purpose of the controls on a CRO.

3. *Professor Messer's full of woe –*
He can't control his CRO.

4. A time-base is adjusted so that it draws a horizontal line, as shown, 50 times per second. Draw diagrams of what is seen when the following are connected:

a) a battery which makes the upper Y-plate positive

b) an a.c. supply of 50 Hz

c) an a.c. supply of 100 Hz

d) an a.c. supply of 100 Hz in series with a diode.

5. Here is a waveform on a CRO screen:

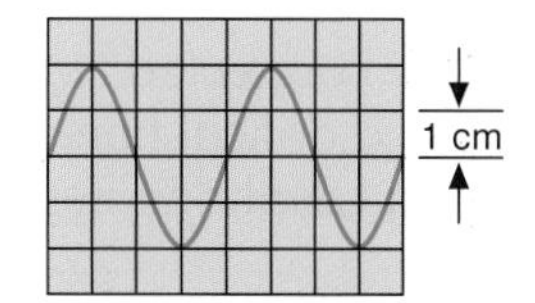

The Y-gain control is set to 2 V/cm and the time-base is set to 5 ms/cm. Calculate:

a) the peak voltage,

b) the time-period of the wave,

c) the frequency of the wave.

With the same input, the controls are set to 1 V/cm and 10 ms/cm. Sketch the screen.

▷ Physics at work: In the office

Computers

Computers use ***digital*** signals (see page 218).

The data is stored in binary code, as a series of 1s and 0s.
Some of the data is stored in integrated circuits ('chips') called ***ROM*** and ***RAM***:

For long-term storage of data you can use:

- ***Floppy disk*** (or disc)
  This stores the data magnetically, as in a cassette tape-recorder (page 307).
- ***Hard-disk*** (or disc)
  This is similar but has many rigid magnetic discs to record the data.
- ***FlashDrive 'memory stick'***
  These use solid-state transistors to store the data.
- ***CD*** or ***DVD*** discs
  These store the data by burning a digital pattern using a laser (see page 306).

Computers can be connected together in a ***network***, using optical fibres (pages 192 and 218) or a telephone network.

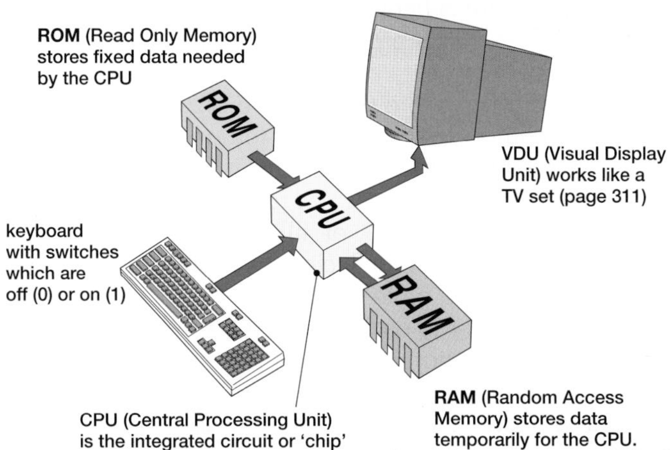
ROM (Read Only Memory) stores fixed data needed by the CPU
ROM
CPU
RAM
VDU (Visual Display Unit) works like a TV set (page 311)
keyboard with switches which are off (0) or on (1)
CPU (Central Processing Unit) is the integrated circuit or 'chip' containing the micro-processor
RAM (Random Access Memory) stores data temporarily for the CPU. It is lost when you switch off.

Telephone networks

Most parts of a telephone network use ***digital*** signals (see page 219 for the reasons).

The digital signals are carried

- by light or infra-red in optical fibres,
- by microwaves for line-of-sight connections, or for beaming up to satellites,
- by UHF radio waves for mobile phones.

All the signals travel at the speed of light (see page 208).

The same network is used to transmit fax messages and emails, and to download web pages.
A key URL (web address) is **www.physicsforyou.co.uk**
It has very useful information for you about your exams.

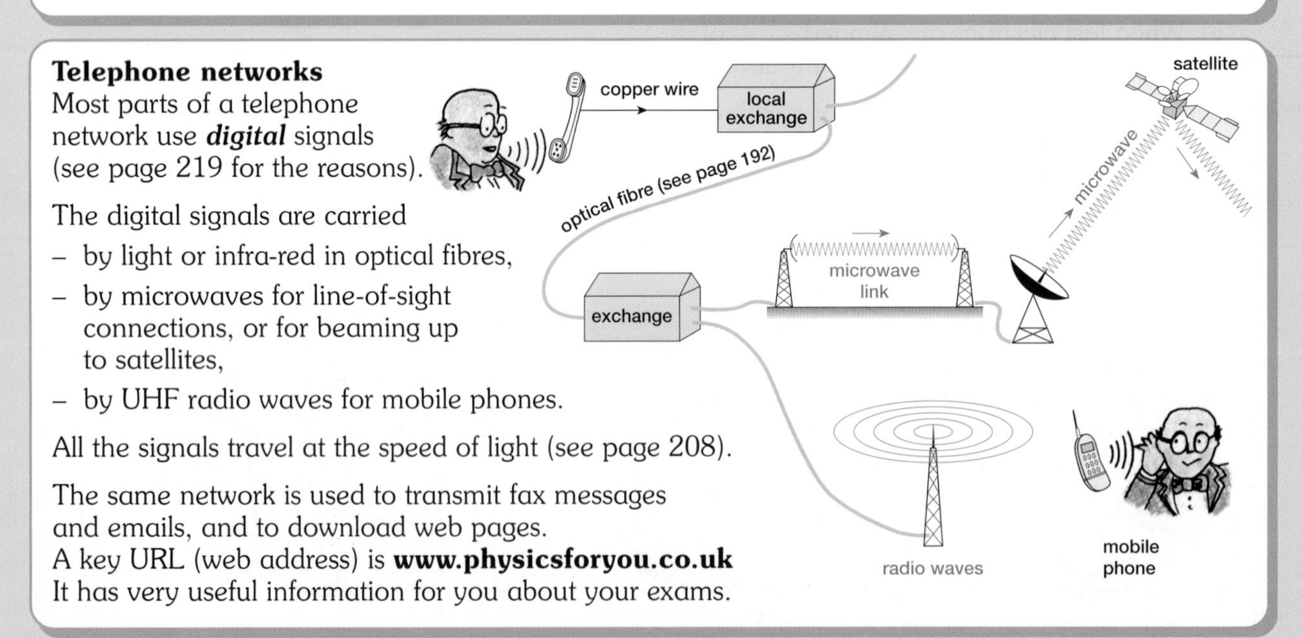
copper wire
local exchange
satellite
optical fibre (see page 192)
microwave
microwave link
exchange
radio waves
mobile phone

Physics at work: In the office

The photocopier

Photocopiers are used in offices and schools to copy pages and documents. They work by using static electricity (see chapter 30). Follow the numbers 1 to 5 to see how:

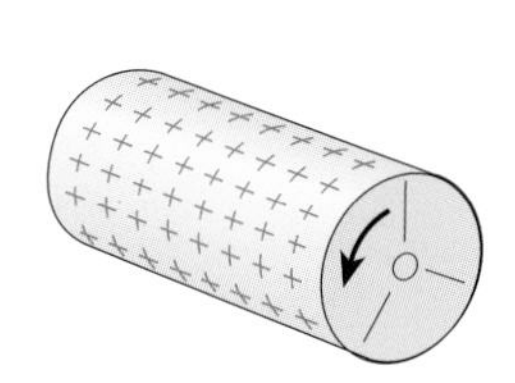

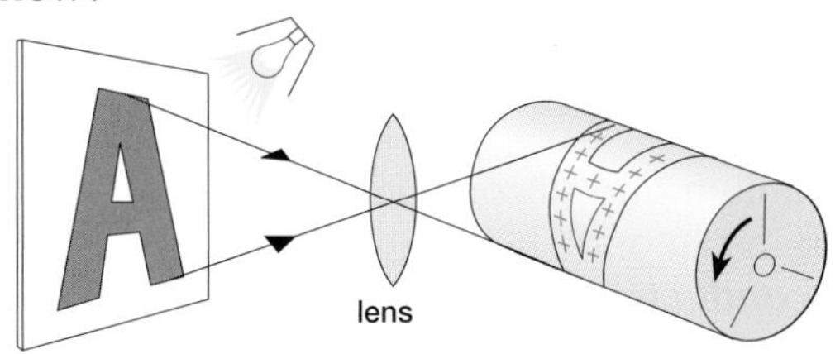

1. Charge
Inside the photocopier, in the dark, a rotating drum is given a positive charge (+). The drum has a special surface which is 'photoconductive'.

2. Expose
A bright light shines on your document. The reflected light is focused on the charged drum. Where the light is bright (reflected by the white paper), the photoconductive surface lets the charge leak away. Only the dark parts stay charged (+).

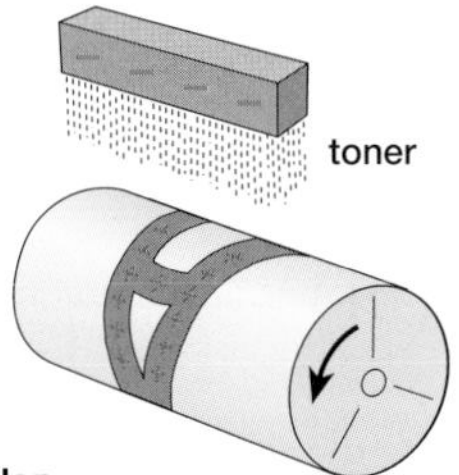

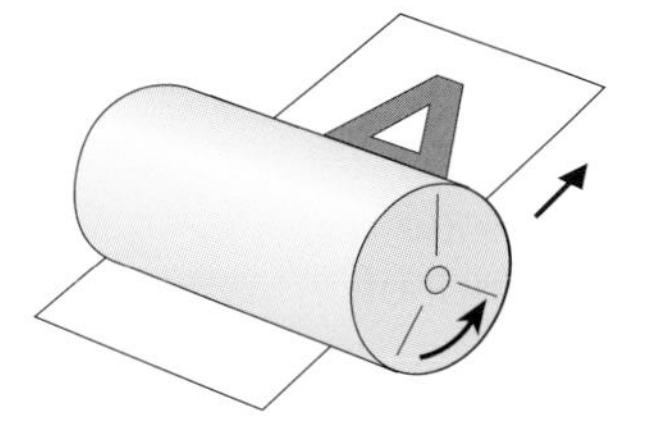

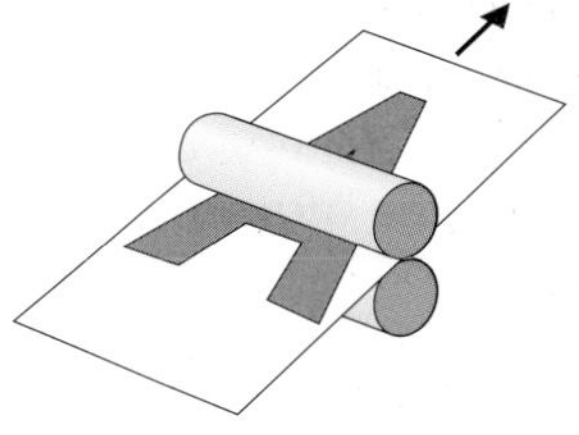

3. Develop
Powdered ink (toner) is negatively-charged and then sprinkled on the drum.
The ink particles (–) stick to the positively-charged parts of the drum (+).

4. Transfer
A blank piece of paper is charged positively. It is pressed against the drum. The toner ink (–) is attracted on to the paper (+).

5. Fix
The paper passes between heated rollers. These melt the toner ink, and fuse it to the paper, to make a permanent copy.
Laser printers work in a similar way.

The ink-jet printer

This is a common kind of printer connected to a computer. One type works in a similar way to the cathode-ray oscilloscope (page 310).

An ink-gun fires tiny droplets of ink out of a nozzle. The droplets are positively-charged (+).
The ink droplets travel between 2 plates as shown:

The ink particles (+) are repelled from the + plate and attracted to the – plate.
The computer controls the voltage on the plates. This controls the deflection of the droplets, to print the shapes of letters on the paper as it goes past.

This method is used to print 'best-before' dates on bottles and cans. For full-colour printing, 4 inks are used (cyan, magenta, yellow, black) just like this book.

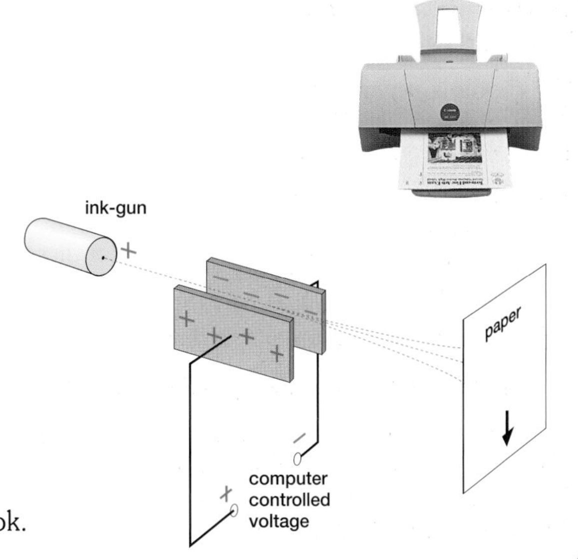

chapter 38

ELECTRONICS

▷ Silicon diodes

Some substances (like germanium and silicon) are neither good conductors nor insulators. These substances are called ***semi-conductors***.

Semi-conductors can be made ***n***-type or ***p***-type, by adding tiny amounts of other substances (***n*** and ***p*** stand for ***n***egative and ***p***ositive charges). If a p-type material and an n-type material are joined together, we get a very useful device, called a p–n junction **diode**.

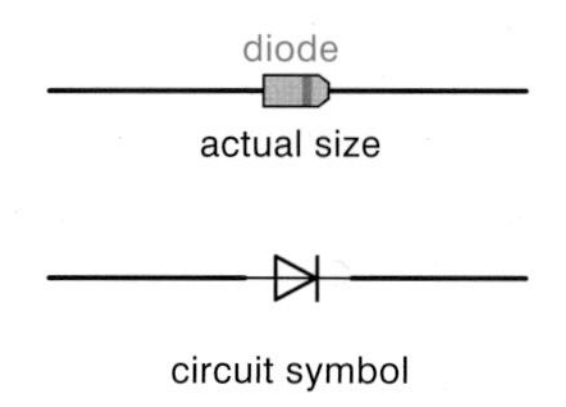

Experiment 38.1 Silicon diode
Connect a battery, a lamp and a diode in series. Connect the marked or narrow end of the diode (the 'cathode') nearest to the negative terminal of the battery. What happens to the lamp?

Now turn the diode the opposite way round in the circuit. Does the lamp still light?

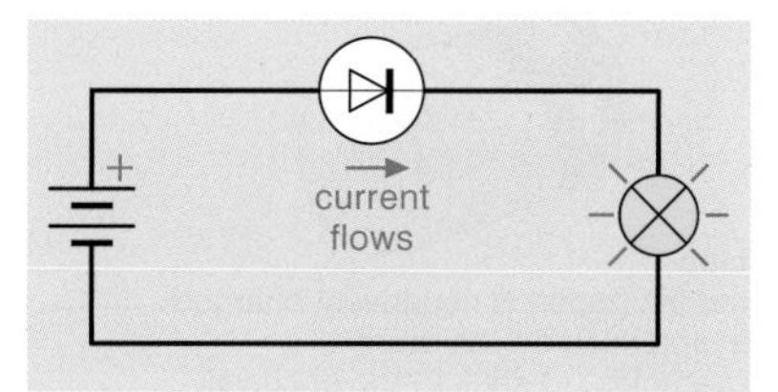

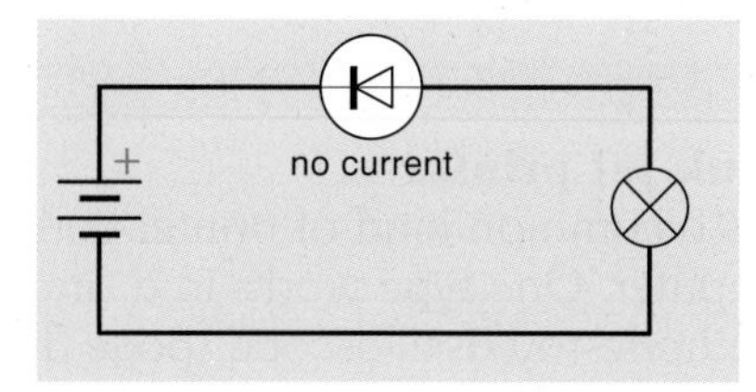

A diode allows a current to flow *only one way*.
It is a ***rectifier***.
The arrow on the diode symbol shows the direction in which conventional current can flow. The diode is then said to be ***forward-biassed***.

If the diode is ***reverse-biassed***, the current is nearly zero.

Experiment 38.2 I : V graph
Repeat experiment 31.23 (page 259) measuring the current ***I*** and the voltage ***V*** for a junction diode.

As the graph shows, almost no current flows if the voltage is applied in the reverse direction.
A current flows in the forward direction whenever the voltage is more than about 0.6 volts.

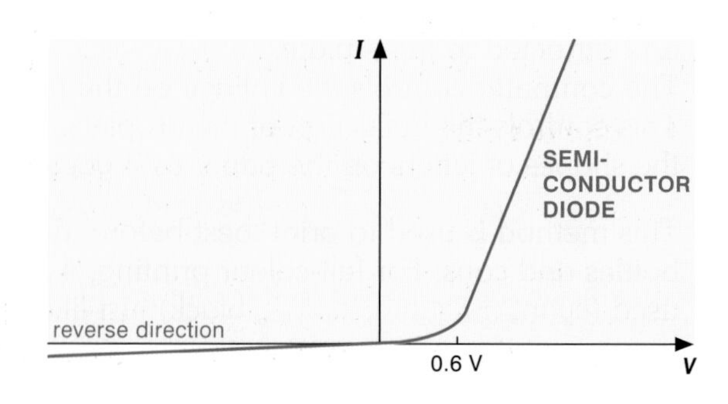

If the reverse voltage or the forward current is increased too much, then the diode will be damaged. See p. 259 for graphs of other ***non-ohmic*** conductors.

Diodes have many uses in radios, TV and computers.

▷ Uses of diodes

1 Protecting equipment

Diodes are used to protect radios and computers which would be damaged if the battery or power supply was connected the wrong way round. If the battery in the diagram was connected the other way round, then no current would flow and no damage could be done.

2 Switching in stand-by batteries

Here is a circuit that could be used on your bicycle:

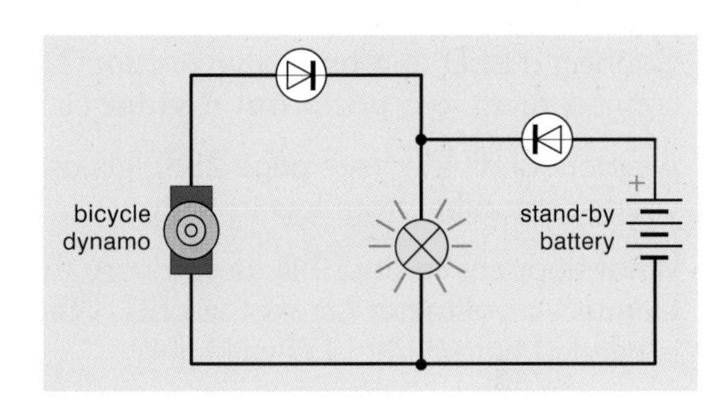

If you are riding the bike, the dynamo will produce a higher voltage than the battery, and light the lamp using the ***left***-hand circuit only, saving your battery.
But if you stop at traffic lights, the dynamo will stop, and then the ***right***-hand circuit will light the lamp.
What does each diode do?
A similar circuit could keep a computer running even if there was a mains power cut.

3 Rectification

The mains supply provides alternating current (a.c.), but often we want direct current (d.c.) as in a battery. How can we use a diode to do this?

Experiment 38.3 Half-wave rectifier ⚠
Connect a diode in series with a 1000 Ω resistor and a low voltage a.c. supply.
Now use an oscilloscope (a CRO, see page 310) to look at the waveform: first connect the CRO across the a.c. supply (the 'input') and sketch the a.c. waveform (see also page 311).
Then connect the CRO across the load resistor as shown in the diagram. Sketch the new waveform:

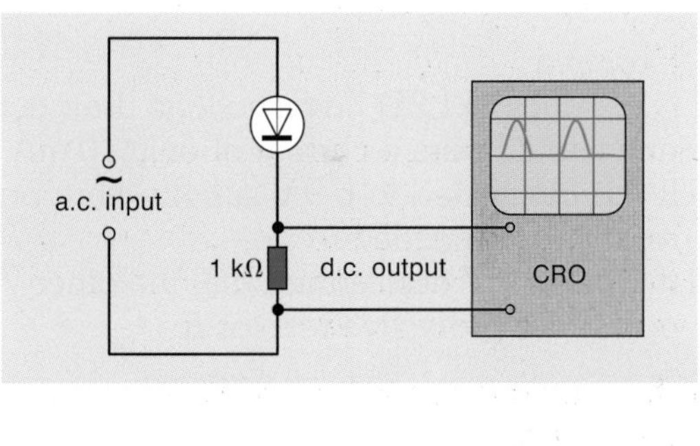

Why is only half the a.c. waveform still present?
The diode conducts only for half the a.c. cycle (only when it is forward-biassed).
The alternating current has been ***rectified*** to an uneven direct current. You are using this circuit whenever you plug your radio into the a.c. mains.

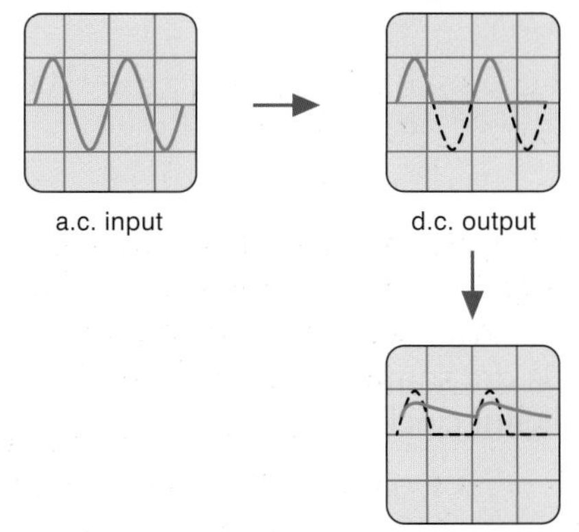

Experiment 38.4 Smoothing
Now connect a 10 μF capacitor (see page 245) in parallel with the resistor (across the input terminals).

Look at the waveform. It has been ***smoothed***.
The capacitor stores some energy when the voltage is high, and then releases it when the voltage falls.

▷ The Light-Emitting Diode (LED)

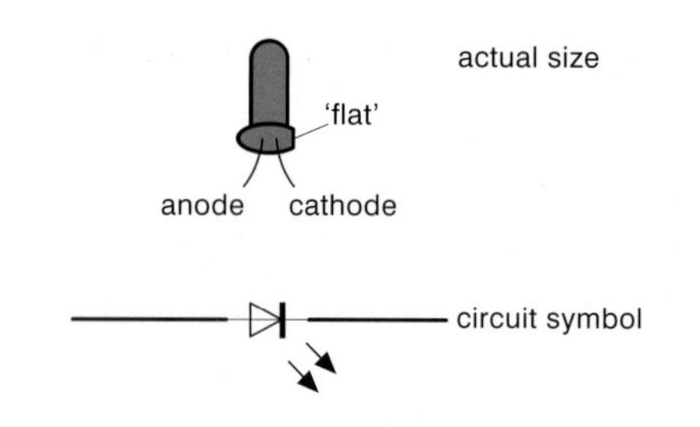

Some special diodes called LEDs will give out light when they are passing a current.
The usual colour is red, but it is possible to get yellow or green or white LEDs.
A light-emitting diode is an example of a ***transducer*** – it transfers energy from one form to another.
What are the two forms of energy in a LED?

Experiment 38.5
Connect a LED to a protective resistor. Then connect them to a **potential divider** circuit:

A potential divider (see page 258) lets us vary a voltage smoothly from low to high.

What happens to the LED as you vary the voltage?
Connect a voltmeter across the LED. What voltage is needed before the LED will light?

Does the diode light if it is reversed in the circuit?

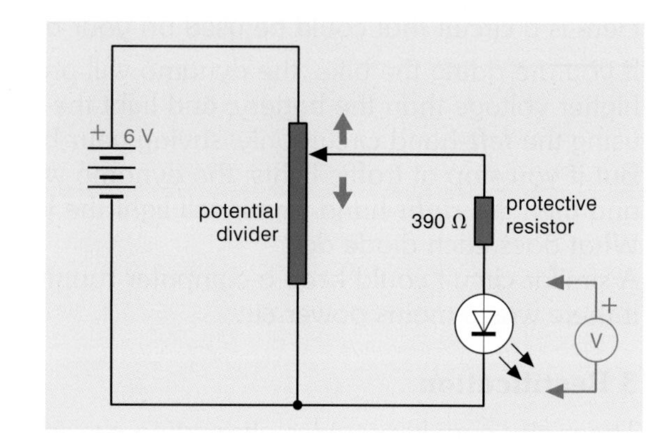

The LED will not light until the applied voltage is greater than about 2 V. However if the full 6 V was applied, the LED would be damaged. A LED must always have a ***protective resistor*** in series with it.

Example
A conducting LED has a voltage drop across it of 2 V and should pass a current of only 10 mA (= 0.01 A).
If it is connected to a 9 V supply, what protective resistance is needed?
If there is 2 V across the LED, the other 7 V must be across the protective resistor R.

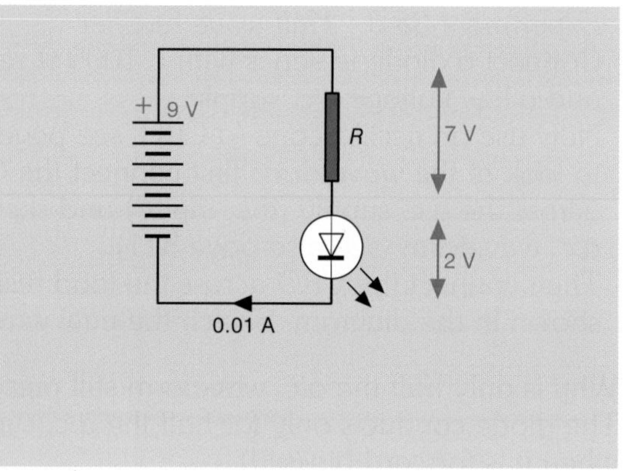

Formula first: $V = I \times R$ (see page 253)

Then numbers: $7 = 0.01 \times R$

$$\therefore R = \frac{7}{0.01} = \underline{700\,\Omega}$$

Uses:
LEDs are often used as indicator lamps, and increasingly for room lighting. LEDs are small, reliable, need only a small current and last much longer than filament lamps.
LEDs are sometimes used in ***seven-segment displays*** in digital clocks, cash registers and calculators:

Professor Messer was using his calculator, sitting opposite Mrs Messer.
Why did she laugh when he typed in 0.7734 and then 5508?
Can you make other words? Try: 0.37108, 5663. Or 35006, 5637. Are they 3781637?

▷ The Light-Dependent Resistor (LDR)

When a semi-conductor has light shining on it, it can conduct electricity more easily.
Its resistance depends on the brightness of the light, so it is called a ***Light-Dependent Resistor*** (or photo-conductive cell).

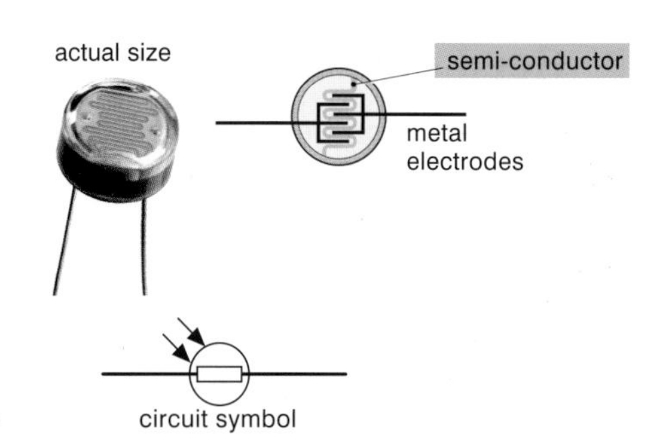

The LDR shown in the photo is made from a semi-conductor called cadmium sulphide (CdS). In the dark its resistance is more than 10 MΩ. In sunlight its resistance is only about 100 Ω.

Experiment 38.6 A light-meter
Connect an LDR in series with a cell, a milli-ammeter and a variable resistor. Adjust the variable resistor to give a deflection on the ammeter:

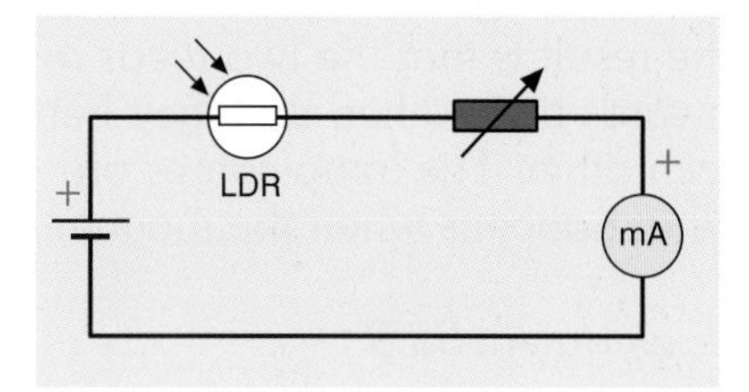

What happens to the current as you cover and uncover the LDR?

Uses:
This circuit could be used as a photographer's light-meter (if it was calibrated). The same circuit can be used to alter the aperture in automatic cameras. More uses are shown later (see page 324).

▷ The thermistor

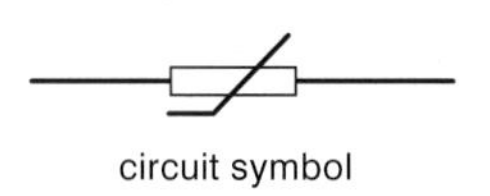

When a semi-conductor is heated, it can conduct electricity more easily. This is because the rise in temperature makes more free electrons available to carry the current.
The higher the temperature, the ***less*** the resistance.

Experiment 38.7 An electronic thermometer
Repeat the last experiment, but with a thermistor in place of the LDR. Heat the thermistor with your hand or a hairdryer:

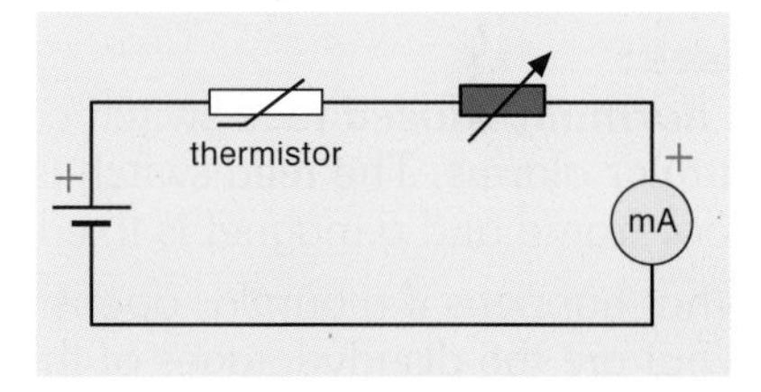

Can you see more current flows as the thermistor gets hotter?

Uses:
This circuit could be used as an electronic thermometer if the ammeter was calibrated in °C (see page 27). It would have a ***non-linear scale***, with ***un***equal divisions.
Thermistors are used to protect the filaments of projector lamps and TV tubes from a current surge as they are switched on. The cold thermistor keeps the current low at first; as the filament heats up, so does the thermistor and it gradually lets more current pass. This extends the life of the filament.
More uses are shown later (page 323 and page 329).

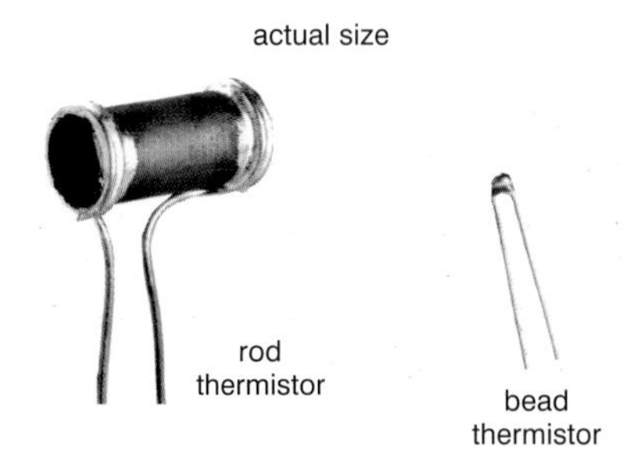

Experiment 38.8 I : V graph
Repeat experiment 31.23 (p. 259) measuring current I and voltage V for a thermistor. The graph is shown on p. 259.

▶ Reed switch

A reed switch can be used in electronic circuits. It consists of a glass tube with two iron reeds sealed inside it:

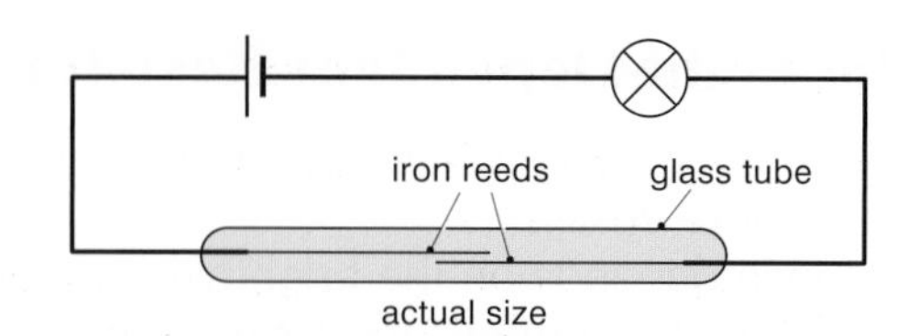

If the iron is unmagnetised, there is a gap between the reeds, and no current can flow. The switch is said to be ***normally-open*** (NO).

If a magnet is brought near the reed switch, it magnetises each iron reed as shown:
(This is 'magnetic induction', see page 283.)
The result is that the two reeds are attracted towards each other, and they bend to touch each other. The lamp comes on.
What happens when the magnet moves away?

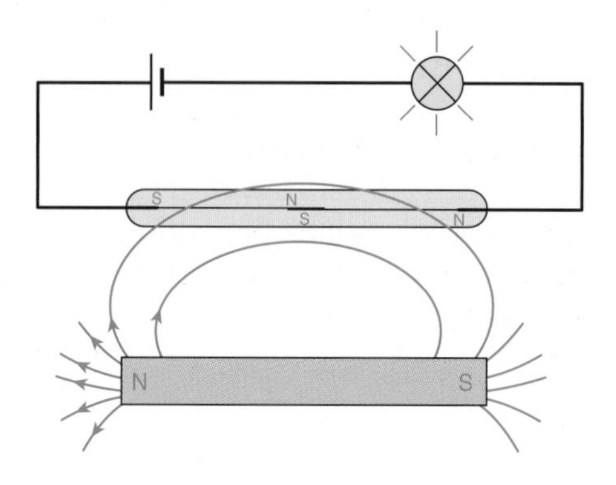

Experiment 38.9
Use a magnifying glass to look closely at the reeds while you move the magnet nearby.

Normally-closed reed switch

If an extra non-magnetic contact is added to the reed switch as shown, it can be used as a ***normally-closed*** (NC) switch.
The current flows from B to C ***until*** a magnet is brought near. What happens then?
How can it be used as a change-over (CO) switch?

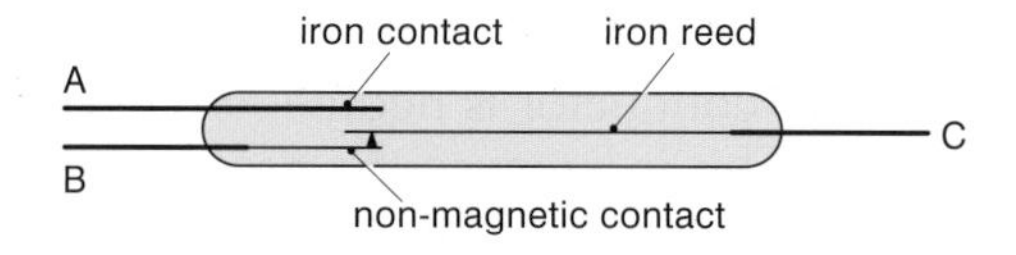

Uses:
A ***normally-closed*** reed switch can be used in burglar alarms. The reed switch is fixed in the door frame and a magnet is fixed to the door:

What happens if a burglar opens the door?
What are the disadvantages of this simple alarm?

A ***normally-open*** reed switch can be used as a safety switch on the door of a microwave oven. Draw a diagram to show how this would work.

Reed relay

If a reed switch is put inside a coil of wire (an electromagnet, see page 287), then it can be operated by a current in the coil.
This is a ***reed relay***.
A small current in the coil can switch on a larger current in the reed switch. (It is a simple kind of amplifier.)

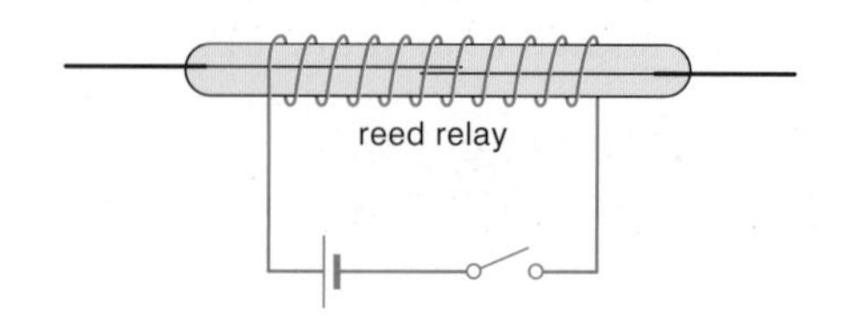

▷ Electromagnetic relay

The diagram shows another kind of relay, which can switch on a larger current than a reed relay.
If enough current flows in the coil it will attract the soft iron armature (see also page 289). The armature is pivoted and so pushes the springy contacts together, so a current can flow through them. What happens if the coil current is switched off?

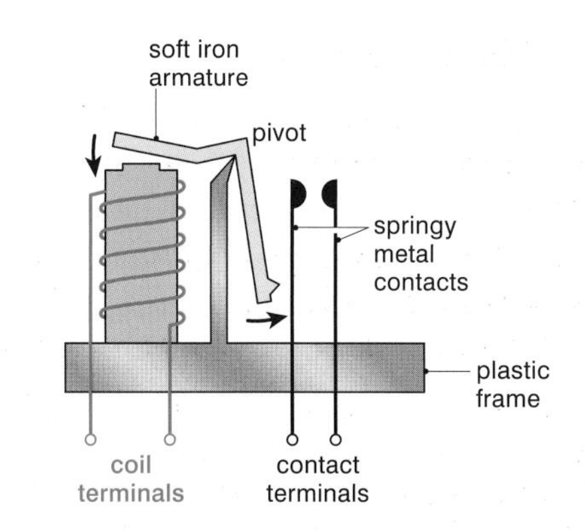

The diagram shows one pair of normally-open contacts. Others can be added side by side so that several circuits can be switched. Normally-closed and change-over contacts can also be used.

Experiment 38.10   A burglar alarm
Connect a relay and a light-dependent resistor (LDR) to a battery and a bell, as shown:
Shine a light on the LDR, so that its resistance falls.

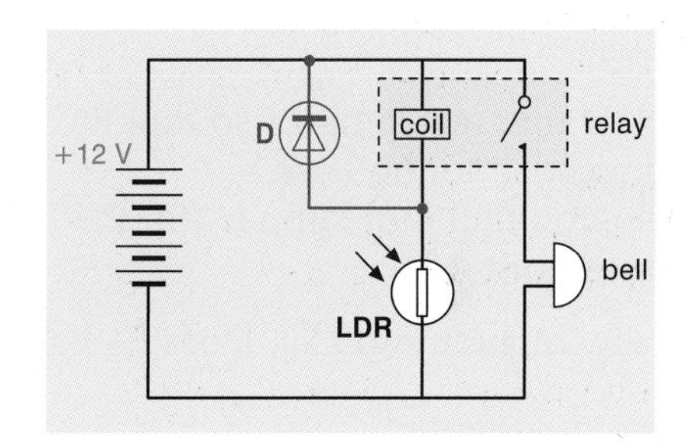

What happens to the current in the LDR?
What happens to the relay? What happens to the bell?
If you put this circuit in a drawer (where it is dark), what would happen if a burglar opened the drawer?

It is important to include a ***diode*** as shown at **D** in the diagram. This is because the magnetic energy stored in the coil can give a large reverse voltage when the relay switches off.
Diode D protects the circuit by shorting out this big voltage.

▷ Transducers and Systems

The LED, LDR, thermistor and reed switch are all examples of ***transducers*** – they transfer energy from one form to another.
The rest of this chapter is about **electronic systems**.
All the systems have three main parts as the diagram shows:

For example, an electric guitar has an input transducer (a pick-up), a processor (the amplifier) and an output transducer (loudspeaker).

Here is a summary of the transducers you have met:

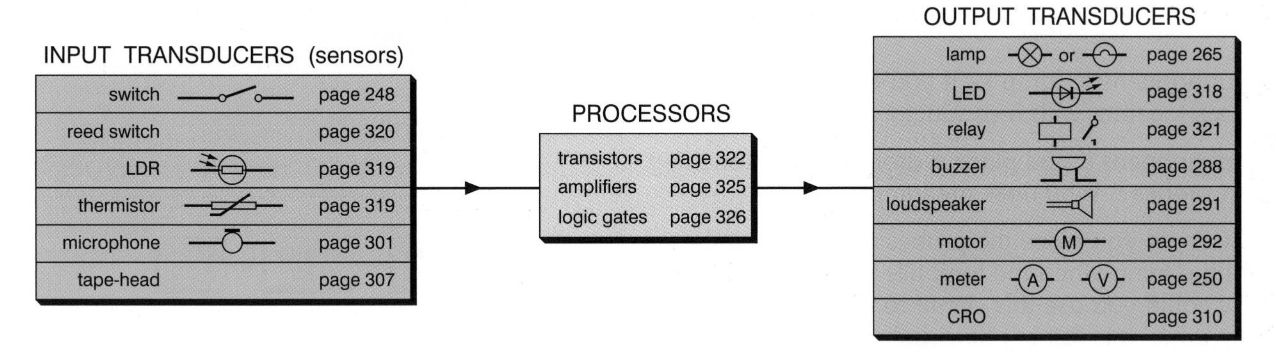

▷ The transistor

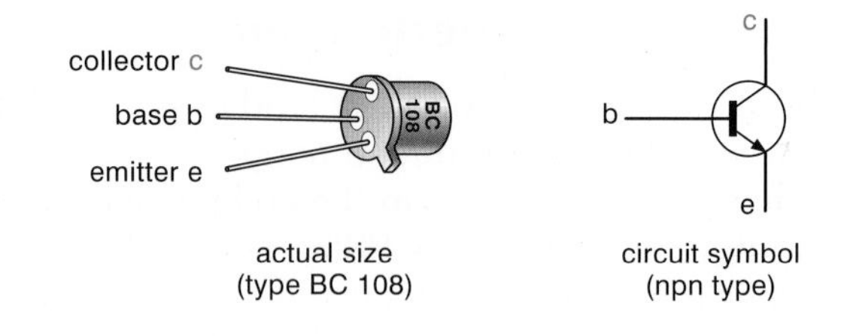

actual size (type BC 108)

circuit symbol (npn type)

This is a device for amplifying small electric currents. It is made of three layers of n, p, n semi-conductor material. The three layers are called the ***collector, base*** and ***emitter***.

The diagram shows one kind of transistor, but there are many different sizes and shapes.

Experiment 38.11 ***Current amplifier***
To investigate a transistor, set up the circuit shown:

Do you find that lamp L_2 lights but lamp L_1 does not?

Now unscrew lamp L_1 so that there is a break in the circuit to the base b.
Do you find that lamp L_2 (in the collector circuit) goes out?

Now replace lamp L_1. Does L_2 light again?

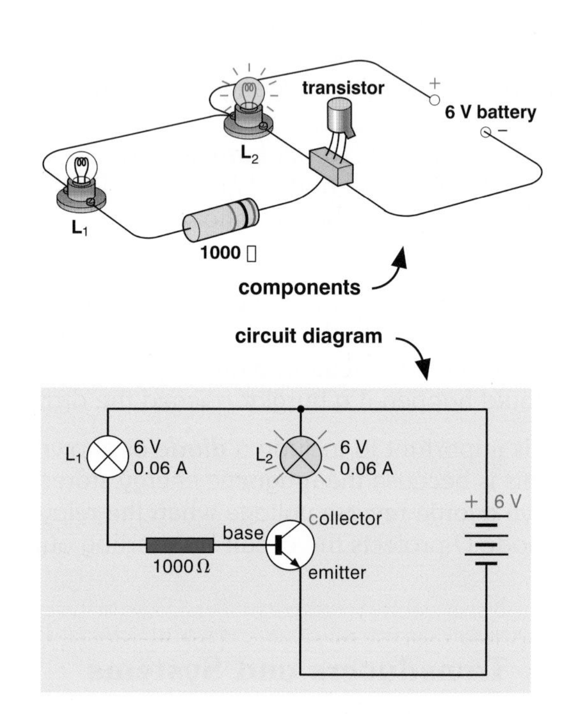

This shows that:
a small current in the base circuit causes a larger current to flow in the collector circuit.

In your circuit, a current of about 1 mA in the base circuit has ***switched on*** a current of about 60 mA in the collector circuit.
A transistor acts as a **current amplifier**.

A transistorised radio contains several transistors to amplify small currents until they are big enough to drive a loudspeaker.

Experiment 38.12 ***Moisture detector***
Change your circuit so that instead of lamp L_1, there are two probes as shown.
The probes are two pieces of copper wire placed close together but not touching.

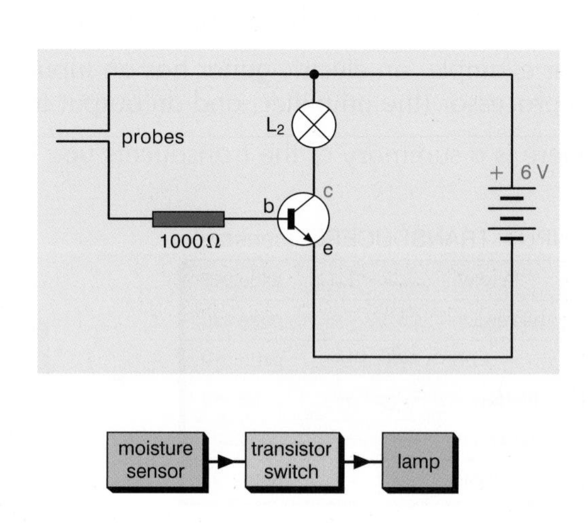

What happens to lamp L_2 if you connect the probes together with a conductor? Why is this?

What happens if you place a drop of water touching both probes? Why does the lamp light?

How could you place the probes in a water tank to light the lamp when the tank fills to a certain depth?
How could you use this circuit to light a warning lamp when it is raining outside?

Experiment 38.13 Help for a blind person
Change your circuit so that it includes a relay to switch on a buzzer:
(Remember to include a diode D to protect the transistor – see experiment 38.10.)
Place the probes so they are inside the top of a cup or beaker.

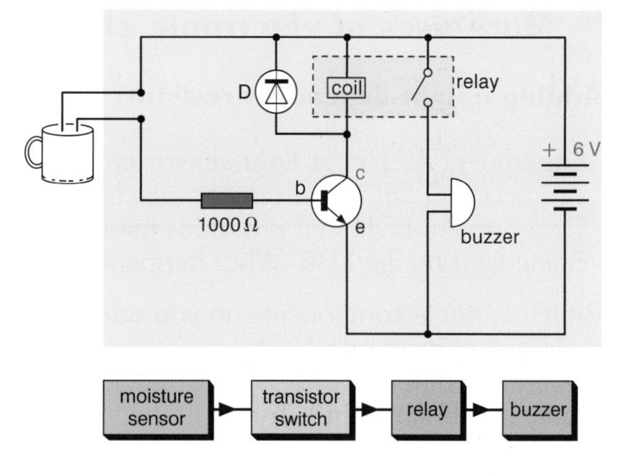

What happens when you fill the beaker with water?

Uses:
How could this be used to help a blind person know when to stop pouring from a tea-pot?
How could it warn when a baby's nappy is wet?
Or switch on windscreen-wipers automatically?

Adding a thermistor

Experiment 38.14 ***A fire alarm*** *(a heat operated switch)*
Connect a thermistor and a variable resistor to a transistor switch as shown. Adjust the resistor until the lamp is just turned off.

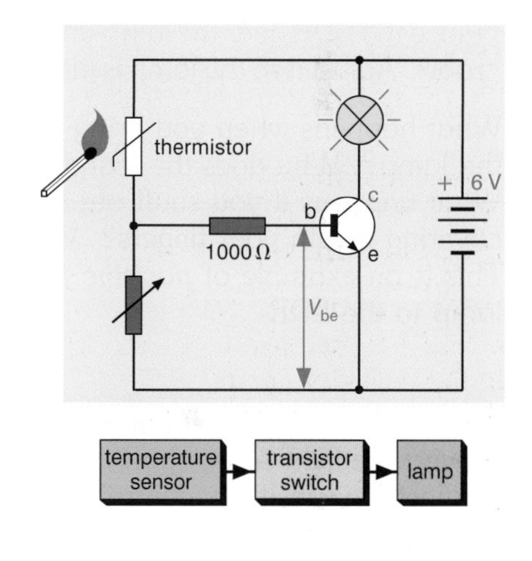

What happens when you heat up the thermistor? (See p. 319.)

The thermistor and the variable resistor are the two parts of a voltage divider or 'potential divider' (see pages 258, 329).
As the thermistor gets hotter and its resistance ***de***creases, its 'share' of the 6 V ***de***creases and so the voltage across the variable resistor ***in***creases. This increases V_{be} (see the diagram) and switches ***on*** the transistor.
If V_{be} is less than 0.6 V the transistor switch is off.
If V_{be} rises above 0.6 V the transistor starts to conduct.
If V_{be} is above 1.5 V the transistor is fully on.

Why is the variable resistor called a 'sensitivity control'?

Uses:
Draw a diagram showing how you could add a relay so that a bell would ring as a fire warning.
How could it be used to warn if a chip pan was getting too hot?

Experiment 38.15 ***A frost alarm***
Invert the temperature sensor part of your circuit, as shown:

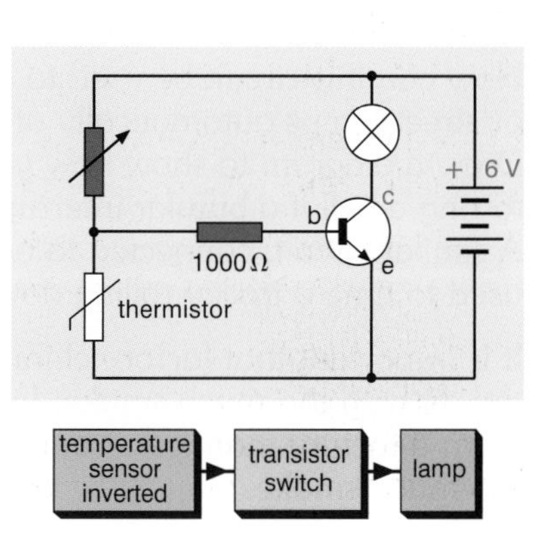

Does the lamp light now when the thermistor is hot or when it is cold? Why?

Uses:
How could this be used as a frost alarm to warn a fruit farmer?
How could this be adapted (using a relay) to keep a tropical fish tank or a greenhouse from becoming too cold?

▶ More uses of electronic circuits

Adding a light-dependent resistor

Experiment 38.16 ***A light-sensor circuit***
Connect an LDR to a transistor switch as shown:
Cover up the LDR so it is dark. Is the lamp off?
Shine light on the LDR. What happens?

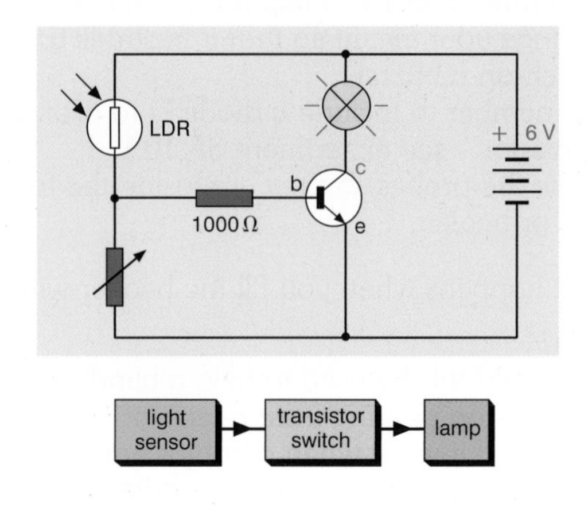

Which variable component can you adjust to make your circuit switch on:
– even in a dim light?
– only in a bright light? Try it.

How could you use this as a burglar alarm for a drawer? (It is usually dark inside a drawer.)

Experiment 38.17 *An electronic candle*
Place the LDR near the lamp and facing it, in a dark room. Adjust it so the lamp is just off.

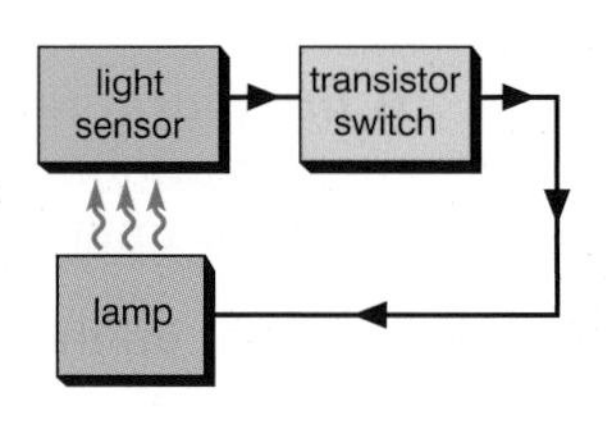

What happens when you bring a lighted match near the lamp? Why does the 'candle' light?
What happens if you snuff out the 'candle' by covering it with your fingers? Why?
This is an example of ***positive feedback*** from the lamp to the LDR.

Experiment 38.18 *A burglar alarm (dark-operated)*
Invert the light-sensor part of your circuit as shown:
Adjust the sensitivity control until the lamp is off in daylight.
Now cover up the LDR. What happens to the lamp?

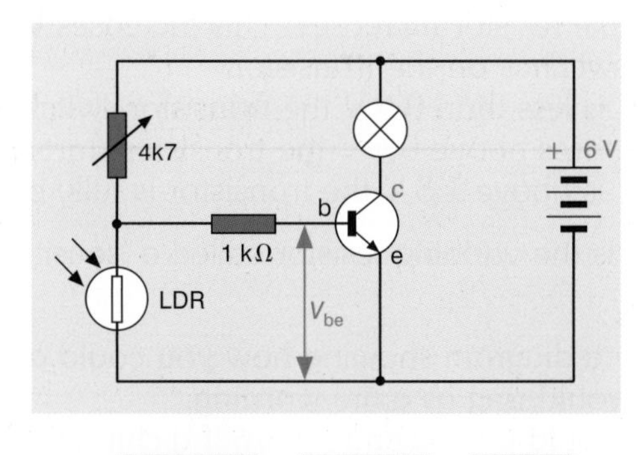

What happens to the resistance of the LDR when it is covered (see page 319)? Does this increase or decrease the voltage V_{be}? Why does the lamp light?

Uses:
How can this circuit be used to switch on car lamps or street-lamps automatically at dusk?
Draw a diagram to show how you could add a relay to ring a bell if a burglar interrupts a beam of light.
A similar circuit, connected to a computer, can be used to time a trolley rolling down a ramp.

It is important that factory chimneys do not pollute the air with too much smoke. Invent a device to warn a factory manager if the chimney is producing too much smoke.

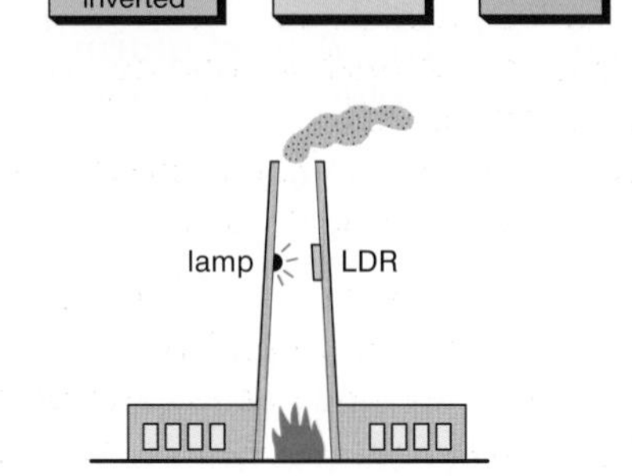

Causing a delay with a capacitor

Experiment 38.19 ***A delay circuit***
Connect a resistor and a capacitor (see page 345) in series with a battery as shown:
Add a high resistance voltmeter and make sure the capacitor and voltmeter are connected the correct way round (correct ***polarity***).
Watch the voltmeter and close the switch. What happens?

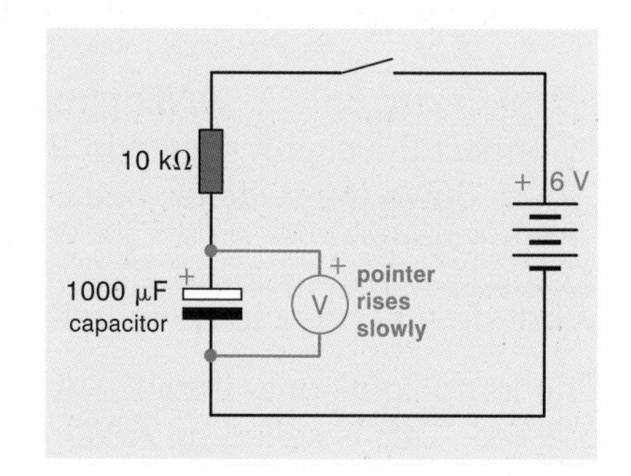

The battery is charging up the capacitor through the resistance. Because it is a large capacitance and a large resistance, it takes a long time. The larger the resistance and the larger the capacitance, the longer the ***time constant*** of the circuit.

Experiment 38.20 *A time-operated switch on*
Connect a capacitor and resistor to a transistor switch as shown:
Press switch S_1 and wait. What happens?

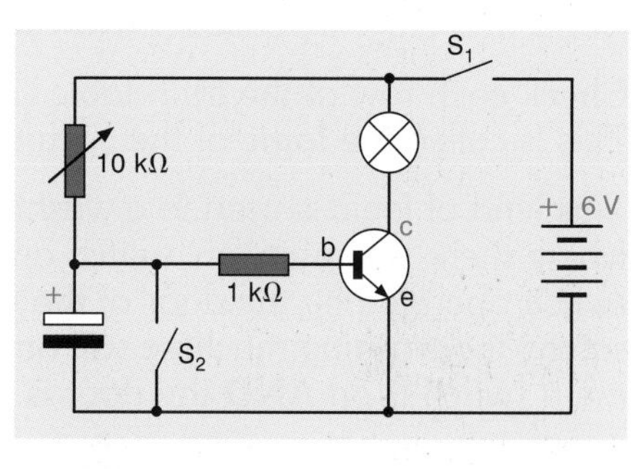

Why is there a delay before the transistor is switched on?
The transistor does not come on until V_{be} rises to 0.6 V.

Press switch S_2 to discharge the capacitor and so re-set the circuit. How can you vary the time delay? Try it.
How could this circuit be used to make an egg-timer for a deaf person? Draw the circuit you would use if the person was blind.

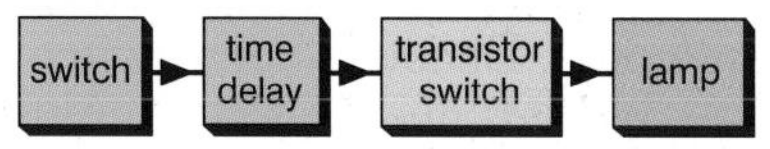

▷ Amplifiers

In experiment 38.11 we saw that a small current in the base of a transistor causes a larger current to flow in the collector circuit. It is a ***current amplifier***.

Amplifiers are needed in radios and CD-players to amplify small signals until they are big enough to drive a loudspeaker.

Experiment 38.21 *One-stage amplifier (an intercom)*
The diagram shows a simple one-stage amplifier:

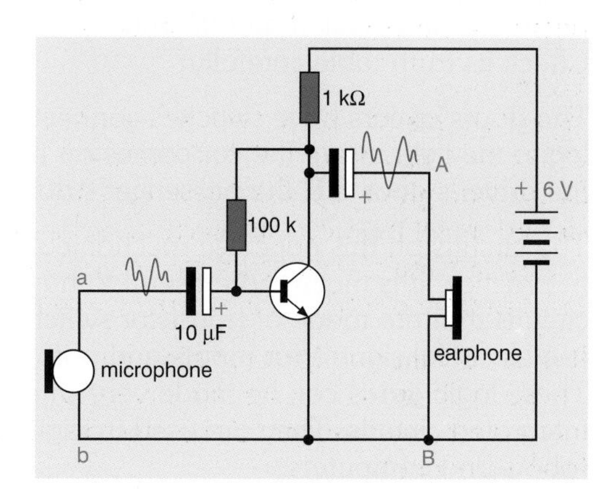

The microphone gives a varying ***input*** voltage across a-b (see also page 232). This is amplified to give a larger ***output*** voltage across A–B.
It is a ***voltage amplifier***.
If the output is 100 times the input voltage, the circuit has a ***voltage gain*** of 100.

In practice, amplifiers usually have 2 or more stages to give greater amplification.
This is done by connecting the output A–B of the first stage to the input a–b of the second stage.

▷ Logic circuits

Experiment 38.22 ***AND circuit***
Connect the circuit shown in the diagram:
Press down the switches one at a time and then both together.

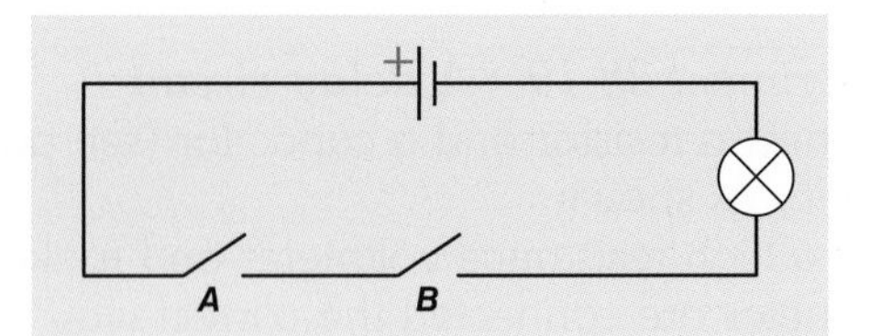

What do you have to do to get the lamp to light?

The lamp lights only if switch ***A*** **AND** switch ***B*** are pressed. This is called an ***AND gate***.
We can describe the way this circuit or 'gate' works in a diagram called a ***truth table***:

switch *A*	switch *B*	lamp
open	open	off
closed	open	off
open	closed	off
closed	closed	ON
inputs		output

Check each row of the truth table carefully.
This is called the ***logic*** of the circuit.

This kind of logic is used in a washing machine where there is a main 'on' switch and another switch operated by the door of the machine. For safety, the washing machine will only work if the main switch is on **AND** the door is closed.

Truth tables are usually shown with numbers rather than words. Because each switch has only two states (open, closed) we use only two digits: **0** and **1**.
If the switch is open and so electricity cannot pass, it is given the digital value **0**.
If the switch is closed so electricity can go through, it is called **1**.
In the same way, a lamp which is off is **0**, and a lamp which is on is **1**.

AND

switch *A*	switch *B*	lamp
0	0	0
1	0	0
0	1	0
1	1	1
inputs		output

Experiment 38.23 ***OR circuit***
Here is a circuit with a different logic:

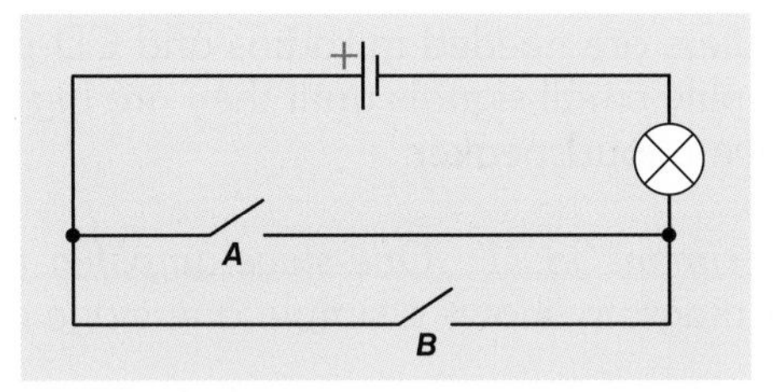

This lamp comes on if switch ***A*** **OR** switch ***B*** is pressed. This is called an ***OR gate***.
Check its truth table carefully:

OR

switch *A*	switch *B*	lamp
0	0	0
1	0	1
0	1	1
1	1	1
inputs		output

The doors of cars have switches connected with OR logic: the light inside the car comes on if either the driver's door **OR** the passenger's door is opened.

The circuits on this page use mechanical switches. In practice, it is much more useful to have logic circuits that are made of transistor switches, and use electrical signals for inputs and outputs.
These logic ***gates*** can be made very small, inside integrated circuits. They are used in digital watches, robots and computers.

AND gate

A tiny integrated circuit can easily contain many AND gates, built from transistor switches (see page 322).
Each transistor switch operates on the ***voltage*** applied to it.
The voltage can be 0 (called logical 0) or it can be higher (called logical 1).
The circuit symbol for an AND gate with two inputs is shown here:
(Remember its shape: it looks like the D of AND.)

The truth table is the same as the AND gate on the opposite page.
There is an output only if input ***A*** **AND** input ***B*** are at a logical 1.
Remember: 0 is zero volts and 1 is a higher voltage (usually 5 V).

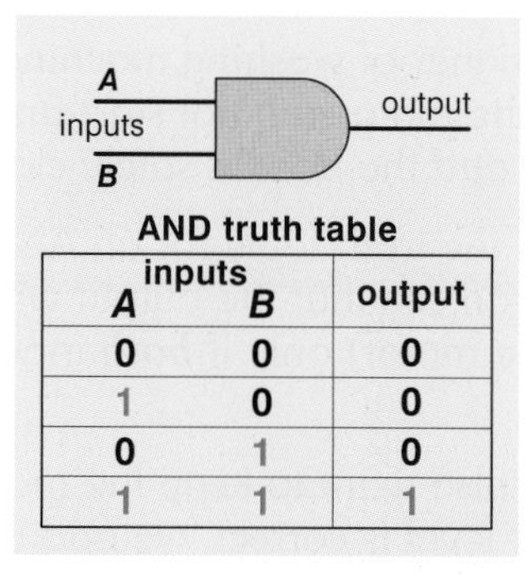

AND truth table

inputs A	inputs B	output
0	0	0
1	0	0
0	1	0
1	1	1

OR gate

The circuit symbol for an OR gate with two inputs is shown here:
Notice the different shape from an AND gate.

The truth table is the same as the OR gate on the opposite page.
There is an output if either input ***A*** **OR** input ***B*** is at a higher voltage.

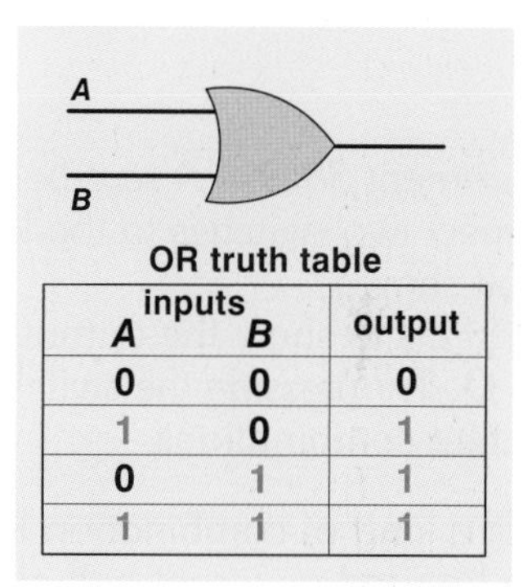

OR truth table

inputs A	inputs B	output
0	0	0
1	0	1
0	1	1
1	1	1

NOT gate

The third kind of logic gate is the NOT gate or inverter.
Its circuit symbol is shown here:

Its truth table is very simple: the output is always the ***opposite*** of the input. It is an ***inverter***.
The output is **1** if the input is **NOT 1**.

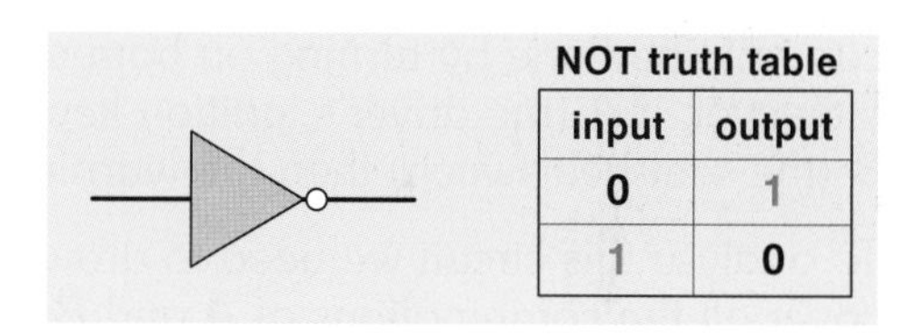

NOT truth table

input	output
0	1
1	0

Logic gates,
Have just two states:
Just off or on,
*With **0** or **1**.*

Sometimes these gates are combined. For example,
AND with NOT = NAND gate.
OR with NOT = NOR gate.
Compare these symbols and truth tables with the AND and OR gates. What do you notice?

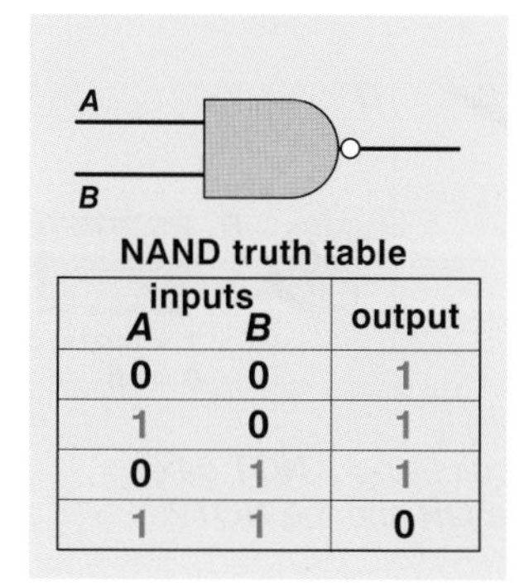

NAND truth table

inputs A	inputs B	output
0	0	1
1	0	1
0	1	1
1	1	0

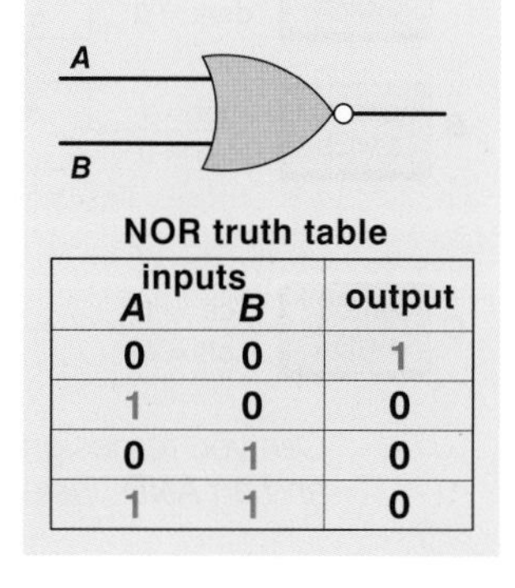

NOR truth table

inputs A	inputs B	output
0	0	1
1	0	0
0	1	0
1	1	0

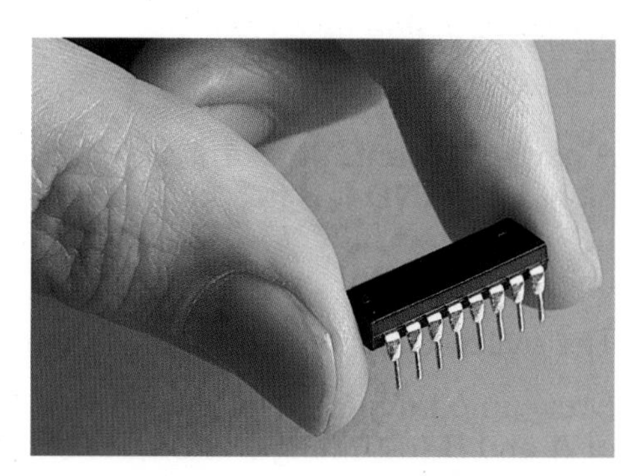

This chip contains four AND gates

▷ Using logic gates

A designer of washing machines wants to ensure that the motor will not start until the on/off switch is on ***and*** the door is safely closed, like this:

From the truth table for an AND gate (page 327), you can see that the output will be a **1** (to switch on the motor) only if ***both*** inputs are **1**.

If we also want to keep the motor off until the water level is correct, we add another AND gate:

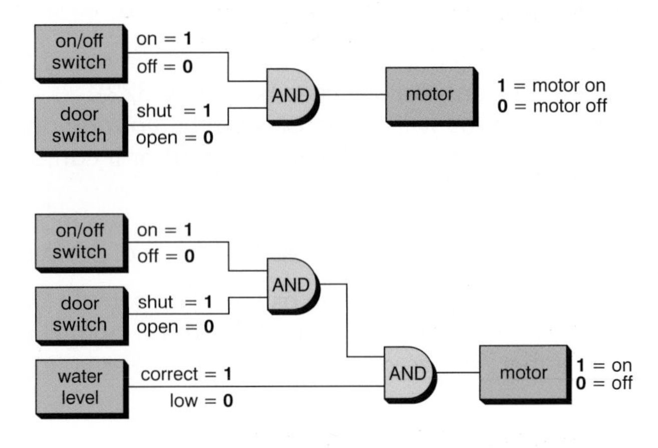

Experiment 38.24 A seat-belt alarm
Connect two switches to the logic circuit shown:
Use LEDs to show the outputs at ***P*** and ***Q***. Try pressing the switches in different combinations.

This is a kind of combination lock.
It can be used as a seat-belt alarm.

To turn on the engine, point ***P*** must be at logical **1** and this can only be done by turning on both switches ***A*** and ***B***:
However, if ***A*** (the driver's ignition key) is turned on before ***B*** (the seat-belt switch) then the warning buzzer sounds.

To analyse this circuit we need to draw up a truth table to cover all the combinations of ***A*** and ***B***:

The third column shows the result at **P** (of ***A* AND *B***), while column ***Q*** shows the result at ***Q*** (of **NOT *B***). You should work through this table carefully, moving down each column in turn.

A	*B*	*P*	*Q*	*R*
inputs		engine *A* AND *B*	NOT *B*	buzzer *A* AND *Q*
0	0	0	1	0
0	1	0	0	0
1	0	0	1	1 = on
1	1	1 = on	0	0

Experiment 38.25 Dark and cold alarm
Connect up this system:

A gardener wants an alarm to sound if it is night-time **AND** it is freezing outside, so that he can protect his crops. There is more than one way to do this. Here is one way:

The switch **C** lets him turn off the alarm.

Draw up a truth table for all the combinations of ***A*** and ***B*** (assuming ***C*** is ***on***), to check that it works correctly.

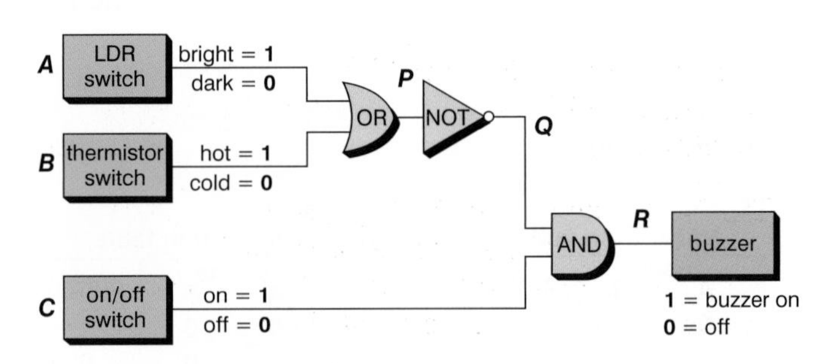

Can you re-design this circuit using 2 ***NOT*** *gates and an* ***AND****, instead of the* ***OR*** *and the* ***NOT****?*

To operate these logic gates, we need **input circuits** that give either 0 volts (logical **0**) or give about 5 volts (logical **1**). This is how we do it:

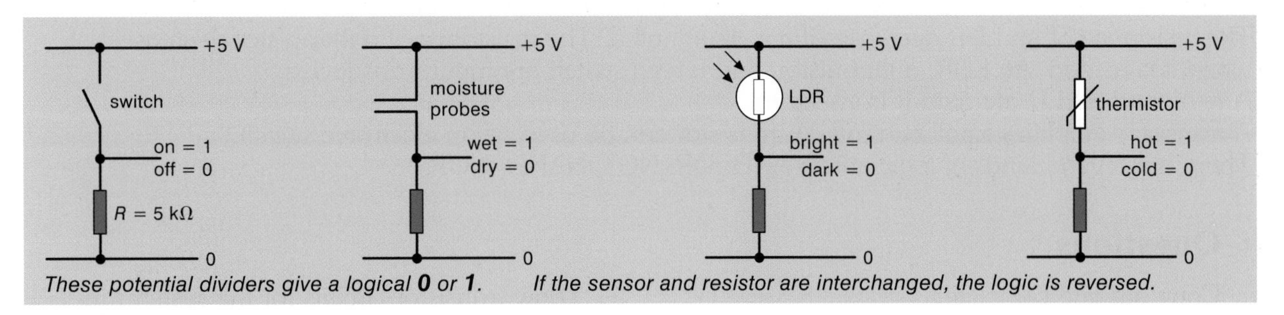

*These potential dividers give a logical **0** or **1**. If the sensor and resistor are interchanged, the logic is reversed.*

Experiment 38.26 Light-dependent switch
Connect up this system:

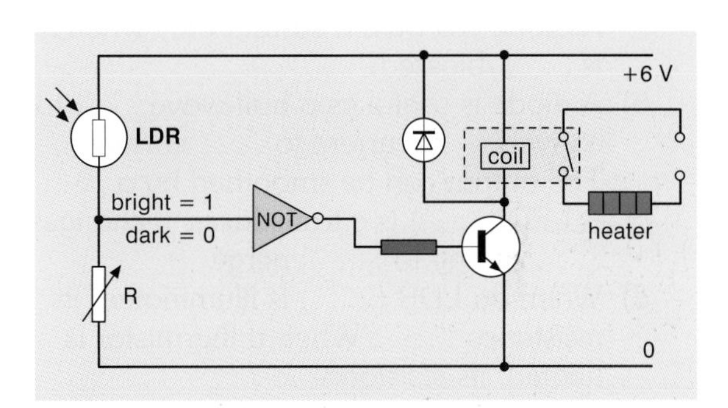

This uses the third circuit shown above.
If it is dark, the input to the NOT gate is **0**, so its output is **1**, which turns on the transistor. The relay then switches on the heater.

Page 321 explains the reason for the diode.
The variable resistor R changes the sensitivity of the circuit to the light.

Experiment 38.27 Wet AND light sensor
Connect up this system:

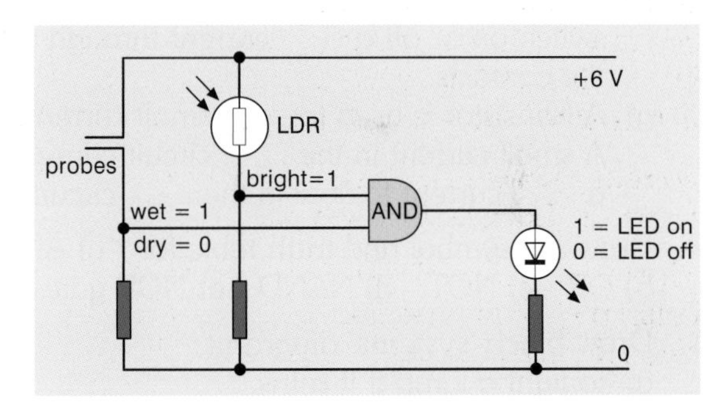

If either input is **0**, the output is **0** and the LED is off. The LED will only light if the probes are wet **AND** light is shining on the LDR.

Use: To warn you (with a bell) if it was raining AND it was day-time – so you could bring in the washing from the line or the baby from the pram (but it wouldn't bother you if it rained at night!).

Experiment 38.28 Wet AND cold sensor
Replace the LDR with a thermistor, and add a NOT gate:

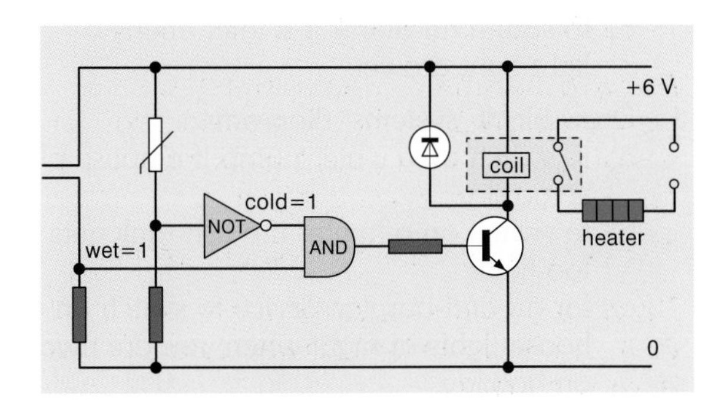

The NOT gate inverts the logic, so that it gives a logical **1** when the thermistor is ***cold***.
If the output of the AND gate is **1**, it switches on the transistor and the relay circuit.

Use: A safety circuit for a washing machine. How can you change the circuit so that it rings a bell if it is either wet OR cold?

Summary

A diode allows a current to flow one way only. This can be used to rectify a.c. to d.c.
The resistance of an LDR decreases if it is illuminated. The resistance of a thermistor decreases when it is heated. An LDR, a thermistor, and a reed switch are input transducers.
A relay and a LED are output transducers.
A transistor amplifies small currents. A transistor can be used as an electronic switch.
There are several kinds of logic gates: AND, OR, NOT, NAND, NOR.

▶ Questions

1. Copy out and complete:
  a) Germanium and silicon are called . . . . . A diode will pass a current only when it is . . . . -biassed.
  b) A diode is useful as a half-wave . . . . to convert . . . . current to . . . . current. The output can be smoothed by a . . . . .
  c) A LED (. . . .) is a transducer; it changes . . . . energy to . . . . energy.
  d) When an LDR (. . . .) is illuminated, its resistance . . . . . When a thermistor is heated, its resistance . . . . .
  e) A . . . . (NO) reed switch does not pass a current until a . . . . is brought near it.
  f) A . . . . current in the coil of a relay can switch on or off a . . . . current through the contacts.
  g) A transistor is used to . . . . small currents. A small current in the . . . . circuit causes a . . . . current to flow in the . . . . circuit.

2. Draw the symbol and truth table for a) AND b) OR c) NOT d) NAND e) NOR gates.

3. Draw block 'systems' diagrams:
  a) to light a lamp if it rains,
  b) to ring a bell if it rains,
  c) to switch on a fan motor if a greenhouse gets too hot,
  d) to sound an alarm if a thief opens a light-tight drawer.

4. Draw block 'systems' diagrams:
  a) to switch on a water pump if a house plant is too dry,
  b) to switch on a heater if a fish tank gets too cold,
  c) for an anti-burglar device to switch on the house lights at night when you are away on holiday.

5. Draw transistor circuits for the systems in question 3.

6. Draw block 'systems' diagrams:
  a) to switch on a water pump if a ditch fills with water,
  b) to switch on a roadworks light at night,
  c) to switch on a fan in a storage heater if it heats up,
  d) to ring a bell if a baby cries,
  e) for a fire alarm to detect smoke ('smoke is more dangerous than fire'),
  f) to switch on a pump to fill a water tank if the level is too low,
  g) to ring a bell when an oven is up to the correct temperature,
  h) for an oven timer.

7. Draw a truth table for this system:

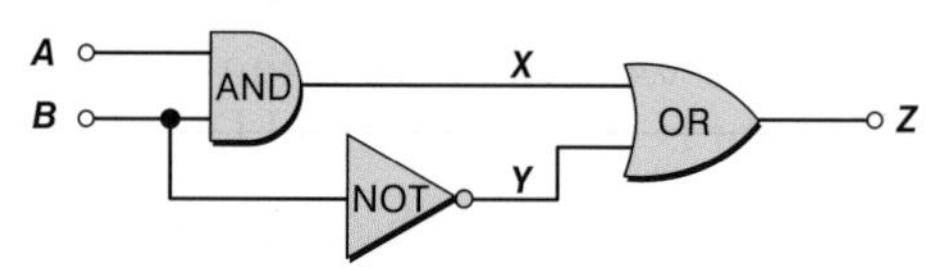

8. Draw block 'systems' diagrams:
  a) to sound an alarm only if the baby cries ***and*** its nappy is wet,
  b) to open a greenhouse window if it is warm in the day-time,
  c) to tell you if it is a good day for gardening because it is warm, dry and light,
  d) to sound an alarm if the tropical fish tank is dark or cold or the water is low,
  e) to allow a drilling machine to run only if the operator presses a foot switch and has not got her hands breaking a beam of light which is shining near the drill,
  f) to show a deaf person if the doorbell has rung since last they looked. (Add a re-set.)

Further questions on page 335.

▷ Physics at work: Radio

A **communication system** consists of:

① ***A method of transferring energy***.
A radio uses electromagnetic waves (page 209).
A telephone (page 289) uses a current in a wire.

② ***Information*** (for example, someone talking into a microphone).

③ ***A way of en-coding the information***, combining it with the energy to be transferred. This is transmitted to the receiver, where it is de-coded.

Radio

The diagram shows how a radio wave (called the ***carrier*** wave) can have its amplitude changed or ***modulated*** by a sound wave (an analogue signal). This is **A**mplitude **M**odulation or **AM**:

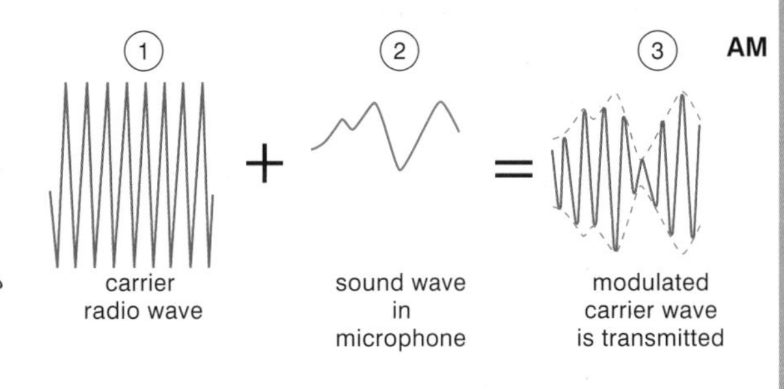

This wave is transmitted from the radio station.

transmitter (see page 211)

aerial

(the alternating radio wave induces an alternating current in the aerial)

radio receiver

volume

tuner

tuner knob

diode

capacitor

de-modulator

sound wave amplifier

loudspeaker

In your radio receiver, a ***tuner*** selects the carrier wave you want (from the hundreds of radio waves passing through your room).

Then a diode rectifies the signal (see page 317), and a capacitor smooths away the carrier wave. Together they act as a ***de-modulator*** (or de-coder).

The signal is then amplified (page 325) and passed to the loudspeaker (page 291), so you can hear it.

An alternative way of encoding the signal is **FM**, **F**requency **M**odulation, as shown here:

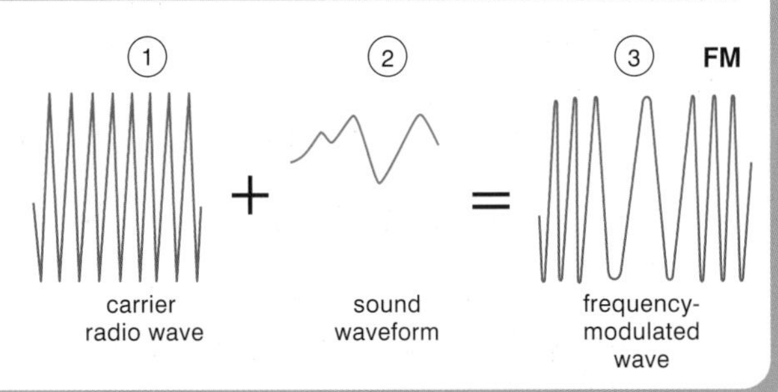

It gives better quality sound, with less interference.

▷ Magnetic effect, motors

1. This question is about an electric **motor**.
a) State two ways of reversing its direction. [2]
b) State two ways to make it spin faster. [2]

2. The diagram shows an incomplete two-pole, single coil, direct current **electric motor** with 50 turns of wire on the armature A.

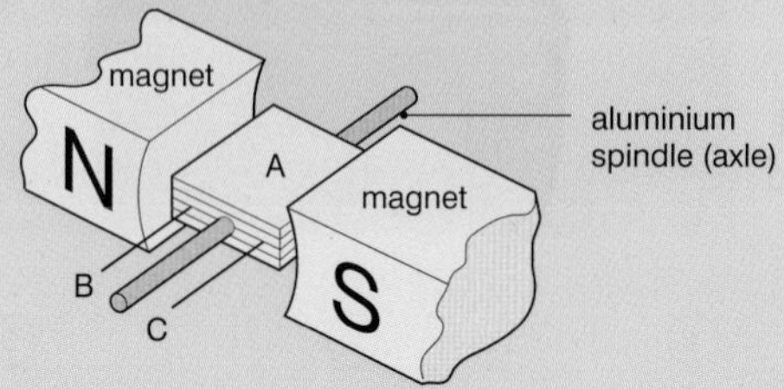

a) Copy and complete the diagram, showing how the ends B and C of the wire wound on A would be connected through a split ring commutator to a d.c. power supply so that the armature would rotate continuously. [4]
b) State what would happen to the direction of rotation of the armature if
i) the polarity of the power supply were to be reversed,
ii) the direction of the current in the armature were to be reversed,
iii) the polarity of the magnetic field were to be reversed,
iv) both the changes described in ii) and iii) were made at the same time. [4]
c) State what would happen to the speed of rotation of the motor if
i) the number of turns on A were to be increased, with the current unchanged,
ii) stronger magnets were to be used with the same coil and power supply. [2]
d) Explain how the commutator makes the armature rotate continuously when a power supply is connected. [6]

3. A student hung a magnet next to a coil of wire to make a door chime.

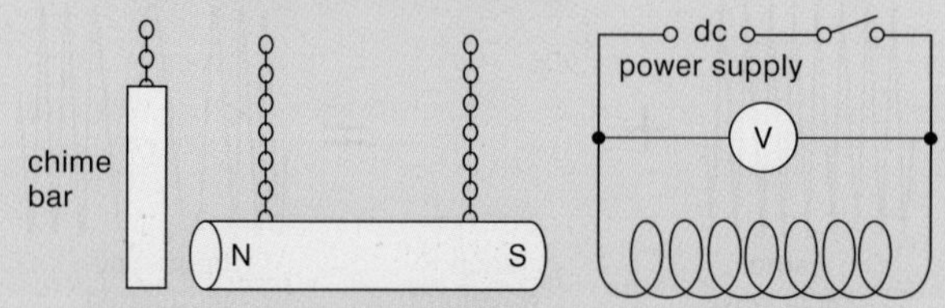

Study the diagram and then explain, step by step, how the door chime works. [2]

4. An electric **motor** in a central-heating pump is supplied with energy at 80 J/s, of which 10 J/s appears as heat and the rest pumps the water. What is the efficiency of the pump? [2]

5. A student was given a small electric **motor** and asked to measure its efficiency when lifting a load of 200 N.
The student connected the motor to an electrical circuit and read the voltmeter and the ammeter while the motor raised the load slowly and steadily. Here are the results:

Load lifted	= 200 N
Distance load was lifted	= 0.4 m
Time taken to lift load	= 5.0 s
Voltage across motor	= 10.0 V
Current through motor	= 4.0 A

a) Calculate the work done on the load. [2]
b) Calculate the power **output** of the motor. [2]
c) Calculate the power **input** to the motor. [2]
d) Calculate the efficiency of the motor. [3]
e) Motors are always less than 100% efficient because of energy losses inside the motor. Give **two** energy losses saying where in the motor they may occur. [2] (AQA)

6. The diagram below shows a **moving coil ammeter**. The current to be measured is passed through the coil.

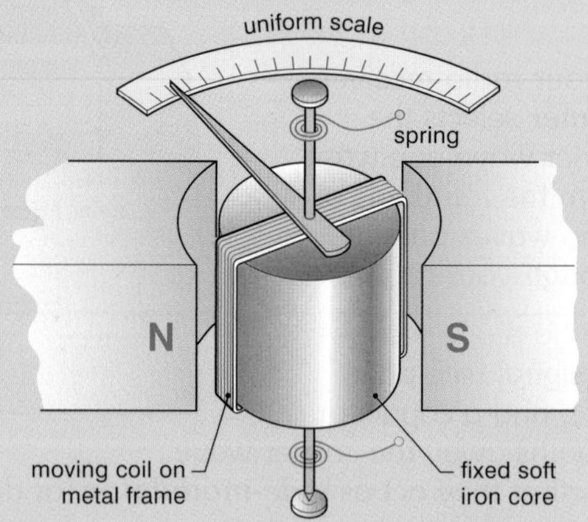

A current of 5 mA is passed through it. Explain why the pointer moves to the 5 mA mark on the scale and no further. [3] (AQA)

7. This question is about a **loudspeaker** (page 291). When a current flows in the coil, a force makes the diaphragm move. Explain why a force is produced. [3]

 Note: although a question may be marked (Edex) etc, it will often be relevant to other examinations as well.

▷ Electromagnetic induction

8. This question is about an electric **generator** (see page 298).
a) What must be done to a generator to enable it to produce electricity? [1]
b) Why is a voltage induced in the coil? [1]
c) Give **four** ways in which the size of the induced voltage could be increased if the generator was changed. [4] (AQA)

9. a) Here is a drawing of a simple **alternator**. A single coil is positioned between the poles of a magnet.

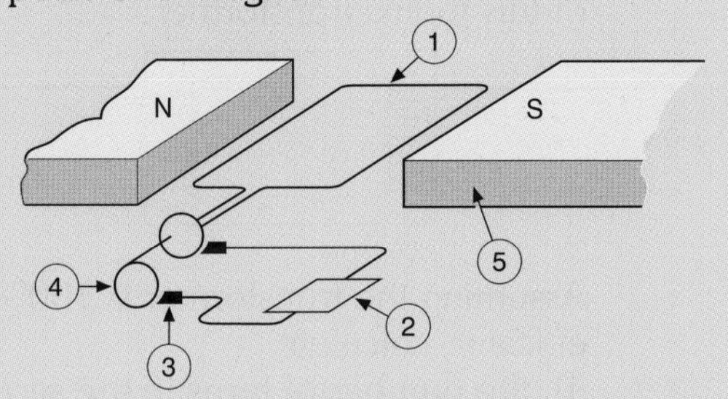

Certain parts of the alternator have been numbered. Write down the correct name for each part 1 to 5. [5]
b) The coil is rotated at a steady speed of 60 rotations per minute.
i) An oscilloscope is connected across the load and adjusted to get on its screen a trace for **one** rotation of the coil. Sketch the trace for one rotation starting and finishing in the position of the coil shown in the diagram. [3]
ii) State **two** ways of **increasing** the voltage produced by the alternator. [2]
c) An example of a practical alternator is a bicycle dynamo. Here, the magnet is rotated with the wheel and the coil is kept still.
i) Give an advantage of rotating the magnet and not the coil. [1]
ii) Give a disadvantage of using a dynamo to power bicycle lights. [1]
(AQA)

10. A simple electrical generator produces an alternating voltage.
a) Explain carefully, using a diagram, what is meant by an alternating voltage. [2]
b) State three ways in which the voltage output could be increased. [3]

11. a) Describe, in as much detail as you can, how the energy stored in coal is transferred into electrical energy in a power station. [5]
b) A **transformer** is used to produce a very high voltage to transmit electrical energy through the national Grid.
Explain why electrical energy is transmitted at a very high voltage. [3] (AQA)

12. Explain why
a) a transformer will only transform alternating voltages, [2]
b) transformers are widely used throughout the National Grid. [2] (WJEC)

13. a) What is the difference between alternating current and direct current? [2]
b) Electric power is distributed in the national Grid at the very high voltage of 400 000 V. Why is this voltage so high? [1]
c) How is the voltage reduced from 400 000 V? [2]
d) Why is it necessary to reduce the voltage to 230 V for our homes? [1] (OCR)

14.

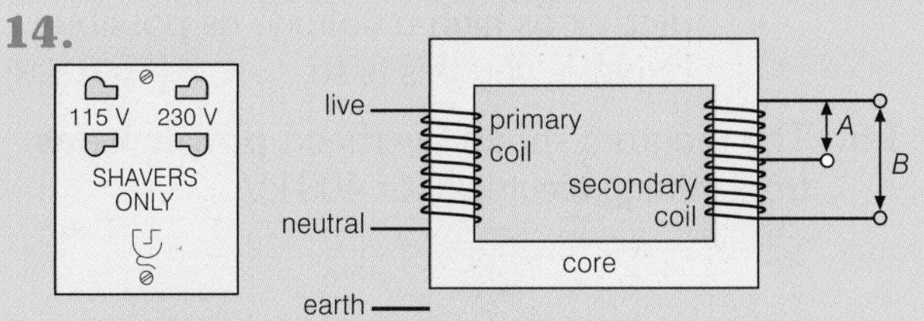

A **bathroom razor socket** contains a transformer. This is to isolate the user from the mains supply.
a) Copy the transformer diagram, and:
i) add an S to show the correct position for a Switch.
ii) add an F to show the correct position for a Fuse.
iii) connect the Earth lead to the correct point on the transformer. [3]
b) The transformer has two outputs, labelled A and B. Copy and complete the table below to show the number of turns and output voltage for each coil. [4] (Edex)

Input voltage	Primary turns	Secondary turns	*A or B*	Output voltage
230 V	5000	2500		
230 V	5000	5000		

▷ Electromagnetic induction (continued)

15. The diagram shows a simple form of **transformer** used for stepping down an alternating voltage supply.

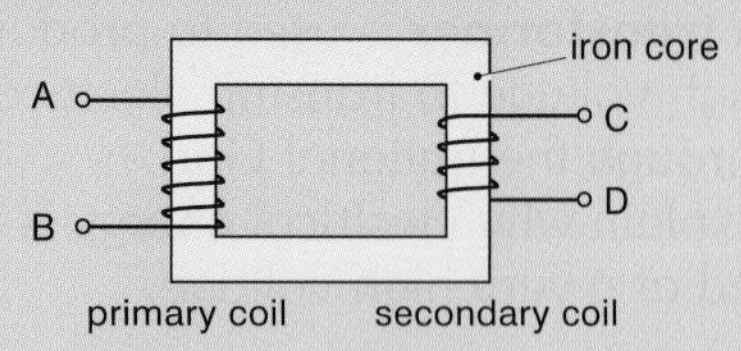

a) By writing about magnetic fields, explain how the transformer works. [3]

b) A power line supplies electrical energy to a transformer in a factory.
The input voltage to the transformer is 11 000 V. The transformer changes this to 415 V for use in the factory.
The power input to the transformer is 800 kW.

i) Calculate the current in the secondary coil of the transformer. [3]
What assumption have you made about the transformer? [1]

ii) The power line to the factory is operated at as high a voltage as possible. Explain why this is so. [4] (OCR)

16. The diagram shows overhead **power lines** transmitting electricity at 400 kV.

a) How is the electricity prevented from flowing to earth through the metal pylon?

b) Some people think that underground cables would be better. What are the advantages and disadvantages of underground and overhead cables? [4]

c) The transmission lines carry 200 MW.

i) Calculate the current in the lines. [3]

ii) If the total resistance of the transmission lines is 10 Ω, calculate the voltage drop when the lines carry 200 MW. [3]

iii) Calculate the power loss in the lines. [3]

iv) What happens to this 'lost' power? [1]

d) Explain why electricity is transmitted at high voltages like 400 kV rather than low voltages like 230 V. [4] (Edex)

17. a) i) Describe, briefly, the energy changes involved in the generation of electrical energy at a **power station**. [2]

ii) State and explain **two** advantages of a hydroelectric power station compared with other types. [2]

b) What are the advantages of transmitting power at i) very high voltage?
ii) alternating voltage? [3]

c) A 6 V, 24 W lamp shines at full brightness when it is connected to the output of this mains transformer:

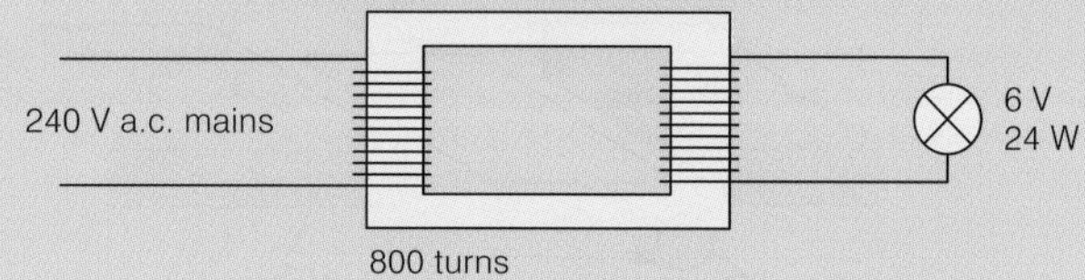

Assuming the transformer is 100% efficient, calculate

i) the number of turns in the secondary coil if the lamp is to work at its normal brightness.

ii) the current which flows in the **mains** cables. [2]

d) Explain whether, and how, the number of secondary turns of the transformer shown in c) should be altered if

i) two 6 V lamps in **series** are to work at normal brightness.

ii) two 6 V lamps in **parallel** are to work at normal brightness. [3] (WJEC)

18. A farmhouse is supplied with electricity from the grid system. Near the farmhouse a transformer steps down an input of 11 500 V to an output of 230 V.

a) Why is the electricity not transmitted all the way from the power station at 230 V? [4]

b) Explain why the 11 500 V is stepped down to 230 V before it is used in the house. [2]

c) Why must alternating current be supplied to the transformer? [1]

d) The number of turns on the secondary coil of the transformer is 2000. Calculate the number of turns on the primary coil. [3]

e) In the farmhouse there is a 690 W electric food mixer which uses the 230 V supply. Calculate the current through its motor. [4]

f) Calculate the resistance of the motor. [4]

▷ Electronics

19. Fiona uses rechargeable batteries for her mp3 player. Her battery charger is solar-powered.

a) Draw a labelled circuit diagram to show how the rechargeable battery, a diode, and the solar cell are connected together in the charger. Show the direction of the current as the battery is being charged. [3]

b) Explain why a diode is used. [1]

c) Suggest two advantages of a solar-powered charger over a mains-powered one. [2]

20. If a variable resistor is connected as shown, it can be used to alter the voltage smoothly.

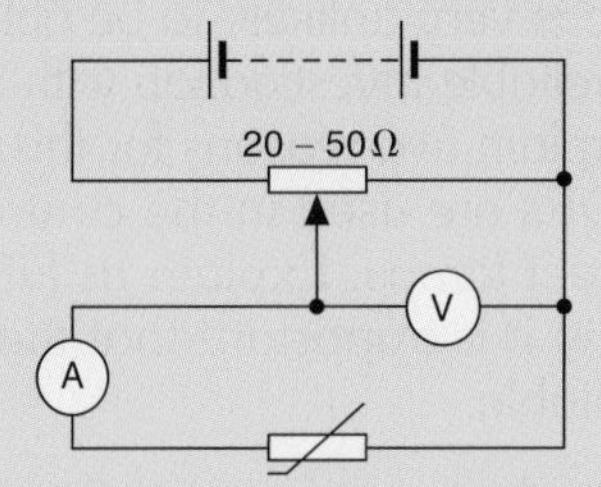

With the temperature of the **thermistor** at 20 °C, the circuit was adjusted until the current was 20 mA and then the voltage noted. This procedure was repeated with the thermistor at different temperatures.

Temperature (in °C)	20	40	60	80	100
Voltage (in V)	12.0	4.2	2.1	1.2	0.8

a) Plot the readings on graph paper. [3]

b) Draw a smooth curve through the points. [1]

c) Use your graph to help you work out the resistance of the thermistor at:

i) 30 °C ii) 70 °C [5] (AQA)

21. This circuit is designed to switch on the beacon automatically when daylight fades.

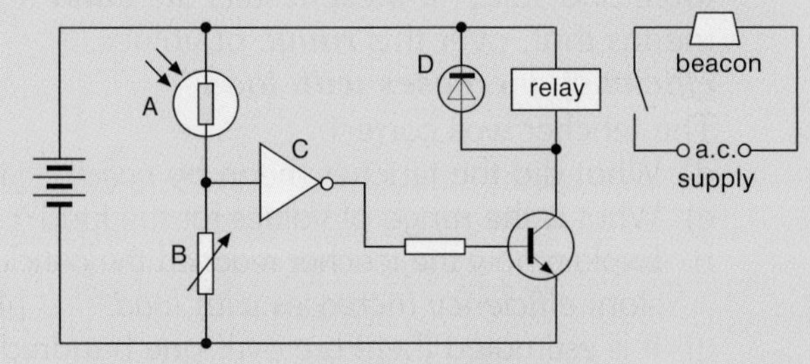

a) Name components A, B, C, D. [4]

b) Explain the function of a relay. [1]

c) Explain in detail how the circuit works. [5]

22. The diagram shows two logic gates.

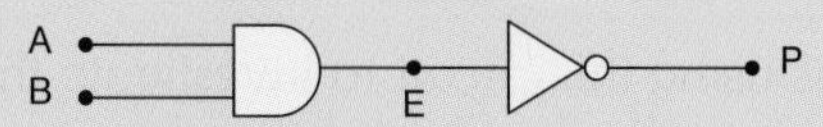

Copy and complete the truth table: [2]

A	B	E	P
0	0		
0	1		
1	0		
1	1		

23. The diagram represents an alarm system.

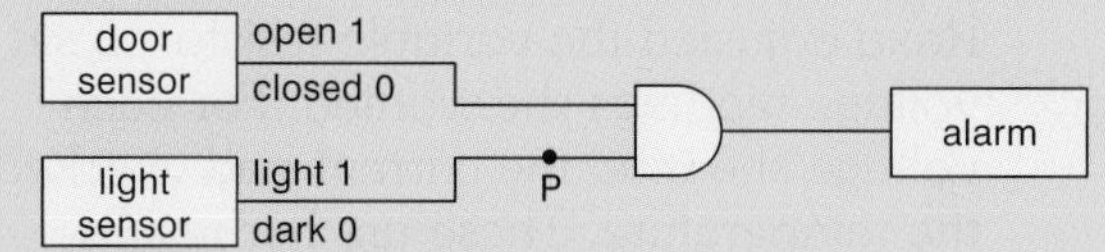

a) When will the alarm sound? [2]

b) A NOT gate is placed at point P. What effect will this have on

i) the input to the AND gate?

ii) the conditions under which the alarm will sound? [3]

24. Plants can be grown in a specially heated box with the correct amount of lighting. Sensors are included in the box and operate as shown below.
Copy and complete the diagram so that the buzzer sounds when the switch is on and when either the light level or the temperature level or both are low.

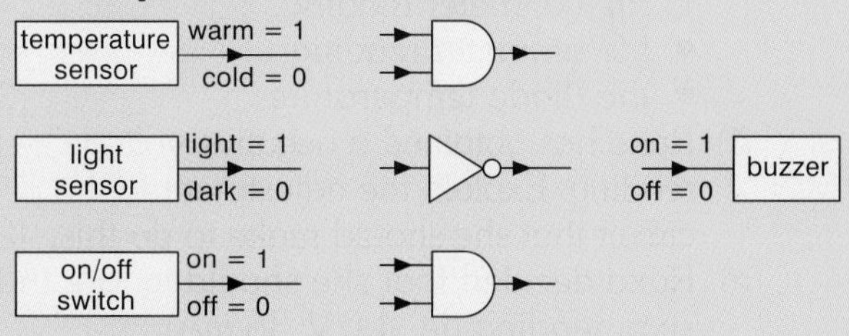

25. The diagram shows a circuit for a combination lock. When a switch is pressed, it changes its input from 0 to 1.

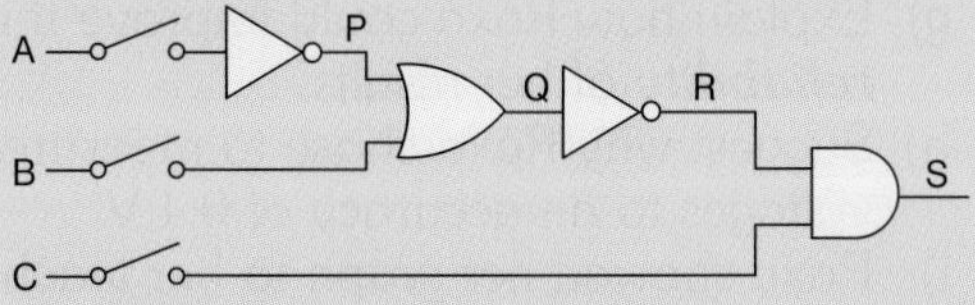

a) Draw up a truth table showing the state of P, Q, R, S for all combinations of A, B, C.

b) If S rises to logical 1, it opens the lock. What is the correct combination?

▷ How Science works

26. Raxa is using this circuit to investigate the current in a **diode**. She has fixed a thermometer to the diode.

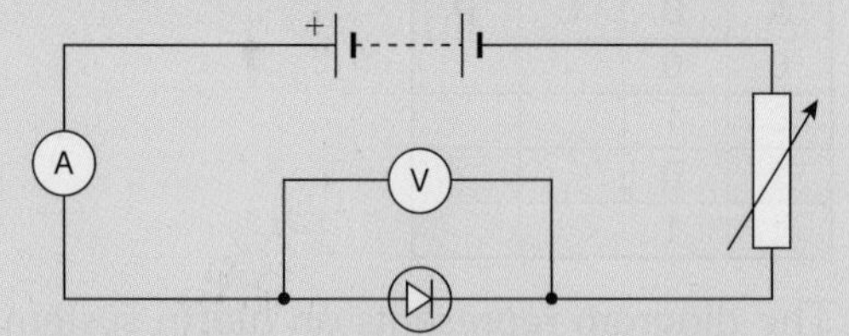

Raxa adjusted the variable resistor to set the voltage readings she wanted. For each voltage she read the current and checked the temperature. These are her results:

Voltmeter reading in V	0	0.3	1.6	2.2	2.7	4.0	−1.2
Ammeter reading in mA	0	0.2	7.0	11.0	14.0	36.0	0
Temperature of diode in °C	22	22	22	22	22	34	22

a) Explain why Raxa fixed a thermometer to the diode. [1]

b) What is the ***range*** of Raxa's voltage readings? [2]

c) State whether each of these variables is a ***dependent*** variable, an ***independent*** variable or a ***control*** variable:
- the voltmeter reading,
- the ammeter reading,
- the diode temperature. [3]

d) Raxa has obtained a negative voltage reading. Explain the adjustment to the circuit that she should make to do this. [2]

e) Raxa decided that she should ignore the sixth reading (ie. 4.0 V, 36 mA). Explain why this was sensible. [2]

f) Plot a graph of the current against voltage. (Ignore the sixth reading.) [4]

g) Explain how Raxa could improve the ***reliability*** of her results. [2]

h) Suggest why Raxa chose to measure the voltages to an accuracy of 0.1 V. [2]

i) Raxa showed her graph to her teacher. The teacher suggested that Raxa should take some extra readings. Explain why and write down a suitable range of voltages for these new readings. [2]

27. Theo has been told the resistance of an **LDR** decreases as the light intensity increases.

a) Sketch a graph of the results you would expect to get if you investigated this. [2]

b) Name i) the independent variable and ii) the dependent variable. [2]

Theo investigates this, using an LDR with an ammeter, voltmeter, 6 V battery, metre rule and a table lamp.

c) Draw the circuit diagram that he should use. [3]

d) Apart from safety considerations to do with the circuit, explain two other safety considerations in Theo's investigation. [4]

e) He is very unlikely to be able to carry out a reliable investigation with this apparatus. Explain two reasons for this. [4]

f) LDRs are used in the control circuits of street lamps. Explain as fully as you can why it is important that these LDRs are reliable. [4]

28. Two students are investigating an **electric motor** using the apparatus shown on page 100. They used different loads and measured the time taken to lift a load through a vertical distance of 0.5 m. The table shows the results:

Load (in N)	1.0	1.5	2.0	2.5	3.0	3.5	4.0
Time taken (s)	1.9	2.7	3.2	3.6	3.9	4.1	4.2

a) Name i) the independent variable and ii) the dependent variable. [2]

b) Choose a word to describe these variables: ***categoric***, ***continuous***, ***discrete***, ***ordered*** [1]

c) Name two of the variables which were controlled for the results in the table. [2]

Their teacher looked at the results in the table and said, "If these results are ***valid*** it means that, over this ***range*** of values, ***efficiency increases with load***."

The teacher was correct.

d) What did the teacher mean by *valid*? [1]

e) What is the *range* of values for the load? [1]

f) Explain how the teacher reached the conclusion: *efficiency increases with load*. [4]

g) It is estimated there are over one hundred million electric motors in use in the UK every day. Explain why it is important that these motors are as efficient as possible. [2]

29. Two students are investigating how the power output from a **solar cell** depends on its area. They use a lamp, shining on a solar cell that is connected to an ammeter and a voltmeter.

a) They keep the distance from lamp to cell constant throughout. Explain why. [2]

b) What is the power output if the voltmeter reads 0.9 V and the ammeter reads 0.8 A? [2]

Here are the results they got for different areas:

Exposed area (cm^2)	10	20	30	40	50	60	70	80	90	100
Power (mW)	60	100	180	260	300	350	490	490	540	600

c) Draw a graph, and a line of best fit. [4]

d) Which result is anomalous? [1]

e) What conclusion can they draw? [1]

Jack says they should have done it outside in the sunshine. Emily disagrees. She says it is easier to get reliable results in the lab.

f) Explain what is meant by ***reliable*** results. [1]

g) State two variables which are relevant to the investigation but which would be difficult to control if you are outside. [2]

30. The table compares the manufacturer's specifications for a small **wind generator** with the results of an engineer's test in *What Turbine* magazine:

	Manufacturer's figures	Magazine test results
Diameter	2.1 m	2.1 m
Maximum power output	900 W	852 W
Voltage	12 V d.c.	12.2 V d.c.
Energy output at max power	700 kWh per month	Not tested
Minimum operating wind speed	3.4 m/s	3.6 m/s
Wind speed for maximum power	14.0 m/s	16.4 m/s
Maximum safe wind speed	55 m/s	Not tested

a) Calculate the maximum current available according to i) the manufacturer and ii) the magazine. [3]

b) Most of the manufacturer's figures appear to be optimistic. Suggest reasons for this. [2]

c) Explain why some of the magazine's results may be unreliable. [1]

d) Suggest why the magazine engineer did not try to test:
- the monthly energy output,
- the maximum safe wind speed. [2]

e) Explain why the wind turbine is unlikely to produce 700 kWh in a month. [1]

31. Calum is investigating how the current in a circuit varies with the applied voltage. He uses this circuit.

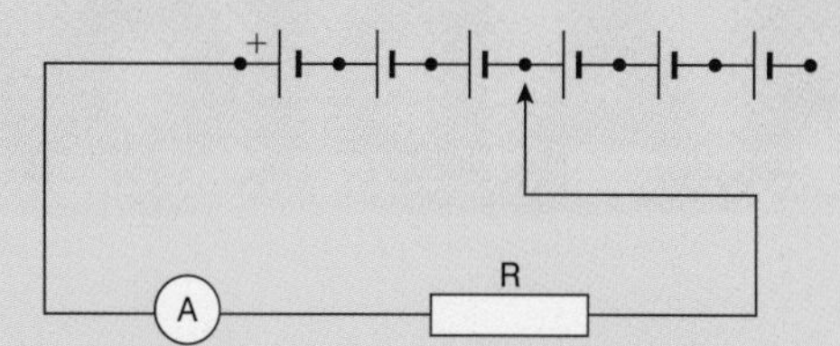

Calum has connected six cells in series to make a battery. Each cell is labelled 1.5 V. He changes the applied voltage by using a different number of cells for each reading. The diagram shows three cells connected. He notes the current for each number of cells:

Number of cells used	1	2	3	4	5	6
Current in mA	13	25	38	42	54	67

a) Calculate the applied voltage in each case. [2]

b) Explain why this voltage is an independent variable, but not a continuous variable. [2]

c) Plot a graph of Calum's results. [2]

d) Calum expected his results to give a perfectly straight line, but they do not. Suggest some reasons for this. [3]

e) Draw a line of best fit and calculate the slope (gradient) of the graph. [2]

f) Use your value for the slope to find the approximate resistance of the circuit. [2]

g) Suggest how the investigation could be improved so as to yield results that are both more ***reliable*** and more ***precise***. [4]

h) Redraw the circuit to show how a voltmeter should be used. [3]

32. Many scientists believe that ***global warming*** is caused by carbon dioxide emissions from burning fossil fuels. Other scientists think that there is not enough evidence for this and global warming might just be part of a natural cycle of temperature change.

a) Explain why some of the evidence, both for and against these views, may be unreliable. [2]

b) Explain how reliable evidence is obtained. [1]

c) Explain why it is likely that people will continue to use fossil fuels for some time to come, even though they may cause global warming. [2]

Further questions on page 368.

chapter 39

In 1896, Henri Becquerel discovered almost by accident that some substances (like uranium and radium) can blacken a photographic film even in the dark. These substances are ***radioactive***.

Radioactivity can be dangerous and experiments in this chapter will be teacher demonstrations only. ⚠

▶ Detecting radioactivity

1. Photographic film

The photograph shows how some film has been blackened by radioactivity except in the shadow of a key.

2. The gold-leaf electroscope

Demonstration 39.1
Charge up a dry gold-leaf electroscope as you did in experiment 30.6 (see page 243).

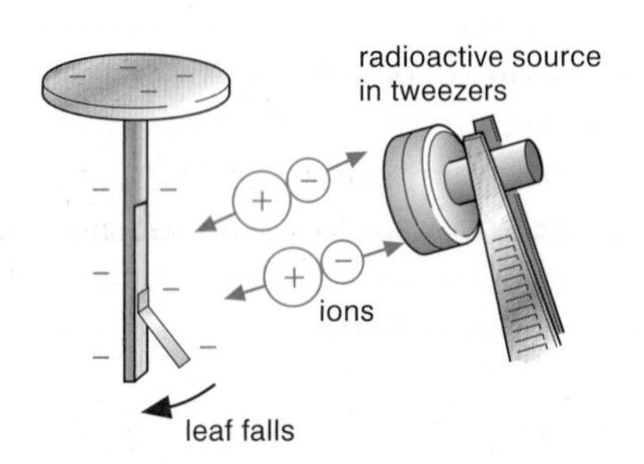

Does the leaf stay up for several minutes?
What happens if a radioactive source (e.g. radium) is brought near the leaf, using tweezers?

The leaf falls because the air is **ionised** by the radioactivity (see experiment 30.11 on page 244). A negative electroscope attracts the positive ions and so it is discharged. This ***ionisation*** is used in other detectors of radioactivity.

3. The cloud chamber

In a cloud in the sky, the water droplets tend to form on dust particles or on charged ions.
In a cloud chamber, a ray from a radioactive source causes a line of ions on which a thin cloud forms (rather like the cloud trail behind a high flying aircraft).

The photograph shows the tracks of 'alpha' particles in a cloud chamber.

4. The spark counter

Demonstration 39.2 ⚠
A spark counter consists of a fine wire stretched just below a piece of metal gauze.
A high voltage between the wire and the gauze is adjusted until it is almost, but not quite, sparking.
Using tweezers, a radioactive source (radium) is brought close to the gauze.

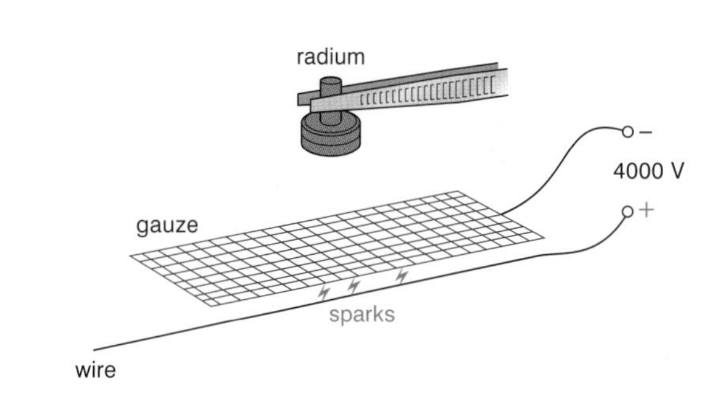

What happens? Why?

Rays from the radioactive source ionise the air so that it is a better conductor, and sparks are produced.

Do the sparks occur regularly or ***randomly***?

Place a thick piece of paper between the source and the gauze. What happens? Why does it stop?

5. The Geiger–Müller tube (G–M tube)

This is a metal tube with a thin wire down the centre. It contains a gas at low pressure. It works on the same principle as a spark counter, but the voltage is lower so that no spark is formed. Instead, a pulse of current is produced. This is amplified and passed to either:

a ***scaler***, which counts the pulses and shows the total number, ***or***

a ***ratemeter***, which shows the rate in 'counts per second'. In addition, a loudspeaker clicks each time a ray from the source passes through the tube.

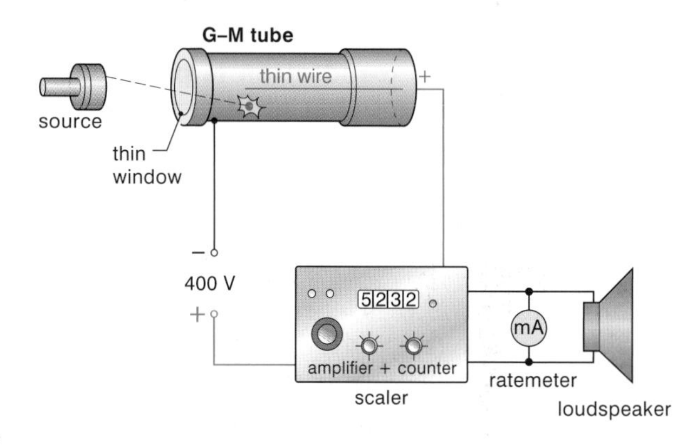

Demonstration 39.3
A radium source is placed near a G–M tube connected to a ratemeter and loudspeaker.

What do you hear as the source moves closer?

Place a thick piece of paper between the source and the tube.
Does the paper stop all the rays?

Because paper stopped the sparks in the last experiment, but not the clicks in this experiment, it seems that there are at least two kinds of radiation from radium.

In fact there are three kinds, called ***alpha***, ***beta*** and ***gamma*** rays.

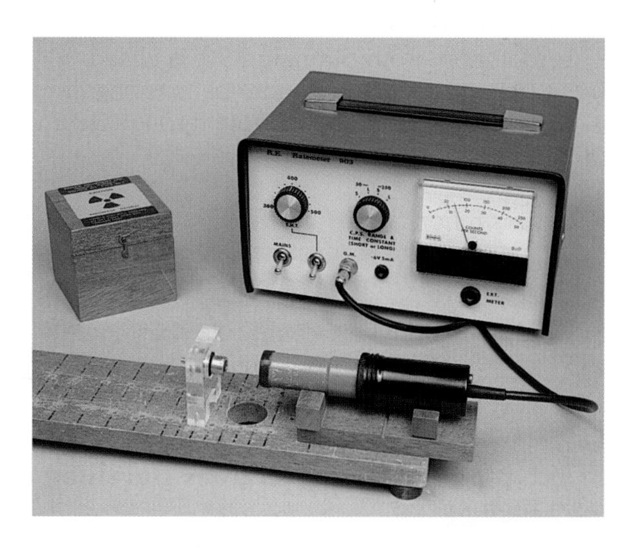

▶ Alpha, beta and gamma rays

Demonstration 39.4 **Background radiation**
Use a Geiger–Müller tube with a very thin window and connect it to a ratemeter with a loudspeaker. What do you hear when it is switched on?

The clicks you hear from the Geiger counter are called the ***background count***. It is due to natural radioactivity, mainly from the rocks of the Earth and cosmic rays from the Sun.

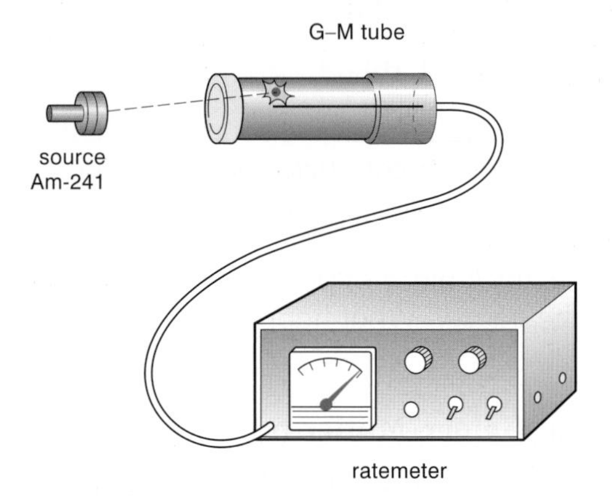

Demonstration 39.5 **Alpha radiation**
Using tweezers, an Americium-241 source is placed close to the window of the tube. Note the count rate on the ratemeter.

Now place a piece of paper between the source and the tube. What happens to the count rate?

This radiation is called **alpha (α) radiation**. It is easily stopped by paper or by your clothing. It can be deflected by electric or magnetic fields. It can be shown that **α-radiation is positively charged particles moving at high speed**.

In fact, an alpha-particle is found to be the same as the ***nucleus*** of a helium atom (see page 342). It consists of 2 protons (+) and 2 neutrons.

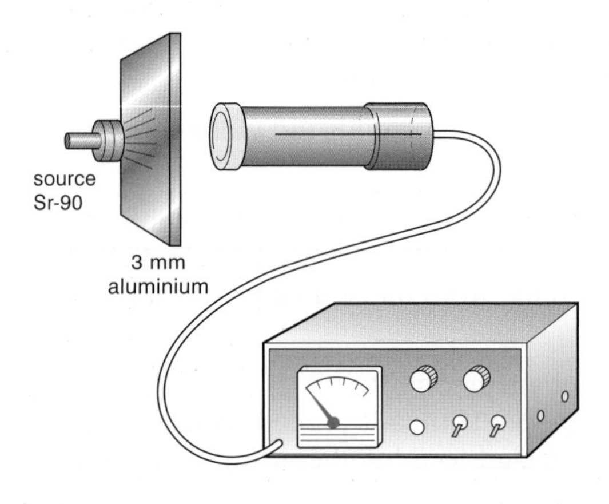

Demonstration 39.6 **Beta radiation**
Using tweezers, the source is replaced by a Strontium-90 source. Note the count rates.

Paper won't stop these rays. Try different thicknesses of aluminium.

This radiation, from Strontium-90, is called **beta (β) radiation**. It is more penetrating than α-particles, but cannot penetrate through 3 mm of aluminium.

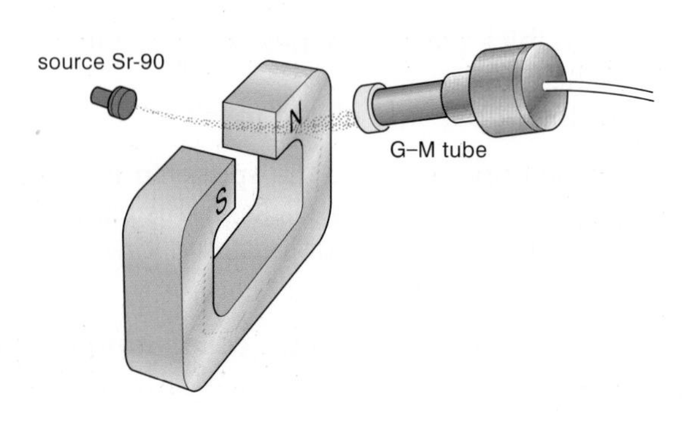

Demonstration 39.7
Place a powerful magnet between the source and the tube. What happens to the count-rate?

Move the tube until the count-rate rises again:

The direction of the deflection shows that beta-rays must be ***negatively-charged particles*** (by using Fleming's Left-hand Rule, see page 290). In fact, **β-particles are electrons travelling at very high speed**.

Demonstration 39.8 ***Gamma radiation***
Use the same apparatus, but with a cobalt-60 source. Do the same tests as before.

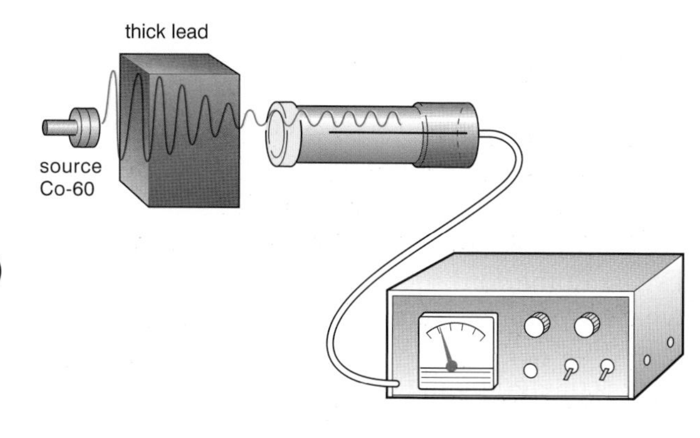

Paper and aluminium won't stop this radiation.

This radiation from cobalt-60 is called **gamma** (γ) **radiation**.
It is not affected by a magnetic field and is very penetrating. Although it is reduced, it is not stopped even by thick pieces of lead.

If the distance between the source and the G–M tube is increased, it is found that at twice the distance the count-rate is only one quarter. At three times the distance, the intensity of γ-rays is only one-ninth.
The intensity of light from a lamp changes in just the same way (see page 214).

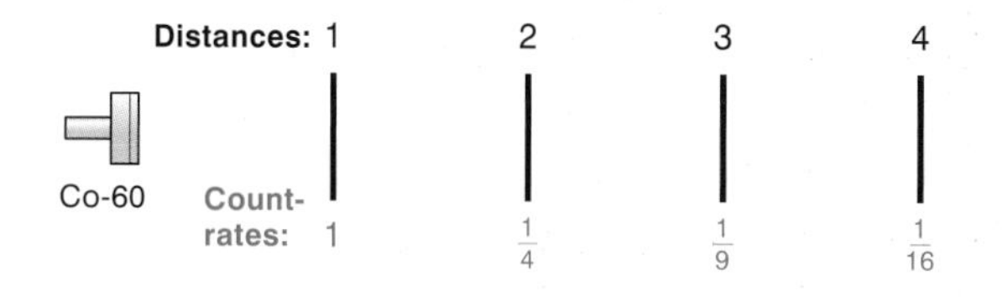

In fact, it can be shown that these **gamma-rays are really electromagnetic waves of very short wavelength** (see page 208).

Repeat all these experiments with a radium source. Do you find that it emits alpha- ***and*** beta- ***and*** gamma-rays?

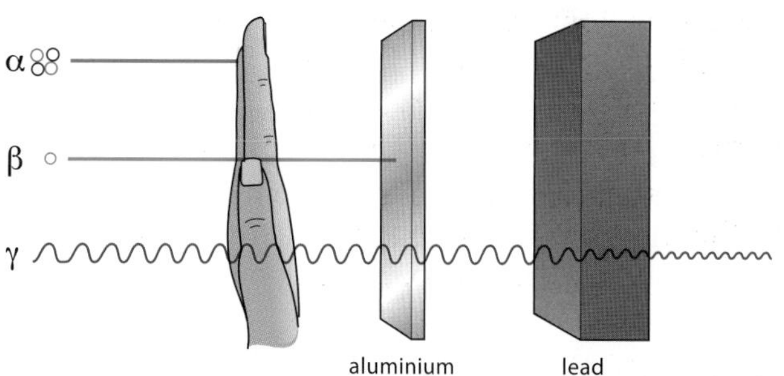

Here is a **summary**:

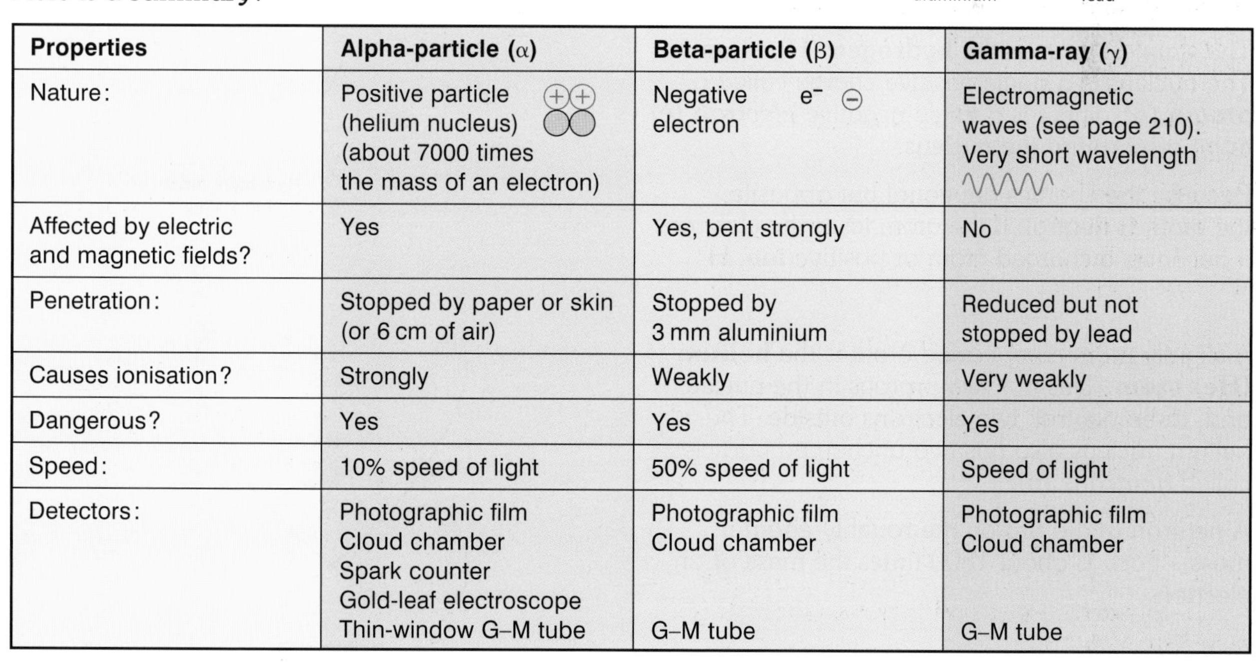

Properties	**Alpha-particle (α)**	**Beta-particle (β)**	**Gamma-ray (γ)**
Nature:	Positive particle (helium nucleus) (about 7000 times the mass of an electron)	Negative electron e^-	Electromagnetic waves (see page 210). Very short wavelength
Affected by electric and magnetic fields?	Yes	Yes, bent strongly	No
Penetration:	Stopped by paper or skin (or 6 cm of air)	Stopped by 3 mm aluminium	Reduced but not stopped by lead
Causes ionisation?	Strongly	Weakly	Very weakly
Dangerous?	Yes	Yes	Yes
Speed:	10% speed of light	50% speed of light	Speed of light
Detectors:	Photographic film Cloud chamber Spark counter Gold-leaf electroscope Thin-window G–M tube	Photographic film Cloud chamber G–M tube	Photographic film Cloud chamber G–M tube

▶ Atomic structure

In 1904, J.J.Thomson suggested that an atom is like a 'plum-pudding', with some negative electrons stuck in a positive blob (see page 372).

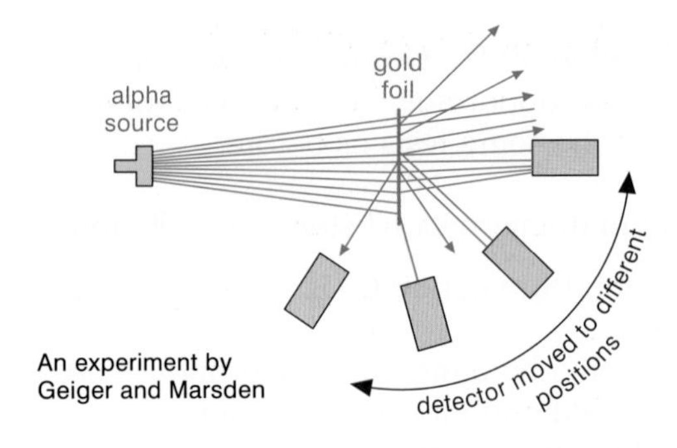

An experiment by Geiger and Marsden

To find more about the structure of an atom, Ernest Rutherford suggested an experiment in which α-particles were fired at thin gold foil. Detectors were used to find how the α-particles were scattered by the gold atoms:
It was found that some α-particles were scattered back ***towards the source*** – rather like firing a machine-gun at tissue paper and finding that some of the bullets bounce back!

In 1911, Rutherford showed that this could be explained if each atom has a tiny core or ***nucleus*** with a ***positive charge***.
A positive nucleus repels the positive α-particles so that they are scattered in different directions:

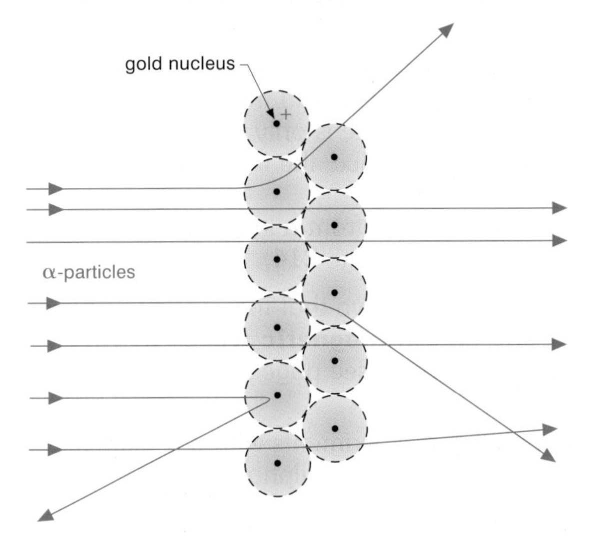

When Rutherford calculated the size of a nucleus, he found that it was very small even compared with a very small atom.
If you imagine magnifying an atom until it is about the size of your school hall, then the nucleus in the centre would be about the size of this full stop.

The simplest atom is the **hydrogen (H) atom**. The nucleus is a single positive charge called a ***proton (p)***. There is a single negative ***electron (e)*** in an orbit round the nucleus.

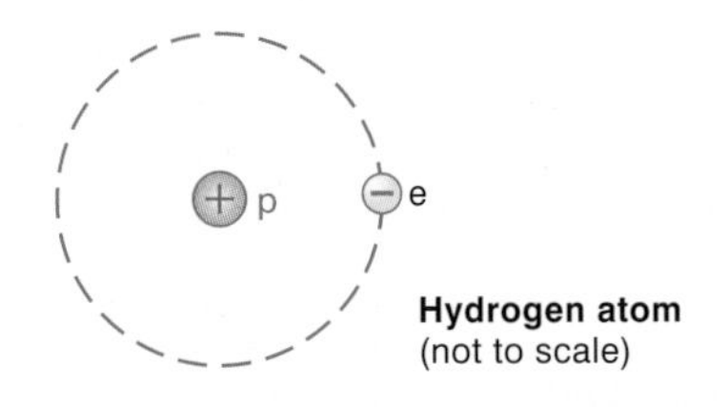

Hydrogen atom
(not to scale)

Because the charges are equal but opposite, the atom is neutral. If the atom loses the electron, it becomes a charged atom or positive ***ion***, H^+.

A slightly more complicated atom is the **helium (He) atom**. This has two protons in the nucleus and, to be neutral, two electrons outside. The helium nucleus also has two uncharged particles, called ***neutrons (n)***.

A neutron and a proton are roughly ***equal*** in mass – each is about 1800 times the mass of an electron.

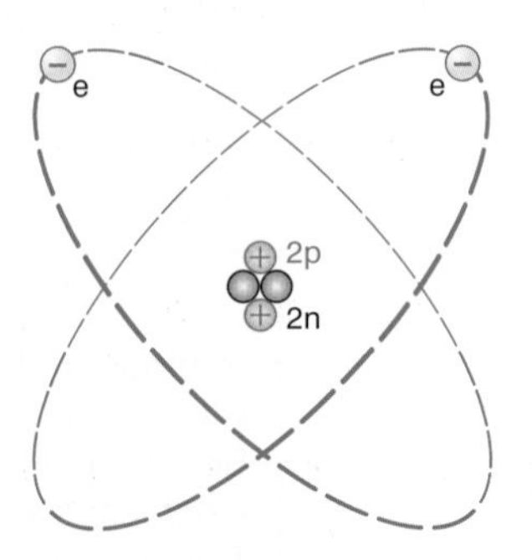

Helium atom

Proton number and nucleon number

Different elements have a different number of protons – for example, a ***lithium*** atom has 3 protons, a ***beryllium*** atom has 4 protons, a ***boron*** atom has 5 protons, a ***carbon*** atom has 6 protons and so on.
The number of protons in an atom is called its **atomic number**.

What is the atomic number of lithium?
A list of some elements in order of atomic number is shown at the bottom of the page.

Another important number is the **mass number**: this is the total number of ***nucleons*** (= protons + neutrons).
For example, the mass number of lithium is 7 (= 3 protons + 4 neutrons).
A shorthand way of describing a ***nuclide*** is shown here:

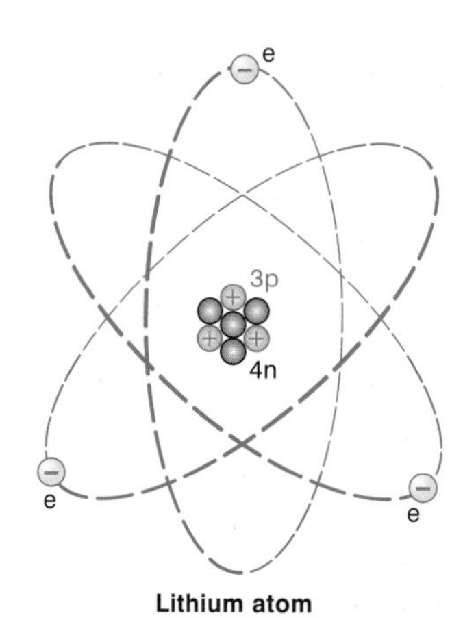

Lithium atom

$^{7}_{3}\mathrm{Li}$

Number of nucleons (protons + neutrons) **(mass number)**

Number of protons **(atomic number)**

Isotopes

All the oxygen that you are breathing has an atomic number of 8. That is, it has 8 protons in the nucleus (and 8 electrons to make a neutral atom).

Most of the oxygen that you are breathing has also 8 neutrons (so the mass number = 8 + 8 = 16).
This is $^{16}_{8}\mathrm{O}$.
Look at the diagram of this atom:

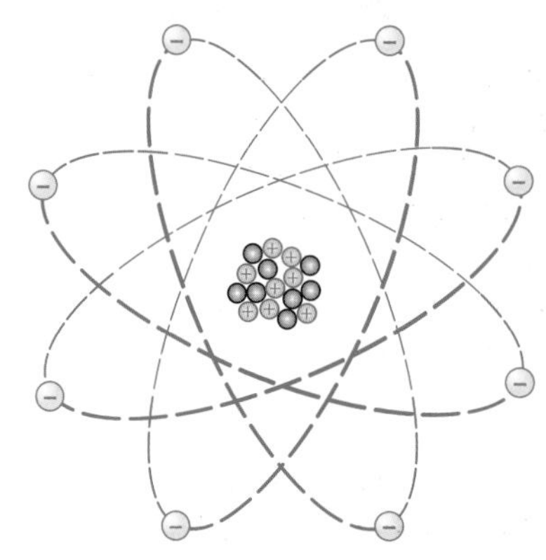

An atom of $^{16}_{8}\mathrm{O}$

Some of the oxygen you are breathing has got 9 neutrons (but still 8 protons, making a mass number of 17). This is $^{17}_{8}\mathrm{O}$.
Draw a diagram of this atom.

$^{16}_{8}\mathrm{O}$ and $^{17}_{8}\mathrm{O}$ are both atoms of oxygen.
They have the same chemical properties but have different masses. They are called **isotopes** of oxygen.

Oxygen has a third isotope, $^{18}_{8}\mathrm{O}$. Draw a diagram of an atom of this isotope.

All elements have more than one isotope. Many of these isotopes are ***unstable*** – the nucleus breaks up into smaller parts.
When this happens α-particles, β-particles or γ-rays may be emitted ***from the nucleus***. This is called ***radioactive decay***.

Atomic number	Element + symbol		Commonest isotope
1	Hydrogen	H	$^{1}_{1}\mathrm{H}$
2	Helium	He	$^{4}_{2}\mathrm{He}$
3	Lithium	Li	$^{7}_{3}\mathrm{Li}$
4	Beryllium	Be	$^{9}_{4}\mathrm{Be}$
5	Boron	B	$^{11}_{5}\mathrm{B}$
6	Carbon	C	$^{12}_{6}\mathrm{C}$
7	Nitrogen	N	$^{14}_{7}\mathrm{N}$
8	Oxygen	O	$^{16}_{8}\mathrm{O}$
92	Uranium	U	$^{238}_{92}\mathrm{U}$

▶ Radioactive decay

Experiment 39.9
For this experiment you need a large number of wooden cubes or dice, each with one face marked.
Imagine each cube is an atom.
If you shake and throw all the dice-atoms, some will land with the marked face upwards.
Let us pretend that these are atoms which have disintegrated and fired out α-particles or β-particles, and γ-rays.
Remove these disintegrated 'atoms' and count how many atoms have survived.

Repeat this process with the surviving dice and continue until all of them have 'decayed'.

Plot a graph of the ***number of dice-atoms surviving*** against the ***number of throws***.

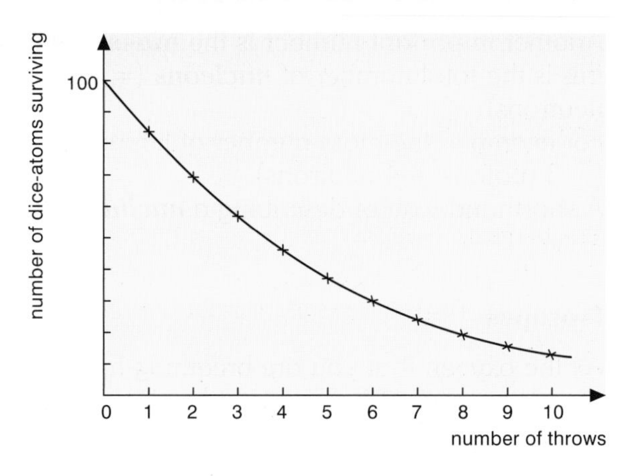

Real atoms behave in a similar way: each unstable atom disintegrates in a ***random***, unpredictable way. A large number of atoms gives a smooth ***radioactive decay curve***.

Half-life

The time taken for ***half*** the atoms to decay is called the ***half-life*** of the substance.

The graph shows a decay curve for a radioactive substance. On the graph you can see that after 1 half-life, half the atoms have disintegrated and half have survived.
After a further half-life, the activity has halved again so that only $\frac{1}{4}$ of the atoms survive. After 3 half-lives it has halved again and only $\frac{1}{8}$th survive.
What fraction survives after 4 half-lives?
What fraction survives after 5 half-lives?

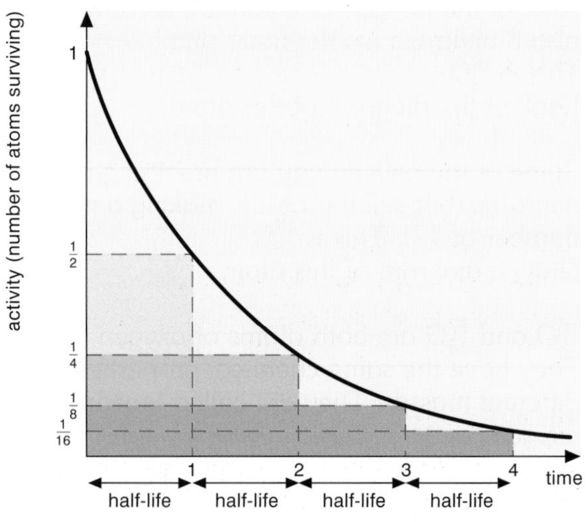

The half-life of different substances varies widely, from fractions of a second up to millions of years.

The half-life of radium is 16 centuries.
This means that in a radium source in school (mass of radium only about five millionths of a gram), the atoms are disintegrating at the rate of 700 million in a Physics lesson, and yet it will do this for 16 centuries before half of the atoms have decayed!

By mistake (in his lunch) Freddy Furze,
Ate radium and died with a curse,
But the point about Fred,
Is that now he is dead,
His half-life is sixteen hundred years!

▶ Unstable atoms

Alpha-decay

An alpha-particle is a helium nucleus (see page 342). It is $^{4}_{2}He$.
It has 4 nucleons: 2 protons + 2 neutrons.

Radium-226 ($^{226}_{88}Ra$) decays by α-emission. When it loses the α-particle, its mass number (226) must decrease by 4 (to 222). Also, its atomic number (88) must decrease by 2, to become 86. It has changed to a ***different element***. It is now ***radon*** (Rn), a radioactive gas.

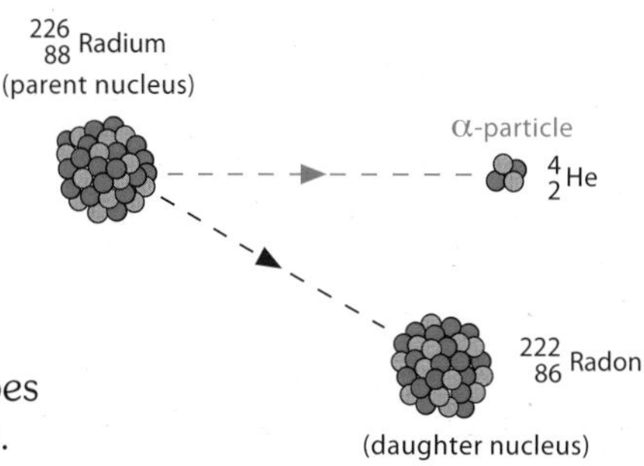

The nuclear equation is: $^{226}_{88}Ra \rightarrow\, ^{222}_{86}Rn + \, ^{4}_{2}He$

Note that the top numbers balance on each side of the equation. So do the bottom numbers.

Radon-222 is itself radioactive and decays by α-emission. What does it become? Write down the equation (element 84 is polonium, Po).

Beta-decay

The $^{218}_{84}Po$ from the last equation can decay by β-emission. This is the emission of an electron ***from the nucleus***. But there are no electrons in the nucleus!
What happens is this: one of the neutrons changes into a proton (which stays in the nucleus) ***and*** an electron (which is emitted as a β-particle).
This means that the atomic number ***in***creases by one, while the total mass number stays the same.
The polonium changes into another element, called astatine (At).

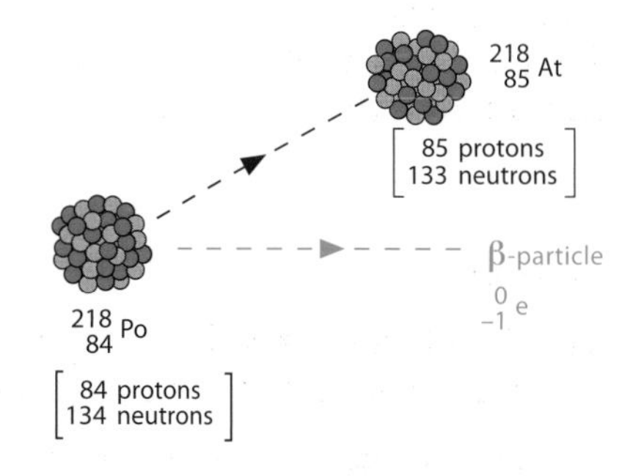

The nuclear equation is: $^{218}_{84}Po \rightarrow\, ^{218}_{85}At + \, ^{0}_{-1}e$

Notice again, the top numbers balance and so do the bottom ones.

Gamma-emission

When an α-particle or a β-particle is emitted, the nucleus is usually left in an 'excited' state. It loses its surplus energy by emitting a γ-ray. See also pages 208, 210.

'Splitting the atom' (transmutation)

In one of Rutherford's experiments, he fired α-particles at the nuclei of nitrogen atoms.
He found that some of the nitrogen atoms turned into oxygen atoms! The nuclear equation is:

$^{14}_{7}N$	+ $^{4}_{2}He$	$\rightarrow$ $^{17}_{8}O$	+ $^{1}_{1}H$
nitrogen nucleus	alpha-particle	oxygen nucleus	hydrogen nucleus (proton)

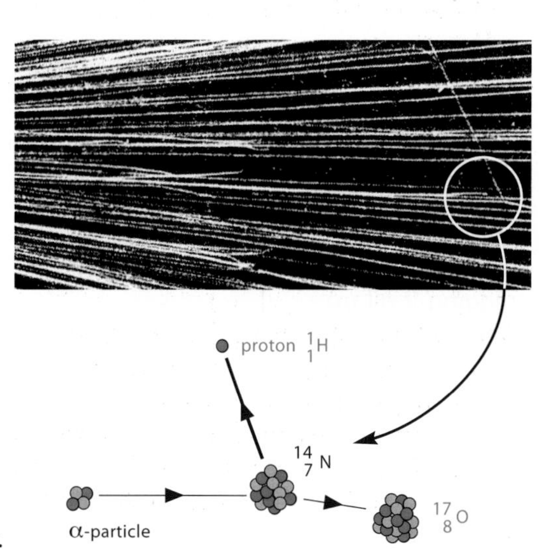

Notice again how the numbers balance.
The photograph shows how this appears in a cloud chamber.

▶ Uses

Radioactive isotopes, or ***radio-isotopes***, can be made by bombarding substances inside a nuclear reactor. Handled with care, they have many uses.

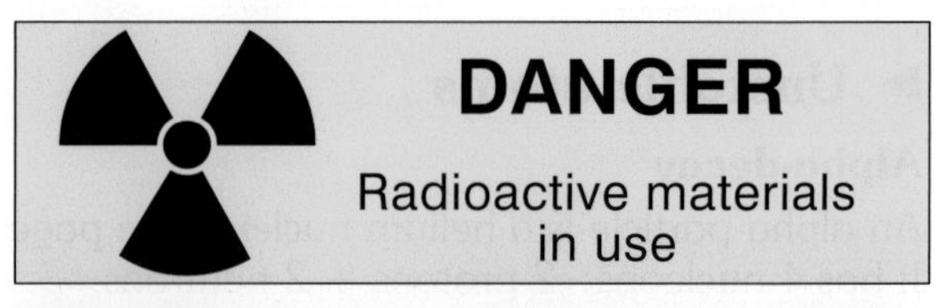

1. Radioactive tracers

Radioactive fertiliser can be fed to plants and then traced through the plant using a Geiger counter. This method is used to develop better fertilisers.

Radioisotopes have important **medical uses**. For example, a doctor might suspect that a patient has a blocked kidney. To find out, the patient sits with a Geiger counter over each kidney:

A small amount of Iodine-123 is injected into the patient. Within 5 minutes both the kidneys should extract the iodine from the blood stream, and then within 20 minutes pass it with urine into the bladder. Which one of these kidneys is blocked?

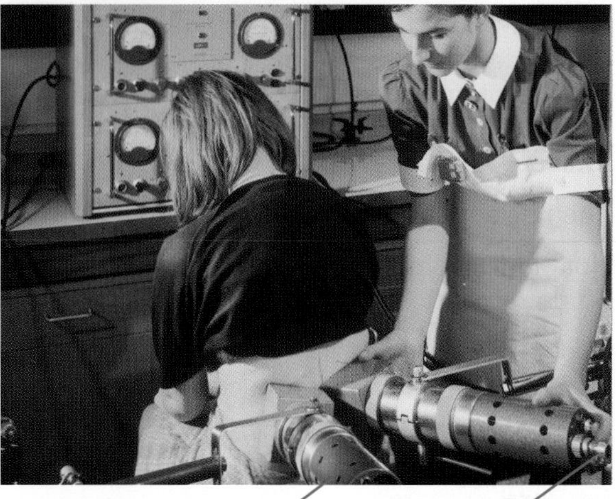

10 mins 20

Chart recorder A

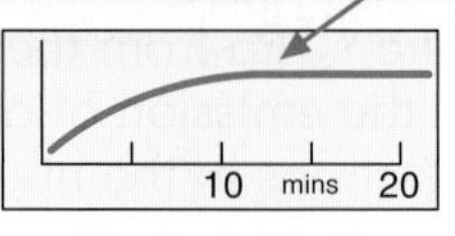

Chart recorder B

The most common tracer is called Technetium-99. This Tc-99 is very useful, and safe, because:

- It emits only gamma-rays. The γ-rays can be detected outside the body by a '***gamma-camera***'. It is also safer because γ-rays do not cause much ***ionisation***. Ionisation can damage cells and cause cancer. (It is alpha-radiation, strongly ionising, that is the most dangerous ***if it gets inside your body***, perhaps from a radioactive gas like radon.)
- It has a short half-life (of 6 hours). How much of it remains after one day?

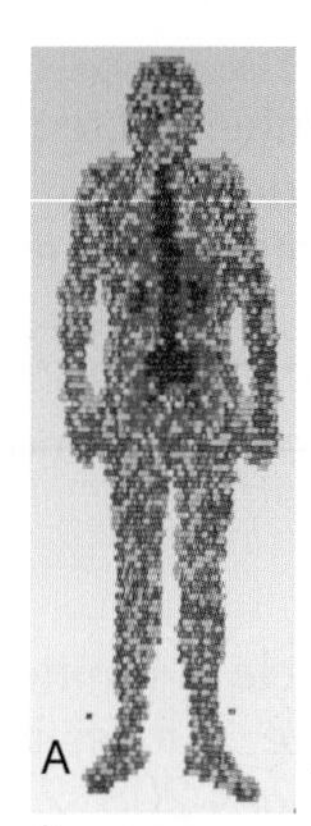

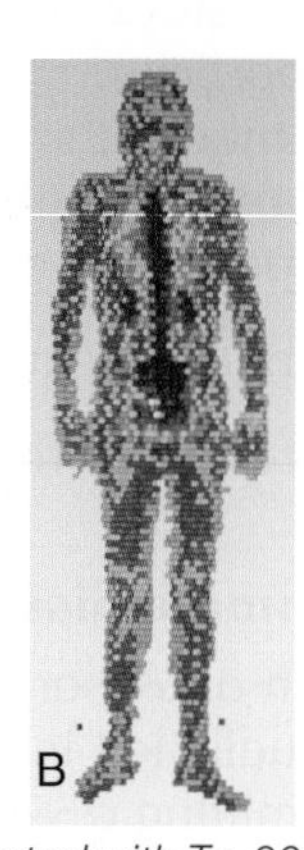

Gamma-ray scans of a man injected with Tc-99
A: initially B: after 6 hours (one half-life)

Tracers are also used in **industry**. If radioactive pistons are fitted to a car, then a Geiger counter can be used to test the oil for traces of wear from the piston. Different piston designs can be tested.

Leaks from a pipeline carrying oil or gas can be traced by injecting a radioisotope into it. This saves digging it all up.
The right isotope is chosen so that:

- It has a half-life of only a few hours or days. This is so that it remains long enough to be detected but not so long that it remains a safety problem.
- It is a beta-emitter. Alpha-particles would be absorbed by the soil, whereas gamma-rays would pass through the metal pipe anyway.

2. Sterilising

Gamma-rays can be used to kill bacteria, mould and insects in foods, even after the food has been packaged. This prolongs the shelf-life of the food, but it sometimes changes the taste.

Strawberries after 7 days

Gamma-rays are also used to sterilise hospital equipment, especially plastic syringes that would be damaged by heating them.

Cancer cells in a patient's body can be killed by careful use of γ-rays (see page 210).

3. Thickness control

In paper mills, the thickness of the paper can be controlled by measuring how much beta-radiation passes through the paper to a Geiger counter:

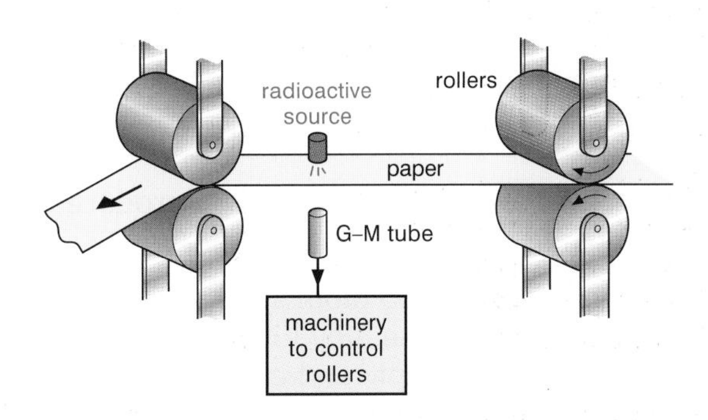

The counter controls the pressure of the rollers to give the correct thickness.
With paper, or plastic, or aluminium foil, β-rays are used. In a sheet-steel factory γ-rays are used. Why?
Why should the source have a ***long*** half-life?

4. Smoke detection

Many homes are fitted with a smoke alarm. This contains a weak source made of Americium-241. This emits alpha-particles which ionise the air, so that it conducts electricity and a small current flows:

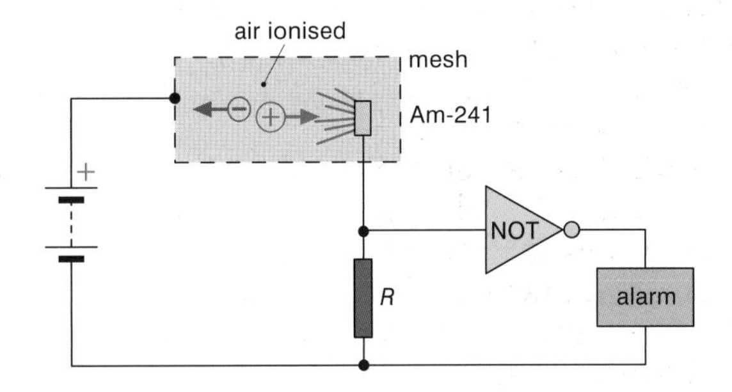

If smoke enters the alarm, it absorbs the α-particles, the current reduces, and the alarm sounds.
Am-241 has a half-life of 460 years. Is this helpful?

5. Checking welds

If a gamma source is placed on one side of the welded metal, and a photographic film on the other side, weak points or air bubbles will show up on the film (like an X-ray, see page 312).

6. Radioactive dating

The Uranium-238 in rocks decays steadily with a very long half-life of 4500 million years. It changes slowly into lead. By measuring how much of the uranium has changed into lead, it is possible to calculate the age of the rock. See page 352.
The U-238 acts as a kind of clock.

With once-living material (like Egyptian mummies), Carbon-14 is used (see page 352).

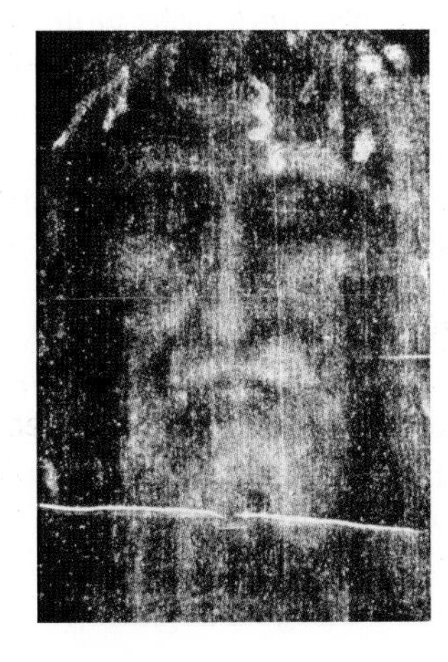

The Shroud of Turin. Carbon-14 dating shows it is 600 years old.

▶ Nuclear energy

If an atom of Uranium-235 ($^{235}_{92}U$) or Plutonium-239 ($^{239}_{94}Pu$) is bombarded with slow-moving neutrons, it breaks up into 2 smaller atoms and 3 neutrons: This process is called **nuclear fission**.

Mass and energy

The mass of the pieces after fission is found to be ***less*** than the mass of the pieces before the collision. The mass that is lost has been converted to energy. Albert Einstein showed how to calculate this:

Energy released (E) = **mass lost** (m) × **velocity of light squared** (c^2) or $E = mc^2$

The fission of 1 kg of Uranium-235 produces more energy than 2 million kg of coal.

When a Uranium-235 atom splits, three neutrons are emitted and these may hit three other uranium atoms so that they split and emit more neutrons which in turn hit other atoms, and so on. This is called a **chain reaction**:

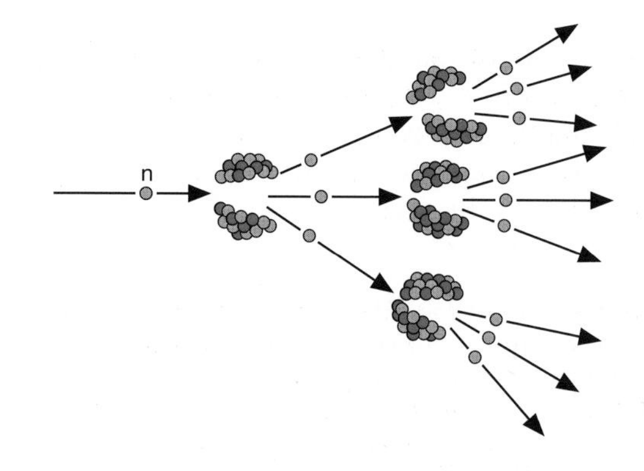

If the uranium is above a certain ***critical size*** (about the size of a tennis ball), then this chain reaction takes place very quickly, as in an atomic bomb.

Nuclear power station

A ***nuclear reactor*** is shown on the opposite page. The energy from the chain reaction makes the uranium **fuel rods** glow red-hot. The heat is carried away by a **coolant** (CO_2 gas, or water) which is pumped through the reactor.
In the **heat exchanger**, this heat boils water to make high-pressure steam, the same as in a coal-fired or oil-fired power station. The steam turns a turbine and a generator makes electricity.

- The chain reaction is controlled by movable **control rods** made of boron or cadmium. These are lowered to absorb neutrons, in order to reduce or stop the chain reaction completely.
- Graphite (or water) is included as a **moderator** to slow down the neutrons. This is because the fission of a uranium atom works more efficiently with slow neutrons.
- The whole reactor is shielded in steel and concrete to absorb the dangerous gamma-rays.

Mrs Messer: "What do nuclear scientists have for dinner?"
Professor Messer: "Fission chips."

The control room of a nuclear power station

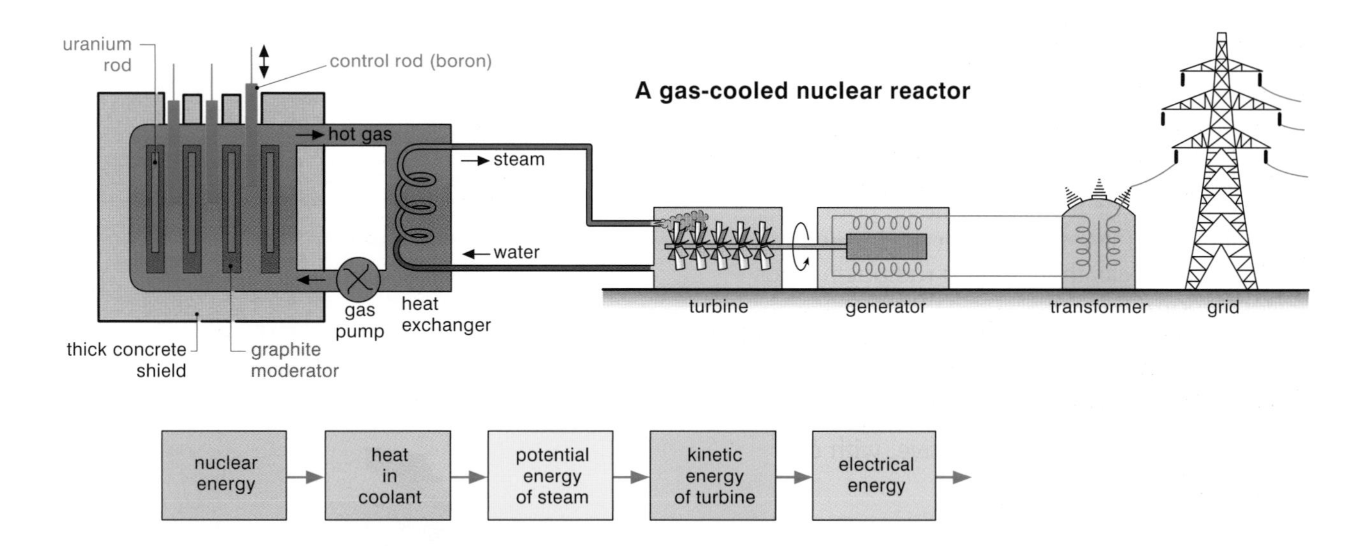

nuclear energy → heat in coolant → potential energy of steam → kinetic energy of turbine → electrical energy →

Nuclear fission raises strong feelings. This is partly because of thoughts of nuclear weapons, and uncertainly about the long-term effects of radiation. The world faces a severe energy crisis (see page 13) and decisions must be made.

Nuclear fusion is another way to get energy. In a hydrogen-bomb and in the Sun, hydrogen nuclei fuse together and release energy (p. 156). This process has not yet been successfully developed for power stations (see page 13).

The nuclear debate. What do *you* think?

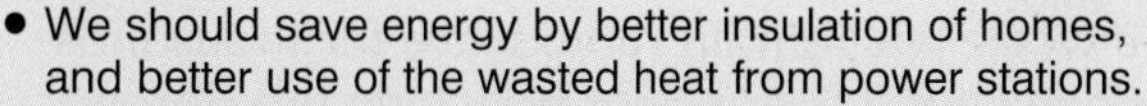

For:	Against:
• We need more nuclear power stations because of future energy demands, to keep our standards of living.	• We should save energy by better insulation of homes, and better use of the wasted heat from power stations.
• Nuclear power can save fossil fuels (coal, oil, gas), which are running out. They could be saved for the Third World, or not used at all (to reduce the Greenhouse Effect and global warming). See also page 107.	• The world's uranium is limited. We must develop renewable sources of energy, like wave power and solar power (see page 14). They do less damage to the environment.
• Nuclear power causes less damage to the environment than do coal-powered stations. Nuclear stations do not produce CO_2 and SO_2, and so do not make acid rain.	• Nuclear power stations produce waste which stays highly radioactive for thousands of years.
• The amount of nuclear waste is small. It can be safely stored by enclosing it in thick glass and burying it.	• Leaving waste is irresponsible – it pollutes the world for our grandchildren. The nuclear waste might leak out.
• Only 0.1% of our background radiation comes from the nuclear power industry (see page 350).	• No level of radioactivity is safe. Statistics show that children nearby are more likely to get leukaemia.
• Nuclear stations are very carefully designed to be safe.	• Chernobyl blew up, due to human error. A lot of people across Europe will die because of it. It is too risky.
• Risks due to the nuclear industry are less than other areas. Look at the figures: 	• Other risks are irrelevant. There is always a risk of a nuclear accident, with enormous consequences. Future generations would not forgive us.

▶ Radiation and You

Your body is receiving ***background radiation*** all the time. Most of this is due to natural sources:

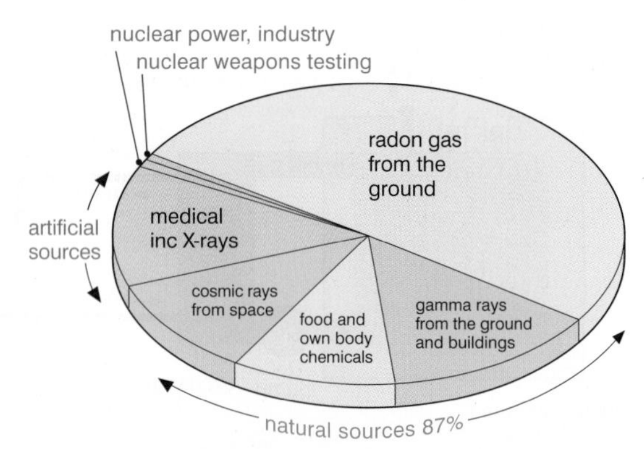

Background radiation in Britain

The amount of radiation you receive depends on the number of X-rays you have, your job, and where you live (it is higher in Cornwall because the rocks there emit a radioactive gas called radon).

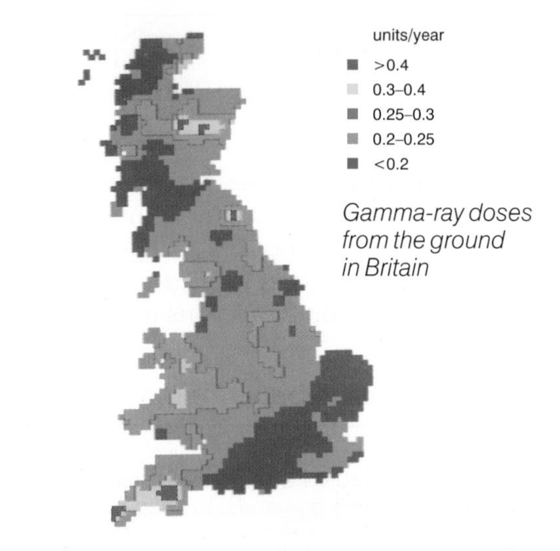

Gamma-ray doses from the ground in Britain

The activity of a source is measured in ***becquerels*** (Bq), where:
1 becquerel = 1 nucleus decaying per second.
Our bodies are slightly radioactive, with an activity of about 4000 Bq!

Disposing of nuclear waste

The fuel rods in a nuclear power station (page 348) need to be replaced every four years. They are highly radioactive with a long half-life. Disposing of them safely is difficult and costly.

This '***High-level***' waste is sealed in glass blocks ('vitrified') and then buried deep underground. '***Intermediate-level***' waste is less dangerous and can be stored in concrete drums underground.

The waste will remain radioactive for thousands of years, but it is not clear how long the storage will remain secure.
It is vital that the waste does not get into our water supplies or our food chain.

Working with sources

A worker using sources should wear a film badge (a ***dosemeter***) on his or her chest:

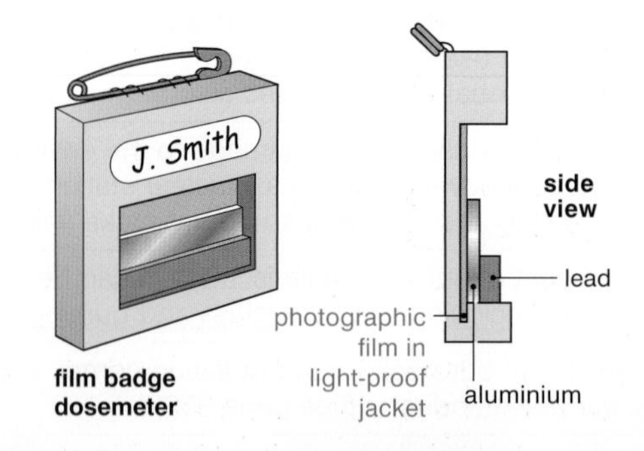

How would this show an exposure to beta rays?

The intensity of gamma radiation decreases with the distance squared (see page 341).
Why does your teacher use tongs to lift sources?

Alpha-particles are easily stopped, but if they get inside your body they do 20 times as much damage to cells as betas, gammas or X-rays. They can all cause cell mutation and cancers.

Workers in the nuclear industry use thick shields and remote-handling equipment:

Summary

The properties and detectors of alpha-particles, beta-particles and gamma-rays are shown in the table on page 341.

An atom has a small, heavy, central nucleus containing protons (+) and neutrons (no charge).

The number of protons is the atomic number. The number of nucleons (= protons + neutrons) is called the mass number.

In a neutral atom, the number of electrons in orbit = the number of protons. If the numbers are not equal, then it is an ion.

An isotope has the same atomic number but a different mass number.

The half-life of a radio-isotope is the time taken for half the atoms to decay. It is also the time for the activity (count rate) to halve.

▶ Questions

1. Copy out and complete:

a) An alpha . . . . is . . . .-charged and has the same charge and mass as the nucleus of a . . . . atom.

b) Beta . . . . are . . . . -charged electrons travelling at very . . . . speed.

c) Gamma rays are . . . . rays of very short . . . . .

d) Beta . . . . are more penetrating than . . . . but less penetrating than . . . . .

e) . . . . and . . . . can be deflected by electric and magnetic fields, but . . . . cannot.

f) Alpha-, beta- and gamma-rays can be detected by . . . . film, a . . . . chamber, or a . . . . tube. A gold-leaf electroscope or a spark counter can be used to detect . . . . .

2. Copy out and complete:

a) An atom has a small heavy . . . . containing . . . . (. . . .-charged) and . . . . (. . . .-charged).

b) A . . . . and a . . . . are each about 1800 times the mass of an electron.

c) The atomic number is the number of . . . . .

d) The mass number is the number of . . . . .

e) An isotope has the same number of . . . . but a different number of . . . . .

f) The half-life of an element is the . . . . taken for . . . . of the atoms to decay, or the count rate (activity) to . . . . . After . . . . half-lives, only one-eighth of the element remains.

3. Copy and complete the table:

	Mass	Charge
proton	1 unit	+ 1 unit
neutron		
electron		
α-particle		
β-particle		
γ-ray		

4. a) Most of the carbon in your body is ${}^{12}_{6}C$. Draw a diagram of this atom.

b) Another ***isotope*** is ${}^{14}_{6}C$ with a ***half-life*** of 5700 years. What are its ***atomic number*** and ***mass number***? Explain the words in italics and draw a diagram of this atom. What fraction of ${}^{14}_{6}C$ remains after 11 400 years?

c) While animals or plants are living, the proportion of ${}^{14}_{6}C$ in them remains constant. But once they die, the Carbon-14 decays. Suppose a modern bone contains 80 units of C-14, and an old bone contains just 10 units. How old is the bone?

5. Strontium-90 has a half-life of 28 years. It is a β-emitter and may be absorbed into human bone. How much time must pass before its activity falls to $\frac{1}{32}$ of its original value? Why would it be dangerous in our food chain?

6. Explain why workers in a nuclear power station wear badges containing film.

7. Explain the problems of nuclear waste.

8. Professor Messer fell asleep and dreamed that the nuclei of atoms could not be changed in any way. Why was his dream a nightmare?

▶ Physics at work: Radioactive dating

The age of rocks

Radioactivity can be used as a kind of clock to find the age of a rock. This is because the half-life of a radio-isotope is a fixed length of time (see page 344).
For example, the half-life of Uranium-238 (^{238}U) has a very long half-life of 4500 million years. It changes slowly into Lead-206 (^{206}Pb). After one half-life, half of it is unchanged and the other half has changed into lead. After two half-lives, $\frac{1}{4}$ of the ^{238}U is left and $\frac{3}{4}$ is lead.

By measuring how much of the ^{238}U in a rock has changed to ^{206}Pb, it is possible to calculate the age of the rock.

A rock from the Moon. Radioactive dating shows it is 4500 million years old, the same age as the Earth.

Example 1
In a rock sample, the proportion of ^{238}U atoms to ^{206}Pb atoms was found to be 4 : 1.
How old is the rock?

This means that, on average, for every 5 atoms of ^{238}U when the rock was formed, 1 atom has decayed and 4 atoms have not decayed yet.
That is, $\frac{4}{5}$ atoms are still radioactive ^{238}U.

So the activity is 80% of the initial activity.
From the half-life graph for ^{238}U we find that this has taken about 1500 million years:

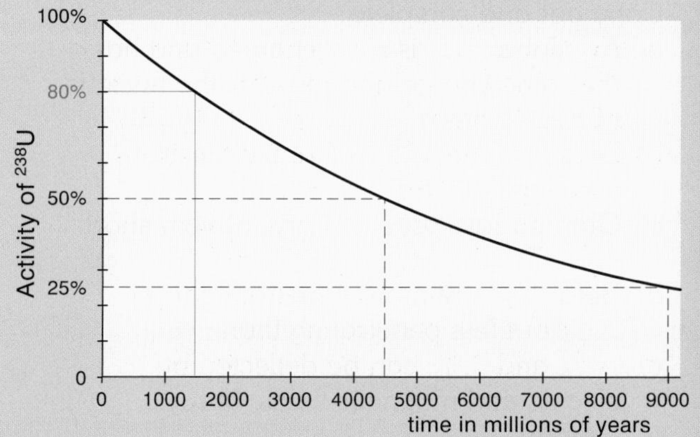

Other radio isotopes are sometimes used.
^{235}U (half-life = 700 million years) also decays to lead.
Potassium-40 decays to Argon-40 with a half-life of 1400 million years.

Carbon-dating

All living things take in some radioactive Carbon-14 (^{14}C) in photosynthesis, as well as the more usual ^{12}C.
When living things die, they stop taking in carbon and the ^{14}C slowly decays, with a half-life of 5700 years.
This can be used to date bones, wood, paper and cloth.

The Dead Sea Scrolls. ^{14}C-dating shows they are 1900 years old.

Example 2
A fresh bone gives a radioactive count of 100 counts per minute.
A bone of the same mass from a skeleton gives 30 counts per minute.
The background count rate is 20 counts per minute. How old is the bone?

Subtracting the background count gives the counts ***due to the bones alone*** as 80 and 10 counts per minute.
The count rate of the ^{14}C has fallen to $\frac{1}{8}$th of its initial value.
This means that ***three*** half-lives have passed (original $\rightarrow \frac{1}{2} \rightarrow \frac{1}{4} \rightarrow \frac{1}{8}$).
So the bone is $3 \times 5700 = 17\,100$ years old.

Alternatively you could read the time from a graph as in Example 1.

▶ Radioactivity

1. The table gives values for the activity of a sample of an isotope at different times:

Time/minutes	0	10	20	30	40	50	60	70	80	90
Activity/counts per second	96	78	62	54	40	32	26	21	15	14

a) On graph paper, plot a graph to show how the activity of the sample changes with time. [3 marks]
b) State a value for the half-life of the isotope. Give a reason for your answer. [2]
(OCR)

2. Two students were set the task of measuring the half-life of a radioactive substance which emits beta particles only. They measured the count rate every 20 minutes and recorded their readings as shown below.

Counts/min	330	231	165	120	90	71	57
Time (min)	0	20	40	60	80	100	120

They then discovered a count rate of 30/min was obtained even when there was no radioactive source near.
a) What was the cause of the count rate of 30/min? [1]
b) Draw up a table showing how the count rate **due to the beta-source only** varies with time. [1]
c) Plot a graph of the beta count rate against time and use it to find the half-life of the source. [4]
d) What effect does the emission of a beta particle have on the structure of the nucleus? [1]
(WJEC)

3. A radioactive mixture is made up of element **X** which gives a count rate of 160 per second and has a half-life of 2 hours and element **Y** which also gives a count rate of 160 per second but has a half-life of 1 hour.
If the initial count rate of the mixture is 320 what will be its value after
a) 2 hours? b) 4 hours? [4]
(WJEC)

4. Nuclear power plants are not always big. Heart pacemakers can be powered from a small piece of plutonium. The isotope used is written as ${}^{238}_{94}\mathrm{Pu}$.
a) In a neutral atom of plutonium,
i) how many protons are there?
ii) how many neutrons are there?
iii) how many electrons are there? [3]
b) Make a **simple** sketch of a plutonium atom. Label the protons, neutrons and electrons. [4]
(Edex)

5. Radioactive iodine is used to investigate tumours of the thyroid gland. It decays by emitting beta particles and gamma radiation. The beta decay process is represented by the following equation:

$${}^{131}_{53}\mathrm{I} \rightarrow {}^{A}_{Z}\mathrm{Xe} + {}^{0}_{-1}\mathrm{e}$$

a) Complete the equation by giving values for A and Z. [2]
Navdeep does an experiment to find the half-life of iodine-131. His first job is to find the level of background radiation.
b) What is background radiation? [1]
c) Write down 2 sources of this radiation. [2]
Iodine-131 has a half-life of 8 days. It is used as a tracer in the human body, to see if the thyroid gland is working properly. It is monitored by a doctor for several days after it has been given to the patient. Another isotope, iodine-128 has a half-life of 25 mins.
d) Why is iodine-128 not suitable here? [1]
e) Using tracers in this way has ***benefits*** and ***risks*** for the patient. Suggest how a doctor could explain these benefits and risks to a patient. [4] (OCR)

6. Carol is a teacher. She is showing the class an experiment using a radioactive source. She insists on these safety precautions:
a) No students may sit in the front row.
b) The radioactive source is kept in a lead box when not in use.
c) She only uses long tweezers to pick up the radioactive source.
d) Everybody must wear plastic goggles.
Explain why each of these precautions is important. [4]

Further questions on radioactivity

7. a) Three radioactive substances have to be stored safely.
Details of the substances are given below.

Substance	Half-life	Type of radiation given out
A	5000 years	alpha (α)
B	4 years	beta (β)
C	156 years	gamma (γ) alpha (α)

Which of the following containers would you use for **each** substance? [3]

i) Aluminium ii) Thin plastic
iii) Lead lined
iv) Give a reason for your answer to part (iii) [1]

b) Copy and complete the table below for substance **B**. [2]

Date	Mass of original radioactive substance left
1 March 2002	8 kg
1 March 2006	
1 March 2014	

c) A Geiger counter was used to measure the activity (in counts per minute) from a radioactive sample in the laboratory over a period of years. Over this period, the background radiation was regularly measured at 4 counts/minute.
The table of results is shown below.

Time in years	0	1	2	3	4	5	6
Recorded activity in counts/min	124	80	52	34	23	16	12
Activity due to sample alone	120						8

i) Copy and complete the table giving the activity of the sample alone. [1]
ii) Explain what is meant by background radiation. [1]
iii) On graph paper, plot the values for the **activity of the sample alone** against time, and join them with a smooth line. [4]
iv) Find the half-life of the substance from your graph. [1] (WJEC)

8. a) The atomic number of americium-241 is 95. Complete the table to give the number of particles in an atom of americium-241.

Particle	Electrons	Neutrons	Protons
Number			

[3]

b) Am-241 decays by losing an alpha particle.
i) Explain why smoke detectors containing Am-241 are not a danger to people.
ii) Complete the equation showing this decay by giving values for ***X*** and ***Y***. [4]

$${}^{241}_{95}\text{Am} \rightarrow {}^{4}_{2}\text{He} + {}^{X}_{Y}\text{Np} + \text{energy}$$

c) The table shows how the activity of a sample of americium-241 changes over a long period of time.

Time/years	0	500	1000	1500	2000
Activity/counts per minute	64	30	14	6	2

i) Use these numbers to draw a graph of activity against time for americium-241.
ii) Use your graph to find the half-life of americium-241. [4]

d) Californium-241 is also radioactive and decays by losing an alpha particle. It has a half-life of 4 minutes.
Suggest why it would be unsuitable for use in a smoke detector. [2] (Edex)

9. Cobalt-60 is used to irradiate food. It has a half-life of 5 years.
a) What useful effect does the radiation have? [2]
b) Why is a source with a relatively long half-life used? [2]

Many people are worried that irradiated food is radioactive. A scientist tests 10 different food samples. Here are her results:

irradiated foods	
food sample	**count rate (per min)**
A	40
B	42
C	37
D	41
E	39

non-irradiated foods	
food sample	**count rate (per min)**
F	43
G	41
H	39
I	37
J	39

c) How could she use this evidence to convince people that the food is safe? [3]
d) Comment on the ***reliability*** and ***validity*** of this evidence. [2] (OCR)

10. a)

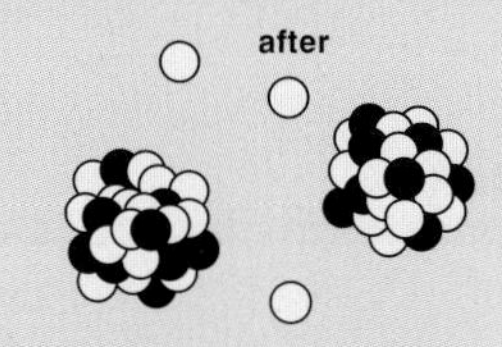

In a nuclear reactor, atoms of uranium-235 split into two nearly equal parts when they capture a neutron. An atom of uranium-235 has 92 electrons and 143 neutrons.

i) How many protons are there in one atom of uranium-235? [1]

ii) What is its proton number? [1]

iii) What is the nucleon number of the uranium-235 atom? [1]

b) It is proposed to build a nuclear power station at Lower Snodbury, Bumpshire. The Electricity Generating Board support the idea but the local residents oppose it. Suggest **three** advantages and **three** disadvantages of the scheme. Write a short paragraph on each. [6] (Edex)

11. In a nuclear reactor, a uranium-235 nucleus undergoes fission after it absorbs a slow neutron.

a) Explain how neutrons are slowed down so that a chain reaction can occur. [2]

b) Explain how the chain reaction can be controlled in a nuclear reactor. [2]

c) Explain why some of the fission products can cause long-term storage problems. [2]

d) Suggest ways in which the storage problems might be overcome. [2]

12. Carbon is an element with an atomic number of 6. Most naturally occurring carbon is carbon-12 but all living material contains a small amount of carbon-14.

a) Draw a labelled diagram to show the structure of a carbon-14 atom. (Label the protons, neutrons and electrons.) [3]

b) Carbon-14 is radioactive and decays by beta-emission. What new element forms when carbon-14 decays? (The periodic table may help you to answer.) [1]

c) In living material there are 96 carbon-14 atoms for every 10^{14} atoms of carbon-12. A fossil gives a count rate showing that it contains 12 atoms of carbon-14 for every 10^{14} atoms of carbon-12.
Calculate the age of the fossil. (The half-life of carbon-14 is 5600 years). [2]
(Edex)

13. Andrew is a medical physicist in a hospital where a machine with a radioactive source is used to treat cancer patients. The source contains ^{60}Co, with a ***half-life*** of 5.27 years. It decays by ***β-emission*** and emits high energy ***γ-rays*** that destroy cancer cells. One of Andrew's jobs is to measure the ***activity*** of the ^{60}Co source every month. Here are his readings:

Month	Jan	Feb	Mar	Apr	May	Jun	Jul
Activity (Bq)	173	171	169	167	166	164	162

a) Explain the meaning of the terms ***β-emission***, ***γ-ray***, ***half-life***, ***activity***. [4]

b) Explain the trend of Andrew's readings. [1]

c) Andrew calculates each patient's dose of γ-rays and he finds that the treatment times are gradually becoming longer. Explain why this is happening. [2]

d) Andrew must replace the ^{60}Co source when its activity has fallen to 50% of the original value. How long will this take? [1]

e) Suggest some safety precautions he should take when replacing the cobalt source. [2]

f) Andrew must measure the activity of the new source precisely. Suggest some experimental precautions that he could take to make sure that his readings are very accurate. [2]

g) Explain why it is important that his measurements and calculations are accurate. Describe possible outcomes if he relies on:
– mistaken readings that are too high,
– mistaken readings that are too low. [3]

14. Some rocks contain the radioactive isotope uranium-235 (^{235}U).
^{235}U has a half-life of 700 million years and, as it decays, lead-207 (^{207}Pb) is eventually formed.

a) On graph paper, draw a decay curve for ^{235}U, showing the proportion of ^{235}U remaining (range 0–1.0) against time (range 0–2500 million years). [4]

b) Samples of an igneous rock gave an average ratio of 70 atoms of ^{235}U to 30 atoms of ^{207}Pb. Use the decay curve you have drawn to estimate the age of the igneous rock. [1]

▶ Physics at work: Measuring your body

Biometrics: iris recognition

Look in a mirror at the details of your iris.
The pattern on it has stayed the same since you were 8 months old, and it will stay the same until 10 minutes after you die. Like your fingerprints, it is unique to you.

An electronic camera (linked to a computer) can scan your eye and recognise the pattern, and so identify you, in about 2 seconds.
This biometric data can be used with or instead of ID cards or PIN numbers. For example, at cash machines or for entry to secure buildings.

In the future, cameras will be able to scan and identify you as you walk down the street.
Discuss the advantages and disadvantages of this.

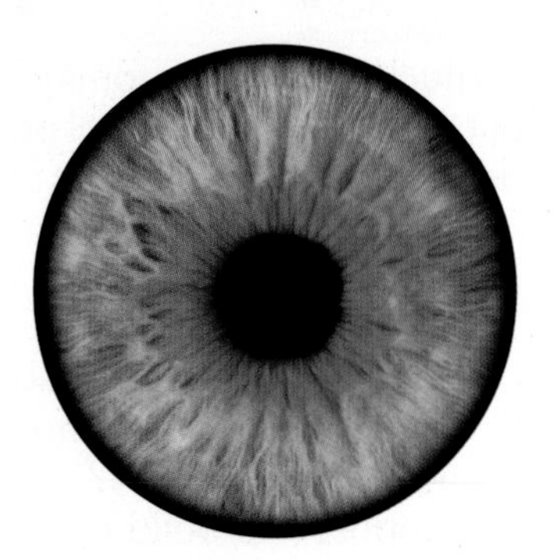

How is your iris different from this one?

Pulse oximetry

A pulse 'oximeter' (oxygen-meter) monitors the amount of oxygen in your blood. This can be very important, eg. after an operation in hospital.

It consists of a small probe attached to your finger (or ear).
It passes 2 beams of light through your finger, to a detector. One beam has a wavelength of 650 nm (red light) and the other is at 900 nm (infra-red).
These 2 beams are absorbed differently, depending on how much the haemoglobin in your blood is saturated with oxygen. (The more oxygen, the redder the blood.)

A small computer calculates the percentage saturation and shows the result on a display.
A reading of less than 95% could be dangerous.

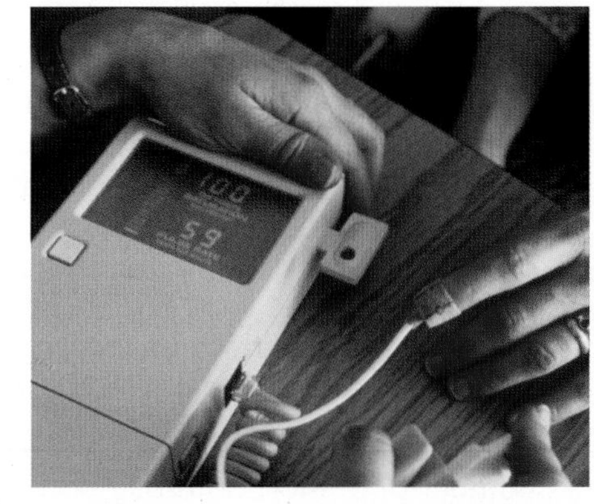

A pulse oximeter

Electrocardiography (ECG or EKG)

Can you feel your pulse? Every time your heart pumps, it produces a small electrical signal or 'action potential' (of several mV).
This can be picked up by electrodes, amplified, and displayed as a graph (eg. on a CRO).
The shape of the ECG wave gives a doctor important information about the heart.

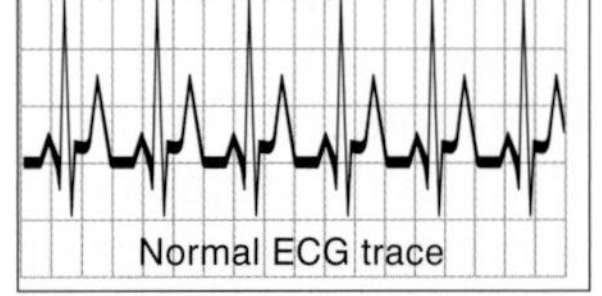

Normal ECG trace

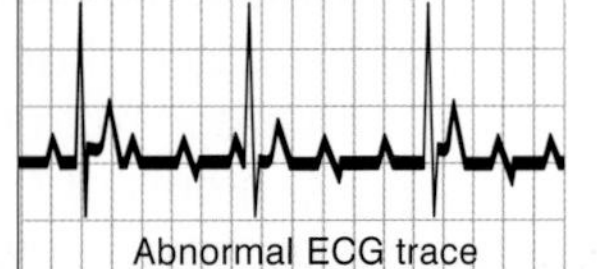

Abnormal ECG trace

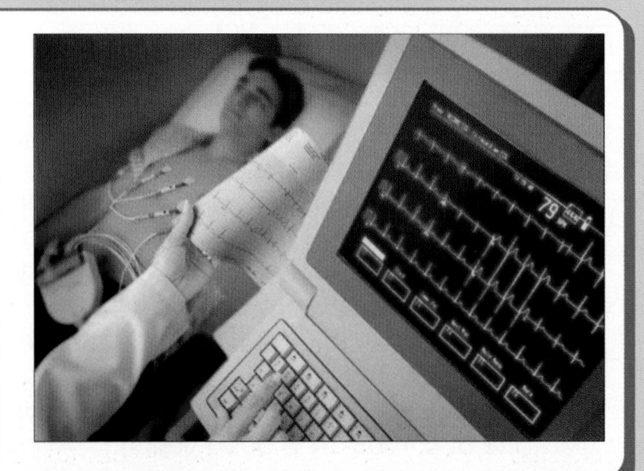

▶ Physics at work: Medical scanning

Scanning

Doctors can gain a lot of information about a body by scanning it with different waves (sound waves, or electromagnetic waves), and by using different wavelengths.

The list here shows examples in other parts of this book:

- ultrasound, eg. for prenatal scans (page 229).
- thermographs, for skin (pages 50, 211).
- X-rays, for bones and teeth (page 312).
  CT or CAT scans use multiple X-rays to build up a detailed image on a computer screen.
- Gamma-rays, in a gamma-camera (page 346).

Magnetic Resonance Imaging (MRI)

This method of scanning uses radio waves.
The patient is placed in a tunnel inside a large superconducting electromagnet. The magnetic field is about 40 000 times stronger than the Earth's field.

The nuclei (protons) of the hydrogen atoms in the body are aligned by the strong magnetic field.

A pulse of radio waves is passed through the body, and this throws some protons out of alignment.
A short time later, when they re-align themselves, they give out radio waves.
These are analysed by a computer to produce a image. MRI scans give very good images of the tissues in a human body (but they are expensive).

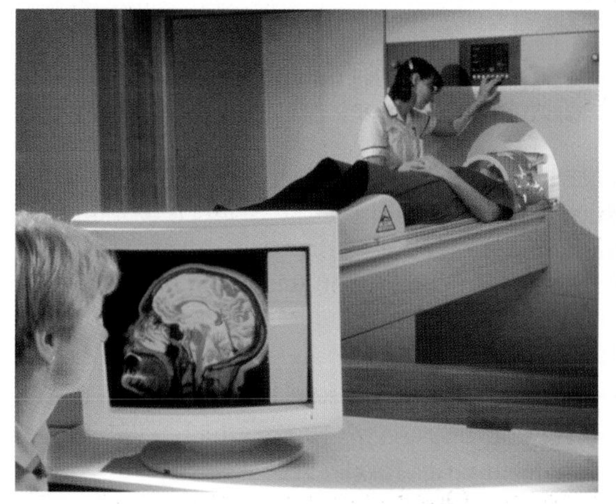

MRI scanner being used to scan a patient's brain

Positron Emission Tomography (PET)

A ***positron*** is a positive electron, or 'anti-electron'! It is an 'anti-particle' or 'anti-matter'.
When a positron meets an ordinary electron, they destroy ('annihilate') each other.
Their mass disappears and they give out energy (by $E = mc^2$, page 348).
The energy is emitted as 2 gamma-rays, which travel in opposite directions (because of the conservation of momentum, page 137).

In a **PET scanner** the patient drinks some water or glucose containing a positron-emitting isotope (with a short half-life). It goes to each part of the body, emitting positrons, which meet electrons.
The scanner detects each pair of γ-rays emitted, and so builds up an image of the brain, etc.

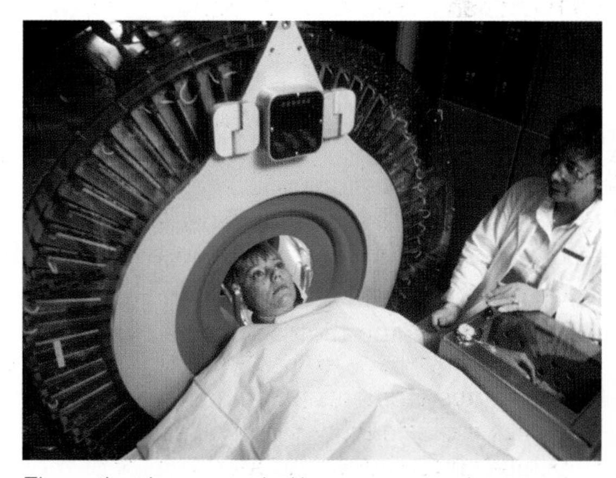

The patient is surrounded by gamma-ray detectors

New medical scanning techniques are expensive and so they may not be available to everyone.
Discuss the social and ethical issues that this involves.

How Science works

Science and your life

Science attempts to explain the universe around us.
It has helped to shape the world that you live in today.
It has provided modern technologies, like mobile phones, cars, TV and the web. These bring great benefits to us, but also problems for our society, such as pollution.

Yes, of course I'm against pollution

Scientists try to explain the world and solve its problems by using ***evidence***. They do this by collecting ***data*** in experiments and investigations, as described in this chapter.

We all have to make decisions about how to lead our lives. Knowing more about 'How Science works' will help you to make decisions based on good evidence.
As a modern citizen you will often have to question the evidence and compare conflicting ideas.
For example:

- Are mobile phones safe for children to use? (page 216)
- Will you buy a car? If so, which will be least harmful to our environment? (page 115)
- Will you vote for a political party which wants to make greater use of nuclear power stations? Or should we use thousands of wind-generators? (pages 106, 349)
- Should money be spent on journeys into space? (page 160)
- Is global warming a danger to us? If so, how can you help? (page 107)

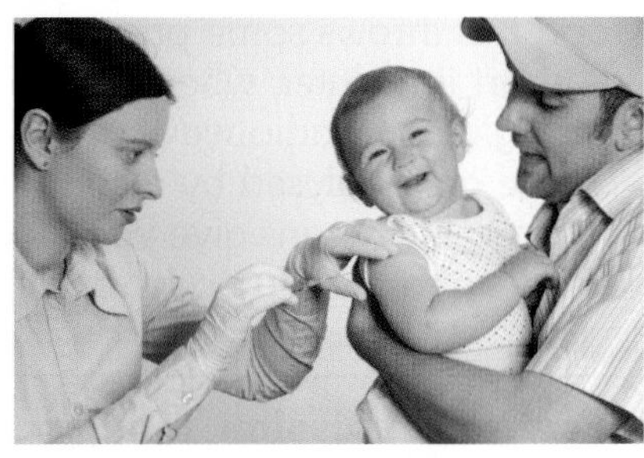

Many people claim that giving a baby an MMR jab can cause autism. However the evidence used to suggest a link came from a small sample of cases. More evidence was needed.

Evidence

To answer such questions you need to consider the ***evidence*** that is available. Then you will need to judge if the evidence is based on good science. Is it **reliable** and **valid**? (see opposite)

In discussions you may find that some people want to use 'evidence' or opinion that is not scientific at all.

Discuss the statements below, and decide :
- Which are scientific statements (for which you know that good evidence exists)?
- Which are opinions, personal views or moral judgements?

1. The spectrum of white light consists of 7 distinct colours.
2. The nicest colour is blue.
3. Cars cause pollution.
4. Bullying people is wrong.
5. Smoking can cause cancer.

Reliability and validity

A key point about the scientific method is that the data must be both ***reliable*** and ***valid***. See also chapter 1, page 7.

Reliable data is evidence that you can trust. If someone else did the same experiment, they would get the same result.

Valid data is evidence from a fair test which is reliable ***and*** which is relevant to the question being investigated.

> *Example 1: Reliability*
> Three students measured the time for one swing of a pendulum. Jo measured 1 swing; Emma measured 1 swing but 20 times and calculated the average time; Jack measured 20 swings and divided the time by 20.
> ***Discuss*** which method is the most reliable and why.
> Can you think of a way making it even more reliable?

> *Example 2: Validity*
> ***Discuss*** which of the following is valid evidence.
> - Measuring the length of a magnet to decide its strength?
> - Measuring the extension of a spring to find the force pulling on it?
> - Measuring the volume of a firework to find the energy in it?

Science and society

Advances in science and technology may be shown in a dramatic way in the media. It sells newspapers! But can you trust them? Is all the evidence given to you? Is it reliable and valid?

Many factors affect the way the evidence is presented to us. *For example*, if it is going to be unpopular with the public, or expensive for the government, the evidence may be watered down or hidden in unpopular publications.

Or the evidence may come from a biassed source. Would you trust evidence about the safety of mobile phones if you knew that the research was funded by a manufacturer of mobile phones?

And famous eminent scientists are more likely to be reported than scientists with less experience. (See also page 370.)

Mobile phone scare

New drug cures cancer

Global warming : new crisis

Headlines send a simple message – but where is the scientific evidence?

Limitations of scientific evidence

Sometimes it is difficult to collect enough evidence to answer a question. *For example*, global warming (page 107) is linked to carbon dioxide in the air as well as other factors, but at present not all scientists agree about the effects of all the factors. The evidence is not yet sufficiently reliable and valid.

And there are some questions that Science cannot answer at all. Science can answer the question '***How** can we produce electricity from nuclear fission?*' (page 349), but the question '***Should** we produce electricity from nuclear fission?*' is for society to decide.

> The rest of this chapter is divided into sections on:
> 1. Planning an investigation
> 2 Making measurements
> 3. Presenting data in graphs
> 4. Finding patterns and drawing conclusions
> 5. Evaluating

▶ Planning an investigation

Sometimes the starting point for an investigation is your ***observation*** of some ***phenomenon***.
For example, you might notice in a sunny car-park that a black car feels hotter than a silver car.
This might lead you to make a ***hypothesis*** that black objects in the Sun heat up faster than silver ones.
This can lead to a ***prediction*** and allow you to devise an investigation like Experiment 9.10 on page 47:

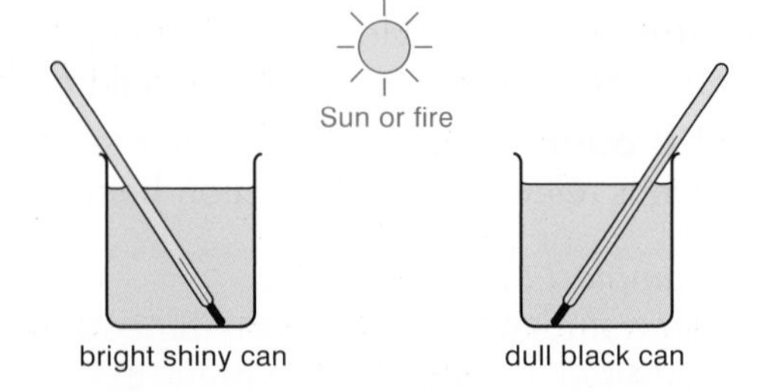

Fair testing

When you are designing an investigation the most important thing is to devise a ***fair test***.

For example, consider the investigation about the black can and shiny can (Experiment 9.10, page 47).
In this experiment it is important that we use two cans of the ***same*** size, and ***same*** shape, with the ***same*** amount of water, at the ***same*** temperature, placed the ***same*** distance from the ***same*** fire, with the temperatures taken at the ***same*** time.
This is called 'controlling the variables' so that only ***one*** variable is changed (the surface of the cans).

Deciding the variables

Look at this investigation into the stretching of a spring when a force (weight) is applied to it (see also page 66):

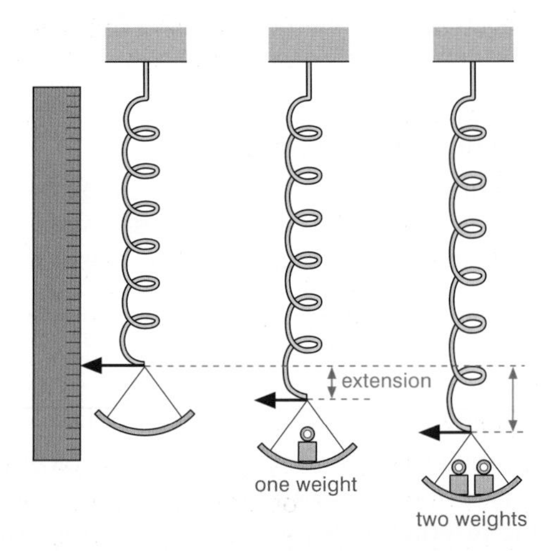

In this experiment, the force applied is the ***independent variable***. This is the variable that you change deliberately, step by step.
As a result of this force, the extension changes.
The extension is the ***dependent variable***. The size of this variable *depends* on the first variable.
All the other variables must be ***controlled*** (kept constant).

Different types of variables

Some of the types of variables that you may meet are:

Categoric variables	These have word labels, eg. iron, steel, glass (in Experiment 9.3, page 41).
Ordered variables	These are categoric variables that can be ranked, eg. small, medium, large.
Discrete variables	These variables can only have whole number values, eg. 1 weight, 2 weights, 3 weights in the spring experiment above. eg. the number of layers of insulation keeping a beaker warm.
Continuous variables	These variables can have any numerical value, eg. the extension of the spring above. eg. the temperature of water in a can.

Trial runs

Sometimes you will need to do a trial run before you do the investigation accurately.
A trial run helps you to decide the best ***range*** of values to use, and the number of readings to take.

For example in the spring experiment on the opposite page, you would need to decide to:

- use sufficient weight to extend the spring,
- use a sensible range of weights,
- have readings closer together (at smaller intervals) where there is a change in the pattern (where the graph is curved).

For a straight-line graph you should take at least 5 sets of results (to give 5 points on the graph).
If the graph is a curve you need more readings.

It's important to get your range right!

Remember: the ***reliability*** and accuracy of your results will be improved if you repeat your readings and then calculate their ***mean*** value (their average).
You can also improve reliability by using another instrument to repeat readings as a cross-check.

Here are 2 sets of repeated readings (in grams) for two objects weighed on the same balance:
Object A: 25.0, 27.0
Object B: 26.0, 27.2, 25.9, 26.4, 24.9, 26.2
For each object, find (a) the maximum and minimum values, (b) the range, (c) the mean (average).
Discuss which result is more accurate and reliable.

ICT
Consider using a sensor and a data-logger or computer if it will help you to get the evidence.
And you can use a spreadsheet to calculate, analyse and display your data.

'Not-so-fair' tests

In most Physics investigations, in the lab, it is easy to control the variables so it is a fair test.

However in other investigations it may be difficult (or even impossible) to control all the variables.
For example, an investigation to test different solar cells for supplying electricity in remote areas in all weathers. You can't control the weather but you can ensure that all the cells have the same weather.

If an investigation uses a large-scale survey you need to try to select similar conditions.
For example, if you were surveying whether the blood pressure of people is affected by the time they spend on their mobile phones, then you should use a group of people with approximately the same weight and diet. And perhaps have a 'control group' that didn't use a phone at all.

Investigating the effect of different fertilisers on areas of wheat.
You can't control all the variables, but all the plants experience the same varying weather.

▶ Making measurements

To provide good evidence, your measurements need to be both ***reliable*** (repeatable) and ***valid*** (relevant).

There will usually be some ***variation*** in the readings you take. *For example*, if you time the fall of a paper parachute over a fixed distance, the times will vary slightly. By repeating the measurement and calculating the mean (average) your results will be more reliable.

You need to consider the ***accuracy*** of your measuring instrument. *For example*, an expensive thermometer is likely to be more accurate than a cheap one. It is also likely to be more ***sensitive*** with a better ***resolution*** (it will respond to smaller changes in temperature).

The ***precision*** is also important. Measuring with a ruler that has a millimetre scale will give greater precision than using a ruler with a centimetre scale.

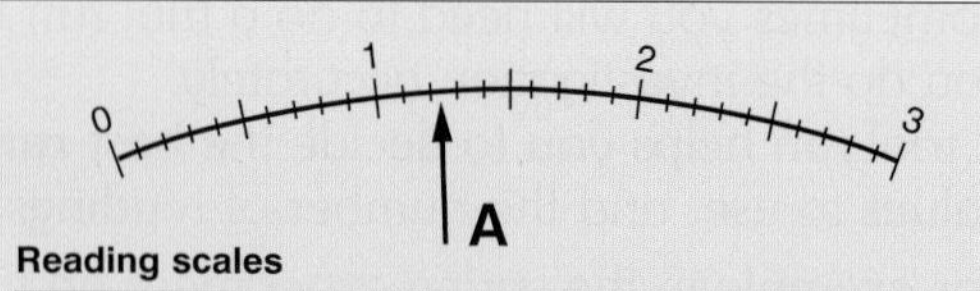

Reading scales

When reading scales, make sure you work out first what each small division stands for. It can help if you make a note of this before you start the experiment.
For example, you could put a note next to your ammeter saying '0–3 A, each mark = 0.1 A'.

Reading scales can be tricky. The makers usually put as many marks on the scale as the accuracy of it will allow. This means that you should only read to the nearest scale division (ie. the nearest mark). For example, from the diagram above, you would write down 1.2 A.

However in some papers the examiners expect you to 'interpolate' and try to judge the fractions of a division. In this case you might write down 1.23 A. Your teacher will advise you which to do.

Example: Precision due to consistent readings
A beaker is weighed on Balance A, 3 times.
The readings are 73 g, 77 g, 71 g. So the range is 77 g – 71 g = 6 g.

It is then weighed on Balance B, 3 times.
The readings are 75 g, 73 g, 74 g. So the range is 75 g – 73 g = 2 g.
Balance B has the greater precision. Its readings are grouped closer together.

The difference between accuracy and precision

The diagrams below show the difference between accuracy and precision when measuring the length of an object. Each red vertical line represents a reading of the length of the object.

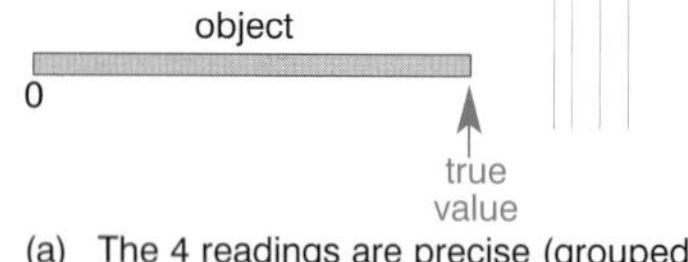

(a) The 4 readings are precise (grouped) but not accurate.

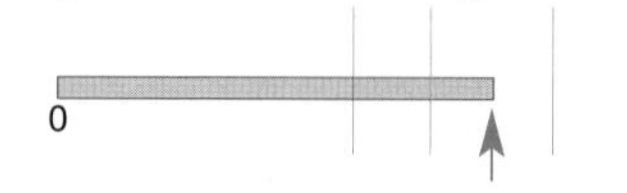

(b) The 4 readings are accurate (when the mean is calculated) but not precise.

(c) The 4 readings are accurate and precise.

Errors

You need to remember to correct for ***zero-errors***.
For example, you should check that your spring-balance or ammeter reads zero before you start to use it.

When reading a scale, make sure you look exactly at right-angles to it, so that you read the correct number:

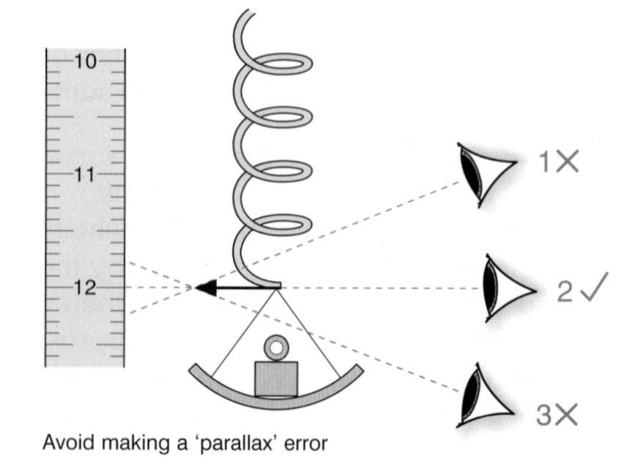

Avoid making a 'parallax' error

Anomalous results

If one of your results seems unusual or 'anomalous', make sure you repeat it. If it was an error you can ignore it when you consider your results at the end.

▶ Presenting your data

As you carry out an investigation, you record your data in a results ***table***. You should label each column with the ***quantity*** you are measuring and its ***unit***.

For example, here is a table for an experiment on stretching a spring (page 66):

Extension of a spring

Stretching force N	Extension mm
0	0
10	3
20	7
30	10
independent variable	**dependent variable**

The first column should always show the ***independent variable*** (the stretching force).
This is what you change deliberately, step by step.

The second column should always show the ***dependent variable*** (the extension).
All other variables must be controlled (kept constant).

If you decide to repeat your readings to get more reliable data, you can split up the second column:

Stretching force N	Extension mm			
	1st reading	2nd reading	3rd reading	Mean (average)

Drawing line-graphs and bar-charts

Having recorded your results, you will usually need to draw a graph (in order to see any pattern in the data).

If the first column in your table (the independent variable) can have a continuous range of values, then use a ***line-graph***. However if the first column can only have certain fixed values, then use a ***bar-chart***.

If you draw a line-graph, do it in 5 steps as shown here:

Five steps in drawing a graph:

1. Choose simple scales.
   For example, 1 large square = 1 newton, (or 2 N, 5 N, 10 N).
   Never choose an awkward scale like 1 square = 3 N or 7 N!
2. Plot the points and mark them neatly. Re-check each one.
3. If the points look as though they form a straight line, draw the best straight line through them with a ruler (and pencil). Check that it looks the best line.
4. If the points form a curve, draw a 'free-hand' curve of best fit. (Don't join the points 'dot-to-dot' with a ruler.)
5. If a point is clearly off the line, you should always use your apparatus to repeat the measurement and check it.

Sometimes you need to find the gradient of your graph.

Finding the gradient of a straight line graph

This is more difficult, but can be done in 3 steps.

a) Draw a large right-angled triangle as shown in red:

b) Find the value of the two sides '***Y***' and '***X***', ***in the units of the graph***. That is, you must find the size of ***Y*** and ***X*** using the scales on the graph (not just measuring with a ruler). For example, in the diagram, Y = 20 mm and X = 60 N. Can you see why?

c) Calculate the gradient (or 'slope') of the graph by dividing ***Y***/***X***. Keep the units with the numbers, so that you find the unit of the gradient.
For example, from the diagram:

$$\text{gradient} = \frac{Y}{X} = \frac{20\text{ mm}}{60\text{ N}} = \underline{0.33\text{ mm/N}}$$

A calculator gave the result as 0.33333333.
Why is it better to write it as 0.33?

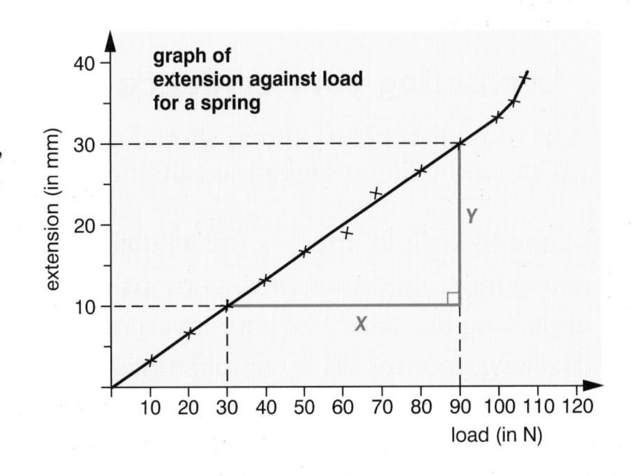

▶ Finding patterns, to draw a conclusion

Once you have drawn your graph, you can see the pattern or relationship between the 2 variables.

Before you can be sure of the pattern, you need to deal with any ***anomalous*** results. If you have repeated the reading then you can decide whether to include it on the graph, or omit it.

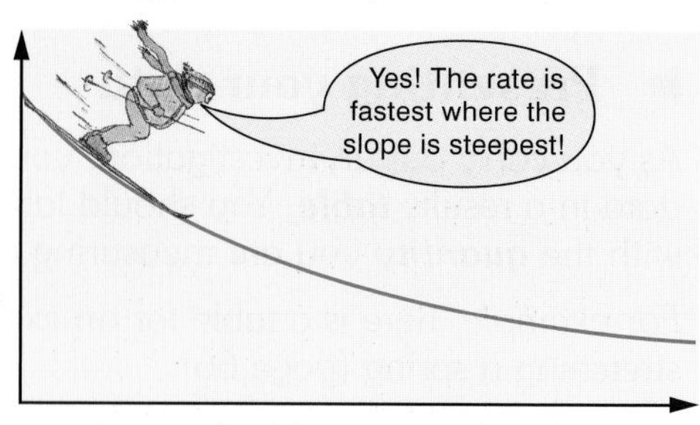

The rate of change depends on the gradient, eg. radioactive decay (page 344).

Examples of the graphs you find in Physics include:

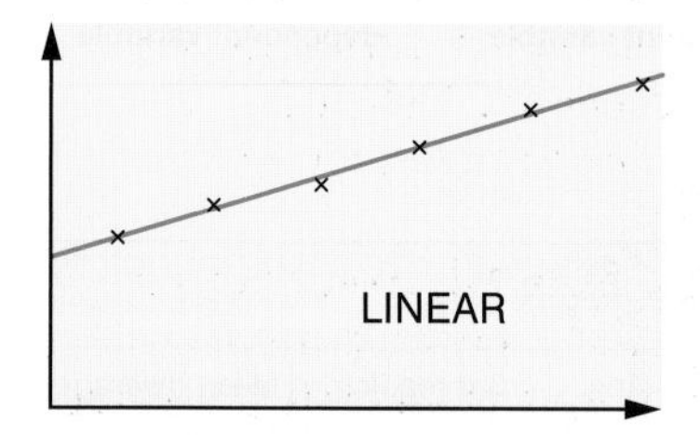

The 'line of best fit' gives an average line.
Any anomalous results have been omitted.
See also page 362.

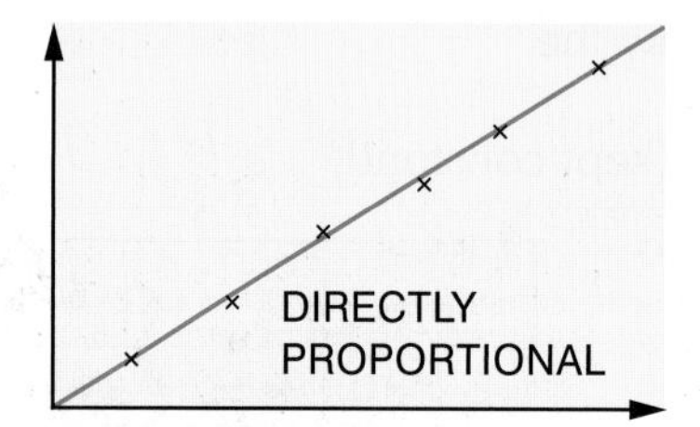

This graph is a straight line through the origin (0,0).
eg. Hooke's Law (page 66).
eg. current–voltage for an ohmic conductor (page 259).

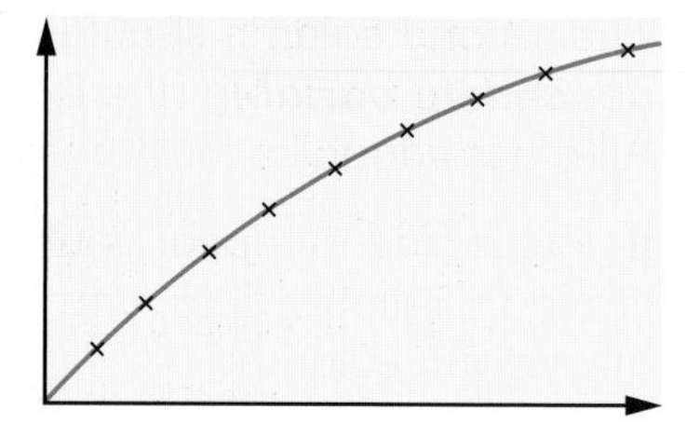

eg. velocity–time for a falling object (page 89).
eg. current–voltage for a filament lamp (page 259).

Drawing a conclusion

Every experiment has a 'conclusion'. This is a summary of what you found out (or sometimes what you didn't find!) Always look at your results or graph or chart to decide what you have discovered.
What reliable and valid deduction can be made from your results? What pattern can you see?

If possible, try to use your scientific knowledge to explain and justify your conclusion.

▶ Evaluating your evidence

As part of your conclusion you should comment on the reliability and validity of the measurements that you made.

You may be able to improve the reliability of your data, by:

- taking more measurements, perhaps changing the range,
- improving the design of your investigation,
- checking your results by an alternative method,
- looking up data from secondary sources (eg. books or the internet),
- seeing if other people following your method get the same results (ie. are your results ***reproducible***?).

▶ Questions

1. Copy and complete:

a) Scientists gather evidence by collecting . . . . . It should be both . . . . and . . . . .

b) Reliable data is evidence that we can . . . . . Valid data is evidence which is . . . . and also . . . . to the investigation.

c) The variable that I change deliberately is called the . . . . variable. As a result the . . . . variable changes. In a fair test it is important to . . . . all the other . . . . .

d) The reliability of my results can be improved by . . . . the measurement and then calculating the mean (. . . .).

e) An . . . . instrument is one which gives readings close to the true value.
A . . . . instrument is one which gives readings which are grouped together.

f) An unusual reading which does not fit the pattern is called . . . . .

g) To see the pattern in my results I can draw a . . . . with a . . . . of best . . . . .

h) After drawing the graph I should draw a . . . . and also . . . . the evidence.

2. For each set of repeated readings below, find:
i) the range, ii) the minimum value,
iii) the maximum value, iv) the mean.

a) 20 °C, 17 °C, 23 °C

b) 75 N, 84 N, 74 N, 73 N, 74 N

c) 17.2 g, 16.0 g, 16.4 g, 15.9 g, 14.9 g, 16.2 g

3. Here are two sets of repeated readings:

a) 1.2 V, 1.7 V, 1.3 V, 1.0 V

b) 1.2 V, 1.4 V, 1.3 V, 1.3 V

For each set, i) calculate the mean,
ii) comment on the precision.

5. The table shows the results of measuring the thickness of a number of sheets of paper:

No. of sheets	15	22	25	35	46	65	75	78	85
Thickness (mm)	1.7	2.4	2.7	4.4	5.1	7.2	8.1	8.6	9.4

a) Plot a graph, discarding any doubtful results.

b) Should it go through the origin? Why?

c) What is the correct thickness of 35 sheets?

d) How many sheets would make 6.6 mm?

e) What is the thickness of 1 sheet?

f) What is the gradient of the graph?

For each of **questions 6 to 12**, plan how you would do the investigation. In each case:

a) Identify the variables, and ensure it is a fair test.

b) Sketch the apparatus you would use, and add notes to explain the method.

c) Draw a table with headings (and, if possible, say what graph or chart you would draw).

6. Investigate the loss of heat from various containers (pages 46, 54).

7. Investigate the efficiency of kettles (page 264).

8. Investigate how the current from a solar cell depends on the surface area (page 103).

9. Investigate the terminal velocity of a paper parachute with different masses (page 89).

10. Investigate how the current through a lamp depends on the p.d. (page 259).

11. Compare the magnifications produced by different lenses (page 195).

12. Investigate the frequencies of a guitar's strings using an oscilloscope (page 232).

4. *Professor Messer's not too bright. It's up to you to put him right.*

More questions on page 336.

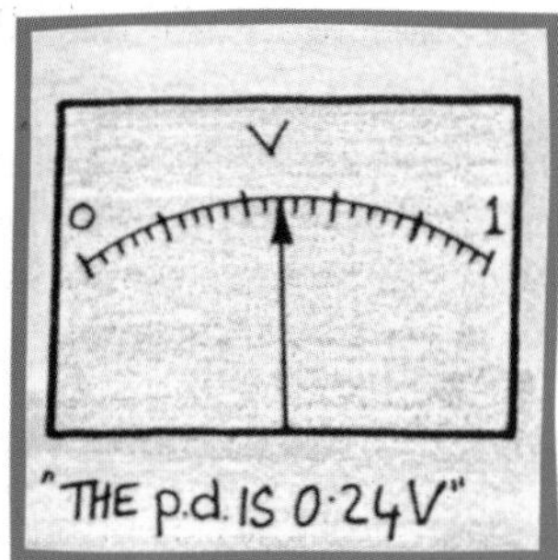

Doing your Coursework

Your coursework or practical assessment is important.
It is an important part of becoming good at Science. And it counts as a significant part of your GCSE or IGCSE grade. Doing good coursework can be one of the best ways to boost your final grade.

'Coursework is important'

The different examination boards vary in the ways in which they assess your coursework. There is a short summary lower down this page.
Your teacher will give you full details of what you need to do for your practical assessment. You can also research the details yourself on the exam board's web-site, see below.

Different examination boards allocate different percentages of marks for coursework or practical assessment.
They may also call it by different names.

For example the following 5 examination boards use the name "controlled assessment". This means that your teacher will observe you while you do some practical work.

- AQA GCSE examination board : controlled assessment = 25%
- Edexcel GCSE examination board : controlled assessment = 25%
- OCR GCSE examination board : controlled assessment = 25%
- WJEC GCSE examination board : controlled assessment = 25%
- CCEA GCSE examination board : controlled assessment = 25%

Other (IGCSE) exam boards use different methods:

- Edexcel IGCSE examination board has an 'assessment of practical skills' within the written papers (see page 368) = 20%
- Cambridge Cam IGCSE examination board has 3 alternative ways 'Coursework' (assessed by your teacher); or a 'Practical Test'; or an 'Alternative to Practical' written paper (see page 368). You do only one of these alternatives =20%

Your teacher will give you more details, or you can find more details by visiting the correct web-site in the list below. Some of these sites have revision materials to help you.

www.aqa.org.uk
www.edexcel.com
www.ocr.org.uk
www.wjec.co.uk
www.rewardinglearning.org.uk (CCEA)
www.cie.org.uk

▶ Helping you to do your practical work

When you are doing your practical work, use the questions in this checklist to help you:

Planning

Before you collect your evidence, think about:

- What is the best way to tackle this particular problem?
- How can you make sure your tests are as ***fair*** as possible?
- Have you identified the independent, dependent, and control ***variables***? (See pages 7, 360.)
- How can you make your tests ***safe***? Do you need to carry out a ***'risk assessment'***?
- Can you make a ***prediction***? If so, can you explain it?
- How many observations or measurements will you make? What ***range*** of values will give suitable results?
- Will you need to repeat results to make them more ***reliable***?
- Which apparatus will you choose to get accurate results? Will datalogging equipment help you to get greater accuracy?
- Can you use other (***secondary***) sources of information to help you plan?
- Should you do a ***trial run*** to check your ideas before you start?
- How will you ***record*** your results clearly and accurately?

Finding a Pattern, Drawing a Conclusion

After you have collected your results, think about:

- Can your results be shown on a ***bar-chart*** or a ***line-graph***?
- Can you see a ***pattern*** in your results?
- Are there any ***anomalous results*** which do not fit the general pattern?
- Do your results support your prediction (if you made one)?
- Can you ***explain*** your results using the work that you have done in Science?

Evaluating

When you have decided your conclusion, think about:

- Can you suggest any ***improvements*** you could make to
  - the way you carried out your tests?
  - the accuracy of your readings?
  - the reliability and validity of your results?
- Are your results good enough to draw a ***firm conclusion***? Do you have enough evidence to be sure? How could you gain ***further evidence***?

▶ Practical work

1. The diagram shows a rheostat (variable resistor) together with its circuit symbol.

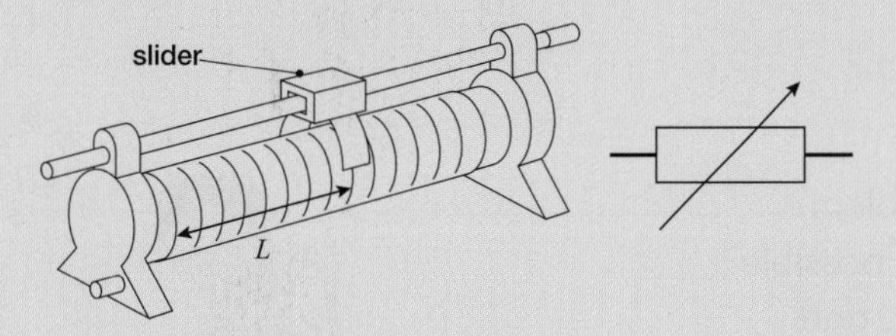

The slider can be moved. When the distance L is changed, the resistance of the rheostat is changed. The maximum value for L is 19 cm. A student investigates how the current I in the rheostat depends on the length L.

a) Draw a circuit diagram for the investigation. [2]

b) i) State one extra item of equipment, not shown on your diagram, that the student would need. [1]

ii) Describe how this investigation would be carried out. [3]

c) During the investigation, the ammeter reading was as shown below.

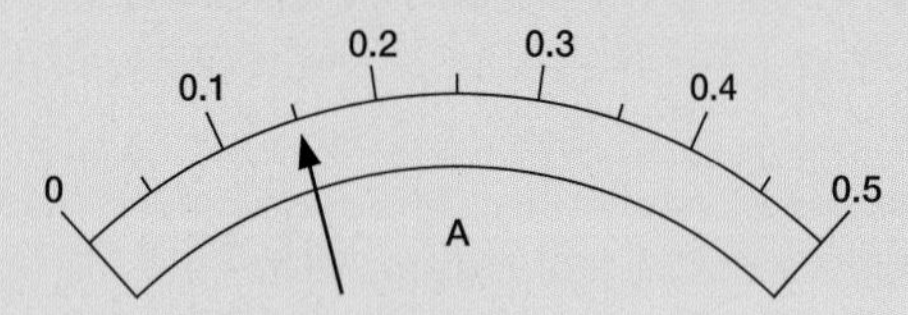

Record this reading. [1]

d) During the investigation the student recorded the following results.

19, 0.11	10, 0.17	7, 0.24
5, 0.33	6, 0.14	16, 0.12

i) Put these results in a table with column headings and units. [3]

ii) Plot a graph of current against length. Label the axes. [3]

iii) Circle the anomalous result. [1]

iv) Draw the best-fit curve for the remaining points. [1]

v) Suggest a reason for the anomalous result. [1]

e) Suggest a reason why the student did not use a value of 2 cm for L in the investigation. [1]

(Edex IGCSE)

2. The IGCSE class is investigating the swing of a loaded metre rule. The arrangement of the apparatus is shown here:

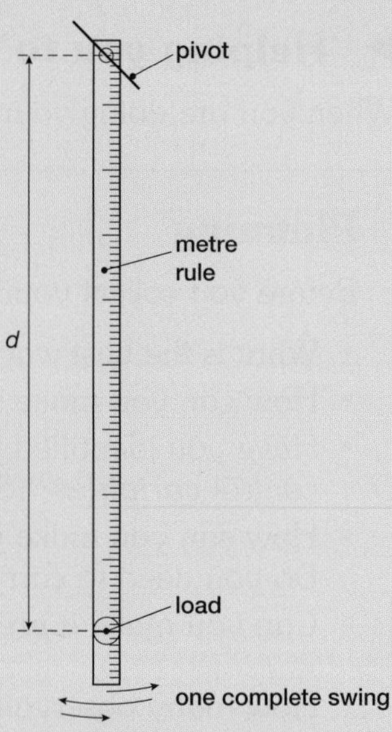

The load is attached to the metre rule so that its centre is 90.0 cm from the pivot. The rule is displaced a small distance to one side and allowed to swing. The time t taken for 10 complete swings is recorded.

This is repeated using different values of the distance d. The readings are shown in the table.

d /	t /	T /
90.0	18.35	
85.0	17.87	
80.0	17.53	
75.0	17.06	
70.0	16.72	

a) Copy the table and complete the headings. [1]

b) Calculate the period T for each value of d to complete the table. The period is the time taken for one complete swing. [2]

c) On graph paper, plot a graph of T /s (y-axis) against d /cm (x-axis). Start the x-axis at d = 70.0 cm and the y-axis at a suitable value of T /s to make the best use of the grid. [5]

d) A student suggests that T is proportional to d. State whether or not the results support this idea and give a reason for your answer. [1]

e) Explain why the student takes the time for ten swings and then calculates the time for one swing (the period), rather than just measuring the time for one swing. [1] (Cam IGCSE)

3. a) A teacher uses apparatus to measure the half-life of a radioactive source.

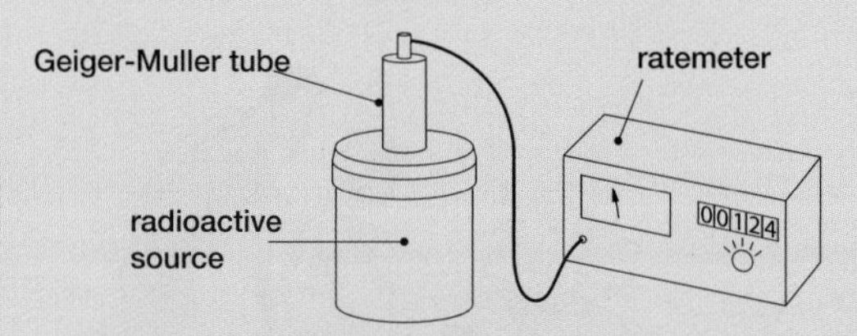

Which part of the apparatus takes safety into account? [1]

b) Before the source is put in place, the teacher takes three readings of count rate, in counts per minute, at one-minute intervals.

23, 22, 27

Calculate the average background rate. [1]

c) At one point during the experiment the ratemeter reading is 78 counts per minute. Calculate how much of this reading is due to the source. [1]

d) Some students record readings from the ratemeter at 5-minute intervals. The data is corrected for background count and placed in a table:

Time (minutes)	Corrected count rate (counts per minute)
0	92
5	54
10	34
15	21
20	14
25	9

i) On graph paper, plot a graph of corrected count rate against time. Draw the best-fit curve for your points. [3]

ii) Half-life is the time taken for the corrected count rate to drop to half of its initial value. Draw lines on your graph to determine a value of half-life. [2]

e) During the experiment another student looks at the ratemeter.

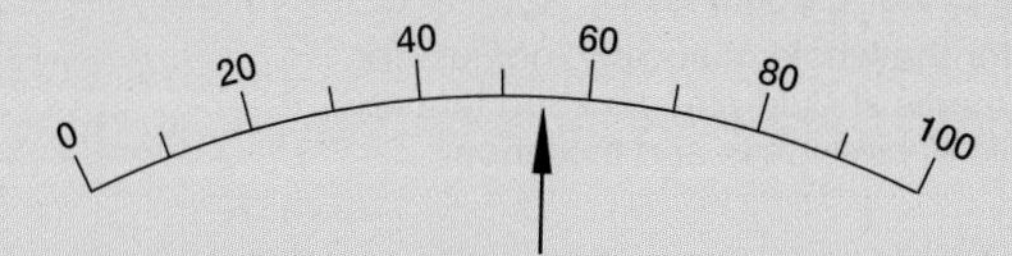

Record the reading. [1]

f) Determine the time after the start of the experiment at which she saw this reading. [3]

g) The teacher took several readings at the beginning. Explain why the students only took one reading for each time interval. [2]

h) Another student comes late to the lesson. She uses her own watch to record the following reading and asks the teacher to include it in the table of data.

489 counts after five minutes

List 3 criticisms of her data. [3]

(Edex IGCSE)

4. The IGCSE class is investigating the change in temperature of hot water as cold water is added to the hot water.

A student measures and records the temperature θ of the hot water before adding any of the cold water available.

He then pours 20 cm^3 of the cold water into the beaker containing the hot water.

He measures and records the temperature of the mixture of hot and cold water.

He repeats this procedure four times until he has added a total of 100 cm^3 of cold water. The temperature readings are shown in the table. V is the volume of cold water added.

V /	θ /
0	82
	68
	58
	50
	45
	42

a) Copy the table and complete the column headings. Enter the values for the volume of cold water added. [2]

b) Use the data in the table to plot a graph of temperature (y-axis) against volume (x-axis). Draw the best-fit curve. [4]

c) During this experiment some heat is lost from the hot water to the surroundings. Also, each time the cold water is added, it is added in quite large volumes and at random times. Suggest two improvements you could make to the procedure to give a graph that more accurately shows the pattern of temperature change of the hot water, due to the addition of cold water alone. [2] (Cam IGCSE)

More questions on page 336.

Changing ideas: How scientists work

In the past, scientists like Newton and Faraday often worked alone.

But today most scientists work in teams, and collaborate internationally.

The world wide web (internet) was originally invented just so that scientists could share their ideas and evidence.

It is important for scientists to report their findings to a science journal such as ***Nature*** or ***New Scientist***.

Experts review and evaluate the work for quality before it is published.

The Scientific Method

Scientists usually begin with an idea or hypothesis in mind, and then:

- research to look for **evidence**,
- look for **patterns** in their evidence,
- see if their hypothesis agrees, or use their imagination to think of a **theory** or **model** to explain the pattern of evidence,
- use the theory to make **predictions** that can be used to test if it is true.

Other scientists then read the journal, evaluate the work, and perhaps repeat the investigation.
The evidence won't be accepted unless the results can be repeated, reliably.

A good theory is one that makes clear **predictions**, that can be tested.

If the tests give new evidence that doesn't fit, then the explanation will have to be modified, or abandoned altogether.
eg. the caloric theory, page 371.

A theory may not be accepted, for different reasons:

- because of insufficient evidence at the time (eg. Wegener's theory of continental drift was not accepted at first, page 373),
- because of different ways of interpreting the evidence that is available (eg. the steady state theory of the Universe v. the Big Bang, page 373),
- because it contradicts some people's strongly-held views (eg. Galileo and the Church, page 376).

The discoveries and inventions of scientists may not seem useful at first.
eg. Faraday's discovery of the dynamo, which was seen as a toy at first, but is now a foundation of modern society.

The **application** of science can be controversial.

eg. the environmental effects of power stations (pages 106–7).
eg. the use of nuclear power (page 349).

These are decisions for the whole of society, not just for scientists. The mass media should play an important role, by informing the public of new ideas and evidence.

Scientists work on a wide range of interesting problems in well-paid jobs, see page 388.

The next few pages describe some of the main ideas of science, and how these ideas have developed.

Ideas about Energy

Ideas about energy started to become important when engineers like **Thomas Newcomen** (in 1705) and **James Watt** (in 1764) tried to improve steam engines. The engines did work, and to do this they had to spend energy (page 99). This was the start of the science of ***thermo-dynamics***.

In 1760, **Joseph Black** showed that temperature and thermal energy (heat) are not the same thing.

He also developed the ***'caloric' theory***. This theory said that heat was a kind of invisible weightless liquid, called *caloric*, that flowed out of hot objects and soaked into cold objects.
At first this seemed a good theory, because it explained most experiments. But it couldn't explain everything — it couldn't explain why your hands get warm when you rub them together quickly.

Nevertheless, scientists kept to this 'caloric' theory as the best theory they had at the time.

In 1798, **Benjamin Thompson** (Count Rumford) was in charge of drilling the holes in cannons.
While a gun was being drilled, he noticed that it got very hot (by friction).
By drilling for a long time and taking measurements, he showed that the gun continued to heat up, long after it should have run out of 'caloric liquid'.
As he wrote: *'It appears to me to be extremely difficult to form any distinct idea of heat, except it be* ***motion****.'*

Benjamin Thompson

James Joule was born in Salford, Lancashire in 1818. As a youth he was taught by **John Dalton** (who had suggested the idea of atoms in 1803).

In 1837, Joule began a series of precise experiments on energy, or *'vis viva'* as it was called then. To do this he had to make his own thermometers, which were accurate to 1/100 °C.
In his electrical experiments he looked at the heating effect of a current in a wire. He found that the heating depends on the *square* of the current (page 266).

In a famous experiment, he built a pulley system so that a falling weight turned some paddles that were in a tank of water.

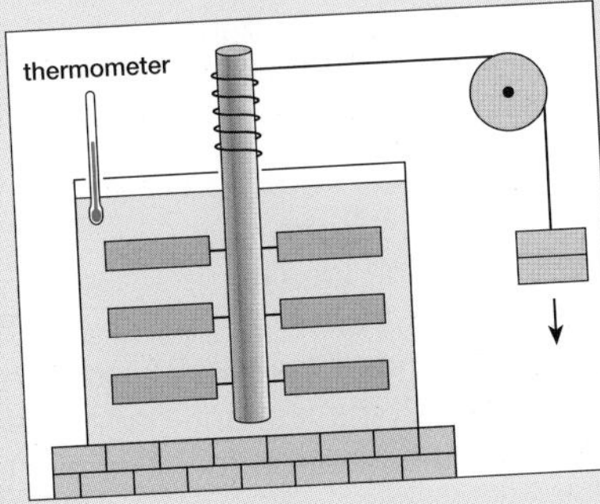

As the weight fell, the turning paddles heated up the water (by friction), and Joule measured the change in temperature.

Joule predicted that the water at the bottom of a waterfall would be slightly warmer than the water at the top (as potential energy changed to kinetic energy and heated it).

James Joule

In 1847 he married Amelia, and they went on honeymoon to the Alps. Joule spent a lot of his honeymoon measuring the temperatures at the top and bottom of waterfalls (in fact he found that the falling water cooled, by evaporation, page 56).

However, from all his experiments, Joule got evidence to convince other scientists that energy is 'conserved'.
This is the famous Law of Conservation of Energy (page 98). This is one of the most fundamental laws of our Universe.
In 1905 **Albert Einstein** extended it to include the idea that mass is a form of energy, $E=mc^2$ (p. 348).

In 1859 **James Clerk Maxwell** showed that thermal energy (internal energy) is really the movement energy of atoms and molecules.
This is the kinetic theory (see page 16).

Ideas about Forces and Gravity

Around 350 BC a Greek called **Aristotle** stated (without proof) that heavier objects fell faster. He believed that objects will only keep moving as long as there is a force exerted on them. (This *appears* to be true in everyday life on Earth, because of friction.)

It wasn't until 2000 years later that **Galileo** actually did some experiments on moving objects.

In 1589 he investigated rolling balls down an inclined plane (a ramp).

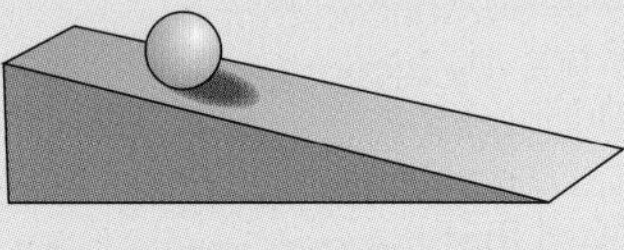

He showed that a force causes a ***change*** in an object's motion. In practice this force is often friction (page 82) which slows down objects. Often the friction is due to air resistance, as with a falling feather or a parachute (pages 89, 128).

In 1657 **Robert Hooke** showed that a feather and a coin fall at the same rate in a vacuum.

Robert Hooke

In 1687 **Isaac Newton** published his laws of motion (pages 69, 130, 84) and his law of gravitation (page 152). He showed how this also applied to the Moon and other satellites (pages 152, 154).

Newton's laws seemed to work perfectly until **Albert Einstein** worked out his ***theory of relativity***. It showed Newton's laws do not work accurately at very fast velocities near the speed of light.

In 1916 Einstein thought about the effect of gravity on space itself. He showed that space (and time) are bent or 'warped' by a mass (rather like a trampoline sags when a weight is on it).

One consequence of this is that light is affected by gravity. If the gravity of a star is so strong that light cannot escape from it, it is a ***black hole***.

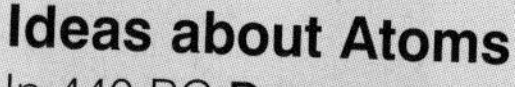

Ideas about Atoms

In 440 BC **Democritus** suggested the idea of atoms, but with no proof.
In 1803, **John Dalton** put forward the idea of atoms to explain differences in the elements.

In 1897, using a cathode ray tube (page 309), **J.J. Thomson** discovered the electron and showed that it is smaller than an atom.

Henri Becquerel discovered radioactivity in 1896 (page 338). Experiments by **Marie Curie** (page 377), **Ernest Rutherford** (page 342) and others, showed the existence of alpha, beta, and gamma rays (page 340), which suggested that atoms have a structure. If so, it would have negative parts (electrons) and positive parts, to make it neutral.

Marie Curie

In 1904, **J.J. Thomson** suggested a 'plum-pudding' model of an atom. He imagined the negative electrons were stuck in a positive blob of matter.

In 1911, **Ernest Rutherford** suggested an experiment to his assistants **Hans Geiger** and **Ernest Marsden**. They fired α-particles at a thin gold foil (see page 342).

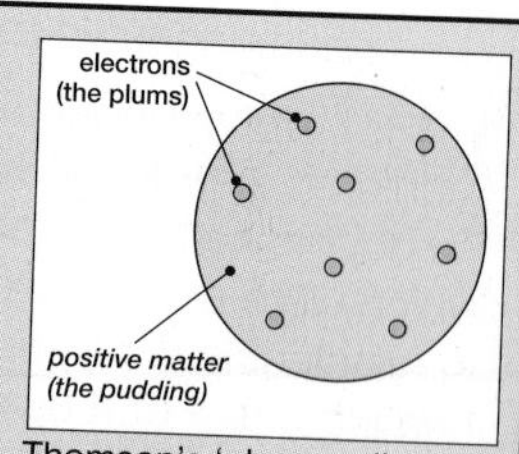

Thomson's 'plum-pud' model

Rutherford used the results to show that atoms have a positive nucleus and this has most of the mass (page 342):

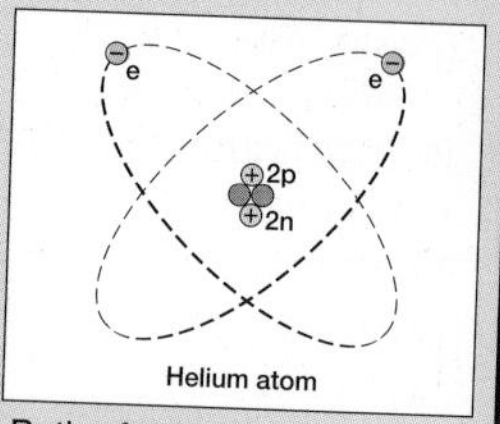

Rutherford's 'nuclear' model

In 1913 **Niels Bohr** developed the idea of electrons in orbits.
In 1932 **James Chadwick** discovered neutrons.
Also in 1932, the first anti-particle was discovered. This was the positron (positive electron).

Since then, many more particles (and anti-particles) have been discovered.
To explain these, **Murray Gell-Mann** suggested in 1961 that protons and neutrons are made up of even smaller particles, called ***quarks***. Experiments have shown this is correct.

Ideas about the Earth

Once upon a time, people believed the Earth was flat. But by 350 BC, **Aristotle** had good arguments for a round Earth. (For example, when a ship sails away, in any direction, you see the mast last of all.)

In 1600, **William Gilbert** (Queen Elizabeth I's doctor) showed that the Earth acts as if it has a magnet inside it (page 282).

From about 1900, the study of earthquakes and seismic waves (page 147) gave evidence for the 'onion-skin' structure of the Earth.
By the 1940s ideas had developed about the birth of the Earth within the Solar System (page 153).

From the first maps, about 1600, people noticed the 'jig-saw' fit of South America and Africa.
In 1915 **Alfred Wegener** suggested his theory of continental drift. However his theory suggested that only the land masses moved, and somehow slid over the ocean floor. This seemed impossible, so his theory was rejected.

In 1928 **Arthur Holmes** suggested that radioactivity inside the Earth causes large convection currents (see page 146).

In the 1950s, the first deep-sea surveys began to investigate the mid-ocean ridges.
Looking at the evidence, **Harry Hess** suggested the 'conveyor-belt' idea of sea-floor spreading.

Harry Hess

In 1963 **Fred Vine** and **Drummond Matthews** used the evidence of magnetic stripes on the sea-floor to confirm the movement.
This led to the idea of tectonic plates, which explains why earthquakes and volcanoes occur in certain places.

However it is still hard to predict when volcanoes and earthquakes will occur.

Ideas about Cosmology

Early astronomers believed that the Earth was at the centre of the Universe, but they found it harder and harder to explain their observations.

In 1543 a Polish monk called **Nicolaus Copernicus** suggested that the Earth moves round the Sun.
In 1609 **Galileo Galilei** found evidence to support this idea, and this led to his famous argument with the Church (page 376).

In 1687 **Isaac Newton** published his laws of motion (pages 69, 130, 84) and gravitation (page 152).
These explained why the planets moved in elliptical orbits to form the Solar System (page 153).

In 1785 **William Herschel** and his sister **Caroline** counted the stars in our galaxy, the Milky Way, and suggested there might be other galaxies.
This wasn't proved until 140 years later.

Caroline Herschel

Albert Einstein's theory of relativity (1917) predicted an expanding Universe.
By 1929 **Edwin Hubble** had found evidence of this, by measuring the red-shift of distant galaxies (page 158).

Edwin Hubble

For some years astronomers split into two camps: those that believed in the ***Big Bang*** theory (page 158), and their opponents who believed in the ***Steady State*** theory.
This Steady State theory suggested that new matter was being created in the Universe all the time, so even though it is expanding, it would look much the same in the past and in the future.

If there had been a Big Bang, then there should be some radio waves left over from the explosion.
In 1964 two physicists using a radio-telescope (page 215) discovered this cosmic background microwave radiation. For most scientists this is evidence of the Big Bang.

Ideas about Light

2500 years ago **Pythagoras** suggested that he saw an object because it gave out a stream of particles which travelled to his eyes, like tiny bullets. The particles were called '***corpuscles***'.

2200 years later (1666) **Isaac Newton** investigated refraction and dispersion by a prism (page 207). In 1675 **Olaus Romer** measured the speed of light.

About the same time, **Christiaan Huygens** suggested that light travels as a ***wave***, not as corpuscles.

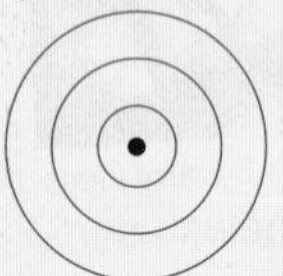

waves spreading out like ripples on a pond?

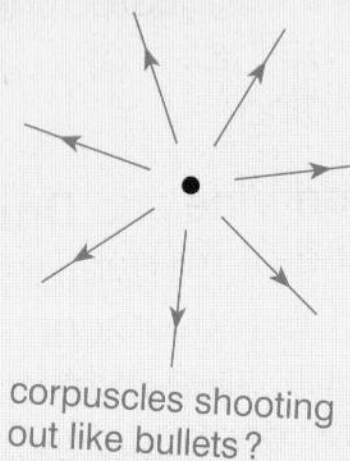

corpuscles shooting out like bullets?

Newton wasn't sure. He tried to use both the wave theory and the corpuscular theory. Later in life he seemed to favour the corpuscular theory.

For the next 150 years, Newton's followers used his name to oppose the wave theory, and most scientists believed the corpuscular theory.

Then in 1801 **Thomas Young** confirmed that light is diffracted (page 169) and discovered 'interference' of light. This was convincing evidence for the wave theory. So for the next 100 years this theory was in favour.

Thomas Young

Then in 1905 **Albert Einstein** produced a theory to explain the photo-electric effect. He used the Quantum Theory devised by **Max Planck**. Einstein showed that light travels as bundles of energy, called ***photons***, rather like corpuscles!

Today Physicists believe that light can behave as both waves ***and*** particles. Sometimes it behaves more like a wave, sometimes more like a particle!

Ideas about Telecommunications

Smoke signals and semaphore flags have been used for a long time, but in 1837 **Charles Wheatstone** patented the first electric telegraph. It used an electric circuit and the Morse Code of dots and dashes devised by **Samuel Morse**.

In 1866 the UK and the USA were connected by the first telegraph cable under the Atlantic Ocean.

In 1876 **Alexander Graham Bell** patented the telephone (page 289).

James Clerk Maxwell developed a mathematical theory of electro-magnetic waves (page 208) which included the prediction that there could be radio waves.

James Clerk Maxwell

In 1888 **Heinrich Hertz** showed how to make radio waves, and in 1901 **Guglielmo Marconi** sent radio signals (page 331) across the Atlantic.

The first communications satellite, Telstar, was launched by rocket in 1962 (page 155).

The transistor (page 322) was invented in 1948 by the team of **John Bardeen**, **Walter Brattain** and **William Shockley**. The invention of the integrated circuit in 1958 led to modern micro-computers (page 314).

Bardeen, Shockley and Brattain

Modern communication has speeded up with the use of optical fibres (page 192) and lasers (page 193), using digital signals (page 218). This has allowed the development of the world wide web on the internet, which was invented by **Tim Berners-Lee** in 1989.

Tim Berners-Lee

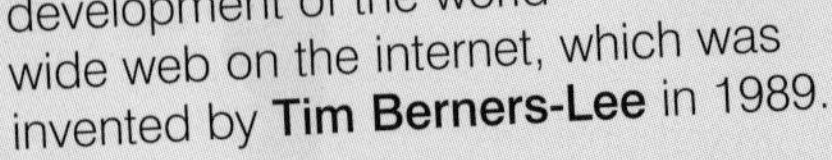

Ideas about Electricity

2500 years ago, the Greeks knew that rubbing a piece of amber resin would make it attract tiny pieces of paper, just as your comb can. (Amber was called elektron in Greek.)

Much later, in 1660, **Otto von Guericke** invented a rubbing machine that would make sparks.

Ben Franklin

In 1752, **Ben Franklin** had an idea while looking at lightning. He suggested that the clouds were charged with electricity, but he had no evidence. In a very dangerous experiment he flew a kite in a storm and found electricity was conducted down the wet string. It was Franklin who first used the words positive and negative for charges, and he invented the lightning conductor (page 245).

In 1791, **Luigi Galvani** noticed that a dead frog's leg twitched if touched by 2 different metals at the same time.

Alessandro Volta followed up this idea, and in 1800 he invented the first battery. Volta imagined that a current was a fluid flowing through the wires.

André Ampère

André Ampère suggested (wrongly) that it was a positive fluid flowing from positive to negative. It wasn't until 1897 that **J.J. Thomson** (see below) showed that it was really negative particles (electrons).

In 1826 **Georg Ohm** discovered Ohm's Law (page 253) and the idea of resistance. **James Joule** found that the heating effect depends on the resistance (page 266). This wastes a huge amount of energy in the national grid.

In 1911, **Heike Kamerlingh-Onnes** discovered superconductivity (***zero*** resistance) at temperatures near absolute zero (page 27). In 1987 it was found at higher temperatures, and the race is now on to find superconducting materials that will work at room temperature.

Ideas about Electromagnetism

Scientists suspected a link between electricity and magnetism. But it wasn't until 1819 that **Christian Ørsted** showed how a current can affect a compass needle (page 286). This was the first ammeter.

In 1821, **Ampère** used a coil to make the first electromagnet (page 287).

Michael Faraday was one of the greatest ever scientists (see page 377). In 1821 he invented the electric motor (page 292). He invented the idea of using 'lines of force' (or 'lines of flux' or 'field lines') to give a picture or 'model' of a magnetic field. This is a very important idea. It is used by modern scientists to picture other fields, such as gravitational fields and electric fields (page 244).

Faraday had an idea (a hypothesis) that it should be possible to do the opposite of an electric motor. After many experiments, in 1831 he discovered the generator (page 298) and the transformer (page 302).

With the invention of the light bulb (by **Joseph Swan** and **Thomas Edison**), this led to the idea of domestic and street lighting, and the development of the National Grid (page 303). Society as we know it could not exist without this way of supplying energy.

J.J. Thomson

In 1897 **J.J. Thomson** did an important experiment with a cathode ray tube (see page 309). By using a magnetic field to bend the beam of current inside the tube, he showed that a current is really a flow of negative electrons, and that the electrons must be smaller than an atom.

The idea of controlling electron beams led to the idea of electronic valves and to the invention of radio (page 331) and television (page 307).

Famous names

All the famous scientists were of course real people, and like all real people they had their hopes and their worries, and their different characters. Some of them were generous and kind; some were rude and unpleasant.
Here are brief biographies of six famous names that changed the world of science by their discoveries.

Galileo Galilei
1564–1642

Galileo was born in Italy on 15 February 1564 (the same year as Shakespeare). His father was a musician. Red-haired Galileo studied medicine at the University of Pisa, but soon became interested in doing experiments – an unusual idea at that time! He investigated the pendulum (page 99) and falling objects (page 128). Using a telescope, he was the first person to see Jupiter's moons and Saturn's rings (page 150). He observed the Moon, the phases of Venus, and sunspots rotating with the Sun. These observations convinced him that Nicolas Copernicus was right: that the Earth was not stationary at the centre of the universe, but that it was spinning on its axis and rotating round the Sun. This was against the views of the powerful Church at that time.
At first Galileo agreed to keep quiet, but later he wrote a sarcastic book supporting Copernicus. The Church brought him to trial in 1633, with ten judges and the threat of torture even though he was ill and 69 years old. He eventually agreed to their opinions and was sentenced to house arrest for the last 9 years of his life.

Galileo was the founder of what we call the scientific method: he believed that theories could be tested or disproved by observations and experiment, not by opinions.

Isaac Newton
1642–1727

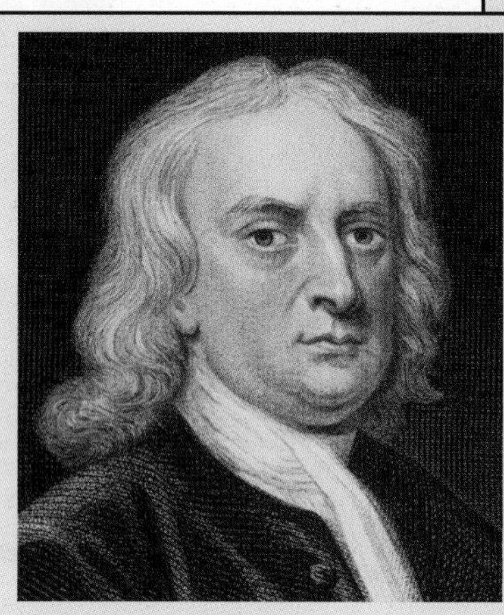

Isaac was born prematurely on Christmas Day 1642, during the English Civil War, in the year that Galileo died. His father died before he was born and his mother soon remarried but left Isaac with his grandmother for 9 years.
He was a very keen reader and went to Cambridge University when he was 19. Because of the plague, the university was closed in 1665–7 and Isaac went home for two years, doing his famous experiments on the spectrum (page 206), inventing the reflecting telescope (page 215), and observing the motions of the Moon and the planets.
He developed his law of gravitation and his laws of motion (pages 69, 130, 84).
Isaac was an insecure man, arrogant and rude.
He suffered two nervous breakdowns and often had flaming rows with other scientists, including Robert Hooke (page 66). He had no women friends and never married.

His great achievement was to show that mathematics can be used to state a set of laws which explain so much about our universe.

Albert Einstein 1879–1955

Albert was born on 14 March 1879 in Germany. At school he was lazy and slow in learning to read, but good at playing the violin. He failed some examinations but eventually got a degree at a polytechnic in Switzerland and then got a job in a patent office.
In 1905 he explained Brownian movement (page 18), published his theory of special relativity, and showed that $E = mc^2$ (page 348). He showed that Newton's Laws are not quite true for objects travelling very fast, at near the speed of light.
When the Nazis came to power in Germany Einstein left, to continue his work in America. He was a wise, cheerful and witty person, who spoke out against war and cruelty, and hated wearing socks!

His theory of relativity included and extended Newton's Laws, and changed the way that scientists look at space and time, energy and gravity.

Michael Faraday 1791–1867

Michael was born on 22 September 1791 near London, the son of a blacksmith from Yorkshire. His family was very poor and at age 13 Michael began work in a bookshop, and became a keen reader. He attended evening classes run by Sir Humphrey Davy and became his laboratory assistant.
Michael was probably the greatest experimental physicist the world has ever known. He discovered benzene, invented an electric motor (page 292), and later the dynamo (page 298) and transformer (page 302). He discovered electromagnetic induction (page 296) and the laws of electrolysis (page 274). He invented the idea of lines of flux (page 282).
He was a gentle, kind and generous man, and an enthusiastic lecturer.

He has been called 'the father of electricity', When Michael started his experiments, electricity and magnetism were seen as toys. When he died, the laws of electromagnetism had been worked out and the technological basis of our electric society had been set up.

Further reading
The Cartoon Guide to Physics
– Larry Gonick & Art Huffman
Mr Tomkins in paperback
– George Gamow & Roger Penrose

Things to do
Find out the main historical events that happened during the lifetime of each of these scientists. Which of them do you think was the 'best' scientist? Can you justify your claim?

The Curie Family
Pierre 1859–1906
Marie 1867–1934
Irene 1897–1956

Maria Sklodowska was born on 7 November 1867 (the year Faraday died), in Poland where her father was a teacher. When she was 24 she went to university in Paris, France where she met and married Pierre Curie.

Together, Marie and Pierre investigated radioactivity (discovered by their friend, Henri Becquerel, page 338). They discovered two new elements, radium (page 344) and polonium (named after her country). Pierre was run over and killed by a truck in 1906, but Marie carried on with the work and was the first person to receive two Nobel Prizes.

Their daughter Irène Curie married Fréderic Joliot, and together they found out how to make artificial radioactivity (which is now used widely to make radio-isotopes, page 346). They were awarded a Nobel Prize also.

Like her mother, Irène spoke out for women's rights, and like her mother she died of leukaemia, caused by working so long with radioactive substances.

Ernest Rutherford
1871–1937

Ernest was born on 30 August 1871 in New Zealand. His father was a Scottish immigrant, an odd-job man and farmer.
Ernest was a keen footballer, clever at school and went to university in England.
He was a great experimental physicist who rather distrusted theoretical physicists like Einstein. He had a good instinct for what should work.
He was a hard worker and covered his nervousness with a loud booming voice. He discovered alpha particles and beta particles (page 340), a radioactive series (page 345), and helped to invent the Geiger counter (page 339). He discovered protons and the transmutation of elements (page 345).

His greatest achievement was to show that atoms have a small nucleus surrounded by orbiting electrons (page 342).

▷ Dates of some inventions and discoveries mentioned in this book

Invention or discovery	Inventor or discoverer	Date	Page number in this book
Wheel and axle	–	before 2000 BC	118
Pulley	–	about 700 BC	119
Pinhole camera	–	about AD 1500	174
Idea of Earth moving round Sun	Nicolaus Copernicus	1543	373, 376
Air thermometer	Galileo Galilei	1597	26, 376
Magnetic poles of Earth	William Gilbert	1599	282
Jupiter's moons	Galileo Galilei	1609	376
Laws of refraction	Willebrord Snell	1621	185
Mercury barometer	Evangelista Torricelli	1644	80
Hooke's Law	Robert Hooke	1660	66
Boyle's Law	Robert Boyle	1662	29
Spectrum	Isaac Newton	1666	206, 374, 376
Reflecting telescope	Isaac Newton	1669	215, 376
Idea of light waves	Christiaan Huygens	1678	207, 374
Pressure cooker	Denis Papin	1680	57
Law of gravity	Isaac Newton	1685	67, 152, 372
Laws of motion	Isaac Newton	1687	69, 84, 130
Steam engine	Thomas Newcomen	1705	371
Mercury thermometer	Gabriel Fahrenheit	1714	27
Conductors and insulators	Stephen Gray	1729	243, 249
Celsius scale	Anders Celsius	1742	27
Capacitor	Pieter van Musschenbroek	1745	245, 325
Lightning conductor	Benjamin Franklin	1752	245, 375
Shape of our galaxy	William & Caroline Herschel	1752	157, 373
Bifocal spectacles	Benjamin Franklin	1760	202
High-pressure steam engine	James Watt	1765	371
Uranus	William Herschel	1781	150
Hot-air balloon	Montgolfier brothers	1783	45
Law of force between charges	Charles Coulomb	1785	241
Parachute	Jean-Pierre Blanchard	1785	89, 128
Gold-leaf electroscope	Abraham Bennet	1787	243
Hydraulic press	Joseph Bramah	1795	79

▷ Something to do

1. Plot a large labelled time-chart of inventions since 1500.

2. Each person in the group chooses one invention, and (after five minutes' preparation) has ***one minute*** to explain to the others how it made life and society different after that date.

Battery of electric cells	Alessandro Volta	1800	252, 260, 375
Magnetic effect	Hans Øersted	1819	286, 375
Electric motor	Michael Faraday	1821	292, 375, 377
Photography	Joseph Niépce	1826	198
Ohm's Law	Georg Ohm	1826	253, 375
Brownian movement	Robert Brown	1827	18
Transformer and dynamo	Michael Faraday	1831	302, 298, 375
Laws of electrolysis	Michael Faraday	1834	274, 377
Refrigerator	Jacob Perkins	1834	56
Neptune	Adams, Leverrier, Galle	1846	150
Conservation of energy	James Joule, Lord Kelvin	1852	98, 102, 371
Telephone	Alexander Graham Bell	1876	289, 314, 374
Four-stroke petrol engine	Nikolaus Otto	1876	114
Gramophone (record-player)	Thomas Edison	1877	306
Microphone	David Hughes	1878	301
Electric lamp	Joseph Swan, Thomas Edison	1879	265
Motor car	Carl Benz	1885	114
Radio waves	Heinrich Hertz	1888	211, 331, 374
Ciné camera	William Friese-Greene	1889	201
X-rays	Wilhelm Röntgen	1895	312
Radioactivity	Henri Becquerel	1896	338, 372
Electron	Joseph Thomson	1897	242, 342, 375
Radium	Marie & Pierre Curie	1898	339, 344, 377
Loudspeaker	Horace Short	1900	291
Tape recorder	Valdemar Poulsen	1900	307
Vacuum flask	James Dewar	1901	49
Diode valve	Ambrose Fleming	1904	308
$E = mc^2$	Albert Einstein	1905	348, 371, 376
G–M tube	Geiger, Müller, Rutherford	1908	339
Idea of nucleus	Ernest Rutherford	1911	342, 372, 377
Sonar	Paul Langevin	1918	226
Television	John Logie Baird	1926	307, 311
Expanding Universe	Edwin Hubble	1929	158, 373
Jet engine	Frank Whittle	1930	114
Pluto	Clyde Tombaugh	1930	150
Electrostatic generator	Robert Van de Graaff	1931	244
Radio telescope	Karl Jansky	1931	215, 373
Neutron	James Chadwick	1932	342, 372
Cat's eye reflectors	Percy Shaw	1934	188
Radar	Robert Watson-Watt	1935	211
Nuclear reactor	Enrico Fermi	1942	348
Transistor	Shockley, Bardeen & Brattain	1947	322, 374
Hovercraft	Christopher Cockerell	1954	82
Integrated circuit	Jack Kilby	1958	374
LASER	Theodore Maiman	1960	193
Idea of quarks	Murray Gell-Mann	1961	372
Optical fibre	Kao, Newnes & Beales	1966	192, 218, 314
Compact disc (CD)	Joop Sinjou, Toshitada Doi	1979	306

Key Skills

As you study Science or Physics, you will need to use some general skills along the way.
These general learning skills are very important whatever subjects you take, or whatever job you go on to do.

To show how important these are, the Government is awarding an extra qualification, which is called the **Key Skills Qualification**.

'Key skills are very important'

There are 6 key skills:

- **Communication**
- **Application of Number**
- **Information Technology (IT)**
- Working with others
- Problem solving
- Improving your own learning

The first 3 of these key skills are assessed by 1-hour Tests and by some evidence that you put together in a ***portfolio***. Your teachers will help you to collect this evidence.

There are five levels but Level 2 is probably the right level for a GCSE student to aim for.

▶ Communication

In this key skill, at Level 2 you will be expected to:

- hold discussions,
- give a short talk,
- read and summarise information,
- write documents.

You will do all these as you go through your courses (and ***not*** just in Science). Look at the criteria in the box below:

'Use an image to make your point clear'

What you must do . . .	. . . to provide evidence that you can:
Contribute to a discussion in class.	Make clear, relevant contributions; Listen and respond to what others say; Help to move the discussion forward.
Give a short talk, using an image*.	Structure your talk, and speak clearly; Use an image* to make your main points clear.
Read and summarise information from two extended documents (which are more than 3 pages long and include at least one image*).	Select and read relevant material; Identify accurately the main points and lines of reasoning; Summarise the information to suit your purpose.
Write two different types of document (one piece of writing should be an extended document and include at least one image*).	Present information in an appropriate form; Use a structure and style of writing to suit your task; Make sure your text is legible and that spelling, punctuation and grammar are accurate, so that your meaning is clear.

* an image can mean a poster, a photo, an OHP slide, a diagram or a graph drawn on a computer, a model, a PowerPoint show, etc., etc.

▶ Application of number

'Interpret information from two different sources'

In this key skill you will be expected to:

- Obtain and interpret information,
- Carry out calculations,
- Interpret and present the results of calculations.

There are many opportunities to do this in Physics, during the course of your lessons and your homework.

What you must do . . .	. . . to provide evidence that you can:
Interpret information from two different sources (including material containing a graph).	Choose how you are going to obtain the information; Obtain the relevant information; Select appropriate methods to get the results you need.
Carry out calculations to do with: a) amount and sizes, b) scales and proportions, c) handling statistics, d) using formulae.	Carry out your calculations, clearly showing your methods and level of accuracy; Check your methods and correct any errors, and make sure that your results make sense.
Interpret the results of your calculations and present your findings. You must use at least one diagram, one chart and one graph.	Select the best ways to present your findings; Present your findings clearly and describe your methods; Explain how the results of your calculations answer your enquiry.

▶ Information Technology

'Using tables to present information'

In this key skill you will be expected to:

- Search for and select information (for example, using CD-ROM databases or web-sites),
- Explore and develop the information for your purpose,
- Present the combined information in the best way.

This will often give you evidence for the Communication skill also.

What you must do . . .	. . . to provide evidence that you can:
Search for and select information for two different purposes.	Identify the information you need, and where to get it; Carry out effective searches for the information; Select information that is relevant to your enquiry.
Explore and develop information, and derive new information, for two different purposes.	Enter and bring together information, using layouts (such as tables) that help to develop and present the information; Explore the information (for example, by entering the data in a spreadsheet or producing a graph); Develop information and derive new information (for example, by calculating a total or an average).
Present combined information for two different purposes. This work must include at least one example of text, one example of images and one example of numbers.	Select and use appropriate layouts for presenting combined information in a consistent way (for example, by use of margins, headings, borders, font size, etc.); Develop the presentation to suit your purpose and the types of information you are presenting; Make sure your work is accurate, clear and saved correctly.

Revision techniques

Why should you revise?

You cannot expect to remember all the Physics that you have studied unless you revise. It is important to review all your course, so that you can answer the examination questions.

Where should you revise?

In a quiet room (perhaps a bedroom), with a table and a clock. The room should be comfortably warm and brightly lighted. A reading lamp on the table helps you to concentrate on your work and reduces eye-strain.

When should you revise?

Start your revision early each evening, before your brain gets tired.

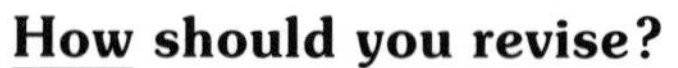

How should you revise?

If you sit down to revise without thinking of a definite finishing time, you will find that your learning efficiency falls lower and lower and lower.

If you sit down to revise, saying to yourself that you will definitely stop work after 2 hours, then your learning efficiency falls at the beginning but ***rises towards the end*** as your brain realises it is coming to the end of the session (see the first graph):

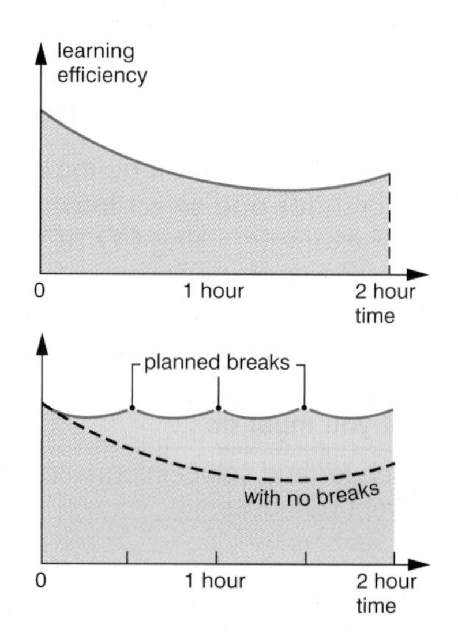

We can use this U-shaped curve to help us work more efficiently by splitting a 2 hour session into 4 shorter sessions, each of about 25 minutes with a short, ***planned*** break between them.

The breaks ***must*** be planned beforehand so that the graph rises near the end of each short session.

The yellow area on the graph shows how much you gain:

For example, if you start your revision at 6.00 p.m., you should look at your clock or watch and say to yourself, 'I will work until 6.25 p.m. and then stop – not earlier and not later.'

At 6.25 p.m. you should leave the table for a relaxation break of 10 minutes (or less), returning by 6.35 p.m. when you should say to yourself, 'I will work until 7.00 p.m. and then stop – not earlier and not later.'

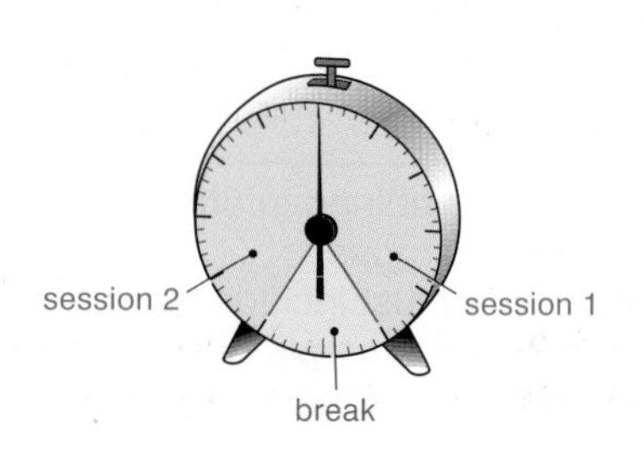

Continuing in this way is more efficient ***and*** causes less strain on you.

You get through more work ***and*** you feel less tired.

How often should you revise?

The diagram shows a graph of the amount of information that your memory can recall at different times after you have finished a revision session:

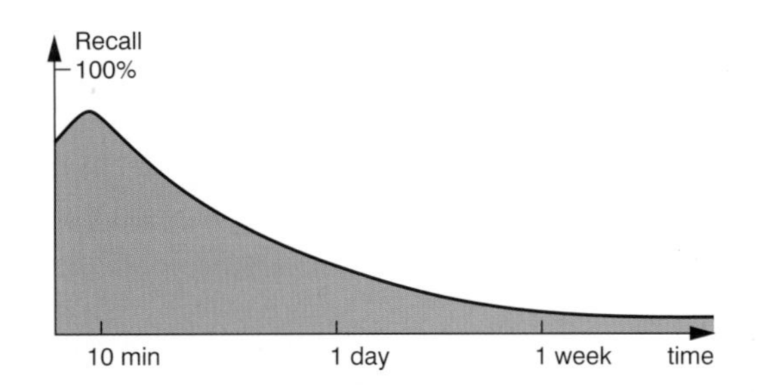

Surprisingly, the graph rises at the beginning: This is because your brain is still sorting out the information that you have been learning.
The graph soon falls rapidly so that after 1 day you may remember only about a quarter of what you had learned.

There are two ways of improving your recall and raising this graph.

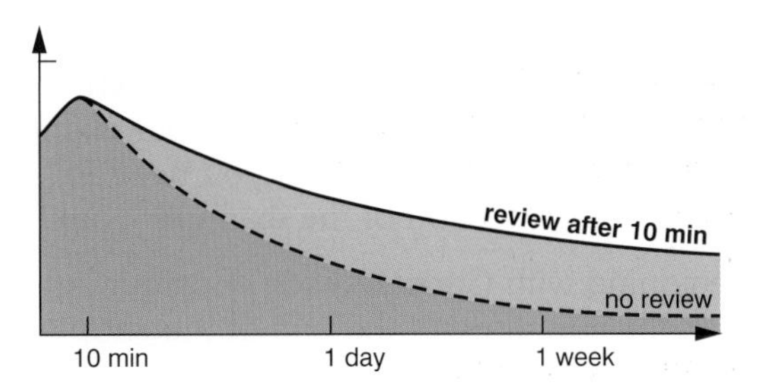

• 1. If you briefly ***revise the same work again after 10 minutes*** (at the high point of the graph) then the graph falls much more slowly:
This fits in with your 10-minute break between revision sessions.
Using the example on the opposite page, when you return to your table at 6.35 p.m., the first thing you should do is ***review***, briefly, the work you learned before 6.25 p.m.

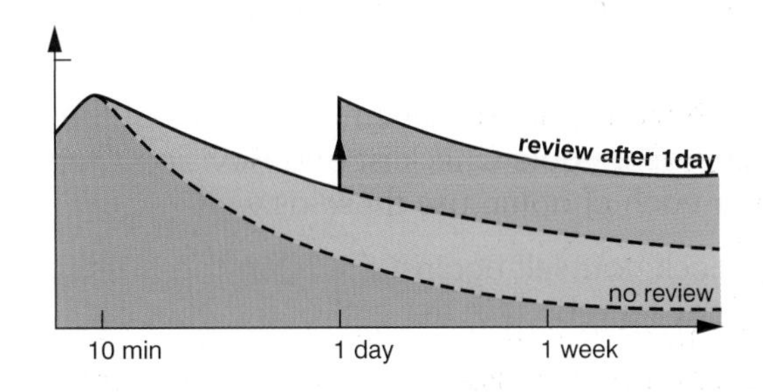

The graph can be lifted again by briefly reviewing the work ***after 1 day*** and then again ***after 1 week***.
That is, on Tuesday night you should look through the work you learned on Monday night and the work you learned on the previous Tuesday night, so that it is fixed quite firmly in your long-term memory.

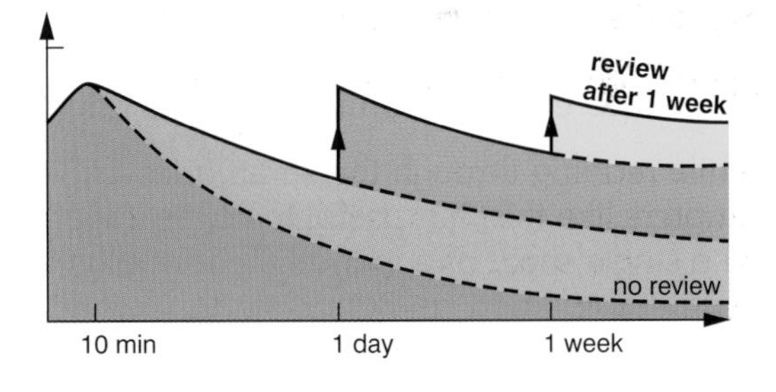

• 2. Another method of improving your memory is by taking care to try to ***understand*** all parts of your work. This makes all the graphs higher.
If you learn your work in a parrot-fashion (as you have to do with telephone numbers), all these graphs will be lower. On the occasions when you have to learn facts by heart, try to picture them as exaggerated, colourful images in your mind.

Remember: **the most important points about revision are that it must occur often and be repeated at the right intervals.**

Suggestions for a revision programme

1. Find out from your teacher which Examination Board you are using. Go to **www.physicsforyou.co.uk** and download the correct specification for your exam. It shows the exact pages that you need to read for each section of the exam syllabus.

2. Beginning with Chapter 3 (on 'Energy'), read the summary at the end of the chapter to gain some idea of the contents. Then read through the chapter looking at the key points in more detail.

   If your teacher has given you copies of the photocopyable Checklist and Revision Quiz for each chapter, use these as well.

3. Check yourself against the fill-in-the-missing-word sentences at the end of the chapter.

   Remember to ***re-read*** the summary and to ***review*** each chapter after the correct revision intervals of 10 minutes, then 1 day, then 1 week (as explained on the previous page).

4. While reading through the summaries and chapters like this, it is useful to collect together on a single sheet of paper all the formulas that you need to know and understand. See also the Checklist on the next page.

5. While doing this for every chapter, you should attempt the questions on the 'Further Questions' pages. These are all taken from previous GCSE exam papers. Your teacher will be able to tell you which are the most important ones for your syllabus.

   Your teacher may be able to give you some 'model answers' from the Support Pack, so that you can see clearly why and where the marks are awarded by the examiners.

6. Read the section on 'Examination technique' on page 386, and check the dates of your exams. Have you enough time to complete your revision before then?

7. A few weeks before the examination, ask your teacher for copies of the 'past papers' from previous years. These will help you to see:
   - the particular style and timing of your exam
   - the way the questions are asked, and the amount of detail needed
   - which topics and questions are asked most often and which suit you personally.

   When doing these past papers, try to get used to doing the questions ***in the specified time***.

   It may be possible for your teacher to read out to you the reports of examiners who have marked these papers in previous years.

8. If you are studying for the AQA exam then the ***'Top Physics Grades for You'*** Revision Guide (see page 2) will be very useful to you.

Professor Messer:
If runners are defeated and cowboys are deranged, are examiners detested?

☑ Revision checklist: definitions, laws and formulas

In some examination papers you may be asked to ***explain*** what is meant by a certain quantity, or to ***state*** a law, or to ***use*** a formula.

You will not need all of these for your exams. **Your teacher will be able to tell you which of them you should tick and then learn about.**

Heat
- Boyle's law 29, 33, 34
- Charles' law, Pressure law 30–34
- Specific heat capacity 37
- Specific latent heat of fusion 53, 54
- Specific latent heat of vaporisation 55

Mechanics
- Force, weight 65, 67, 131
- Hooke's Law 66
- Newton's First Law 69
- Newton's Second Law 130
- Newton's Third Law 84
- Density 74
- Pressure 77
- Pressure in a liquid 78, 79
- Scalar and vector quantities 86
- Parallelogram of forces 86
- Moment, law of moments 90–91
- Centre of mass (of gravity) 92
- Work and energy, joule 97, 99
- Principle of conservation of energy 98, 102
- Gravitational potential energy 108
- Elastic potential energy 108
- Kinetic energy 109
- Power, watt 110
- Efficiency 102, 116, 264
- Average speed, velocity 122, 126
- Acceleration 122, 124, 128, 130
- Constant acceleration equations 127
- Terminal velocity 89, 128
- Momentum 136–138

Astronomy
- Day, month, year, seasons 148, 149
- Planet, comet, star, galaxy 150, 153, 157
- Law of gravitation 152
- Polar, geostationary satellites 154

Waves and sound
- Wavelength, frequency, amplitude 167, 225
- Transverse and longitudinal waves 166, 146
- Speed, frequency, wavelength 167, 225, 228
- Pitch, loudness, quality 232–233

Light
- Law of reflection 177
- Real and virtual images 174, 179, 195, 196
- Focal length of a curved mirror 181
- Refractive index and Snell's Law 185
- Critical angle 187
- Principal focus and focal length of a lens 194
- Spectrum and dispersion 206
- Electromagnetic spectrum 208–217
- Analogue, digital signals 218, 306
- Filters, colours of light 220

Electricity
- Law of electrostatics 241
- Capacitance 245, 325
- Resistors in series 250, 256, 261
- Resistors in parallel 251, 257, 261
- Ohm's Law and resistance 253
- Micro-, milli-, kilo-, mega- 9, 252
- Potential divider 258, 323, 329
- Current : voltage graphs 259, 316, 319
- Coulomb, ampere, volt 260, 261
- Electric energy (in joules) 261, 266
- Electric energy (in kW h) and cost 267
- Electric power 111, 266
- Anode, cathode, law of electrolysis 273, 274
- Law of magnetism 280
- Magnetic field and line of flux 282, 286
- Right-hand Grip Rule 286
- Rule for polarity of a coil 287
- Fleming's Left-hand Rule 290
- Fleming's Right-hand Rule 296
- Faraday's Law and Lenz's Law 296, 297
- Effective value of a.c. 299
- Transformer 302
- Truth tables, logic gates 326

Nuclear physics
- Alpha, beta and gamma rays 341
- Alpha-decay, beta-decay 345
- Atomic number, mass number, isotope 343
- Half-life 344, 352
- Energy and mass 348, 357
- Becquerel 350

Examination technique

In the weeks before the examinations:

Attempt as many 'past papers' as you can so that you get used to the style of the questions and the timing of them.

Note which topics occur most often and revise them thoroughly, using the techniques explained on previous pages.
Read through your list of essential formulas as often as possible if you need to memorise them (see page 385).

Just before the examinations:

Collect together the equipment you will need:

- Two pens, in case one dries up.
- At least one sharpened pencil for drawing diagrams.
- A rubber and a ruler for diagrams.
  Diagrams usually look best if they are drawn in pencil and labelled in ink.
  Coloured pencils are usually ***not*** necessary (but may sometimes make part of your diagram clearer, eg. when drawing a spectrum).
- A watch for pacing yourself during the examination. The clock in the examination room may be difficult to see.
- A calculator (with good batteries).
- For some examinations you may need special instruments eg. compasses or a protractor.

Use your lists of essential formulas for last-minute revision.

It will help if you have previously collected all the information about the length and style of the examination papers (for ***all*** your subjects) as shown below:

Date, time and room	Subject, paper number and tier	Length (mins)	Types of question: – structured? – single-word answers? – longer answers? – essays?	Sections?	Details of choice (if any)	Approximate time per mark (minutes)
5th June 9.30 Hall	Science (Additional) Paper 3 (Physics) Foundation Tier	45 min	Structured questions (with single-word answers and longer answers)	1	no choice	1 min.

In the examination room:

Read the front of the examination paper carefully. It gives you important information. How is your examination paper different from this one:

Some hints on answering questions are given in the box below.

NATIONAL EXAMINING BOARD

Science: Physics
Foundation Tier
5th June
9.30 a.m.

Time: 1 hour 30 minutes

Answer **all** the questions.

In calculations, show clearly how you work out your answer.

Calculators may be used.

Mark allocations are shown in the right-hand margin.

In what ways is your examination paper different *from this one?*

Answering 'structured' questions:

- Read the information at the start of each question carefully. Make sure you understand what the question is about, and what you are expected to do.
- Pace yourself with a watch so you don't run out of time. If you have spare time at the end, use it wisely to check over your answers.

How much detail do you need to give?

- The question gives you clues:
  - Give short answers to questions which start: '***State ...***' or '***List ...***' or '***Name ...***'.
  - Give longer answers if you are asked to 'Explain ...' or '***Describe ...***' or asked '***Why does ...?***'.
- Look for the marks awarded for each part of the question. It is often given in brackets, eg. [2] This tells you how many points the examiner is looking for in your answer.
- The number of lines of space is also a guide to how much you are expected to write.
- Always show the steps in your working out of calculations. This way, you can gain marks for the way you tackle the problem, even if your final answer is wrong.
- Try to write something for every part of each question.
- Don't explain something just because you know how to! You only earn marks for exactly what the question asks.
- Follow the instructions given in the question. If it asks for one answer, give only one answer. Sometimes you are given a list of alternatives to choose from. If you include more answers than asked for, any wrong answers will cancel out your right ones!

If your exam includes 'multiple-choice' questions:

- Read the instructions very carefully.
- If there is a separate answer sheet, mark it exactly as you are instructed, and take care to mark your answer (A, B, C, D, E) opposite the correct question number.
- Even if the answer looks obvious, you should look at all the alternatives before making a decision.
- If you do not know the correct answer and have to guess, then you can improve your chances by first eliminating as many wrong answers as possible.
- Ensure you give an answer to every question.

Careers

Have you thought what you want to do when you leave school?

After English and Mathematics, Physics is the most important qualification for a great many careers.

In the list below: ★ = Physics is usually essential
+ = Physics is an advantage

Employers rate Physics qualifications very highly, particularly if you study it to A-level or higher.

And studying Physics can open the door to a surprising variety of jobs.

Aeronautical Engineer ★
Agricultural Scientist +
Air-traffic Controller +
Architect +
Army ★ or +
Astronomer ★
Audiologist ★
Automobile Engineer ★
Biomedical Engineer ★
Biophysicist ★
Building Technologist ★
Civil Engineer ★
Civil Service Scientific Officer +
Computer-aided Design +
Computer Programmer +
Dental Technician +
Dentist +
Doctor +
Draughtsperson +
Electrical Engineer ★
Electrician ★
Electronics Engineer ★
Environmental Health Officer +
Ergonomicist ★
Flight Engineer ★
Food Scientist +
Forensic Scientist ★
Geophysicist ★
Health & Safety Officer +
Industrial Designer +
Information Scientist +
Journalist (science) +
Laboratory Technician ★
Lighting Technologist ★

Marine Scientist +
Materials Scientist +
Mechanical Engineer ★
Medical Physicist/Technician +
Merchant Navy, deck,
engineer, or radio officer ★
Metallurgist ★
Meteorologist ★
Mining Engineer ★
Motor Mechanic +
Nuclear Scientist ★
Optician ★
Patent Agent/Examiner ★
Pharmacist +
Physicist ★
Physiotherapist +
Pilot +
Production Engineer ★
Quantity Surveyor +
Radio and TV repair ★
Radiographer ★
Radio Studio Manager +
Recording Engineer ★
Royal Air Force ★ or +
Royal Navy ★ or +
Space Scientist ★
Structural Engineer ★
Systems Analyst +
Teacher (science) ★
Technical Writer +
Telecommunications (radio,
telephone, satellite) ★
TV Camera Operator +
Veterinary Surgeon/Assistant +

using electronics

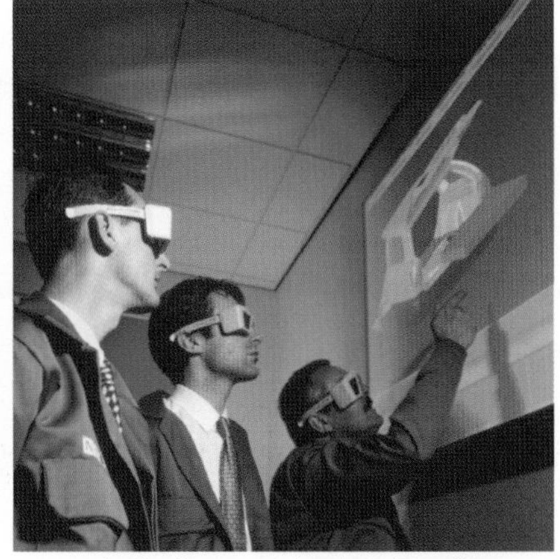

designing cars in virtual reality

a civil engineer

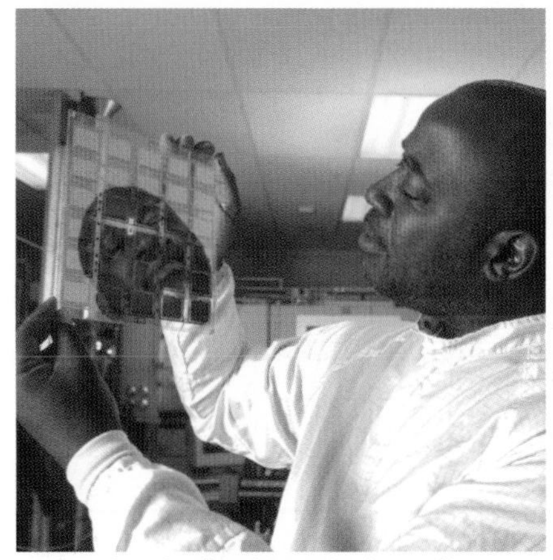

developing better solar cells

crime investigation

teamwork is important

a quantity surveyor

adjusting a jet engine

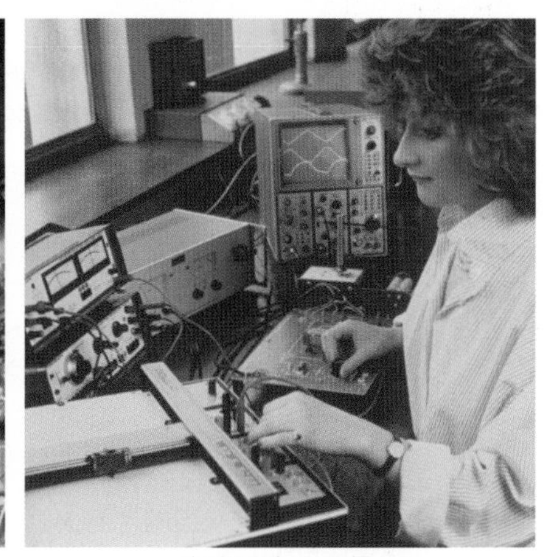

electronic testing

Of course, for many of the careers listed or illustrated on these pages, you will need further study at school or in a College of Further Education. You should consult your careers teacher for more detailed information.

To find out more, visit the Institute of Physics web-site at: ***http://learningphysics.iop.org*** and select 'Beyond School'.

Check your maths

Directly proportional

Look at the table. Do you see how $\mathbf{Y}$ and $\mathbf{X}$ are connected?
If $\mathbf{X}$ doubles, so does $\mathbf{Y}$. If $\mathbf{X}$ halves, so does $\mathbf{Y}$, etc.

Y	X
3	1
6	2
9	3
12	4

We say: $\mathbf{Y}$ is ***directly proportiona***l to $\mathbf{X}$

In symbols: $\mathbf{Y} \propto \mathbf{X}$

For an example, see Hooke's Law (page 66) or Ohm's Law (page 253).

Making an equation

We can always change a proportionality into an equation by putting in a constant, $\boldsymbol{k}$:

$$\boldsymbol{Y} = \boldsymbol{kX} \quad \textit{or} \quad \frac{\boldsymbol{Y}}{\boldsymbol{X}} = \boldsymbol{k}$$

In the example in the table, $\boldsymbol{k} = 3$.

Inversely proportional

Look at the table. Do you see how $\boldsymbol{P}$ and $\boldsymbol{V}$ are connected?
If $\boldsymbol{V}$ doubles, $\boldsymbol{P}$ halves. If $\boldsymbol{V}$ halves, $\boldsymbol{P}$ doubles, etc.

P	V
12	1
6	2
4	3
3	4

We say: $\boldsymbol{P}$ is ***inversely proportional*** to $\boldsymbol{V}$

In symbols: $\boldsymbol{P} \propto \dfrac{1}{\boldsymbol{V}} \quad \textit{or} \quad \dfrac{1}{\boldsymbol{P}} \propto \boldsymbol{V}$

We make an equation in the same way as before:

$$\boldsymbol{P} = \boldsymbol{k} \times \frac{1}{\boldsymbol{V}} \quad \textit{or} \quad \boldsymbol{P} = \frac{\boldsymbol{k}}{\boldsymbol{V}} \quad \textit{or} \quad \boldsymbol{P} \times \boldsymbol{V} = \boldsymbol{k}$$

In the example in the table, $\boldsymbol{k} = 12$.

Changing the subject of an equation

As long as you do the same thing to ***both*** sides of an equation, it will still balance:

Example 1

From page 74: $\boldsymbol{D} = \dfrac{\boldsymbol{M}}{\boldsymbol{V}}$

a) To change the subject to $\boldsymbol{M}$:
Multiply both sides by $\boldsymbol{V}$, then cancel: $\boldsymbol{D} \times \boldsymbol{V} = \dfrac{\boldsymbol{M}}{\cancel{\boldsymbol{V}}} \times \cancel{\boldsymbol{V}}$

$\therefore \underline{\boldsymbol{M} = \boldsymbol{D} \times \boldsymbol{V}}$

b) To change the subject to $\boldsymbol{V}$:
First multiply by $\boldsymbol{V}$: $\boldsymbol{D} \times \boldsymbol{V} = \boldsymbol{M}$

Then divide by $\boldsymbol{D}$: $\dfrac{\cancel{\boldsymbol{D}} \times \boldsymbol{V}}{\cancel{\boldsymbol{D}}} = \dfrac{\boldsymbol{M}}{\boldsymbol{D}}$

$\therefore \underline{\boldsymbol{V} = \dfrac{\boldsymbol{M}}{\boldsymbol{D}}}$

Example 2

From page 127: $\boldsymbol{a} = \dfrac{\boldsymbol{v} - \boldsymbol{u}}{\boldsymbol{t}}$

To change the subject to $\boldsymbol{v}$:

First multiply by $\boldsymbol{t}$: $\boldsymbol{a} \times \boldsymbol{t} = \dfrac{\boldsymbol{v} - \boldsymbol{u}}{\cancel{\boldsymbol{t}}} \times \cancel{\boldsymbol{t}}$

$\therefore \boldsymbol{at} = \boldsymbol{v} - \boldsymbol{u}$

Then ***add u***: $\boldsymbol{at} + \boldsymbol{u} = \boldsymbol{v} - \cancel{\boldsymbol{u}} + \cancel{\boldsymbol{u}}$

$\therefore \underline{\boldsymbol{v} = \boldsymbol{u} + \boldsymbol{at}}$

Often it is easier if you put in your numbers ***before*** changing the equation.

Graphs

a) ***Directly proportional***

For an equation like $\boldsymbol{Y = kX}$ (see the opposite page), the graph is a ***straight*** line, ***through the origin***.

The gradient (or slope) = $\boldsymbol{k}$ (see page 363).
For an example, see Hooke's Law (page 66).

b) ***Linear but not directly proportional***

Consider an equation like $\boldsymbol{Y = kX + c}$ or like $\boldsymbol{v = at + u}$ (see example 2 opposite and page 127).
For these equations the graph is a straight line but ***not*** through the origin.

c) ***Inversely proportional***

A graph of $\boldsymbol{P}$ against $\boldsymbol{V}$ (see table opposite) would give a curve.

The equation is $\boldsymbol{P} = \boldsymbol{k} \times \frac{1}{\boldsymbol{V}}$, so to get a straight line, we must plot $\boldsymbol{P}$ against $\frac{1}{\boldsymbol{V}}$.

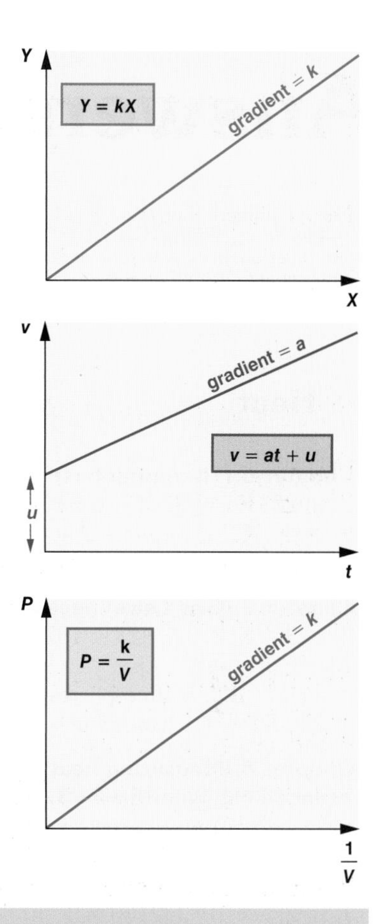

Dividing by fractions

Remember that dividing by $\frac{1}{2}$ is the same as multiplying by 2.
Dividing by $\frac{1}{3}$ is the same as multiplying by 3, etc.

Many people make mistakes with the parallel-resistor formula on page 257. Read the example on that page carefully.

Large and small numbers – indices

The small ***index*** tells us how many decimal places to move.
The + or − sign tells us which direction to move.
That is, for 10^n, move the decimal point $\boldsymbol{n}$ places to the right.

Examples:

$2 \times 10^6 = 2\,000\,000$
$2.1 \times 10^6 = 2\,100\,000$
$0.21 \times 10^7 = 2\,100\,000$

For 10^{-n}, move the decimal point $\boldsymbol{n}$ places to the left.

Examples:

$4 \times 10^{-1} = 0.4$
$4 \times 10^{-3} = 0.004$
$4.1 \times 10^{-3} = 0.0041$

Seven steps in solving a Physics problem

1. Draw a diagram showing all the information given in the question.
2. Decide which ***formula*** to use. (If you are not sure, write down the possibilities and then decide.)
3. Decide the ***units*** for the formula. If necessary, change your information to the correct units (eg. mm to metres).
4. ***Substitute*** the numbers in the formula.
5. ***Calculate*** the answer.
   Check that your answer is a sensible size.
6. Decide the ***units*** for your answer.
7. Read the question to see if there is ***another part*** to do.

Answers

These are numerical answers only.

Note: fuller answers, with a Mark Scheme, are included in the Teacher Support Pack.

Heat

Chapter 6 (Thermometers) page 28
7. a) 283 K b) 27 °C c) 310 K
8. b) 40 °C c) Between 2 hours and 4 hours
d) 4 hours e) 9 hours f) 0.5 h, 6.4 h g) 5.5 h

Chapter 7 (Gas Laws) page 34
2. a) 300 K, 270 K, 423 K, 183 K b) 100 °C, –73 °C 727 °C
3. 4 cm^3 **4.** 1.5 atmospheres
5. 2500 cm^3, 1.1 atmospheres **6.** 273 cm^3

Chapter 8 (Measuring heat) page 39
2. Jack 4 MJ, Wife 18 MJ **3.** Low specific heat capacity, poisonous, expensive, heavy
4. a) 42 000 J b) 8400 J c) 880 J d) 3800 J
5. 500 J/kg °C
7. a) 2.1×10^9 J (2100 MJ) b) £63
c) 7.2×10^8 J (720 MJ) d) 1.7 K (1.7 °C)

Chapter 9 (Conduction) page 52
3. Roof (cost regained in 2 years)

Chapter 10 (Changing state) page 59
7. a) 680 000 J b) 170 000 J c) 6 900 000 J
d) 230 000 J
8. 2 400 000 J/kg **9.** 6 120 000 J

Further questions on Heat

Gas Laws (page 61)
11. b) iii) 104.5 kPa v) −273 °C

Conduction, Convection, Radiation (page 62)
15. d) 4 years
17. a) 2600 J, 1200 J c) 65%, 45%
19. a) 100 kJ b) i) 29 kJ/minute ii) 29 kJ/minute
22. a) 53 200 J b) 67.2%
23. a) 450 000 J b) 336 000 J

Mechanics

Chapter 11 (Pushes and pulls) page 72
2. 40 N (39.2 N) **3.** a) 600 N (588 N)
b) Approx. 100 N c) 60 kg **4.** a) 10 cm b) 15 cm
5. a) 50 mm d) 70 mm length should be 66 mm
e) 5 N f) 3.75 N g) 86 mm
7. b) 6.6 cm should be 6.4 cm c) 9.0 cm d) 2600 N

Chapter 12 (Density) page 76
1. 5 g/cm^3 (5000 kg/m^3) **2.** 8000 kg/m^3
3. 18 000 kg **4.** a) B b) C c) A and C d) D
5. 40 000 kg **6.** 5 m^3 **7.** 3 g/cm^3 (3000 kg/m^3)
8. a) 2 m^3 b) 0.002 m^3 c) 5000 kg d) 200 bricks
9. 130 kg **11.** b) Density = 11 g/cm^3 (lead?)

Chapter 13 (Pressure) page 81
2. 50 N/m^2 **3.** 1 000 000 N (1 MN)
7. a) 100 N/cm^2 (1 000 000 N/m^2) b) 2000 N

Chapter 14 (More about forces) page 88
3. b) 4.7 m/s

Chapter 15 (Turning forces) page 96
3. 40 N m (4000 N cm) **4.** a) 5 N b) 4 N c) 1 N
5. c) 5 m

Chapter 16 (Work, energy and power) page 112
2. 10 J **4.** 20 000 J (20 kJ)
10. a) 60 J b) 5% (0.05) **11.** d) 28.5 m
14. 100 W **15.** 500 W
16. a) 7500 W b) 7.5 kW
20. a) 100 J b) 10 000 J (10 kJ)
21. a) 450 000 J b) 50 000 J c) 400 000 J d) 5000 N
22. 5 m

Chapter 17 (Machines) page 120
2. 25%, 35%, 10%, 20% **3.** 2%, heat **4.** 300 N
6. a) 600 J b) 800 J c) 75% (0.75) **7.** 10 rev/s anticlockwise, 15 rev/s clockwise **9.** i) 20 J, 30 J, 67%, (0.67) ii) 600 J, 800 J, 75% (0.75)

Chapter 18 (Velocity and acceleration) page 133
1. a) 20 m/s b) 2 m/s^2 **2.** 20 m/s
4. b) i) Zero ii) 10 m/s^2
5. From rest, it accelerates uniformly for 4 s, at 2.5 m/s^2, up to 10 m/s. Travels at constant velocity of 10 m/s for 10 s. Then constant deceleration, of 2 m/s^2, for 5 s, to stop after a total time of 19 s.
6. a) 0.7 s c) 4.7 m/s **8.** 40 m/s
10. b) 500 N c) 50 kg **11.** 7.7 s **12.** 5 m
13. a) 20 m/s b) 2 s c) 10 m **14.** 8.45 m
15. 5 m/s^2, 3 m/s^2
16. a) 3.0×10^7 N b) 0.3×10^7 N (3×10^6 N)
c) 1 m/s d) zero **17.** a) 15 m/s^2 b) 1500 N
18. a) 1000 N b) 160 N

Chapter 19 (Momentum) page 139
2. 6 kg m/s
4. a) 2 m/s b) 64 J, 16 J c) Heat, sound
5. 2 m/s **6.** 40 m/s **7.** 0.1 m/s
8. a) 32 500 000 kg m/s b) 32 500 000 N (3.25×10^7 N)

Further questions on Mechanics

Hooke's Law (page 140)
1. c) i) 24 mm ii) 5 N

Moments and machines (page 140)
3. 33.3 g **4.** a) 300 Nm b) 100 N
5. 4 Nm (400 Ncm)

Pressure (page 140)
6. b) 3000 N c) 37.5 N/cm^2 (375 000 N/m^2 or 375 kPa)
7. a) i) 25 N/cm^2 ii) 25 N/cm^2 iii) 2500 N

Forces (page 141)
10. a) 700, 700, 700, 700 d) 40 m/s

Momentum (page 141)
13. a) 8 kg m/s b) $2\frac{2}{3}$ m/s c) 16 J d) 10.7 J
14. a) 0 kg m/s b) 80 m/s

Energy sources (page 141)
18. b) 0.25 (25%)

Energy (page 142)
21. b) i) 1.728×10^{14} J ii) 4.86×10^{14} J
iii) 35.5% (0.355)
22. a) i) 20 J ii) 60 J iii) 20 J b) 25% (0.25)
23. b) 28% (0.28)

Energy, work and power (page 143)
26. b) 4 m c) 800 J d) gravitational potential energy
e) 840 J f) Food
27. a) 300 m b) 180 000 J (180 kJ)
c) 18 000 W (18kW) e) 30% (0.3)
28. a) 150 m b) 120 000 J c) 12 000 W (12 kW)
29. a) 900 N c) 630 000 J (630 kJ)
d) 720 000 J (720 kJ) e) i) 90 000 J (90 kJ)
ii) 45 m/s
30. a) 14 400 J b) 600 W **31.** a) 3000 J b) 600 W
32. 1500 W, 600 joules/second (W)

Velocity and acceleration (page 144)
33. b) Accelerating at 1 m/s^2; constant velocity = 15 m/s; decelerating at –0.75 m/s^2 c) 1 m/s^2; d) 112.5 m, 225 m, 337.5 m **34.** c) i) 9 m ii) 6 m
35. a) 200 m b) 11 s c) 80 m d) 10 s e) 8 m/s
f) Constant g) Zero
36. b) i) 36 s ii) 750 m iii) –2.5 m/s^2
iv) 250 000 N c) Look at areas

Force and acceleration (page 145)
38. a) 18 m b) 15 m/s^2 c) 13 500 N
d) 405 000 J (405 kJ) e) 405 000 J (405 kJ)
39. a) i) 1.0 s ii) 4 m iii) 20 m/s^2 iv) 1400 N
v) 1.6 m b) i) 1 m ii) 700 J iii) 560 J
40. b) 1000 kg c) i) 400 N ii) 0.4 m/s^2
d) See pages 69, 138
41. a) 10 m/s^2 b) 10 000 N
42. a) 14 m/s b) 140 m c) 2 m/s^2
d) 10 N e) 120 kg m/s f) 40 J g) 160 J

▶ The Earth and beyond

Chapter 20 (The Earth and beyond) page 162
7. g) i) ~500 000 000 km iii) ~6 years

Further questions (page 163)
3. b) i) Uranus ii) Pluto iii) 228 000 000 km
12. b) i) 1 c) i) 14.1 **15.** c) 12 000 million years

▷ Waves: Light and Sound

Chapter 21 (Waves) page 170
3. 340 m/s
4. a) 300 000 000 m/s (3×10^8 m/s) b) 600 m **5.** 0.5 s
6. a) 2 cm b) 3 Hz **8.** b) 15 mm c) 4 Hz

Chapter 22 (Light) page 175
4. 3.9×10^8 m (390 000 km) (see page 122)

Chapter 23 (Reflection) page 180
1. a) 60° b) 60° c) 60° **3.** 2 m behind mirror, 5 cm wide, 1 m/s (2 m/s relative to him)

Chapter 25 (Refraction) page 191
13. b) 15° c) 23°

Chapter 26 (Lenses) page 197
6. 7.5 cm from lens; 2 cm high; real, inverted, diminished
7. a) 100, 60, 40, 30, 24 c) 26 cm d) 25 cm e) 20 cm

Chapter 27 (Optical instruments) page 205
7. a) $\frac{1}{1000}$ s at ***f***/5.6 b) $\frac{1}{1000}$ s at ***f***/2
c) $\frac{1}{60}$ s at ***f***/16 d) $\frac{1}{1000}$ s at ***f***/4

Chapter 28 (Colour) page 223
14. 15 km **15.** 0.3 s
22. a) 0.25 W/m^2 b) 0.11 W/m^2 c) 0.010 W/m^2

Chapter 29 (Sound) page 235
2. 6800 m (6.8 km) **3.** 3.4 m **4.** a) 20 000 Hz, 20 Hz
b) 1.7 cm, 17 m **5.** 1430 m **7.** 680 m
8. 360 m/s **9.** 2250 m **13.** 104 dB, not safe

Further questions on Waves: Light and Sound

Waves (page 236)
1. b) i) 1.5 cm ii) 15 cm/s
3. a) 300 000 km/s (3×10^8 m/s) b) 1200 kHz

The electromagnetic spectrum (page 238)
18. c) i) 0.03 m (3 cm) **22.** a) 0.25 m (25 cm)

Sound (page 239)
25. a) 150 m b) 75 m **26.** b) 721 m c) 0.36 s
29. a) 0.0005 m (0.5 mm)

Chapter 35 (Magnetic effect) page 295
14. b) i) 7.7 N ii) 15.0 N
15. a) 30 J b) 15 W c) 20 W d) 75% (0.75)

Chapter 36 (Electromagnetic induction) page 305
9. d) 10 V e) 4 A
10. a) 10 : 1 b) 46 W c) 46 W if 100% efficient d) 0.2 A
11. 1000 V a.c., 10 V a.c., 10, 1000
14. a) i) 1000 A ii) 1 A

Chapter 37 (Electron beams) page 313
5. a) 4 V (8 V peak to peak) b) 20 ms (0.020 s) c) 50 Hz

▶ Electricity

Chapter 31 (Circuits) page 262
3. 3 A out b) 1 A in
5. a) 30 Ω b) 1500 Ω c) 1 Ω d) 5 Ω
8. a) 1 A, 0.5 A, 1 A **9.** a) 10 V b) 3 A c) 10 Ω
10. a) 2 A b) 2 A c) 6 V d) 14 V
11. a) 5 Ω b) 2 A c) 4 V d) 6 V **12.** 1200 J
13. a) 2 Ω b) 12 V c) 12 V d) 4 A e) 2 A
14. 1000 A, 100 MJ (100 000 000 J)

Chapter 32 (Heating effect) page 271
2. a) 20 000 J (20 kJ) b) 5 kJ c) 25%
3. 2 A
4. a) 0.25 A b) 960 Ω **6.** £1.20 (120 p)
7. a) 1968 Units b) £157.44 **8.** 80 p
10. 4 A, 5 A; 1 A, 3 A; 460 W, 3 A; 4 A, 5 A; 2300 W, 13 A

Chapter 33 (Chemical effect) page 275
4. 6 g **5.** a) 0.33 g b) 0.66 g

Further questions on Electricity (1)

Circuits (page 276)
4. a) 3000 Ω b) 0.003 A (3 mA) c) 3 V
5. a) i) 0.01 A (10 mA) ii) 10 V b) 2000 Ω
7. a) See page 259 c) i) 2.5 V ii) 10 Ω
8. a) 7 Ω b) 2 A c) 10 V d) 1.3 A
9. c) 0.064 K/min A^2 d) 0.051 K/min A^2
11. a) 2 A b) 6 Ω c) 24 W d) i) Increases
ii) Decreases
12. a) 6 Ω b) 1.5 W **13.** £4.20

Electric power (page 278)
14. a) 2300 W b) 13 A d) 23 Ω
15. b) i) S_3 ii) S_3 and S_1 c) i) 120 Ω ii) 4.5 A iii) 7 A
16. b) i) 1.5p ii) 2p **17.** a) 48.0 W b) 12.0 V
18. a) 10 A b) 13 A d) £3.68 **19.** b) 720 J
20. a) 15 Ω b) 0.6 W c) 120 C d) 360 J

Further questions on Electricity (2)

Magnetic effect, motors (page 332)
4. 87.5% (0.875)
5. a) 80 J b) 16 W c) 40 W d) 40% (0.40)

Electromagnetic induction (page 333)
14. b) A, 115 V; B, 230 V **15.** 1930 A
16. c) i) 500 A ii) 5000 V iii) 2 500 000 W (2.5 MW)
iv) Heat **17.** c) i) 20 turns ii) 0.1 A
d) i) Doubled to 40 turns ii) 20 turns
18. d) 100 000 e) 3 A f) 76.7 Ω

Electronics (page 335)
20. c) i) About 340 Ω ii) 80 Ω
22. a) E: 0,0,0,1 P: 1,1,1,0
25. A = 1; B = 0; C = 1

How Science works (page 336)
29. b) 0.72 W **30.** a) i) 75 A ii) 69.8 A

▶ Radioactivity

Chapter 39 (Radioactivity) page 351
4. b) 6, 14, $\frac{1}{4}$ c) About 17 100 years **5.** 140 years

Further questions on radioactivity (page 353)
1. b) approx 32 minutes
2. a) Background radiation c) 35 min d) See p. 345
3. a) 120 per second b) 50 per second
4. a) i) 94 ii) 144 iii) 94 **5.** a) i) 131 ii) 54
7. b) 4 kg, 1 kg c) i) 76, 48, 30, 19, 12 iv) 1.5 y
8. a) 95, 146, 95 b) ii) 237, 93
c) ii) 460 years (smaller counts are less accurate)
10. a) i) 92 ii) 92 iii) 235
12. c) 16 800 years
13. d) 5.27 y
14. b) about 400 million years

For help with your mathematics see page 390

Dotty Definitions

joule	– fight to the death
satellite	– burning in the chair
unit	– you do it with wool
centigrade	– scale for perfumes
change of state	– emigration
insulate	– getting home after time
ionise	– and a steel nose
newton	– up-to-date weight
watt	– question
power	– hit the lady
sunspots	– his acne
maximum	– large mother
principle	– royal tug of war
inclined plane	– there's writing on the walls of the aircraft
velocity	– but we didn't lose the coffee
hertz	– painful
amplitude	– well and truly eaten
dispersion	– or that person
circuit	– for making a teacher?
anode, cathode	– female debtors
armature	– has strong teeth
a.c./d.c.	– I view the ocean
dynamo	– eat briefly

Something to do

For each of these terms give the correct definition or write as much as you can to describe them.

Eg., joule: unit of energy; 1 kJ = 1000 joules; work done when 1 newton moves through 1 m.

Wanted

A reward is offered for information leading to the arrest of Eddy Current, charged with assault and battery on a teenage coil named Milli Amp. He is also wanted in connection with the parallel theft of valuable joules from a bank volt.

Milli Amp tried to run but met series resistance and was overpowered. Later she was found by her friend Dinah Mo. After the couple had had a torque for a moment, Dinah said, "She almost diode but conducted herself well." Milli said, "Anode I would survive but it still hertz. It's enough to make a maltese cross."

Police say that the unrectified criminal escaped from a dry cell where he had been clamped in ions. First he had fused the electrolytes and then squeezed through a grid system. He was almost run to Earth in a magnetic field by a line of force, but he has been missing since Faraday.

Watt seems most likely is that he stole a d.c. motor. He may decide to switch it for a megacycle and return ohm by a short circuit.

How many puns can you find? (Over 30?) Write a sentence to explain the correct meaning of each one.

Index

A

B

C

D

E

F

G

H

I

J

K

L

M

N

O

P

Q

R

S

T

U

V

W

X

Photograph acknowledgements

AEA Technology: 346T, 350; Air Pictures: 169; Alex Segre/Alamy: 182TL; Bettmann/Corbis: 377B; Blind Mobility Research Unit Nottingham: 228B; Bosch: 293; BoxMag Rapid: 288; British Aerospace: 51BR, 189B; British Rail Research: 22; Burstein Collection/Corbis: 371T; Photoshot/Bruce Coleman: 213TL; Camera Press: 376B; Castle Associates: 234; Colorsport: 95TL, 108B, 134R, 135T, 135BR, 137; Corbis: 64BR Eddy Lemaistre/Photo & Co, 138T Parrot Pascal, 215T, Jonathan Blair, 356B Lester Lefkowitz; Corel (NT): 6 C418, 42L C127, 106 C94, 114T C62, 122 C494, 208C C285; Digital Vision (NT): 4B, 150, 173, DV9, 82 DV13, 109 Karl Ammann DVAA, 128T DVXA, 149 DV6; Elcomer Instruments: 284; Fischer Scientific: 255; Ford Motor Company Ltd: 89; Format Photography: 389BL Brenda Prince, 389TL Maggie Murray; FLPA/Nigel Catlin: 361; GEC: 300; Getty Images: 64TR, 70, 95BR, 108T, 125, 128B, 130, 135BL; 291 John Stanton; iStock: 13, 42R, 47L, 71, 79, 114B, 116, 120, 138B, 182TC, 199C, 199B, 199A, 209D, 228T, 240ML, 240BL, 240T, 252T, 260, 297, 306B, 309, 314, 315T; Joel Finler Collection: 201; John Bailey: 217T; Keith Johnson: 12, 64bl, 172, 174, 180, 188, 196, 199D, 199E, 220, 338T; Last Resort Picture Library: 83B; Leyland DAF: 97; London Buses: 95BL; London Fire Brigade: 213TR; Martyn Chillmaid: 49, 86, 105, 182TR, 184, 186, 204, 203, 209A, 213BC, 213BR, 229B, 232, 245, 250L, 250R, 252BL, 252BR, 275, 283, 302, 319T, 319B, 325, 327, 331, 339, 344; Mary Evans Picture Library: 47BR, 95TR; 373BL Alamy; Barnaby's Picture Library; MEMTEK: 231; NASA: 152, 153, 154, 158, 161T, 161BL, 161BR, 215B; National Gallery London: 212 UML, 212 UMR; National Power: 15B; National Remote Sensing Centre: 157T, 157B; Nokia: 216; OMRON: 27B; Ontario Science Centre: 244; PhotoDisc (NT): 200 PD40, 208A PD54, 208B PD18, 208E PD2, 209B PD22, 389TR PD72; Photolibrary: 146B; Racall: 311B; Ripley's Believe It or Not!: 312B; Robert Harding Picture Library: 45, 47TR; Rolls Royce: 389BC; Royal Astronomical Society: 151; RS Components: 258; Sally & Richard Greenhill: 210B; Science and Society Picture Library: 338B, 345; Science Photolibrary: 4t Dr Mitsuo Ohtsuki, 14T, 311T, 312TL, 371B, 374BL, 375B, 376TR SPL, 14B, 50T, 103B, 115T Martin Bond, 27T Chris Priest & Mark Clarke, 46L, 51BL, 182B, 212B, 217B, 240MR, 347T, 250B Cordelia Molloy, 46R Martyn F Chillmaid, 50BL Dr Ray Clarke & Mervyn Goff, 51TR Dr Ray Clark, 64TL Alex Bartel, 104T Kaj R. Svensson, 103T Tony Wood, 115M Peter Menzel, 127 Renee Lynn, 134L Jerry Wachter, 156, 159T, 160, 213BL, 352T NASA, 159B Robin Scagell, 155T David Ducros, 155BL, 155BR, 146T NRSC Ltd, 166 Martin Dohrn, 173B George East, 182TL Francoise Sauze, 189T Edelmann, 192 Steve Horrell, 193 Rosenfeld Images Ltd, 206 Simon Fraser, 207 Fred Burrell, 210T Martin Dohrn, 210M Phillipe Plailly, 211T Dr R. Clark & M.R. Goff, 211B Agema Infrared Systems, 212TL Erich Schrempp, 212TC Phil Jude, 212lML, 212lMR, 267, 268R, 268C, 268L, 270, 374T, 375TL, 376TL Sheila Terry, 229T Saturn Stills, 229UM Cnri, 229LM Alexander Tsiaras, 306T Dr Jeremy Burgess, 312TR Stammers/Thompson, 346BL 346BR Elscint, 347B Gianni Tortoli, 348T Hank Morgan, 356T David Parker, 356M James Prince, 357T Geof Tompkinson, 357B Hank Morgan, 358 Samuel Ashfield, 372T Adam Hart-Davis, 372B National Library Of Medicine, 374M Emilio Segre Visual Archives/American Institute Of Physics, 374BR Sam Ogden, 375TR Jean-Loup Charmet, 377TR Physics Today Collection/American Institute Of Physics, 389BL James King-Holmes, 389ML Volker Steger, 389 BR Physics Department/Imperial College London, 389MC Mauro Fermariello, 389MR Maximilian Stock Ltd; Scottish Power: 15T; Shell: 74; Spectrum Colour Library: 53; Stone/ Getty Images: 104B; The Print Collector/Alamy: 377TL; Time & Life Pictures/Getty Images: 373BR; Topfoto.co.uk: 83T, 115B Rachel Epstein; Transport Research Laboratory: 69T USGS: 373T; Volvo: 69B; Yamaha: 233; ZEFA: 135M, 212TR.

Every effort has been made to trace and contact all copyright holders, but if any have been overlooked, the publisher will be pleased to make the necessary arrangements at the first opportunity.

Picture research by johnbailey@ntlworld.com and Sue Sharp
Illustrations by IFA Design Ltd, Tony Wilkins Illustration, Jordan Publishing Design, Jane Cope and Ann Johnson

Thanks to Chris and Rachel Johnson for checking the answers section.

Acknowledgement is also made to the following Examining Groups for permission to reprint questions from their examination papers. The questions are not necessarily from examinations for the current specification but are believed to be relevant. The Examining Groups do not take responsibility for the answers provided.

AQA Assessment and Qualifications Alliance
Edex Edexcel Foundation
OCR Oxford, Cambridge and RSA Examinations
NIS Northern Ireland Schools Examinations Council
WAEC West African Examinations Council
WJEC Welsh Joint Education Committee
Cam IGCSE University of Cambridge Local Examinations Syndicate (Cambridge IGCSE Physics Paper 6 Nov 06 Q2; Nov 08 Q2)
Edex IGCSE Edexcel IGCSE Examinations

Websites: www.physicsforyou.co.uk and www.physics4u.co.uk
From these you can download exactly which pages in this book you need to study for your particular examination course.

Other books by Keith Johnson

Advanced Physics for You
with Simmone Hewett, Sue Holt, and John Miller
This is written in the same friendly style as the GCSE book, and covers the core of AS and A-level Physics, with over 200 worked examples.

Timetabling: A Timetabler's CookBook
This book is a complete and practical guide for those staff responsible for timetabling in schools.

Spotlight Science 7, 8 and 9
with Sue Adamson, Gareth Williams, Lawrie Ryan
This is a flexible and accessible science course for KS3, for students at all attainment levels.
The Teacher's Support Packs contain an enormous amount of valuable support material to support differentiation in your teaching and learning.
There are 2 versions: the original 'Spiral' version and the newer 'Framework' version.